SOLUTE MOVEMENT IN THE RHIZOSPHERE

TOPICS IN SUSTAINABLE AGRONOMY

Series Editors

Rattan Lal
Pedro Sanchez
Malcolm Sumner
Marilyn E. Swisher
P. B. Tinker
Robert E. White

Volumes

T.R. Yu *Chemistry of Variable Charge Soils*

M.E. Sumner and R. Naidu *Sodic Soils: Distribution Properties, Management, and Environmental Consequences*

L. Upton Hatch and Marilyn E. Swisher *Managed Ecosystems: The Mesoamerican Experience*

P. B. Tinker and P. H. Nye *Solute Movement in the Rhizosphere*

SOLUTE MOVEMENT IN THE RHIZOSPHERE

P.B. TINKER, M.A., D.Sc.
Senior Visiting Fellow
Department of Plant Sciences
Oxford University

P.H. NYE, M.A., FRS
Emeritus Reader in Soil Science
Emeritus Fellow, St. Cross College
Oxford University

New York Oxford
Oxford University Press
2000

Oxford University Press

Oxford New York
Athens Auckland Bangkok Bogotá Buenos Aires Calcutta
Cape Town Chennai Dar es Salaam Delhi Florence Hong Kong Istanbul
Karachi Kuala Lumpur Madrid Melbourne Mexico City Mumbai
Nairobi Paris São Paulo Singapore Taipei Tokyo Toronto Warsaw

and associated companies in
Berlin Ibadan

Published by Oxford University Press, Inc.
198 Madison Avenue, New York, New York 10016

Library of Congress Cataloging-in-Publication Data

Tinker, P.B. (Philip Bernard)
Solute movement in the rhizosphere/P.B. Tinker, P.H. Nye.
p. cm. — (Topics in sustainable agronomy)
Includes bibliographical references and index.
ISBN 0-19-512492-8
1. Crops and soils — Mathematical models. 2. Plant-soil relationships — Mathematical models. 3. Soils — Solute movement — Mathematical models. 4. Rhizosphere — Mathematical models. 5. Roots (Botany) — Physiology — Mathematical models. 6. Plant nutrients — Mathematical models. 7. Crops — Nutrition — Mathematical models. 8. Plants — Nutrition — Mathematical models. I. Nye, Peter Hague. II. Title. III. Series.
S596.7 T55 2000
631.8 — dc21 98-39136

9 8 7 6 5 4 3 2 1

Printed in the United States of America
on acid-free paper

Preface

In this book, we describe in detail how plant nutrients and other solutes move in the soil in response to leaching and plant uptake. The plants we consider may grow in isolation, or as a crop, a mixture of crops, or a natural community. The way their roots interact with the soil is not so fully understood as the way their shoots respond to the atmosphere, because the root–soil system is both complex and too inaccessible to study easily. But our aim is to understand processes in the rhizosphere so fully that we can model them realistically, and predict the effects of variations in natural conditions or in our own practices. Although our aim is not fully achieved, we think the approach is correct, and has great potential.

At present, the world's developed countries have a vast experience of the effects of the nutrient elements on important crops, based on repeated field trials. But experience is confined to existing or past conditions, and new varieties, cultural practices, and environmental conditions bring with them the need to reassess former conclusions. Often, resources do not match the range of crops or vegetation, or the diversity of soil, climate, and treatment. In these circumstances, advice on practice can best be given by combining fundamental insight with the information given by field trials.

These phenomena are dynamic: soil solutes move, and plants grow; yet, as we show in chapter 1, the intimate connection between the two has only recently been understood. Therefore, an expanded account of solute transport processes in the rhizosphere is timely. Until relatively recently, it has been difficult to link all the separate steps involved in the movement of solutes through the soil and their uptake by extending roots, because the mathematics was too difficult or tedious;

and hence simplifying assumptions had to be made: fluctuations of external variables, such as rainfall, were difficult to include in any detail. The universal availability of computers has removed these obstacles, and provided the essential tool in modelling the various pieces of the system that comprises growing roots in soil. The scientist studying a single process can therefore be fairly confident that if he can describe it quantitatively, it can readily be incorporated in a larger system.

Most of the mechanisms described here have been worked out for the major nutrient elements, simply because they were first in the field. But they are equally relevant, with modification, to other solutes, whether beneficial or harmful.

A good model or theory will suggest questions, and combinations of treatments and effects, that had been overlooked or undervalued. It will help to decide whether new trials are worthwhile. Thus, essential measurements in an experiment have often been omitted for lack of a coherent model. We have been unable to use many experimental results achieved at great expense, simply because a measurement that could readily have been made, such as root length, has been omitted. In addition, it does not seem to be sufficiently appreciated that statistical techniques, such as multiple regression, do not prove causal relations between variables.

Most models of natural ecosystems or crop production are coarse grained, the object being to establish a framework and fill in the details later. Our approach is different in that we analyse the working of small-scale, often simplified, systems first, before combining them into a more complicated one.

This book retains the general approach of the authors' *Solute Movement in the Soil–Root System* published in 1977. It takes account of research done since then to the extent that it has been expanded from 8 to 11 chapters and largely rewritten. Much of the new material concerns the rhizosphere. We have retained text and figures that are still relevant, though we have profited from advice by critics and clarified or corrected passages they found obscure or wrong. We have omitted material that is out of date; taken account of new developments or insights, such as in water and nutrient uptake; included areas of increased interest, such as environmental pollution, crop modelling, and mycorrhizas; and revised much that can now be more clearly expressed. Since our target is understanding, we have not used new examples or diagrams simply because they are more recent, but only if they better illustrate our points.

The general outlines of the foundation chapters, 1–6, are retained, though updated and clarified in details:

- Chapter 1 outlines the history of ideas on the subject and concludes with a simple account of the continuity, or mass-balance, equation, which underlies most quantitative treatments of transport.
- Chapter 2 deals with water movement and uptake by plants in sufficient depth to show how it should be introduced into models of solute movement; and it explains uptake into and passage through roots, where there have been considerable advances.
- Chapter 3 describes how solutes are distributed between the solid, liquid, and gas phases of the soil, with greater emphasis on the often slow transfers between these phases in undisturbed soils.
- Chapter 4 describes local solute movements, particularly by diffusion.

- Chapter 5 concentrates on plant roots, the knowledge we have gained from solution culture about the uptake rate of solutes, and much new information on the molecular and cell processes involved.
- Chapter 6 examines the solute movements occurring in the soil around a single root of an intact plant due to its uptake of water and solutes.

The next chapters contain much new material on processes that affect ion fluxes into and near roots, followed by chapters on the whole plant, and field vegetation.

- Chapter 7 deals with solubilization of sparingly soluble nutrients in the rhizosphere caused by root-induced changes in pH and by chemical agents such as siderophores and phosphatases.
- Chapter 8 contains a much expanded treatment of mycorrhizas and rhizosphere microorganisms in relation to nutrient uptake.
- Chapter 9 considers carbon allocation to roots and rhizosphere in understanding plant and root growth, and the architecture of the root system and the factors influencing it.
- Chapter 10 describes the modelling of the growth and nutrient uptake of a single whole plant in homogeneous soil, including the interactions within the root zone caused by competition for nutrients between individual roots.
- Chapter 11 confronts the complexities of modelling the growth and nutrient uptake of crops and mixtures of plants in the field. We describe the approximations needed to take account of variable weather, soil heterogeneity, and bypass flow; and examples of how some of the numerous crop and vegetation models deal with nutrient uptake processes. There is still much to do on this subject. This book is intended to summarize the present position and help in further developments.

We have used the International System of Units and its abbreviations generally, but since so many of the original results we quote are in centimetres, we normally use this as the most convenient unit of length.

P.B.T.
P.H.N.

Acknowledgements

We wish to thank scientific colleagues who have helped with discussions, information and advice, and in particular Dr. P.B. Barraclough, Professor D. Clarkson, Professor N. Comerford, Dr. P. Darrah, Professor J. Farrar, Professor A. Fitter, Dr. Melanie Jones, Dr. C. Mullins, Professor J. Ritchie, Professor J. Porter and Professor D.J. Read. We thank Professor J.P. Lynch for figure material.

We also thank Professor J.F. Nye, FRS, for help with proof-reading of mathematical material, and Dr. P. D. Nye for assistance with computer wordprocessing problems. In particular, we thank our publishers for their patience.

We are grateful for permission to make use of material from the following sources for the figures and tables listed. Additional information can be found in the reference list of this book, from the reference given in each caption.

Journals

Advances in Agronomy. By permission of Academic Press. Figures 9.3; 9.7; 9.15. Table 9.5.

Advances in Plant Pathology. By permission. Figure 8.10.

Agroforestry Systems. By permission of Kluwer Academic Publishers. Figure 11.29.

Agronomy Journal. By permission of the American Society of Agronomy. Figures 10.2; 10.4; 10.12.

Annals of Botany. By permission. Figure 11.28.

Aspects of Applied Biology. Figure 9.8.

Australian Journal of Agricultural Research. Figure 9.20.

Australian Journal of Botany. Figure 11.31.

British Journal of Applied Physics. By permission of the Institute of Physics. Figures 4.7; 4.8.

Canadian Journal of Soil Science. Figure 2.2.

Chemistry and Industry. Figure 9.14.

Clay Minerals. Figure 3.11.

European Journal of Agronomy. Figures 11.18; 11.19.

Forest Ecology and Management. By permission. Figure 11.23.

Journal of Agricultural Science, Cambridge. By permission of Cambridge University Press. Figures 9.19; 10.3; 11.16; 11.17; 11.25.

Journal of Applied Ecology. By permission of Blackwell Science Ltd. Figures 2.5; 2.10; 6.18. Tables 2.1; 6.4; 8.8.

Journal of Experimental Botany. By permission. Figures 5.7; 9.9. Table 5.4.

Journal of Field Crops Research. By permission of Elsevier Science. Table 9.4 (1998, *55*, 34, Table 3).

Journal of Soil Science. By permission of Blackwell Science Ltd. Figures 3.1; 3.2; 3.3; 3.5; 3.10; 4.1; 4.3; 4.5; 4.6; 8.11; 11.7. Table 3.2.

New Phytologist. By permission. Figures 8.9; 8.15; 8.16; 9.5; 9.16; 9.18. Tables 7.2; 8.7.

Pesticide Science. By permission of the Society of Chemistry and Industry. Figures 6.2; 8.8.

Philosophical Transactions of the Royal Society. By permission of the Royal Society. Figure 2.11 (1994, *B345*, 400, Fig. 5). Table 2.3 (1976, *B273*, 464, Table 1).

Physiologia Plantarum. By permission. Figure 8.14.

Plant and Soil. By kind permission of Kluwers Scientific Publishers. Figures 5.11 (1990, *124*, 179, Fig. 3); 6.4a (1973, *38*, 166, Fig. 1); 6.11 (1972, *37*, 637, Fig. 8); 6.13 (1966, *25*, 99, Fig. 11); 6.14 (1973, *38*, 170, Fig. 4); 6.15 (1969, *30*, 464, Fig. 1); 8.3 (1994, *163*, 4, Fig. 1); 8.5 (1977, *48*, 32, Fig. 4); 8.6 (1961, *15*, 174, Fig. 4); 9.1 (1996, *185*, 13, Fig. 1); 10.1 (1990, *124*, 180, Fig. 5); 10.8 (1972, *36*, 852, Fig. 1); 10.9 (1972, *36*, 705, Fig. 5a); 10.11 (1994, *165*, 166, Fig. 5); 10.14 (1975, *42*, 164, Fig. 1); 10.15 (1975, *42*, 177, 178, Figs 3 and 13); 10.16 (1975, *42*, 218, Fig. 18); 11.11 (1989, *119*, 67, Fig. 4); 11.13 (1990, *124*, 305, Fig. 1); 11.21 (1997, *190*, 156, Fig. 6); 11.24 (1972, *37*, 211, Fig. 2). Tables 7.1 (1992, *139*, 121, Table 1); 8.2 (1961, *14*, 220, Table 2); 8.13 (1983, *70*, 204, Table 2); 9.6 (1989, *113*, 112, Table 1); 10.2 (1990, *124*, 178, Table 1).

Plant Cell and Environment. By permission of Blackwell Science Ltd. Figure 9.13.

Plant Physiology. By permission of American Society of Plant Physiologists. Figure 2.8. Tables 2.2; 2.4.

Planta. By permission of Springer-Verlag GmbH & Co. Figure 5.12 (1984, *160*, 494, 495, Figs 3 and 4).

Proceedings of the Fertilizer Society. Figure 3.9.

Royal Institute of Chemistry. Table 3.3.

Science. By permission of *Science*. Figure 11.30 (1985, *229*, 385, Fig. 1).

Society of Experimental Biology Symposium No. 19, 1965. Figure 2.9.

Soil Biology and Biochemistry. By permission of Elsevier Science. Figure 8.2 (1994, *26*, 177, Fig. 4). Table 9.2.

Soil Science. By permission from Lippincott, Williams and Wilkins. Figures 2.4 (1952, *74*, 338, Fig. 3); 3.4 (1964, *197*, 377, Fig. 1); 4.9 (1971, *111*, 377, Fig. 3); 6.12 (1969, *107*, 338, Fig. 3).

Soil Science Society of America Proceedings (or Journal). By permission. Figures 2.3; 3.8; 9.10; 9.17; 10.7; 10.13; 11.6; 11.8; 11.12; 11.14; 11.26. Tables 10.4; 11.5; 11.6.

Soil Use and Management. By permission of the British Society of Soil Science. Figures 11.3; 11.4. Table 11.1.

Weed Research. Figure 11.9.

Books

Atkinson, D. (ed)., *Plant Root Growth*. By permission of Blackwell Science Ltd. Table 9.8.

Baver, L.D., Gardner, W.H. & Gardner, W.R., *Soil Physics*. By permission of John Wiley and Sons, Inc. Figure 2.1.

Esau, K., *Plant Anatomy*, 2nd Edition. By permission of John Wiley and Sons. Figure 5.2.

Fahn, A., *Plant Anatomy*, 4th Edition. Pergamon Press. By permission of the author. Figure 5.3.

Feddes, R.A., Kowalik, P.J. & Zaradny, H., *Simulation of Water Use and Crop Yield*. Centre for Agricultural Publishing and Documentation. Figures 11.1; 11.2.

Hanks, J. & Ritchie, J.T. (eds.), *Modelling Plant and Soil Systems*, pp. 287–321. By permission of the American Society of Agronomy. Figure 11.20.

Haverkort, A.J. & MacKerron, D.J.L. (eds.), *Potato Ecology and Modelling of Crops under Conditions Limiting Growth*. By kind permission of Kluwers Academic Publishers. Figure 11.22 (1995, 78, Fig. 1).

Kramer, P.J. & Boyer, J.S., *Water Relations of Plants and Soils*. By permission of Academic Press. Figures 2.7; 9.12.

Lynch, J.M., *The Rhizosphere*. By permission of John Wiley and Sons. Figure 8.1. Tables 9.1; 9.3.

Marschner, H., *Mineral Nutrition of Higher Plants*. Academic Press. Figures 5.6; 5.8a; 5.9; 5.14. Tables 5.1; 5.2; 5.5; 10.3.

McMichael, M. & Persson, H. (eds.), *Plant Roots and Their Environment*. By permission of Elsevier Science. Figure 11.10. Tables 9.4; 9.9.

Monteith, J.H., Scott, R.K. & Unsworth, M. (eds.), *Resource Capture by Crops.* Nottingham University Press. By permission of the author. Figure 9.6.

National Agricultural Advisory Service, *Residual Value of Applied Nutrients.* By permission of UK Ministry of Agriculture, Fisheries and Food, Tech. Bulletin No. 20. Figure 3.7.

Nye, P.H. & Tinker, P.B., *Solute Movement in the Soil–Root System.* Blackwell. Figures 5.4; 5.5; 6.3; 8.4; 8.7a; 10.5; 10.6; 10.7a; 11.15. Tables 5.4; 8.1.

Proceedings of the 9th Congress of the International Society of Soil Science, Adelaide, 1968. Figure 3.6.

Proceedings of the 3rd International Conference on Plant Protection in the Tropics. Figure 11.27.

Read, D.J., Lewis, D.H., Fitter, A.H. & Alexander, I.J. (eds.), *Mycorrhizas in Ecosystems.* By permission of CABI Publishing. Figure 8.12. Tables 8.4; 8.5; 8.10.

Robb, D. & Pierpoint, W.S. (eds.), *Metals and Micronutrients.* By permission of Academic Press. Table 8.12.

Roy, J. & Garnier, E. (eds.), *A Whole-Plant Perspective on Carbon–Nitrogen Interactions.* SPB Academic. Figure 5.13.

Sanders, F.E.T., Mosse, B. & Tinker, P.B., *Endomycorrhizas.* By permission of Academic Press. Table 8.3.

Scott Russell, R., *Plant Root Systems.* McGraw-Hill. By permission of the author. Figure 8.7b.

Smith, J.A.C. & Griffiths, H. (eds.), *Water Deficits.* By permission of Bios Scientific Publishers, Oxford. Figures 2.6; 9.11.

Smith, S.E. & Read, D.J., *Mycorrhizal Symbiosis.* By permission of Academic Press. Figure 8.13. Table 8.6.

Thornley, J.H.M. & Johnson, I.R., *Plant and Crop Modelling—a Mathematical Approach to Plant Physiology.* By permission of Oxford University Press. Figures 9.2; 9.21.

Tinker, P.B. & Lauchli, A., *Advances in Plant Nutrition 1.* By permission of Greenwood Publishing Group. Figure 5.10. Table 5.6.

Transactions of the Society of Soil Science, Commissions IV and V. Soil Bureau, Lower Hutt, New Zealand, 1962. Table 6.1.

Troughton, A., *The Underground Organs of Pasture Grasses.* By permission of CABI Publishing. Figure 5.1.

Waisel, Y., Eshel, A. & Kafkafi, U. (eds.), *Plant Roots—the Hidden Half.* By courtesy of Marcel Dekker, Inc. Figure 10.10.

Walker, A. & Allen, R. (eds.), *Pesticide Movement to Water.* By permission of British Crop Protection Council. Figure 11.5.

Weaver, J.E., *Root Development of Field Crops.* McGraw-Hill. Figure 9.4.

Doctoral Dissertations

Al-Najafi, M.A., Root shrinkage in relation to water stress. D.Phil. thesis, Oxford, 1990. Figure 7.1.

Barraclough, D., The diffusion of macromolecules in the soil pore space. D.Phil. thesis, Oxford University, 1976. Figure 4.4.

Kirsch, B.H., Solute movement in soil under conditions of evaporating water. D.Phil. thesis, Oxford, 1992. Figure 4.11.

Sanders, F.E.T., Effects of soil and root properties on the uptake of nutrients by competing roots. D.Phil. thesis, Oxford, 1971. Figures 6.8; 6.9; 7.2.

Wray, F.J., Changes in the ionic environment around plant roots. D.Phil. thesis, Oxford, 1971. Figures 6.16; 6.17.

Zriek, R.A., Concentration changes of sodium and chloride associated with roots in saline environments. D.Phil. thesis, Oxford, 1987. Figure 6.10.

Contents

Main Symbols

Note: if a symbol is used in another sense this is noted in the text.

Symbol	Definition	Units
a	radius of root axis	cm
a_h	radius of root hair	μm
A_L	leaf area	cm^2
A_R	root surface area	cm^2
b	solute buffer power of soil $\mathrm{d}C/\mathrm{d}C_L$	
C	concentration of diffusible solute in soil	mol cm^{-3} soil
C_g	concentration of solute in gas phase	mol cm^{-3} gas
C_L	concentration of solute in liquid phase	mol cm^{-3} liquid
C_s	concentration of solute in solid phase	mol cm^{-3} solid
D	diffusion coefficient of solute in soil	cm^2 s^{-1}
D^*	dispersion coefficient of solute in soil	cm^2 s^{-1}
D_g, D_L, D_s	diffusion coefficient of solute in gas, liquid, solid phases	cm^2 s^{-1}
D_L^*	longitudinal dispersion coefficient of solute	cm^2 s^{-1}

E	net assimilation rate	g cm^{-2} day^{-1}
f_g, f_L, f_s	diffusion impedance factor in gas, liquid, solid phases	
F	flux of solute	mol cm^{-2} s^{-1}
G	Gibbs free energy	J mol^{-1}
I	inflow (net absorption rate by unit root length)	mol cm^{-1} s^{-1}
I_W	absorption rate of water by unit root length	cm^3 cm^{-1} s^{-1}
K	hydraulic conductivity	cm s^{-1}
K_m	Michaelis–Menten constant	mol cm^{-3}
L	root length	cm
L_A	root length under unit area land surface	cm cm^{-2}
L_P	root length per plant	cm
L_V	root density (length per unit soil volume)	cm cm^{-3}
LAR	leaf area ratio (A_L/W)	cm^2 g^{-1}
M	molecular weight	
M_t, M_∞	amounts absorbed at times t and ∞	mol
N	equivalent fraction	
P	external pressure	MPa
q	water uptake rate per unit crop area	cm^3 cm^{-2} s^{-1}
r	radial distance	cm
R_W	relative growth rate ($1/W(\mathrm{d}W/\mathrm{d}t)$)	day^{-1} or s^{-1}
S	reaction (loss or gain) rate in soil	mol cm^{-3} s^{-1}
t	time	s or day or year
U, U_R, U_S	solute content of plant, root, shoot	mol
v	water flux	cm^3 cm^{-2} s^{-1}
V	uptake rate per unit fresh weight	day^{-1}
W, W_R, W_S	dry weight of plant, root, shoot	g or mg
W_{RF}	root fresh weight	g
X	concentration of solute in dry plant material	mol g^{-1}
z	ionic charge	
α	root absorbing power (F/C_{La})	cm s^{-1}
$\overline{\alpha a}$	root demand coefficient	cm^2 s^{-1}
β	gas solubility coefficient (C_L/C_g)	
λ	packing factor	
θ	soil moisture fraction by volume	
θ_g, θ_L, θ_s	volume fraction of gas, liquid, solid	

μ	chemical potential	J mol^{-1}
μ_e	electrochemical potential	J mol^{-1}
ϕ	water potential	MPa
ψ	water matric potential	MPa

Subscripts

a at root surface
g gas phase
h root hair
i initial value
L liquid phase
s solid phase
t at time t
w water

SOLUTE MOVEMENT IN THE RHIZOSPHERE

1

Introduction

The art and study of plant nutrition go back at least to Roman times, as essential parts of the business of producing food. This long historical perspective can usefully be studied now, when plant nutrition is largely a matter of science in its principles, but still, to a surprising extent, an art in its application, even in developed countries. In the past, the delay between a scientific advance and the application in practical agriculture was usually many decades. Thus, the rates of fertilizer used by Lawes (Johnston 1994) in experiments in 1850 were not applied widely in practice until after 1950. The movement to precision agriculture may now take the final step to a full science-based nutrition of plants in the field. For these reasons, we have thought it worthwhile to give a highly condensed outline of the history of scientific advance in our subject.

It is now generally accepted that under given growth conditions, uptake of a solute by roots is related to its concentration in the soil solution and the extent to which this, in turn, is buffered by the soil. Though these apparently simple ideas were advanced more than a century and a half ago, only recently have they been defined clearly enough to form a basis for detailed understanding of the effect of solutes on plants grown in the soil. These ideas have, in particular, been obscured by specific effects of roots with their associated rhizosphere organisms: for roots not only vary widely in their response to solute concentration, but also alter near them the soil properties we measure in the bulk of the soil. Thus, it is only since the 1950s that we have come within reach of the objective clearly set us by Liebig in 1840 when he wrote: 'A rational system of agriculture must be based on an exact acquaintance with the means of nutrition of vegetables, and with the influence of soils and action of manure upon them'.

1.1 The Origin of Current Ideas

The history of ideas about soil and plant relations has been well described by Russell (1937) and Wild (1988) for the period up to the beginning of the twentieth century. Even though Jethro Tull (1731) could not decide whether nitre, water, air, fire, or earth was the food of plants, John Woodward (1699), 30 years earlier, seems to have grasped the essential idea of a flow of solutes to roots induced by the transpiration stream when, having grown spearmint in a primitive 'nutrient' culture solution, he wrote:

> It has been shown that there is a considerable quantity of (terrestrial) matter contained in rain, spring and river water; that the greatest part of the fluid mass that ascends up into plants does not settle there but passes through their pores and exhales up into the atmosphere: that a great part of the terrestrial matter, mixed with water, passes up into the plant along with it, and that the plant is more or less augmented in proportion as the water contains a greater or lesser quantity of that matter: from all of which we may reasonably infer, that earth, and not water, is the matter that constitutes vegetables.

Then in 1804, De Saussure showed that the root was selective and could absorb the salts in water at different rates, excluding some so that they became more concentrated in the external solution. De Saussure isolated soil solution in the soil pores by displacing it with water coloured with carmine. The idea that the soil solution was the main source of plant nutrients was further advanced by Whitney & Cameron (1903).

A fascinating history of solute–solid adsorption studies has been chronicled by Forrester & Giles (1971). The capacity of the soil to adsorb substances from solution — so rendering barnyard liquor and other coloured waters clear and odourless — had long been known, but a clear statement of the buffering role of the soil lies in the following quotation from Gazzeri (1823): 'Loam and especially clay take possession of soluble matters which are entrusted to the soil, and retain them in order to give them by degrees to plants, conformably with their needs'. Such ideas as these were the forerunners of experiments made between 1845 and 1852 by Huxtable, Thompson, and particularly Way (1850, 1852) that established the process of cation exchange.

Meanwhile, Daubeny (1846) decided that the active fraction of the total soil phosphorus available to plants was that dissolved in water impregnated with carbon dioxide. He recognized that 'if the results of soil phosphorus studies are to be meaningful, the soil must be kept as chemically intact as possible . . .' The acid and alkali extractions in the following century violated his principle. Other ideas considered that the feeding power of plants could be attributed to the total acidity of sap in their roots (Dyer 1894); and that carbon dioxide excreted by the roots forms carbonic acid which can make available a fraction of the nutrients in 'readily available' minerals (Czapek 1896).

It is remarkable that there were no really new developments in this position for the first 40 years of the twentieth century except in the understanding of the exchange chemistry and structure of soil clays. The agricultural chemists and the plant physiologists scarcely seemed to communicate with each other. The former concentrated on finding a suitable extractant for 'available' nutrients in the soil, the first example being Dyer's 1% citric acid (1894), to be followed by numerous others, all based on a search for the 'available' fraction. The main method of investigation of soil nutrient questions during and since this period was field experiments, designed and interpreted by the powerful statistical methods developed by R. A. Fisher (1925). In practical terms, this approach has been extremely successful, and forms the basis of modern fertilizer practice and the great post-war increases in crop yields. However, for scientific purposes, it was overemphasized, since it stressed quantity of nutrients rather than their intensity of supply, and led agricultural chemists to be satisfied with correlations and regressions between fertilizer responses and chemical extracts, and so inhibited the search for more fundamental and detailed explanations of their results.

Meanwhile, plant physiologists, eschewing such a complex medium as soil in favour of nutrient culture solution, established the 'essential' nutrient elements; and determined the main metabolic factors affecting salt absorption. Unfortunately, their experiments were usually made at concentrations very much higher than that of the soil solution, so that the quantitative aspects of their work are all too often irrelevant to soil conditions. During this period, California workers (Hoagland 1944) were exceptional in attempting to bridge the gap between soil and nutrient solution-cultured plants — possibly because salt concentrations in their soils are unusually high.

Much of the early history of agricultural chemistry was dominated by the search for the role of humus in plant nutrition, and the source of plant nitrogen (Russell 1937). In one of the major triumphs of the last quarter of the nineteenth century bacteriologists discovered the main outlines of the nitrogen cycle and particularly the sources of ammonium and nitrate in the soil. During the twentieth century, the incredible diversity of the whole soil population of microbes, fungi and animals has been recognized, and a new branch of the subject — soil ecology — has come into being. It was also recognized that there was a particularly large population in the root–soil interfacial region, to which the 'rhizosphere' was given in 1904 by Hiltner (Russell 1973, p. 241). The special properties of the rhizosphere have subsequently received much attention (see chapters 7 and 8).

It has proved very difficult to relate the numbers and types of organisms to the rates of mineralization of organic matter or microbial metabolism. This is due to the difficulty in identifying the many organisms and to the spatial complexity of soils. We now have a wide experience of the rates of mineralization of litter and humus and also of specific pesticides, gained in a variety of environments, but know little about the details of the mechanisms involved. We have, in general, therefore to be content with empirical rate functions.

1.2 The Beginning of the Modern Period (c. 1940–1960)

In 1939, Jenny & Overstreet amplified the concept of 'contact exchange', which was suggested by Devaux (1916). He noted that roots, like the soil, had cation exchange properties, and thought that since they were in such close contact they could readily exchange cations. Jenny & Overstreet (1939a) held that the nutrient cations were absorbed in this way in exchange for hydrogen ions produced by the root at its surface. They further designated 'contact exchange' as an overlapping of the volumes within which the individual adsorbed ions oscillated at their adsorption sites, and added that 'the ions do not enter the soil solution *per se*, but, in the moment of contact they jump directly from one particle to another'. So stated, contact exchange, while it may exist, is quite inadequate to account for the number of cations absorbed by plants. However, Jenny & Overstreet (1939b) further claimed that cations could migrate along clay surfaces, and could pass from one clay surface to another if they were in close contact with one another. Hence, the roots could draw on a much greater number of cations than they could by immediate contact (Albrecht *et al.* 1942). Here, we can recognize the beginnings of our current ideas about the zones around the root from which nutrients are diffusing. However, as Russell (Russell & Russell 1950, p. 444) realized: 'The process of contact exchange does not, in fact, differ fundamentally from uptake from the solution, for in either case the ion can only be transferred through the water film that surrounds the root, and the composition of the soil solution is controlled by the solid phase of the soil as well as by the uptake of nutrients by the crop'. Another problem with 'contact exchange' is that it seems to assume that active uptake occurs only at the root surface, which, as we discuss in chapter 5, is not so.

It proved difficult to demonstrate that plants would grow as well in dilute nutrient solution, similar to typical soil solution, as they did in the same solution buffered by a suspension of clay or ion exchange resin. However, when precautions were taken to ensure that the solution was well stirred and that the supply of nutrient ions was maintained as they were depleted by uptake (Loneragan & Asher 1967). Lagerwerff (1960) was able to conclude, 'It seems clear that uptake rates from solution, sand, and sand and resin are the same provided the soil solution is renewed sufficiently often'; and Olsen & Peech (1960) concluded that their results of rubidium uptake from clay suspensions 'fully support the classical soil solution theory of mineral nutrition of plants proposed long ago by Cameron'.

There is another difficulty about the extended theory of contact exchange. It had long been generally accepted that plant roots make the soil more acid. Jenny & Overstreet (1939a) suggested that they exchange hydrogen ions. Other sources of hydrogen ion have been thought to be carbon dioxide and organic acids excreted by the roots (Schander 1941). However, as we discuss in chapter 7 roots in soil normally take up more anions than cations, and consequently have to liberate bicarbonate ion if a high electrical potential difference is not to develop between the root and the soil. Thus, as a rule, the soil near them becomes more alkaline.

If the concentration of an element in a plant is only a minute fraction of its concentration in the soil, then it is plausible that the layer of mucilage, a few micrometres thick, often observed on the root surface, may contact or envelop sufficient soil to contain enough of the element required. Passioura (1966) concluded that finely divided manganese and iron oxides (Jones 1957; Jenny 1960), and molybdenum adsorbed on ferric hydroxide might provide elements in this way. However, as we discuss in chapter 7, there is little evidence that mucilage can solubilize nutrient elements.

The next notable advance was made independently by Schofield (1955) and Woodruff (1955). They attempted to define more precisely the idea of the 'availability' of an ion, and hypothesized that a measure of this was the work needed to withdraw it from an equipotential pool in the soil. This work was to be measured as the chemical potential of the ion referred to the calcium ion, which is usually the dominant ion in the soil solution. Schofield, writing of phosphate, drew a very clear distinction between the intensity of its supply, to be measured by the chemical potential of $Ca(H_2PO_4)_2$, and the quantity in the exchangeable pool of phosphate that he considered maintained it. He used the analogy of a well:

> It is the depth to water which determines the work needed to get the water to the top. Similarly [the chemical potential of $Ca(H_2PO_4)_2$] is a measure of the 'depth' of the 'level' of the pool of soil phosphate. As water is taken out of the well it is replaced by lateral movement of ground water, and as phosphate is taken from the soil solution it is replaced by desorption. In neither case is replacement quite complete; withdrawals generally cause some lowering of the 'level', depending on the 'capacity' of the system.

These thoughts led to a great deal of valuable work on the relation between changes in chemical potential of a nutrient ion in solution and related changes in the amount adsorbed on the soil, which we describe in chapter 3.

Nevertheless, it had not been shown that nutrient uptake was controlled by the work the root had to do to absorb a nutrient: and, indeed, Nye (1968b) thought it unlikely that it should be since he considered the energy available as a result of root respiration greatly exceeded that required over the same period to accumulate ions such as potassium, which are at a higher concentration in plant cell vacuoles than in the soil solution. However, this argument is probably mistaken since Lambers *et al.* (1996) have shown that the energy demand associated with ion uptake is considerable (section 5.2.3). Wild (1964) and Wild *et al.* (1969) have shown that uptakes of phosphorus and of potassium are more closely related to their concentrations in solution than to their chemical potentials referred to the calcium ion potential.

Meanwhile, plant physiologists concentrated their attention on salt absorption by portions of plant tissue, rather than by whole plants. Radioisotopes, now readily available, greatly assisted them. Much work was done with roots excised from their shoots, with barley seedlings being particularly suitable (Epstein 1972). In addition, uptake by slices of storage tissue; the nutrient-absorbing leaves of aquatic plants like *Elodea*; single and multicellular algae; cell organelles, for example mitochondria; and ectotrophic mycorrhizal tissue were all studied (Sutcliffe 1962). For the purpose of our theme, many of these experiments are important because they were carried out at lower solution concentrations than hitherto. They distinguished between passive and active uptake, and revealed that the initial rates of active uptake were usually related to the solution concentration by a diminishing returns form of curve. They also revealed quantitatively the effects of competition between ions on their rates of uptake. Consequently, it became clear that to interpret the effect of applying a salt to a soil on the composition of a plant, for example the effect of adding potassium chloride on the magnesium composition, one had to know both the change produced in the composition of the soil solution and the consequent effect on the plant's absorption mechanism. However, in spite of these advances, the experimenters with plant parts did not recognize that cell uptake is controlled by the whole intact plant (see chapter 5).

During this period, the importance of knowing the rate of uptake of solutes per unit of root in intact growing plants was not widely appreciated, and in the discussion of this question in chapter 5 there are few references to work before 1960.

The approach of soil scientists to the problem of nutrient uptake from the soil had so far been essentially static. It considered that as roots ramified through the soil they took up nutrients at particularly active zones just behind the root tips, 'tapping' the available nutrients or rendering them 'available' as they progressed. The question of how the individual ions reached the surface of the root does not seem to have been raised, and even the role of transpiration in sucking the soil solution to the root surface seems to have been neglected. A possible reason is that the most studied crops, the cereals, have very dense root systems. The first signs of a more dynamic view emerged from a seminal paper by Bray (1954). He wrote:

> The mobility of nutrients in soils is one of the most important single factors in soil fertility relationships. The term 'mobility' as used here, means the overall process whereby nutrient ions reach sorbing root surfaces, thereby making possible their sorption into the plant. Thus the term involves the solution or exchange of the nutrient as well as its movement to the root surfaces. A correlative process, just as important, is the growth of the roots and the extension of the [working] root surfaces into areas where the nutrients occur. These two processes, complementing each other, largely [determine] the soil fertility requirements of a plant.

Bray also recognized that individual roots would be more likely to compete with each other for mobile nutrients, such as nitrate, than for relatively immobile ones, such as phosphate.

Tepe & Leidenfrost (1958) also expressed the new dynamic view:

> To obtain a realistic estimate of the availability of plant nutrients in soil due consideration should be given to the degree of mobility of ions and their proximity to an absorbing surface (root hair), and to the fact that the uptake of nutrients on the one hand and the growth of plant roots on the other prevents the plant–nutrient system from ever reaching equilibrium.

These ideas soon bore fruit and the subject was put on a quantitative basis. The 'mobility' of ions was resolved into two processes: mass flow of the soil solution to the root induced by transpiration, and diffusion of ions to the root induced by lowering of their concentration by uptake at its surface. Barber (1962) calculated the proportion of nutrients in a plant that could be supplied by mass flow of the soil solution, and Walker & Barber (1961) also demonstrated by autoradiography zones of nutrient depletion and of accumulation around roots. Bouldin (1961) and Olsen *et al.* (1962) applied the mathematics of diffusion theory to ions diffusing to the root. It is interesting to note that the first theoretically sound treatment of water movement to roots was made at about the same time by Philip (1957).

Measurements of ionic diffusion in pure clay systems and ion exchange resins had been made for many years previously, but it was Porter *et al.* (1960) and Schofield & Graham-Bryce (1960) who showed how the diffusion coefficients could be measured in soil.

This book is largely an account of the quantitative work over the past half century that sprang from these beginnings.

1.3 Wider Perspectives

So far, we have given an account of the development of ideas about the localized movement of ions in the immediate neighbourhood of the absorbing surfaces of roots. To apply these ideas to problems of crop production and ecology we need a model of uptake by whole plants or communities of plants. Thus, we have to show how behaviour of solutes at the root–soil interface may be integrated over whole root systems; and to do this we have to understand, in detail, the grosser movements of solutes within the profile.

A broad-scale approach to this question has been adopted in innumerable studies, ever since the main plant nutrients derived from the soil were clearly recognized in the middle of the nineteenth century. In 1868, Augustus Voelcker reported analysing the water draining from variously fertilized plots of Broadbalk Field, Rothamsted, anticipating that his investigation was likely to open up 'quite a mine of theoretical enquiry' (Russell 1966, p. 126); and the relative ease of leaching of various nutrients from the topsoil was early appreciated, for example by Dyer (1901). Nutrient balance sheets for cropped soils have been drawn up, and their importance in assessing long-term changes in fertility recognized — a subject thoroughly reviewed by Cooke (1967, 1969). Foresters, particularly in Germany and France, measured the nutrient uptake of mature trees, and published numerous papers between 1876 and 1893. Rennie (1955) has reviewed this and later work, and discussed the long-term effects of timber production on the nutrient balance in the soil.

Ecologists were slower to study the wider aspects of the nutrient balance in the ecosystem. Sampling of mixed communities is laborious, and natural vegetation does not provide the commercial incentive of a crop. Where the natural vegetation is relied on to restore fertility of worn out soils, as in the practice of shifting cultivation, considerable information does exist (Nye & Greenland 1960). Information on nutrient contents and cycles in a range of natural vegetation from tundra to tropical forest was assembled by Rodin & Bazilevich (1965). The subject is now part of 'geochemical cycling' and much information is available.

In the exploitation of the soil by plant roots there is a marked difference between a root system that develops from seed in a fertilized soil, and progressively exploits deeper layers of the soil — characteristic of an annual crop — and the established root system of perennial plants, such as occurs in many natural communities, for example a humid woodland. In these, nutrient levels in the soil solution are usually much lower, and the overall rate of uptake will depend on rates of mineralization of humus, release by weathering, and leaching from living vegetation and litter by rain, so that competition between roots is intense. Factors that may aid this competition, such as mycorrhizal hyphae or root surface enzymes, are likely to be particularly important in explaining the success of a particular species. Since the action of these factors in competitive situations is still poorly understood, our understanding of such systems is correspondingly much less complete.

The possibility of a much more exact prediction of movement of solutes in the whole root zone arose when the ideas of ion exchange chromatography were applied to leaching of ions through columns of soil by Ribble & Davis (1955). A soil profile can, in fact, be viewed as a large, irregularly packed, erratically eluted, multilayer ion exchange column. Needless to say, no exact predictions of its behaviour are possible, but, as long as this is recognized, the principles of exchange chromatography provide a sound theoretical starting point for models, such as those described in section 11.2, that can be increasingly refined as the complexities are understood.

In recent years, the impact of agriculture on the environment and its sustainability has increased the need for more detailed understanding of the likely consequences of different systems of land management. Though, in the early days, inorganic nutrients received most attention, concerns about pollution of food and groundwater have enormously widened the range of solutes to be considered. Some pollutants are, indeed, common inorganics in excess. Many newer ones are organics. In addition to being leached, they may be decomposed by soil organisms. However, as our examples show, the general principles of solute movement, established over the years, apply equally to them, so we have not needed to create any separate chapters to cover them.

In chapter 11, we aim to show how the detailed understanding that we are now gaining of solute movement around roots and in the profile can give greater insight into larger and complex systems in the field.

After a few decades in which the nutrition of plants in soil has been regarded as a rather outmoded or dull subject, there are now many exciting new approaches. In fundamental biology, the mechanisms of uptake and control of nutrients in the roots and the whole plant at last look soluble, partly because the tools of molecular biology give us a much clearer idea of the mechanisms, so that we are no longer dealing with a 'black box'. The genetic engineering techniques will be applied to plant nutrition directly, and plant nutrition will also have to be adjusted to meet the needs of the very different crops that will be grown in the future. A better understanding of the complexities of the rhizosphere should allow more exact and more economical plant nutrition. In the more applied subjects, the modelling of soil systems will allow better — but never perfect — prediction of the nutrient needs of plants, and the engineering systems of precision farming will allow these better predictions to be more accurately followed.

All possible agricultural improvements will be essential over at least the first 50 years of the twenty-first century. With population and wealth increasing faster than ever before, agricultural outputs will have to escalate enormously. This will have to be done in a world with an increasing level of carbon dioxide in the atmosphere, and a shortage of water, both of which have nutritional implications. This will call for more basic science, but, even more, for flexible and responsive applied research, so that agricultural systems can meet changing demands and conditions.

Another avenue is opening up in the natural vegetation of forests and grasslands. Until quite recently, ecologists and soil scientists went their separate ways. The science of plant nutrition has now reached a stage where it can give real insights into the complexities of mixed-species natural vegetation, even if it cannot deal with them as precisely as with agricultural crops. Natural vegetation is governed by competition for resources, and nutrients are the most important growth-limiting resources in very many ecosystems.

1.4 The Continuity Equation

Although our preliminary survey of solute movement has ranged from small- to large-scale processes, their quantitative treatment, which is developed in subsequent chapters, is usually based on some form of the 'continuity' equation, described below, which expresses the change in the mass of a substance in a small volume over a small time. According to the problem, this may be formulated in Cartesian, cylindrical or spherical coordinates; and solved to satisfy various boundary conditions, for example that there is a specified concentration of solute at the boundary between a root surface and the adjacent soil. It is therefore convenient to explain this simple concept here, and to present together, for future reference, a variety of forms in which it appears.

1.4.1 Solute Transfer

We may illustrate the principle involved by considering a movement of a solute through soil (e.g. by diffusion) in the direction of the x axis. Consider two imaginary similar planes of unit cross-section normal to the axis, and distance δx apart. The volume enclosed is $\delta x \times 1 = \delta x$. Within this volume we have the following:

Rate of gain in a solute in volume δx = rate of entry across plane at $x -$ rate of exit across plane at $x + \delta x$, that is,

$$\begin{aligned} \delta x(\partial C/\partial t)_x &\approx (F_x - F_{x+\delta x})_t \\ &\approx -\delta x(\partial F/\partial x)_t \end{aligned} \tag{1.1}$$

where

$F_x, F_{x+\delta x}$ = flux of solute at x, $x + \delta x$,
C = amount of diffusible solute per unit volume of soil.

Note that $C =$ all mobile solute.

As $\delta x \to 0$,

$$(\partial C/\partial t)_x = (-\partial F/\partial x)_t \tag{1.2}$$

Equation (1.2) is the continuity equation in one dimension. (For those unfamiliar with partial differential equations, $(\partial C/\partial t)_x$ means the rate of change in C with t, when x is held constant.)

If movement of solute is by diffusion alone, by Fick's First Law (section 4.1),

$$F = -D(\partial C/\partial x)_t \tag{1.3}$$

where D is the coefficient of diffusion. Therefore,

$$\partial C/\partial t = \partial/\partial x(D\partial C/\partial x) \tag{1.4}$$

an equation known as Fick's Second Law.

If the solvent (water) is also moving, solute is carried by convection, or mass flow, as it is sometimes termed:

$$F = -D^* \, \partial C/\partial x + \nu C_L \tag{1.5}$$

where D^*, the dispersion coefficient, differs from the diffusion coefficient because the water movement itself causes some dispersion of the solute molecules (section 4.3); ν is the water flux in the direction x; and C_L is the concentration of solute in the soil solution.

For combined convection and diffusion, the continuity equation, obtained by substituting F in equation (1.5) into equation (1.2), becomes

$$\partial C/\partial t = \partial/\partial x(D^* \, \partial C/\partial x) - \partial(\nu C_L)/\partial x \tag{1.6}$$

Notice that ν must occur within the differential unless the water flux is constant. It is evident that the flux of solute often depends to a great extent upon the movement of water, which we consider in chapters 2 and 11.

If the solute is volatile, we have to consider its movement through soil air, usually by diffusion, though convection may also contribute in some instances. These complications are considered as they arise. We do not treat the movement of gases or problems of soil aeration in detail.

1.4.2 Solute Reaction

The level of solute may also change as a result of processes occurring within the volume δx. To allow for the rate of such reactions we add a term $S(C, x, t)$, often called a 'source' term, to equation (1.2) to give

$$(\partial C/\partial t)_x = (-\partial F/\partial x)_t + S(C, x, t) \tag{1.7}$$

Note that $S(C, x, t)$ means that the source term, S, is a function of C, x and t.

Examples of processes that are specified by this rate function are:

(a) Slow release of ions from non-exchangeable to exchangeable form or the reverse.
(b) Release or fixation of solutes by organic matter transformation.
(c) Degradation of organic solutes by purely chemical, or microbial action.

We do not attempt to cover all this ground in detail, though we aim to mention any quantitative generalizations that can be made. A function of this form is also particularly useful for expressing the rate of uptake by a system of roots in a small finite volume of soil, and we develop this approach for a single plant in chapter 10, and for plant communities in chapter 11.

1.4.3 Generalization of the Continuity Equation

1.4.3.1 Cartesian Coordinates

When movement may occur in three dimensions, equation (1.7) becomes

$$(\partial C/\partial t)_{x,y,z} = -[(\partial F_x/\partial x)_{y,z} + (\partial F_y/\partial y)_{x,z} + (\partial F_z/\partial z)_{x,y} + S(C, x, y, z, t)] \quad (1.8)$$

1.4.3.2 Cylindrical Coordinates

If movement occurs in a direction normal to a cylinder, such as to a root, equation (1.8) may be expressed as

$$(\partial C/\partial t)_r = -(1/r)\partial/\partial r(rF_r) + S(C, r, t) \quad (1.9)$$

where r is the radial distance from the axis of the cylinder.

1.4.3.3 Spherical Coordinates

For some problems, such as movement into a soil aggregate, movement may be considered normal to the surface of a sphere and equation (1.8) may be expressed as

$$\partial C/\partial t = -(1/r^2)\partial/\partial r(r^2 F_r) + S(C, r, t) \quad (1.10)$$

where r is the distance from the centre of the sphere.

In simple cases, these equations may be solved analytically. Most of these solutions will be found in the books of Carslaw & Jaeger (1959) and Crank (1975). More often, numerical methods must be used, and many examples of these will be encountered in subsequent chapters. Crank (1975, chapter 8) gives an introduction to them, and multidimensional problems are treated more fully by Mitchell (1969, chapter 2) and Smith (1978).

2

Soil and Plant Water

Water is of central importance in the transport of solutes, whether by diffusion or mass flow, and whether in soils or plants (Lösch 1995). It is also extremely important for the biota that live in the soil (Parr *et al.* 1981). Water is an unusual component of the environment, because its structure suggests it should be a gas at normal temperatures rather than a liquid, and it is the only common compound in the biosphere that occurs to a significant extent in the vapour, liquid and solid phases.

We begin this chapter with a very brief statement of the thermodynamic approach to the study of water, which defines the water potential. Without an understanding of chemical potentials, it is difficult to deal with the relationships of ions and water in the soil and the plant. Therefore, in this chapter we give an introduction to this subject with special reference to water, which we then take further in chapters 4 and 5. A clear exposition of this is given in Nobel (1991).

2.1 Water Potential

2.1.1 Basic Theory

The concept of chemical potential is fundamental. It is a measure of the energy state of a particular compound in a particular system, and hence of the ability of a unit amount of the compound to perform work and thereby cause change. In particular, the difference in potential at different points in a system gives a measure of the tendency of the component to move from the region with the high potential to the region with the low potential. A component of a system can have various forms of potential energy in this sense, all of which contribute to the total chemical potential. Here, we exclude chemical reaction energy and kinetic energy.

The main forms of energy that contribute to the chemical potential of a specified compound or material are due to its concentration (which may release energy on dilution), to its compression (which may perform work on expansion), to its position in an electrical field (which may release energy if the component is electrically charged and moves within the field), and to its position in the gravitational field (which may release energy as the component moves downwards). This is best stated in equation (2.1) (Nobel 1991):

$$\mu_j = \mu_j^o + RT \ln a_j + \overline{V}_j P + z_j F^* E + m_j g h \tag{2.1}$$

Here, the subscript j indicates the terms that apply specifically to the compound j in the system we are interested in, such as the chemical activity a, the molar volume $\overline{V}$, the valency z, and the mass m. Other terms either describe the parameters for the system, (height h, electrical field E, pressure P and temperature T), or are universal constants (gas constant R, Faraday constant F^* and acceleration due to gravity g). The dimensions of each term are in energy per mole, in units of J mol^{-1}. All changes in chemical potential are measured from some standard state for that compound, where the chemical potential is defined as μ_j^o.

With charged entities, such as ions, for which the term $z_j F^* E$ in equation (2.1) is significant, the total chemical potential is usually termed the electrochemical potential μ_e.

When vapour and liquid are in equilibrium, the chemical potential of water has to be the same in both. It is convenient to define the chemical potential in the vapour phase for systems that contain both liquid and vapour. From the gas law, by considering the energy that a volume of gas can deliver on expansion, the energy per mole, or chemical potential, can be derived (Baver *et al.* 1972), as in equation (2.2):

$$\mu_j = \mu_j^o + RT \ln P_j/P_j^o + m_j g(h - h^o) \tag{2.2}$$

where μ_j is the chemical potential of vapour, m_j is the molecular weight, and P_j is the pressure of vapour, with superscript o relating to the reference state of the compound. A reduction of chemical potential of liquid water, as by addition of solute to water, will therefore reduce the partial pressure of water vapour over the solution, and hence the vapour phase chemical potential.

2.1.2 Soil and Plant Water Potential

2.1.2.1 Definition

Based on equations (2.1) and (2.2), a practical simplification can be made, in which we define the term water potential φ, rather than the chemical potential of water, by moving to a volume rather than a molar basis. Water potential is obtained by dividing the chemical potential by $\overline{V}_w$, the molar volume of water, as in equation (2.3):

$$\varphi = (\mu_w - \mu_w^o)/\overline{V}_w = (RT/\overline{V}_w)\ln(P_{vw}/P_{vw}^o) + \rho g(h - h^o) \qquad (2.3)$$

where μ_w and μ_w^o are the chemical potentials of water, and of water in the standard state, φ is the water potential, P_{vw} is the water vapour pressure, P_{vw}^o is the water vapour pressure at the reference point and state, and ρ is the density of water ($m_w/\overline{V}_w = \rho$). The logarithmic term is the relative humidity divided by 100.

The advantage of this transformation for soil and plant studies is that energy per unit volume is equivalent to pressure. Hence, φ is approximately also energy per unit weight, in units of $\mathrm{J\,kg^{-1}}$ for SI. Water potential can therefore be expressed in a variety of virtually equivalent units, which are noted here for convenience: 10 bar = 1 MPa = 10^6 N $\mathrm{m^{-2}}$ = 1000 J $\mathrm{kg^{-1}}$ = 10 × 750 mm Hg = 10.2 m H_2O head.

It is very useful to be able to define the chemical potential of water in an equilibrium system by reference to the vapour phase, if the non-vapour part is a complex system. It is easy to see from equation (2.3) that the relation between water potential and vapour pressure is non-linear. A potential of −1 MPa is equivalent to a relative humidity of about 99%. The soil air can therefore almost always be regarded as saturated. Conversely, the potential difference between the moist surfaces within a leaf and the atmosphere outside it at a relative humidity of 50% is −93.6 MPa.

Within a body of pure water, the chemical potential is therefore defined by the height above a reference level, and by the pressure at a point. The latter needs to be defined carefully with reference to the standard state, because it will depend both upon pressures outside the water, the hydrostatic pressure due to overlying water, and upon the pressures that may arise from surface tension effects in small water–air interfaces. These may be positive or negative depending on the nature of the solid material.

The height term is included in the water potential, because it is possible that water in soils or plants at different heights are being compared, and the term could be significant. It is certainly important in water transport in the xylem of trees and deep roots or in soil profiles.

In a solution, water potential is usually defined by physical pressure P, osmotic pressure Π and height, as expressed by equation (2.4):

$$\varphi = (\mu_w - \mu_w^o)/\overline{V}_w = P - \Pi + \rho g h \qquad (2.4)$$

In equation (2.4), the water potential effect due to the presence of solutes is expressed directly in the osmotic pressure form, from equation (2.6).

2.1.2.2 Matric Potential

The pressure term P in equation (2.4) includes positive pressures, such as would arise in a saturated soil, where water in the lower layers is under a pressure head, or in plants with a positive pressure in the xylem. In some soils, part of the weight of the overlying mass of soil itself is carried by the pore water, and this is called 'overburden potential' (Baver *et al.* 1972). This is easily understood by watching a swelling soil imbibing water, when the entering water raises the soil weight.

However, in an unsaturated soil or plant tissue, water is held in a microporous or capillary system. If the system is not saturated, some pores are only partly filled with water, and surface tension at the many air–water interfaces in the pores means that the water bodies within the soil are under tension, or a negative pressure. This suction is called matric potential ψ. The expression for the pressure across a curved interface is $(2\gamma \cos \beta)/r$, where γ is the water surface tension, r the radius of curvature and β is the angle of water–soil contact. Most surfaces in the soil are hydrophilic, so the angle of contact is usually 0° and $\cos \beta$ is 1, though there are non-wetting soils in which β can exceed 90°.

The term P in equation (2.4) must therefore be separated into two components: the external pressure P to which the system is subject, as in a pressure plate apparatus, or a hydrostatic pressure in a saturated soil; and the matric potential which is the negative pressure within all the distributed water bodies in an unsaturated soil. Consequently, there cannot be a hydrostatic pressure of a water column and a matric potential at the same time. In soils, the matric potential is often the most important part of water potential. The use of pressure units therefore allows equation (2.4) to be stated in a particularly convenient form as equation (2.5):

$$\varphi = P + \psi - \Pi + \rho g h \tag{2.5}$$

As water is lost from a soil, by evaporation or suction, the larger pores empty first, because the capillary tension there is smallest. As the limiting radii of the soil–water interfaces diminish, the pressure difference increases, and water potential becomes steadily more negative (figure 2.1). The relationship between the water content and the matric potential is the 'moisture characteristic', and defines a basic property of soils.

However, the moisture characteristic is not a unique relationship, but varies with the previous wetting–drying history of the soil, and particularly with whether the last change was towards wetting or drying. Simply, on wetting it is the narrowest empty pores that fill first, whereas on drying it is the widest full pores that empty first. This causes differences in water distribution in the body of the soil, at the same water content, so that φ, K and D_w depend upon the wetting–drying history. This is known as hysteresis (figure 2.2).

To help in understanding water potential, it may be useful to consider a tall vessel largely filled with soil, and with a pure water table above the bottom of the vessel. In this system, water potential is defined by relative humidity in the air at the top, by matric potential in the upper unsaturated layers, and by positive pressure in the saturated layers at the bottom. When equilibrium has been established within the column, the sum of each of these terms with the height term (all in the same units) is the water potential, and it has to be the same at every level, even though the individual values change continuously with height.

2.1.2.3 Solutes and Water — Osmotic Pressure and Osmotic Potential

All forms of chemical potential are, in principle, interchangeable, but in practice there are many restrictions over which forms are operative in particular situations. The effects of solute concentration on the chemical potential of water can be particularly confusing, so we will explain them in detail here. If pure water and

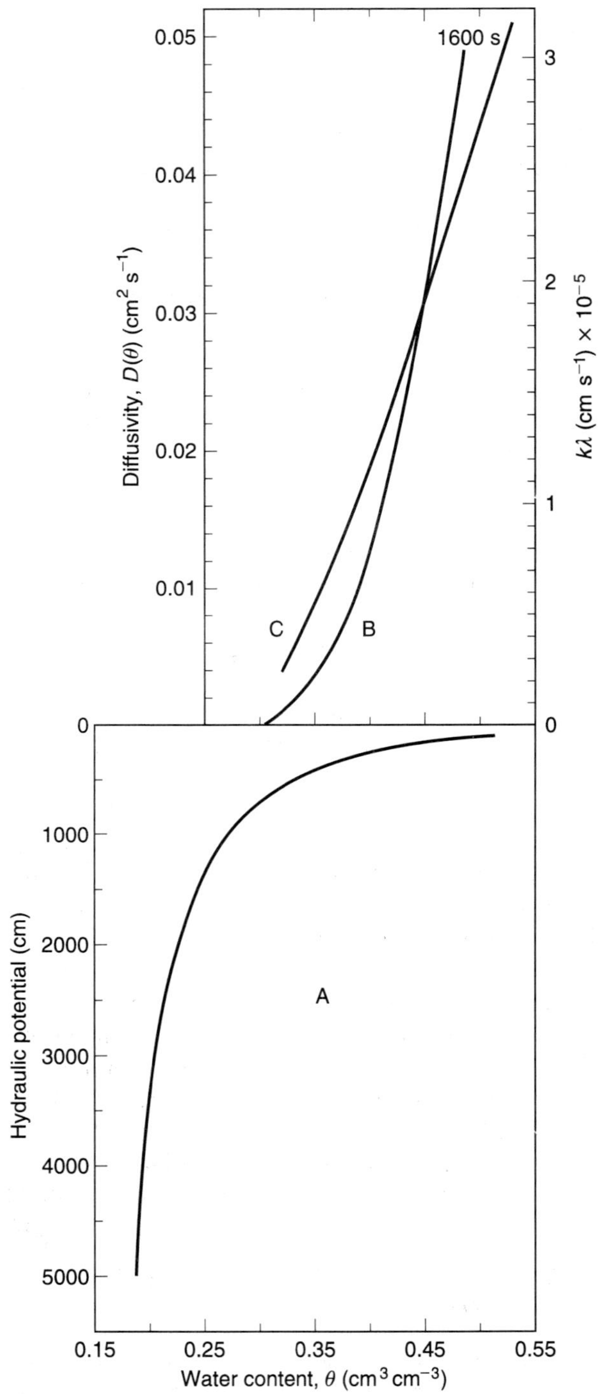

Figure 2.1 The relationship of the fractional moisture content θ of a Salkum silty clay loam with (A) the hydraulic potential (the moisture characteristic), (B) the hydraulic conductivity, and (C) the soil water diffusivity. The last two were measured after 1600 s, and depended upon the duration of the method (after Baver *et al.* 1972).

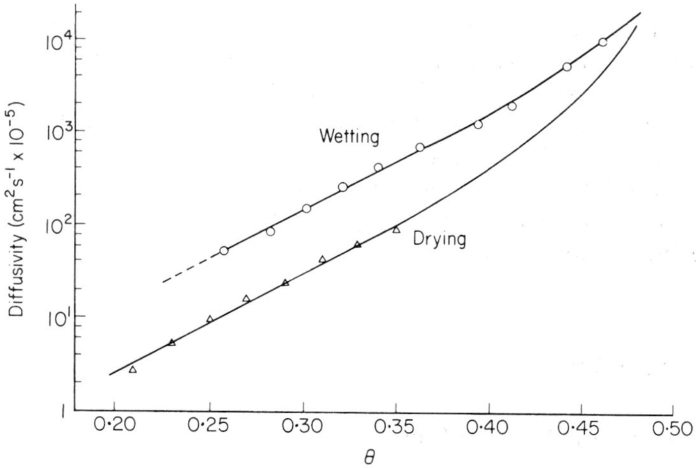

Figure 2.2 The diffusivity of a Grenville silt loam in relation to the fractional water content. Hysteresis causes the curves obtained for soil in the process of wetting or drying to differ markedly (after Staple 1965).

a solution are separated by a perfect semipermeable membrane (which allows water, but not the solute, to cross it), the water will move into the solution. Equilibrium can only be attained if the solution is placed under an external pressure. This is due to the fact that the chemical *activity* of water in the solution is lowered by the presence of the solute (equation (2.1)), so that the resulting chemical potential difference tends to drive water from the pure water into the solution and so to dilute it. However, this can be balanced by an external pressure — the osmotic pressure Π — on the solution, which increases the potential of water in it. It is therefore an excellent example of a system that balances one form of chemical potential against another. This equivalence is given in equation (2.6):

$$RT \ln a_w = -\overline{V}_w \Pi \tag{2.6}$$

where a_w is the activity of water in the solution and $\overline{V}_w$ is the molar water volume. Osmotic pressure is therefore another manifestation of, or another way of expressing, the reduction in chemical potential of water by the presence of solutes (equation (2.4)). The activity of water is not a simple concept to handle in practice, and it is easier to express the osmotic pressure as a function of the solute concentration, as in equation (2.7), which is applicable for dilute solutions (Kramer & Boyer 1995, p. 38):

$$\sum \Pi_j \approx RT \sum C_j \tag{2.7}$$

where $\sum \Pi_j$ is the total osmotic pressure due to all the solutes present and ΣC_j is the molar concentration, summed over all solutes. If a salt dissociates into ions, it is the total molar concentration of all particles — ions or molecules — which is effective.

The osmotic pressure states the effects of solutes on the water potential, but a physical pressure can be generated only if there is a semipermeable membrane separating two bodies of water with different solute concentrations. If there is not, then there is simply a concentration gradient within the aqueous phase, which will drive a diffusion flux of water into the more concentrated solution and of solute into the more dilute solution, until they are uniformly mixed. Thus, osmotic pressures do not normally develop into physical pressures in soil, because efficient semipermeable membranes are rare.

2.2 Transfer of Water

2.2.1 Transfer of Water in Soil

2.2.1.1 Saturated Soils

Soil water moves in response to a difference of water potential, as expressed by Darcy's law, in equation (2.8):

$$\nu = -K\,\mathrm{d}\varphi/\mathrm{d}x \tag{2.8}$$

where ν is the flux of water in $\mathrm{cm^3\,cm^{-2}\,s^{-1}}$, and K is the hydraulic conductivity, in $\mathrm{cm\,s^{-1}}$ if the water potential φ is in centimetre head of water. In a saturated soil, K will be a constant so long as the structure remains stable, because the water flow pathway will be unchanged. However, in practice many soils are not stable when saturated, and their structures are sensitive to small stresses, so it cannot be assumed that all saturated soils have constant hydraulic conductivities. The source of the potential gradient in saturated soils will normally be in the pressure potential (i.e. the depth of the overlying water column).

Because the actual flow rates in different water-filled pores and in different parts of the same pore differ widely, a mixing occurs in the direction of flow, which tends to disperse any sharp changes in solute concentration. The effects are seen most clearly in a strongly structured soil, with large cracks or faunal passages that allow very high flow speeds when they are water-filled (section 11.2.2).

2.2.1.2 Unsaturated Soils

In unsaturated soil, K varies widely with the water content, because the latter defines the total cross-section for water flow, the mean effective water-filled pore radius, and the effective pathlength. The larger water-filled pores will empty first as the soil dries (figure 2.1), because of the larger radius of curvature of the soil–water interface. This has a large effect on the flow rate, because the Poiseuille equation (equation (2.9)) indicates that the flow rate in a tube depends upon the fourth power of its radius, for a constant pressure gradient:

$$f = (\pi r^4/8\eta)\,\mathrm{d}P/\mathrm{d}x \tag{2.9}$$

where f is the volume flow rate in a tube (in $\mathrm{cm^3\,s^{-1}}$), r is the radius, η is the viscosity, and $\mathrm{d}P/\mathrm{d}x$ is the pressure gradient.

As water is lost from the soil, the continuity between water-filled pores also decreases, and small water bodies can become isolated. A soil with water-filled volume fraction θ less than 0.1 will normally have a very low value of K for this reason (figure 2.1). Several empirical equations have been put forward that give a relationship between K and either θ or ψ, and that can be useful if only approximate values are required. Campbell (1991) suggested equation (2.10):

$$K = K_{sat}(\theta/\theta_{sat})^m \tag{2.10}$$

in which K_{sat} is the hydraulic conductivity of a saturated soil, θ_{sat} is the water content at saturation, and m is a constant.

In some special cases, there can be flow in unsaturated soils with constant θ, ψ, and K, as during infiltration (figure 2.3), because there is a gravitational gradient which drives the flow (section 2.1.2). It is more usual for the driving force to be due to a difference of matric potential, resulting from a difference in the water content. As the matric potential and the water content for a soil are related by the 'moisture characteristic' curve (figure 2.1), the driving force can be expressed as a gradient of either. By using the slope of the characteristic, equation (2.11) is obtained, in which the flow in unsaturated soil can be expressed in terms of the water content gradient and the 'water diffusivity'. However, the real driving force is the water potential gradient, and it should not be forgotten that the use of the water content gradient is a convenience, and is physically incorrect. The flow of water is a mass flow under a pressure difference — it is not a diffusion process driven by random thermal motion (chapter 4).

$$v = -K\partial\psi/\partial x = -K(\partial\psi/\partial\theta)(\partial\theta/\partial x) = -D_w\partial\theta/\partial x \tag{2.11}$$

where $D_w = K\partial\psi/\partial\theta$.

The advantage of using D_w instead of K is that the great majority of situations with unsaturated flow are 'transient states'; that is, where the redistribution of water means that θ, K, and D_w are varying continuously with time and position in the soil. It is then necessary to combine the continuity equation (1.2) and the water flow equation (2.11), to give equation (2.12):

$$\partial\theta/\partial t = \partial/\partial x\{K\,\partial\psi/\partial x\} = \partial/\partial x\{D_w\,\partial\theta/\partial x\} \tag{2.12}$$

The diffusivity formulation in equation (2.12) has one less variable than that using the water potential, and it is in the same mathematical form as Fick's second law (see chapter 4). There are many analytical solutions to the latter, which can in principle be applied to water transport through equation (2.12). However, the variability of D_w (and K) with θ makes D_w analogous to a concentration-dependent diffusion coefficient. Analytical solutions to the equation are then difficult to obtain, and numerical solutions or approximations may be necessary (section 4.2.1) (Philip 1973). Flow in unsaturated soils is also complicated by the phenomenon of hysteresis, which causes K and D_w to vary even at the same θ (figure 2.2), though D_w may have a more linear relation with θ than K (Baver *et al.* 1972) (figure 2.1).

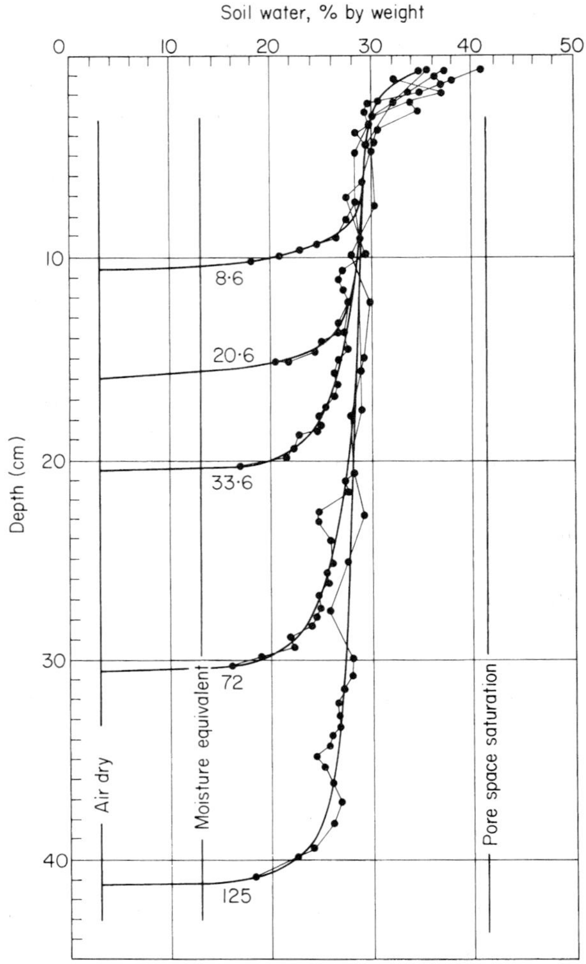

Figure 2.3 Sequential curves of soil moisture against depth in a Yolo sandy loam during the process of water infiltration. Infiltration times are marked on the curves, in minutes. Points are the observed values, with lines interpolated and extrapolated (after Bodman & Colman 1943).

2.2.2 Infiltration

Infiltration and the movement of water down soil profiles under gravity are very important for our subject, because so much of the transfer and redistribution of solutes in soils occurs by this mechanism (section 11.2). If water is supplied at a steady rate to the soil surface, for example by rain that does not run off, the water profile in the surface layers must eventually settle to a steady state in which the water moves under a gravitational gradient only, and the same flux of water has to pass through each layer, irrespective of differences in soil texture and structure. Consequently, the water content at each level adjusts itself to be such that K has the same value at all levels, according to equation (2.10) (figures 2.1 and 2.3). This assumes that all levels will be able to pass the constant flux of water in a barely unsaturated state. However, if K cannot reach the required value for a soil horizon, this soil layer will become saturated, and a perched water table will form immediately above it. The pressure potential caused by the water in this layer of saturated soil will then tend to drive water more rapidly through it, and an equilibrium but dynamic state will be re-established.

The situation is more complicated during the initial stages of water infiltration into a dry soil. The descending water then forms a 'wetting front', which is in a transient state. After this front has passed, a 'transmission zone' forms above it, which contains the steady state described above. A simple empirical equation (2.13) describes the infiltration rate:

$$\nu = \tfrac{1}{2}st^{-1/2} + a \tag{2.13}$$

where ν is the infiltration rate in length per unit time as measured above the ground and s is a constant. After long times, the infiltration rate becomes constant at a. This varies widely with texture and structure: Slatyer (1967) gives values of 1.0 cm h^{-1} for sandy or well-aggregated soils, 0.25 cm h^{-1} in aggregated soils of high clay content, and less than 0.1 cm h^{-1} in swelling soils. This corresponds to a mean actual flow velocity of some 3–4 cm h^{-1} in the pores of the sandy soils on average; the actual value will, of course, vary enormously between pores, depending upon the pore radius, as noted above.

In practice, this idealized situation is not often observed, which causes many practical problems. Some soils disperse and form crusts under direct rainfall, when infiltration may become very variable with position, leading to very localized ponding and runoff. Moren *et al.* (1989) discuss various pre-wetting treatments aimed to get better infiltration into such a soil. Irregular penetration of water into strongly structured soils is frequent, for reasons discussed above, but localized rapid flow or 'finger' formation can also occur in homogeneous sandy soils (Selker *et al.* 1992). Glass *et al.* (1989) presented a general theory of the cause of this non-homogeneity of flow in non-saturated conditions based upon the idea that the 'finger' tips are close to saturation, but, as they pass on, water moves slowly outwards into the surrounding drier soil. However, hysteresis causes the centre of the finger always to remain wetter than the outer fringes; hence, it will continue to form a flowpath of higher conductivity, and the finger remains stable. It is unusual to find absolutely regular infiltration into any natural soil, and the accurate measurement of soil hydrological parameters in the field is never easy.

2.2.3 Osmotic Potential, Semipermeable Membranes, and Water Flow

It was pointed out in section 2.1.2 that moderate solution concentration differences rarely develop physical pressures and bulk flows of solution in soils. However, membranes that are semipermeable to varying degrees are ubiquitous in biological material, and the turgor pressures of all tissues and cells, and most of the fluxes of water, depend upon this property (sections 2.3.3 and 5.2.3). The scale of osmotic pressures is indicated by the fact that a 0.1 M solution of a non-dissociating compound is about 0.24 MPa, that for the fluid in young plant tissue is about 0.73 MPa, and for sea water is about 2.5 MPa.

Most real semipermeable membranes are to some degree leaky (section 2.3.3), in which case some solute moves through them, and the full osmotic pressure is never developed. This leakiness is defined by the reflection coefficient σ. By definition, there can be no real equilibrium in systems with a leaky membrane, because solute will continue to move through it into the water until the concentration is uniform throughout. The reflection coefficient can therefore be defined only in terms of the fluxes generated in this dynamic situation, by using the methods of irreversible thermodynamics (Nobel 1991), as in equation (2.14). This equation expresses the transfer when both hydraulic and osmotic pressure differences occur across a membrane:

$$J_v = L_p\,\Delta P + L_{pd}\Delta\Pi = L_p(\Delta P - \sigma\Delta\Pi) \tag{2.14}$$

where $\sigma = -L_{pd}/L_p$. In a general form, with multiple solutes, this becomes

$$J_v = L_p(\Delta P - \Sigma\sigma_j\,\Delta\Pi_j) \tag{2.15}$$

where J_v is the flux of solution volume (water plus some solute), L_p is a conductivity coefficient, ΔP is the hydrostatic pressure difference across the membrane, $\Delta\Pi_j$ is the osmotic pressure difference generated by each species and σ is the reflection coefficient. This last term therefore states the relative effect of the two forms of pressure on flow. If σ is 1, the membrane is perfectly semipermeable so no solute can pass, and gradients of osmotic or hydrostatic potential will have the same effect on flux of solvent. If σ is 0, there is no semipermeability, so osmotic pressure differences will have no effect on water movement.

2.2.4 Vapour-Phase Transport of Water

In principle, water may obviously be transferred in the vapour phase in soils, but there has been a long debate about whether significant fluxes of water are possible, particularly in relation to supply of water to roots. The driving force for diffusive vapour fluxes is the gradient of a gas-phase concentration, and the very small humidity gradients that correspond to large water potential differences (see section 2.1) suggest that vapour-phase fluxes must be very small compared with the flux that would be generated in the liquid phase by the same water potential difference. In any case, the saturated vapour pressure sets an upper limit to the size of the concentration difference. Rose (1963) has given data for the fraction of the total water flux that moves through the vapour phase in various materials and conditions. At relative humidities below 60%, essentially all the flux in soil moved through the vapour phase, but the total flux was then very small. The vapour-phase flux was unimportant relative to the liquid-phase flux when the water content was greater than 15% of saturation. This almost certainly means that a long pathway that is solely through the vapour phase is insignificant normally, but just might be important under drought conditions (Nobel & Cui 1992).

It has been suggested that the vapour-phase pathway could act as a short-circuit where the liquid-phase pathway is interrupted by very short air-filled gaps, and hence increase the unsaturated conductivity significantly. This could occur where a wider, empty pore lies between narrower, water-filled pores, or between soil and root if the contact is imperfect. There is still debate about whether this process could contribute significantly to the supply of water to plants in some circumstances (section 2.3.4), but it seems unlikely.

2.2.5 Water Movement under Temperature and Solute Gradients

The parameter temperature occurs in equation (2.1), but it was not then explicitly considered as a component of water potential. The driving forces for transport under a temperature gradient are rather more complex than the simple matric and gravitational potential gradients discussed above. However, these gradients are of practical importance and occur frequently in soil in the field, and it is probable that they are ignored far too often. Temperature gradients are almost ubiquitous in soils due to the diurnal and seasonal temperature cycles, and can be particularly steep under bare topsoils.

The vapour pressure is increased and the surface tension of water is decreased by increasing temperature. A gradient of temperature will therefore alter both the matric potential and the vapour pressure of the soil water, so that both will drive water from hot to cold, but both the driving forces and the conductivities will differ. The behaviour of these two fluxes can be almost independent, leading to some very complex situations. For example, if a steady state is set up within a cylinder of moist soil containing a labelling solute, and is maintained for a long time with different temperatures at the two ends, the final state is as in figure 2.4. Here, the vapour pressure difference drives a flux of water to the cool end, where the vapour condenses. Irrespective of changes in matric suction due to temperature, this sets up a counterflow of liquid water, with its salt concentration, to the hot end, leading to a quasi-steady state (Gurr *et al.* 1952). Nassar & Horton (1992) and Nassar *et al.* (1992) also found that water moved to the cold regions of a soil column, and solute to the hot region. The final effect of a varying temperature gradient on solute movement in the field is likely to be complicated.

Philip & De Vries (1957) and Taylor & Cary (1964) have given quantitative theories to predict thermally driven water fluxes. The former was found most

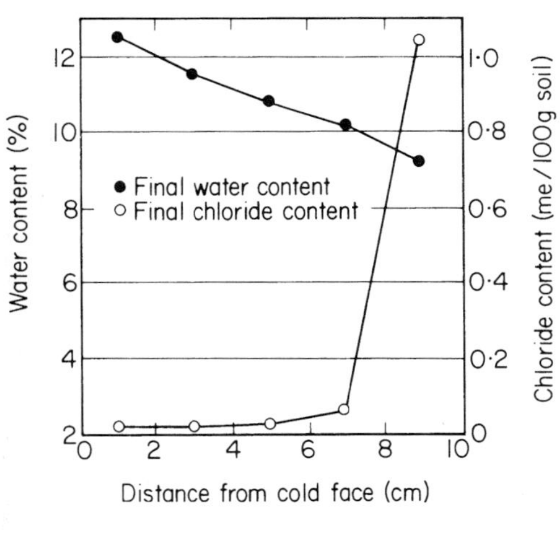

Figure 2.4 Steady-state distribution of chloride and water in a horizontal column of soil subject to a steady temperature difference at the two ends for 18 days (after Gurr *et al.* 1952).

reliable in a test by Cassel *et al.* (1969). Philip & De Vries (1957) regard the water potential and the thermally driven water fluxes as separate and non-interacting at the microscopic level. The flux may be stated as

$$v = -D_w \, \partial\theta/\partial x - D_T \, \partial T/\partial x \tag{2.16}$$

where D_w is the diffusivity for a water content gradient and D_T is the constant for a thermal gradient. Both D_w and D_T have liquid-phase and vapour-phase components.

The effects of very high solute concentrations on water transport are also complex, and can express themselves without the presence of a semipermeable membrane (section 2.1.1). Very large solute concentration differences can often occur, by deposition of salt crusts from evaporating soil solution or from added fertilizer. The distribution of fertilizer on the soil surface is a fairly common process in the field that could produce complex distributions of water and of salt (Scotter & Raats 1970) (section 11.2.6). This results from condensation of water vapour on the salt, and formation of a salt solution with high surface tension and low vapour pressure, to which soil water moves by liquid and vapour transport, as described above.

A gradient of solute concentration (i.e. of osmotic pressure) in soil will normally cause diffusive counterfluxes of salt and of water. There are, however, other effects that become important at high solute concentrations. Solutes alter the surface tension and vapour pressure of water; thus, the tension of water at 20°C is 72.7×10^{-3} N m^{-1}, but 1.8 M sodium chloride has a tension of 76.0×10^{-3}. Water will therefore tend to move by mass flow into the zone that contains salt, in addition to the diffusive movement and the vapour-phase movement, so the total flow may be considerable, even with no semipermeable membrane. Anion exclusion (section 3.5) in a fine-textured soil could increase the water flux by salt-sieving. Nassar *et al.* (1992) concluded that salt-sieving and the osmotic effects of solute could be neglected at concentrations below 0.1 M, but were important above 0.5 M. The full explanation of transfers that result from strong salt gradients in soil is mathematically complex (Bresler 1972; Nassar & Horton 1992).

2.3 Water Use by Plants

2.3.1 Soil–Plant–Atmosphere Pathway

The demand of plants for water depends upon the weather conditions that determine the potential evapotranspiration, the size and arrangement of the leaf canopy, and external shading. Plants also have internal mechanisms whereby they can avoid meeting this demand in full, mainly through their ability to alter the size of the stomatal apertures in the leaf surfaces. For terrestrial plants, by far the greater amount of this water is supplied via the soil, though water intercepted by the canopy from rain, mist or dew can, in effect, meet part of the total demand. There are many close analogies between the supply of water and nutrients to plants, but there is no nutrient analogy to this externally driven demand for water, which may not be closely linked to the growth rate of the plant, so that water use efficiencies (plant mass/transpired water) vary widely (Kramer & Boyer 1995).

The driving force for this transfer from soil to atmosphere via the plant is a gradient of water potential. The water potential of the atmosphere can be extremely low; for example, a relative humidity of 50% corresponds to −93.6 MPa in water potential. However, the normal effective limit of soil water potential beyond which the supply to the plant ceases, the permanent wilting point (PWP), is −1.5 MPa or about 99% relative humidity. From this, it is seen that the potential difference from soil to atmosphere may be very large, and that almost all of this difference occurs between plant and atmosphere. The plant needs to be maintained at relatively small negative water potentials to survive, so the stomatal control system has to be at the plant–atmosphere interface. Osmotic regulation in plants can allow them to survive at lower water potentials than PWP (Morgan 1984), and genes that control osmoregulation are now being investigated (Zhu *et al.* 1997).

Despite the fact that they form such a small fraction of the total potential difference from atmosphere to soil, the gradients within the soil and plant and the fluxes that they produce are extremely important. There is always some loss of water from the canopy, even if the stomata are fully closed, and permanently closed stomata would preclude entry of carbon dioxide and hence further plant growth. The potential differences and the fluxes of water in soils and plants under dry conditions are thus critical for the ability of plants to survive and grow (Nobel 1991, p. 521).

The network of pathways through soils and roots with a potential gradient that drives a flux at once calls to mind an electrical resistance network, and this has been described as the 'Ohm's law analogy'. The pathways can be described as in figure 2.5. This analogy is useful for explanatory purposes, but the processes are not as simple as this implies. Most fundamentally, the part of the flow in soil is probably driven by a matric potential gradient, that in the plant by an osmotic/matric gradient and that in the atmosphere by a vapour pressure gradient, and these gradients vary in very different ways with water content. Second, the resistances of both soil and plant are variable, in physically different ways, and the analogy with a constant ohmic resistance is poor (Weatherley 1982). With these provisos, it is possible to write a resistance equation (equation (2.17)) for water transport from the bulk soil to the stomatal membrane where the water evaporates (Lösch 1995) (figure 2.5), which is at least a useful conceptual tool, and has been used as the basis for effective models of crop water uptake (Campbell 1991).

$$q = (\varphi_s - \varphi_l)/(R_p + R_s) \tag{2.17}$$

where subscript s refers to soil, p to plant and l to the leaf surface, q is rate of flow, and R is resistance to flow. This equation gives useful information, as shown in table 2.1 (section 2.3.3).

The parts of the pathway that need detailed discussions are (i) the control mechanisms of the stomata, because this sets the rate that the roots are called upon to supply; (ii) the water uptake properties of the root and the internal hydraulic resistances of root and stem; and (iii) water transport in the rhizosphere, because the properties of this zone and the root surface are still debated. None of

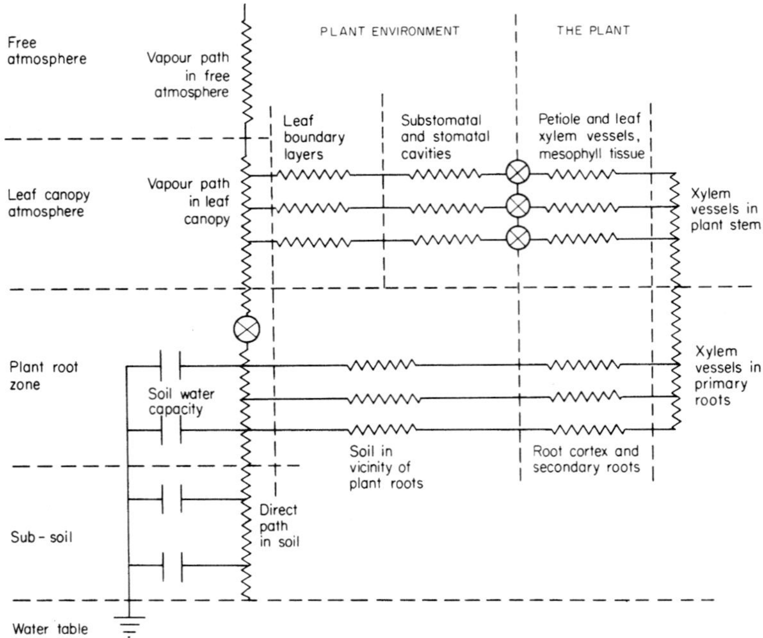

Figure 2.5 The 'Ohm's law analogy': the pathway of water from the bulk soil to the free atmosphere regarded as an electrical circuit. ⊗, Crosses mark the points where there is a phase change from liquid to gas (after Cowan 1965).

these problems is new (Boyer 1985), but all have seen development in the 1980s and 1990s.

2.3.2 The Stomatal Mechanism

We will not discuss here the climatic and environmental factors that affect the rate of water loss from leaves and plants; this information is available in many good

Table 2.1 Calculated values for resistance in the plant (R_p) and in the soil (R_s) for ryegrass growing in progressively drier soil.

Day	Water loss (cm^3 day^{-1})	ϕ Soil (bar)	ϕ Root (bar)	R_s (bar day cm^{-3})	R_p (bar day cm^{-3})	R_p/R_s
1–5	25	−0.20	−0.21	0.0004	0.10	250
5–8	16	−0.25	−0.26	0.0006	0.17	283
8–11	26	−1.0	−1.02	0.0007	0.10	143
11–13	9	−5.0	−5.10	0.011	1.03	94
13–18	12	−8.0	−8.30	0.025	0.60	24
18–22	6	−11.0	−12.00	0.160	0.83	5.2

Source: after Lawlor (1972).

textbooks and is based on well-defined physical theory. We are concerned with the feedback processes whereby the plant can control its rate of transpiration depending upon the supply of water available to it from the soil (Schulze 1986). The stomata are the apertures on the leaf surface through which water vapour diffuses out. The size of the aperture can be regulated, and closure is initiated both by a low relative humidity in the air (Beerling *et al.* 1986) and by a decline in the water potential in the leaf that follows the onset of drought (Hsiao *et al.* 1976; Kramer & Boyer 1995). The level at which the latter process is initiated was assessed at −0.5 to −1.0 MPa for mesophytic plants and −1.0 to −2.0 MPa for xerophytes (these are, respectively, non-tolerant and tolerant of drought).

It used to be believed that these mechanisms, together with others dependent on changes in carbon dioxide and light, fully explained the stomatal reactions of plants. More recently, it was realized that many species consistently reduce their stomatal conductance when drying conditions are imposed, but before the plant or leaf water potential changes significantly (Steudle & Jeschke 1983). The primary effect is probably a chemical signal transmitted from the roots to the leaves in plants, the signal being triggered by drying out of the surrounding soil (Davies & Zheng 1991; Jackson 1993; Schulze 1993; Tardieu & Davies 1993). This signal appears to be the concentration of abscisic acid (ABA) in the xylem flow (figure 2.6). However, the sensitivity to this chemical message seems to be a function of leaf water potential, so that a type of dual control operates. Abscisic acid has other relevant properties; thus, Saab *et al.* (1990) reported that increased production of ABA maintains primary root growth and inhibits shoot growth of maize at low water potentials, which also will delay droughting of the plant.

This process appears to give a direct linkage with soil water status, but the synthesis of ABA must be controlled by the root water potential rather than the bulk soil water potential. As discussed below, the relationship of the root water potential to the bulk soil water potential depends on whether the rhizosphere soil around the roots becomes dried out. This may occur in some circumstances, and will depend upon soil water content, soil type, root distribution, water flux, and other factors (section 2.3.4). On this argument, the more rapidly the soil ceases to supply water to a root because of local drying out, the more sensitive will be the chemical signalling.

The concentration of ABA in the leaves will itself depend both upon its rate of synthesis and upon the flux of water up through the xylem. The flux of water thus partly regulates the signal, but is itself regulated by the signal, both by this dilution mechanism and by the root water potential mechanism described above, giving scope for complicated feedback behaviour. A model of the ABA concentration in the xylem has been produced (Tardieu 1993). However, there are differences in the findings in laboratory and field research, suggesting that the understanding is still incomplete.

There are particular difficulties in explaining the behaviour of plants that have part of their root systems in dry and part in wet soil. Blum *et al.* (1991) measured plant parameters in barley grown in three realistic situations: irrigated from the soil surface, with a permanent water table, or in a drying soil profile. Growth differed markedly between all three groups, despite the fact that the plants supplied by irrigation or by the water table were both well supplied with water

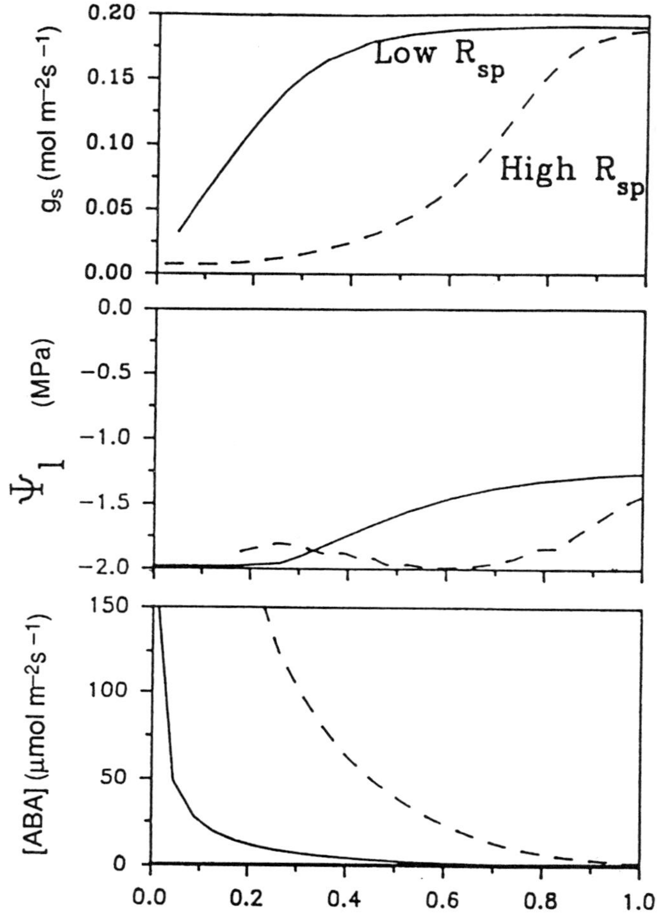

Figure 2.6 Development of the stomatal conductance (g_s), the leaf water potential (ψ_1) and the abscisic acid concentration in the xylem sap ([ABA]) in relation to the soil water level, expressed as the fraction of the transpirable soil water remaining. R_{sp} is soil–root resistance (after Tardieu & Davies 1993).

throughout (table 2.2). Blum *et al.* (1991) explained this as being due to a chemical signal of developing drought that came from the roots in dry soil near the surface in the water table treatment. However, this has be reconciled with field results, such as those of Weir & Barraclough (1986) (section 11.1.3), and the total picture is still not clear. The purely physically based models of the control of water extraction from soil by plants are almost certainly inadequate.

2.3.3 Transport of Water in Root and Stem

The movement of water across the epidermis of the root, into the stele and up the xylem of the plant sets the boundary condition for the transport of water in the soil. It also has a bearing on whether water uptake occurs at different rates in

Table 2.2 Wheat grown with a supply of water from irrigation, a water table at 20 cm, or stored water in the profile plus watering at 57 days after emergence.

			Water table (I)		Stored water (II)	
Variable	Unit	Irrigated (control)	Value	% Control	Value	% Control
Total tillers/plant	number	15.4	15.2	98.9	13.3	86.2
Total ears/plant	number	13.6	9.2	67.5[a]	9.5	69.7[a]
Tiller mortality rate	%	11.4	39.5	—	28.3	—
Shoot dry weight	g	20.0	13.7	68.8[a]	12.9	64.6[a]
Biomass/plant	g	31.9	25.2	79.0[a]	24.8	77.9[a]
Flag leaf area	cm^2	63.9	48.4	75.7[a]	38.2	59.8[a]
Plant height	cm	75.1	65.2	86.8[a]	64.9	86.4[a]
Grain yield/plant	g	11.9	11.4	96.2	12.0	100.4
Kernels/plant	number	426	407	95.6	417	97.9
Kernel weight	mg	28.6	27.5	96.3	28.7	100.4
Harvest index	$g\ g^{-1}$	0.37	0.45	120.8[a]	0.48	128.9[a]
Days of heading	number	82.8	77.1	93.0[a]	62.1	74.9[a]

Source: after Blum *et al.* (1991).

Grain yield per plant was little affected, but significant differences in several parameters were attributed to a non-hydraulic signal from the roots. Superscript a indicates significant ($P >$ 0.05%) differences from control.

different parts of the root system, and how this is related to the water distribution in the soil (root anatomy in chapter 5 and root distribution and density are discussed in chapter 9). Calculations of the hydraulic conductivities of the xylem vessels from their diameters indicates that the greater part of the total root and stem resistance to water flow usually lies in the radial pathway across the parts of the root outside the xylem.

The flows of ions and of water into and up the xylem are not closely linked. The simplest interpretation is that ions are pumped into the xylem (section 5.2.1) at a rate determined by the nutrient status of the plant, its growth rate and the level of external supply. The flow of water into the xylem is determined by the plant transpiration, so long as it can be supplied by the soil. The concentration of the xylem flow can therefore vary widely with the transpiration rate (Munns & Passioura 1984). The structure of the root (chapter 5) is so complex that understanding water transport across it has not been easy. The apoplasmic pathway should follow the water-filled intercellular and cell wall spaces (section 5.2.1). However, the intercellular spaces are of the order of 10 μm diameter, and hence they empty at around -30 kPa, which is a very small lowering of potential. The pores in the cell wall are of the order of 8 nm, which is so narrow that a high potential gradient would be needed to force water along the walls. Even a speed of 1 cm day^{-1} would require the considerable pressure gradient of -30 MPa m^{-1} (Newman 1976; Nobel 1991). Since the cross-section of cell wall available as a direct route from soil to xylem is very small, this suggests that a continuous pathway through cell walls is not very useful. However, a fully apoplasmic pathway exists where water may flow through young roots and root tips before the endodermis is laid down, or where a new lateral root is emerging (section 5.2.1). It is therefore not unexpected that different plant species may show very different conductance (Newman 1976) (table 2.3) and apoplasm/symplasm water flux ratios (Steudle *et al.* 1993). The root conductances may also vary with plant age (Fiscus & Markhart 1979) (figure 2.7).

The reaction to changes in temperature or the use of metabolic inhibitors (Brouwer 1965) shows that simple flow through a porous medium with a single semipermeable membrane at the endodermis could not represent the root correctly. The argument above suggests that most water flow is through the symplasm in the cortex and the pericycle, as well as in the endodermis. The current view is that most water traverses the root radially through the cells and their plasmalemma and tonoplast membranes, because of the low impedance in the

Table 2.3 Permeability values per unit surface area for entire root systems of selected crop plants.

	Permeability (nm s^{-1} MPa^{-1})	Standard error
Broad bean (*Vicia faba*)	5.4	0.5
Dwarf bean (*Phaseolus vulgaris*)	5.6	0.2
Sunflower (*Helianthus annuus*)	7.1	0.6
Maize (*Zea mays*)	22	2.7
Tomato (*Lycopersicon esculentum*)	61	15

Source: after Newman (1976).

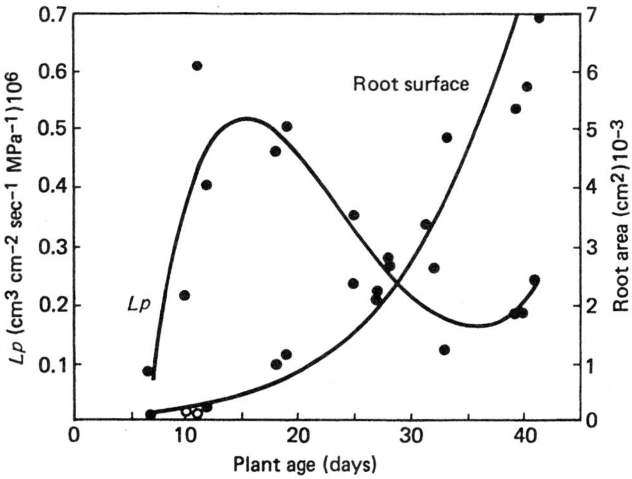

Figure 2.7 Changes in mean hydraulic conductance (L_p) of intact bean root systems with growth of the root system and changing proportions of root surface at different stages of development (after Kramer & Boyer 1995).

latter and the negligible impedance of the cytoplasm and vacuole. This is called the cell-to-cell pathway, and may include both this direct route, which is mainly through the vacuole, and also a purely symplasmic component in which water moves via cytoplasm and plasmodesmata, in a manner analogous to the symplasmic transport of ions (section 5.2.4). These two symplasmic routes cannot in practice be distinguished (Steudle 1994).

A recent representation of flow in a root is the 'composite membrane' model (Steudle 1993; Steudle *et al.* 1993) (figure 2.8). Axial resistance along the root vessels is ignored. Most flow is assumed to be through the symplasm, largely via the cell-to-cell pathway. This pathway traverses several semipermeable membranes, so if it were the only way for water to move from soil to xylem, the root should be an almost perfect osmometer. However, in parallel with this is the fully apoplasmic route described above, which is not interrupted by the endodermis, so there is a pathway right from the external surface to the xylem elements that does not involve crossing a membrane.

The consequences of this are best understood in considering the different fluxes of water traversing the root when osmotic and hydraulic driving forces of equal value are applied. The symplasmic pathway would have a reflection coefficient of near to 1, the apoplasmic component one of 0. An osmotic driving force, such as an externally applied salt solution, would drive water out through the symplasmic pathway, but the negative hydraulic pressure developed by this in the root would then drive solution inwards along the apoplasmic route. On the other hand, an externally applied hydrostatic driving force would drive water through both pathways. The different effects of the differently caused water potential gradients, and the variety of different combinations of pathways and pathway properties, explain why this problem has remained difficult for so long.

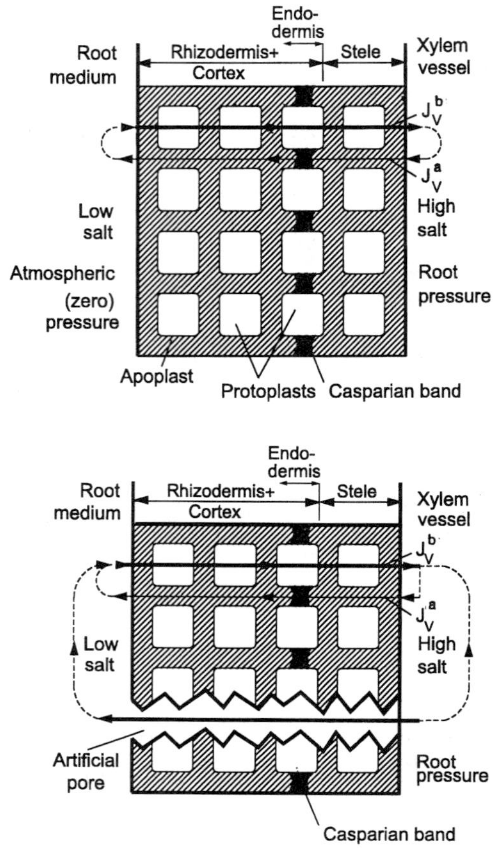

Figure 2.8 Representation of composite transport of water and ions in a root with high salt concentration in the xylem due to ion uptake from the external solution at low transpiration rates. A flow J_v of water from the external solution to the xylem through the cells (symplasm route, superscript b) results from this, causing a positive root pressure. This causes a reverse flow of xylem solution outwards along the apoplasm route (superscript a), that is, through natural or artificial breaks in the Casparian band. The net uptake of ions and water is the resultant of these flows (after Steudle *et al.* 1993).

Steudle *et al.* (1993) tested these ideas with roots that had been radially pierced with very fine glass tubes to within the endodermis. The roots behaved in a manner consistent with the idea that the apoplasmic route had simply been enlarged by the presence of these fine holes. The full mathematical treatment of this model uses irreversible thermodynamics to provide formulae for the effective reflection coefficient and radial hydraulic conductivity. In one case, the latter was 10 times larger for a hydraulic force than for an osmotic driving force. This theory also explains the frequently observed fact that the root resistance to water entry varies with the water flow rate (Weatherley 1982; Passioura 1988) (figure 2.9). At low water flow rates, the ion uptake process produces a concentrated solution in the xylem. This produces a hydraulic pressure that will drive solution back along the apoplasmic route. At high flow rates into the xylem, the solution there is dilute, and the hydraulic pressure will normally be negative; hence, both hydraulic and osmotic potentials cause an inward flow, and both pathways will deliver water into the root. This will appear as a high resistance at low flows, and a low resistance at high flows. Such a backflow at low flow rates, which did not traverse a membrane, would allow a leakage of ions from the plant into the external solution, which may be one source of the leakage that causes C_{min} (the minimum concentration below which no net uptake occurs) to be above zero (section 5.3.2).

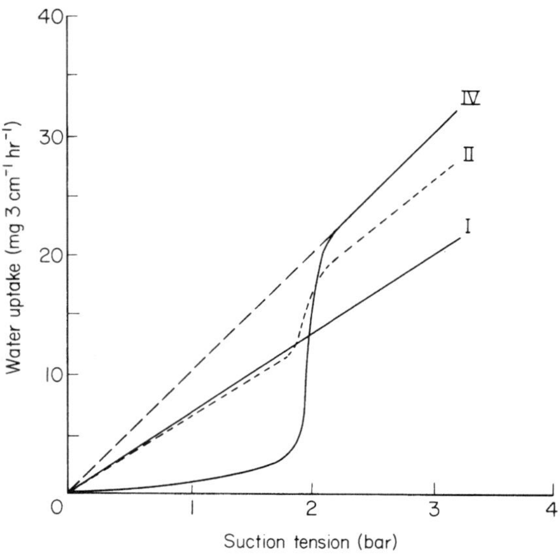

Figure 2.9 The relationship between hydraulic potential (suction) and the water uptake rate for 3-cm-long sections of root of a broad bean plant. I is nearest the tip, and shows a constant conductivity across the root. Further back (IV), the conductivity is low at small potential differences, but decreases sharply as these increase, so that the Ohm's law analogy breaks down (after Brouwer 1965).

Other explanations have assumed that the Casparian band is a complete barrier for solutes, so that if the root is not a perfect osmometer, this is due to the inherently low reflection coefficient of the endodermal plasma membranes (Baker *et al.* 1992). The methods of irreversible thermodynamics have been used to analyse the coupled flows of water and solute through such membranes (Dalton *et al.* 1975), but this attribution of all the properties of the root to a single pathway has not been widely used subsequently. Steudle (1993) concluded that this and other earlier methods do not explain the observed effects.

In general, the flow rate into a given segment of root declines as it ages, possibly because of increasing suberization of the endodermis. Clarkson *et al.* (1974) found that it changed by a factor of 8 in one case, but contradictory results have been obtained (Newman 1976). Haussling *et al.* (1988) also found that the entry rate of water declined sharply from the root tip backwards in Norway spruce. In agreement with Brouwer (1965), they found that when the total uptake rate of water by the plant was increased, the effect on the local rate was much more marked in old than young root, that is the resistance changed more sharply with flow rate in the old roots (figure 2.9).

It is normally stated that water does not move backwards from the root into the soil to any significant extent, that is the root acts as a 'rectifier', though various contradictory results have been published (Passioura 1988; Baker *et al.* 1992). Thus, *Agave* roots have 'rectifier' properties, with a resistance to movement from root to soil some 60 times larger than in the reverse direction (Rundel & Nobel 1991). On the other hand, Caldwell *et al.* (1991) showed with a deuterium tracer that water entering the root system of *Artemisia tridentata* at depth could later be detected in nearby shallow-rooted *Agropyron*, having moved from lower in the profile by a process they call 'hydraulic lift'. Tyree *et al.* (1995) also found strong backflow into the soil from the upper roots of an arid-zone *Artemisia*. However, they also supported Passioura's (1988) explanation of 'rectification' as being due to strong solute concentrations forming on the inward side of membranes if outward-directed solution flow resulted from a positive hydraulic xylem pressure. The osmotic pressure of this solution prevents further flow outwards. It seems that both rectification and hydraulic lift are possible under different conditions. A large apoplasmic pathway across the root and a low solute status in the plant would tend to prevent rectification, and should allow outward flow of water (figure 2.9). A possible advantage to the plant of an outward flow of water would be its subsequent ability to obtain nutrients from the rhizosphere soil that had been wetted.

The boundary condition for water uptake at the surface of the root is therefore not simple. First, there will be some more or less random variation along the root, as shown by the dye perfusion experiments of Canny & Huang (1994), in addition to the more regular effect of ageing. Second, the relationship of the water potential in the xylem and at the root surface may be difficult to define. The potential in the xylem can be reduced to relatively low values, say −2 MPa, and this potential is theoretically available at the root surface. However, part of this may be present as a positive or negative hydrostatic pressure, due to the water status of the plant, and the rest will be an osmotic potential due to the continuing delivery of ions into the xylem. These potentials will have different effective hydraulic conductivities, and will therefore have different effects at the root surface. Third, the root water potential can directly influence the stomatal conductance, and therefore the transpirational flux. It is not surprising that the root resistance (figure 2.9) is highly non-ohmic.

However, a full model has been developed based on these ideas, with five basic elements (Tardieu 1993; Tardieu & Davies 1993; Davies *et al.* 1994): (1) an equation that expresses the stomatal conductance as a function of the concentration of ABA in the xylem and the leaf water potential; (2) the water potential of the water-absorbing roots derived from a resistance equation (figure 2.5) with the soil water potential and the water uptake flux; (3) an equation relating the concentration of ABA in the xylem to the water potential in the roots and the water flux up through the plant; (4) Newman's (1969) model that provides the soil resistance to water uptake around the root,

$$R = 1/(2\pi K\theta L_a)\ \ln\,(d/a) \tag{2.18}$$

where R is resistance, d is half the mean distance between roots and a is mean root radius; (5) the transpiration flux estimated from a simple Penman–Monteith model, using the stomatal conductance from (1).

These five equations with five unknowns (stomatal conductance, transpirational flux, ABA concentration, and root and leaf water potentials), can in principle be solved for a series of times. It was concluded that purely physical or chemical regulation systems for the stomata are not tenable, and that a more complex interactive system is needed. Abscisic acid has a number of other roles in plant development, which need to be integrated with this direct signalling system.

Campbell (1991) has also produced a useful model based on the resistance analogy of plant water uptake along classical lines, with non-uniform distribution of roots with depth. Despite the fact that a simple physical relationship between leaf water potential and stomatal resistance was assumed, with no root control factors, fair agreement with soil profile water distribution was obtained. Most crop models (section 11.3.3) avoid these complications by working with much simpler assumptions, and ignoring mechanistic interpretations.

Environmental and nutritional factors can also affect the hydraulic conductivity of the root. Sodium chloride strongly reduces the conductivity of maize root cells, but Azaizah *et al.* (1992) found that adding calcium chloride, while it slightly reduced the conductivity itself, prevented any effect of sodium. However, as maize has a large apoplasmic pathway, the effect of these salts on the root radial conductance was less than on cell conductance. Radin & Matthews (1989) investigated the hydraulic parameters of N- and P-deficient cotton plants. The cell conductance was very sharply diminished in plants transferred into N- and P-free solution. The root radial conductance was also reduced, but remained in absolute terms twice as large, because of the apoplasmic pathway. These effects on root cell conductivity are claimed to be amongst the earliest consequences of nutrient deficiency (table 2.4). Anaerobiosis similarly reduced root radial conductance (Everard & Drew 1987), and it also decreased the active pumping of nutrient ions, so that the osmotic pressure in the xylem was small.

2.3.4 Water Movement in the Rhizosphere

Much attention has been given to the physical and biological water relationships in the rhizosphere, in the belief that movement of water in the rhizosphere could be a rate-limiting step in water supply to the plant. Gardner (1960) addressed this by treating water uptake of a root from soil as a water diffusivity problem, assuming a single root in an infinite volume of soil. He avoided the problems of defining the plant's water demand and root water potential discussed above by simply assuming that a root has a constant water potential at the surface. He used a solution of the diffusion equation adapted for water flow (equation (2.19)):

$$\psi_s - \psi_a = (I_w/4\pi K)[\ln (4D_w t/a^2) - 0.577] \tag{2.19}$$

where subscripts s and a designate soil and root surface, a is root radius, D_w is the water diffusivity, and I_w is water inflow, in $cm^3\ cm^{-1}\ s^{-1}$.

Table 2.4 Effects of nutrients on water relations of root cortical cells, as tested in water culture. Note the decline in L_p with nutrient deficiency. For comparison, L_p for intact roots was 2.3 for full nutrients, and 1.0 for −P (both m s^{-1} $MPa^{-1} \times 10^{-7}$). ε, Modulus of elasticity; $t_{1/2}$, half time for water exchange; L_p, hydraulic conductivity.

Treatment	Turgor (MPa)	ε (MPa)	$t_{1/2}$ (s)	L_p (m s^{-1} $MPa^{-1} \times 10^7$)
Full nutrients	0.41 ± 0.07	6.6 ± 0.7	11 ± 2	1.19 ± 0.25
Intermediate N	0.44 ± 0.03	5.9 ± 0.7	22 ± 3	0.67 ± 0.12
−P	0.47 ± 0.03	7.3 ± 0.7	29 ± 4	0.45 ± 0.08
−N	0.63 ± 0.04	12.1 ± 1.5	33 ± 5	0.22 ± 0.04

Source: after Radin & Matthews (1989).

Other authors have used different approximations, including steady state (water content remaining constant in an equivalent soil cylinder around the root, with water supplied at the periphery) (equation (2.17)) and steady rate (similar removals of water from all points in a cylinder around the root) (Passioura & Cowan 1968). The simplest statement of the steady-state treatment gives the equation (2.17), which is analogous to the steady-state solution for ion uptake (Tinker 1976) (section 10.5.2). The basic equation assumes water supply from an annular soil cylinder of inner and outer radii a and b, respectively, and where the water enters at the outer radius b and exits at the root radius a — that is, a steady-state solution in which the constant I_w is the water demand set by the environmental conditions. The basic equation is

$$I_w = 2\pi r D_w \, \partial\theta/\partial r \tag{2.20}$$

giving by integration, if D_w is constant,

$$\theta_b - \theta_a = I_w/2\pi D_w \, (\ln \, b/a) \tag{2.21}$$

In all these equations, I_w has a very important effect. The larger the inflow, the greater the potential gradient in the soil, and the lower the water content will be in the rhizosphere near the root surface. This causes difficulties if I_w varies widely along the roots, because normally the mean value over a whole root system or along an individual root has to be used. Figure 2.9 suggests that I_w increases proportionately most rapidly at positions well behind the tip. This agrees with the conclusions of Varney & Canny (1993) that the position of maximum I_w moved further back along maize roots as its absolute value increased, to a point 20–30 cm behind the tip, before decreasing again beyond 60 cm from the tip. The distribution of water uptake along the root is thus significantly different from that of the nutrients, so real values of mass flow will differ greatly along the root (section 5.4.3).

The value of K or of D_w is also very important in equations (2.19, 2.20) and (2.21), in which they are often assumed to be constant. If the inflow of water is sufficient to decrease θ_b, then D_w will decline and the water content gradient must become still steeper. As the drying-out process continues, a point is reached at which a further decrease in the root water potential has no further effect on the inflow, because this merely reduces D_w still further. There are analogies with the 'inflow to a zero sink' found with ion uptake (section 6.1.2(3)). It is, of course, by no means certain that a given plant can produce a sufficiently large negative potential in the root to reach this limiting inflow.

By integrating equation (2.20) with respect to r and to θ, and using an empirical relation between θ and D_w, the equation can be solved to show this drying-out process (Lang & Gardner 1970; Whisler *et al.* 1970). Inserting typical values for the variables for a loam soil gives a maximum inflow of water of the order of 0.1 $cm^3 \, cm^{-1} \, day^{-1}$ for I_w. At this I_w value, the drying out around the root is very marked, and if it is used in model calculations it is not surprising that they predict drying out around the roots. This figure is, in fact, very large compared with the normal mean values found for plants in the field, and it is unlikely that this value is closely approached except in special situations (Newman 1976).

If the soil is dried out locally around the root, this would reduce the ion diffusion coefficients there, as well as affecting water supply and rhizosphere microbiological and faunal populations. There have, indeed, been suggestions that one consequence of drought is phosphorus deficiency induced by the low diffusion coefficient in dry soils (Safir *et al.* 1972), though this is certainly not always true. Such dried-out zones have been produced in specially designed experiments (Dunham & Nye 1973) (figure 2.10). The results from these agreed with theory, as developed for the linear geometry they used. It is not possible to give an exact equivalent to the I_w value in cylindrical geometry, but the flux at the root mat surface was 0.17 cm^3 cm^{-2} day^{-1}.

A similar flux at the surface of a root of 0.05 cm diameter would give I_w of 0.025 cm^3 cm^{-1} day^{-1}. In comparison, Lawlor (1972) found a $\bar{I}_w$ of 0.006 cm^3 cm^{-1} day^{-1} or less for ryegrass growing in large pots. In this work, the calculated ratio of the resistance within the plant to that in the soil, using the Ohm's law analogy, was always well above 1 (table 2.1). That this was so even with the mean soil water potential as low as −1.1 MPa argues that soil resistance rarely becomes important in practical conditions for densely rooted species. This conclusion may differ for sparsely rooted species (chapter 9).

The weakness of these arguments against local drying out around roots, when applied to the field, is, of course, that water does not flow uniformly to the root surfaces at all depths in the profile. Surface soil will normally dry out first (see figure 11.3) (Weir & Barraclough 1986), followed by lower layers, in turn. The variation in root hydraulic conductivity at different points will also affect I_w, and values of I_w several times the mean may therefore apply in parts of a root system. There are also results that suggest important soil resistance to uptake, despite the fact that theory indicates that it should be insignificant in most cases. Thus, Harmsworth & Aylmore (1989) used computerized tomography to show dried-out zones around radish tap roots in soil, but the I_w that can be calculated from the data reported was several ml per centimetre per day (cm^3 cm^{-1} day^{-1}). It is not surprising that dried-out zones appeared, and the system was clearly not typical of fibrous roots. Tinker (1976) discussed the considerable effects of clumping of roots, and Tardieu (1993) made simulations to show that considerable differences between potentials in bulk soil and those near clumped roots can be predicted, with a high value for I_w in a clay. He suggested that roots in heavy and compacted soils are strongly clumped, and that this effect explains the frequently low stomatal conductance in plants growing in such soils.

The possibility of poor root–soil contact, especially after root shrinkage, has had much attention (Nye 1994). This effect and its consequences are discussed in more detail in chapters 7 and 9. There is a good theoretical case for the importance of root hairs in uptake of water, which has been proven to occur in potometer experiments (Rosene 1943; Cailloux 1972). Jones *et al.* (1983) showed with the pressure probe that the hydraulic conductivities of the membranes of root hairs and cortex cells were similar. Similar arguments have been advanced for mycorrhizal hyphae, but the evidence in this case is conflicting (section 8.3.9.2). Given the irregularities and voids in normally structured soils with a faunal population, and the tendency of roots to seek out planes or points of weakness,

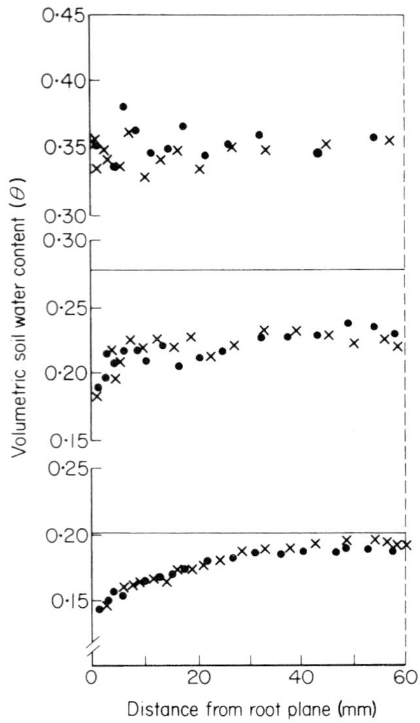

Figure 2.10 Water content gradients in root blocks after 6 days with one end in contact with a layer of absorbing onion roots. The three blocks had different original moisture contents. Note the effect on size of the gradient. • and × indicate replicate blocks on opposite sides of the root curtain (after Dunham & Nye 1973).

total root–soil contact may be a rather rare situation. Sanders (1971) noted that 43% of roots in a rhizotron were not in full soil contact.

A simple indication of poor root–soil contact was given by Faiz & Weatherley (1982), who found that the plant water potential of sunflowers could be sharply increased by shaking the pots, or squeezing the plastic bags in which they were growing in soil. They concluded that this disturbance improved the root–soil contact, but disturbing the soil would, of course, alter the soil water potential also.

Most roots with diameters smaller than the holes they occupy will touch one side of the void, because of surface tension or random curvature. The effect of this on resistance to water flow has been considered in two main ways. Tinker (1976) used an electrical analogue, and the results indicated that as long as contact was maintained over at least 50% of the circumference, the effects on the water content at the root surface were not large, as compared with total contact, even with an I_w as large as 0.015 cm^3 cm^{-1} day^{-1}. However, this method considered only the resistance to water flow in the soil, and not that within the cortex. Nye (1994) developed a full simulation model, allowing for resistance outside and inside the root surface (see chapter 7). When contact was limited, the main effect on I_w was due to the resistance in the root cortex rather than in the soil. The conclusion was that the effect of incomplete contact on I_w could be considerable, even when the fraction of the circumference in contact was as high as 50 or 60% (figure 2.11) (Nye 1994).

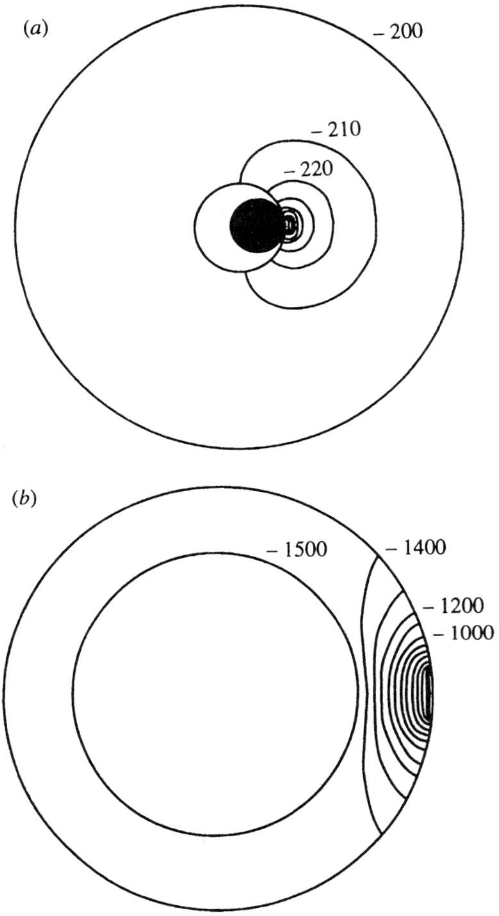

Figure 2.11 Resistance to movement of water in roots and in surrounding soil, including the effect of root–soil contact. (a) Black circle is shrunken root; next circle defines surrounding void, outermost circle is limit of soil. (b) Inner circle is stele of root, outer one the epidermis. Contour lines of equal water potential are numbered in J kg^{-1} (after Nye 1994).

It is possible that vapour-phase transport across the total or partial airgaps could be important. Most workers have concluded that such a transfer must be insignificant (Tinker 1976), especially if most of the effect of partial contact occurs through the root resistance. However, Nobel & Cui (1992), working with desert succulents over a drying period, concluded that the limiting factor for I_w was, in time sequence, the hydraulic conductivity of the root, then the airgap between the shrinking root and the soil, and finally the soil conductivity. In their rather specialized conditions, they considered that vapour-phase transport was important when the airgap formed the main resistance.

The possibility that mucigel (section 8.1.3) can act as a bridge between root and soil has often been suggested. However, Guinel & McCully (1986) found that mucigel produced from maize root tips had a very high water content (99.9%) and water potential (−7.3 kPa). When tested in agar of different concentrations, a mucigel droplet shrank from 10 mm^3 in water to 0.5 mm^3 at −9.7 kPa, and to almost zero volume at −24.4 kPa. Mucigel can therefore be effective as a bridge only at small negative water potentials, where this is probably not useful to the plant.

The present state of this subject is confused. The classical theory shows quite clearly that I_w has to be well above the usual mean values for crops in the field before a significant soil resistance and dried-out zones are credible, though it may happen in desert environments, with very sparsely rooted plants or in plants with very clumped root systems in heavy soils. If stomatal closure is under the control of chemical messengers (section 2.3.2), this conclusion is supported, because drying of soil and shrinkage of roots will certainly then cause a decrease in the transpiration rate and hence in the maximum value of I_w that has to be met. However, there are various points that cause uncertainty, mainly the variability of water inflow along a root, the poor contact between soils and roots, and the non-uniform extraction of water from the soil profile. The main conclusion seems sound, but it may not apply in special cases. Oertli (1996) reviewed this question recently, but was not able to state a firm conclusion.

2.4 Conclusion

As the problems in water uptake have been investigated, those related to soil have generally been solved, at least in principle. The main problems outstanding for soils are not the basic processes, but the heterogeneity of soils from the rhizosphere to the field scale, and the variation of soil parameters across these scales. The less tractable basic problems are associated with the behaviour of plants, or plant–soil interactions. Stomatal control can no longer be seen as a wholly physical mechanism, and precise and proven models for this control are needed. The apparently simple process of uptake of water by roots is being seen as more and more complicated, with the opportunity for root property variation in response to many plant and soil variables. A system to measure the real value of I_w at different points on the root system is the biggest need at present, because a solution to this problem will cast much more light onto the questions of the root surface and rhizosphere impedance to water movement.

3

Solute Interchange between Solid, Liquid, and Gas Phases in the Soil

We noted in chapter 1 that the concentration of solute in the soil solution is buffered by solute adsorbed on the soil surfaces. We also show in chapter 4 that the overall mobility of ions is related to their amounts and mobilities in the solid and solution. In this chapter, we focus on the soil solution concentration, primarily to show how the factors controlling it can be incorporated in models of the growth of crops and the leaching of nutrients or pollutants, such as those described in chapters 10 and 11. We examine the general principles governing the interchange of solutes between all phases in the soil, dealing first with inorganic ions, especially plant nutrients and heavy metals; and later with organic solutes, including biocides, which may also occur in the vapour phase. We also consider the reactions between metal ions and other organic or inorganic ions in solution to form complexes, such as $CuOH^+$.

3.1 Composition of the Soil Solution

3.1.1 Measurement

The method of displacing the pore solution from a column of soil with ethanol, introduced by Ischtscherikow (1907), has been examined by Moss (1963, 1969). He found, in accord with theory (section 3.1.3), that the activity ratios $(K)/(Ca + Mg)^{1/2}$ and $(K)/(Ca)^{1/2}$ determined in the displaced solutions remained constant over considerable changes in soil moisture level to the point of saturation. He also found that the activity ratio $(K)/(Ca + Mg)^{1/2}$ in the extracts from a wide range of soils agreed well with the activity ratio determined by the null point method of Beckett & Craig (1964). In this method, the soil is shaken with dilute $CaCl_2$ solution containing graded amounts of potassium, and the activity ratio at which the soil does not gain or lose potassium to the solution is determined. Ethanol appears to displace solution from the fine as well as the coarse pores, and successive fractions, devoid of alcohol, have the same composition. For small samples of soil, it is more convenient to add a heavy liquid that is immiscible with water, and extract the solution by centrifuging (Kinniburgh & Miles 1983). Suction methods are useful for following changes in composition of moist soils. They should be used with care since they change the pressure of CO_2 and hence the concentration of the bicarbonate ion. These and other methods have been discussed by Nielsen (1972) and Adams (1974).

Some typical soil solution compositions are given in tables 3.1 and 3.2.

3.1.2 Exchangeable Cations — their Concentration in the Soil Solution

The next sections refer frequently to the different types of clay minerals found in soil. White (1979, chapter 2) gives a brief and sound introduction to these minerals. All are aluminosilicates that consist of layers of aluminium oxides and silicon oxides with some of the aluminium and silicon atoms isomorphously replaced by other atoms, so as to leave a net negative charge on the lattice; this is balanced by cations that readily exchange with cations in the soil solution. There are three main types of mineral:

(1) The 1:1 type kaolinites (e.g. kaolin); one layer of silicon/oxygen, and one layer of aluminium/oxygen.
(2) The expanded 2:1 type smectites (e.g. montmorillonite); one layer of aluminium/oxygen sandwiched between two layers of silicon/oxygen — this unit is separated from the next by layers of water molecules.
(3) The non-expanded 2:1 type hydrous micas (e.g. illite).

Humus, a weak acid, can also be negatively charged and act as an exchanger by losing a proton. The hydrous oxides of iron and aluminium can become positively charged, by adsorbing a proton, and then have anion exchange properties.

The total strength of the soil solution depends upon the concentration of unadsorbed anions that it contains. Among these, chloride, nitrate and bicarbonate are not adsorbed by the negatively charged soil colloid surface, unless it also contains positively charged sites, which are often associated with iron and aluminium oxides at pH below about 6. Sulphate also is not usually strongly adsorbed. These ions control the overall strength of the soil solution. The proportions in the soil solution of the different cations that balance these anions is determined by the ionic charge of the adsorbed cations, their proportions on the exchange complex, their ionic size and the properties of the exchanger. These factors are considered, in turn, in the next section. Thus, to predict the effect of a treatment, such as drying the soil, on the solution concentration of a cation, the first question is, What will be the effect on the anion concentration? and then, What proportion of the anions in solution will be balanced by that cation? It is often not appreciated

Table 3.1 Representative soil solution compositions (meq $litre^{-1}$).

	pH	Ca^{2+}	Mg^{2+}	K^+	Na^+	NH_4^+	Total cation	NO_3^-	Cl^-	HCO_3^-	SO_4^{2-}	Total anion	Reference[a]
Acid soil	4.2	1.0	1.4	0.4	0.4		3.2	3.8	0.2	—	0.8	4.8	Vlamis (1953)
Sandy loam	7.2	21.0	1.2	0.7	1.8	0.4	25.1	15.6	2.2	1.1	7.0	25.9	Eaton *et al.* (1960)
Average of 8 cropped soils	7.3	10.1	7.1	0.68	1.8		19.7	3.7	—	1.8	12.5	18.0	Burd & Martin (1923)
Average of 8 fallowed soils	7.0	27.9	10.9	1.6	2.8		43.2	29.6	1.4	1.0	9.7	41.7	Burd & Martin (1923)
Saline soils	8.3	43.5	48.0	9.6	21.7		114.8	(31.2) by difference	20.1	7.2	56.3	114.8	Reitemeier & Richards (1944)

Table 3.2 Soil solution composition (μM).

	pH	Na	K	Mg	Ca	Sr	Ba	Mn	Fe	Cu
Grassland/arable[a]	7.7	465	390	135	2120	1.9	0.16	0.52	3.4	0.25
Woodland[b]	4.8	335	284	104	592	0.63	0.41	69	24.1	49
	TON[c]	Cl	S	P	Si	B	DOC[d]			
Grassland/arable[a]	860	1590	327	64	220	5.1	4250			
Woodland[b]	991	749	398	17	255	6.6				

Source: Campbell *et al.* (1989).

[a]18 grassland and 6 arable topsoils. Median throughout the year.
[b]5 woodland topsoils. Geometric mean.
[c]Total oxidized nitrogen.
[d]Dissolved organic carbon.

that to understand the amounts of cations that may be leached from the soil by rain, it is essential to understand first what controls the concentration of anions.

Typically, a large range of cation and anion species coexist in natural soil solutions. In addition to the common cations, in acid soils there is a significant concentration of aluminium and hydrogen ions. Many cations, such as those of the heavy metals, form complexes in solution, such as Cu^{2+}, $CuOH^{+}$, and $CuCl^{+}$. If the form in which a root may take up an element from the soil solution, or the sorption–desorption reactions of the element with the soil, are to be understood, then it is necessary to decide what the actual concentrations of each possible complex is. The computer program GEOCHEM (Sposito & Mattigod 1979) contains the necessary stability data about the different complexes to effect this task.

3.1.3 Exchangeable Cations — Effect of Ionic Charge

Solute Movement in the Soil–Root System (Nye & Tinker 1977) gave a thermodynamic treatment of this material. Here, we present a simplified account that is sufficient for the purposes of this book.

Consider the exchange reaction

$$\text{soil–Ca} + 2K^{+} = \text{soil–2K} + Ca^{2+}$$

This is governed by the mass action equation

$$\frac{(\text{soil–2K})(Ca^{2+})}{(\text{soil–Ca})(K^{+})^{2}} = K_e \tag{3.1}$$

where K_e is the equilibrium constant for the reaction and the terms in parentheses are activities. Rearranging the terms gives

$$\frac{(\text{soil–2K})}{(\text{soil–Ca})} = \frac{(K^{+})^{2}}{(Ca^{2+})} K_e \tag{3.2}$$

For a small exchange, such as might be caused by drying the soil, the ratio on the left-hand side (LHS) will be little changed since there is usually a large reserve of exchangeable ions adsorbed on the solid. Hence the 'reduced activity ratio' $(K^+)/(Ca^{2+})^{1/2}$ tends to remain constant. Introducing solution concentrations, $[K^+]$ and $[Ca^{2+}]$, and activity coefficients, f_K and f_{Ca}, the reduced activity ratio equals $[K^+]f_K/([Ca^{2+}]f_{Ca})^{1/2}$. In dilute solutions, less than *c.* 0.003 M, the activity ratio may be replaced by the concentration ratio $[K^+]/[Ca^{2+}]^{1/2}$ since the activity coefficient ratio $f_K/(f_{Ca})^{1/2}$ is nearly unity. In stronger solutions, but less than *c.* 0.5 M, the Davies' (1962) equation is commonly used to calculate the activity coefficients:

$$\ln f_i = -az_i^2[I^{1/2}/(1 + I^{1/2}) - 0.3I] \tag{3.3}$$

where

z_i = the charge on the ion,
I = the ionic strength of the solution,
a = 0.509 at 25°C.

For further discussion of activity coefficients in general, see Robinson & Stokes (1959), and in soil solutions, Sposito (1981).

As a rule, calcium is the predominant cation in the soil solution. Thus, if half the water in the soil solution is removed its concentrations of calcium ions will roughly double; and in order that the reduced activity ratio $(K)/(Ca)^{1/2}$ should remain steady, we may predict that the activity of potassium will increase by a factor of $\sqrt{2}$. Moss (1963) has verified on three soils that the reduced activity ratio does remain constant over the field moisture range.

Schofield & Taylor (1955) suggested adopting calcium as a reference ion, and measuring the concentration of other ions in M/100 $CaCl_2$. This should cause only a slight disturbance to the adsorbed ions. Hence, soil pH is best measured in this solution rather than in water which gives rise to a variable calcium concentration, depending on the soil/solution ratio and the initial salt concentration in the soil solution. For soils with weak soil solutions, 0.003 M $CaCl_2$ is often preferred. For very acid soils in which Al^{3+} dominates, even 0.003 M $CaCl_2$ may be inappropriate (Tinker 1964a, b).

3.1.4 Relative Proportions of Exchangeable Cations

The relation between the proportions of two cations on the exchange complex and their reduced concentration ratio in the soil solution has to be determined experimentally, since there are no general equations that account accurately for the complexity of real soils. The more theoretical aspects of exchange reactions are full of subtle questions of definition and interpretation. We recommend Bolt's account (1982, chapters 2, 3 and 4) as being one of the few that treat the subject fundamentally, and White (1979, chapter 7) for a simpler account.

3.1.4.1 Cation Exchange Equations

Most soils have a permanent negative charge, mainly originating from isomorphous substitution of a cation in the clay mineral lattices by another carrying a smaller positive charge: for example Al^{3+} by Fe^{2+}. The negative charge is satisfied by cations that are partly held very close to the mineral surfaces in the 'Stern' layer, a few water molecules thick, and partly in a diffuse layer in which their concentration declines roughly exponentially with distance to that of the surrounding soil solution. There are numerous theories about the details of the distribution of ions at and near charged surfaces. Sposito (1984) gives a full account, and a clear summary is given in Mott (1988).

Of the many exchange equations proposed, the Gapon (1933) equation is probably the most used. It is

$$K_g = \frac{N_A}{N_B} \times \frac{(B)}{(A)} \tag{3.4}$$

where K_g is the Gapon constant, A is the concentration of a monovalent cation, B of a divalent cation, and N_A and N_B are fractions of the total charge on the exchanger. Though first proposed as an empirical equation, Eriksson (1952) has derived it theoretically. The Gapon equation predicts the activity ratio in solution from the proportions of ions on the exchanger in terms of the single Gapon constant that may be determined experimentally. Bolt (1955) showed that the value of K_g for the exchange between sodium and calcium on illite remained constant from 1 to 70% sodium saturation. For many saline soils, N_{Na} has been found to be linearly related to the activity ratio $(Na)/(Ca + Mg)^{1/2}$ (Richards 1954). The exchange of potassium for calcium, however, behaved irregularly.

In the theoretical derivation of the Gapon equation, K_g is derived from the surface charge density of the exchanger, but this can only be determined reliably on pure clays. Beckett & Nafady (1967a, b) have shown that on the planar adsorption sites of several 2:1 type clay minerals, N_K is linearly related to $(K)/(Ca + Mg)^{1/2}$. This is not true of the non-expanded 2:1 type illites whose edge sites show a strong preference for potassium. Since the derivation of the equation assumes a uniformly charged plane surface, it is not surprising that it fails to explain behaviour in soils that contain a significant proportion of their negative charge on clay mineral edge sites, or those sites formed by ionization of a proton from acid groups on inorganic or organic compounds.

Examples of exchange isotherms for K–Ca and Mg–Ca, covering the whole range of saturation from 0 to 100%, are shown in figures 3.1 and 3.2. It should be noticed that when the ions have different charges, the position of the curve depends on the total concentration of the soil solution. Bruggenwert & Kamphorst (1979) have summarized the extensive literature of 700 titles on the exchange constants for various cation pairs on both pure clays and soils. Unfortunately, it is difficult to make useful generalizations from all this data.

The exchange constant for a pair of ion species is affected by the presence of a third species. Bond & Verburg (1997) have shown how such ternary exchanges, such as those between K^+, Na^+, and Ca^{2+}, can be approximately predicted from the three binary exchange constants $K^+ - Na^+$, $Na^+ - Ca^{2+}$, and $K^+ - Ca^{2+}$.

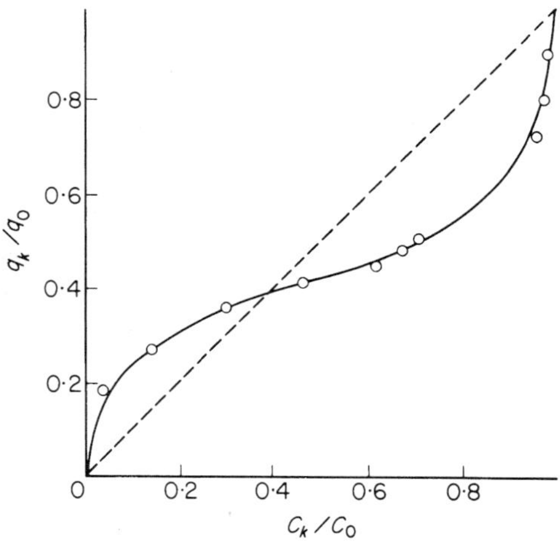

Figure 3.1 Isotherm for exchange between K^+ and Ca^{2+} on Harwell soil. q_K/q_0 is the fraction of the total negative charge satisfied by K^+ on the exchanger, and C_K/C_0 is the fraction of K^+ in the solution (after Deist & Talibudeen 1967).

3.1.5 Effect of Ionic Size and Type of Exchanger

Martin & Laudelout (1963) studied the complete exchange isotherm between the alkali cations and ammonium on montmorillonite. Their affinity relative to the ammonium ion increased with the volume of the ion, in the order Li < Na < Rb < Cs. Laudelout *et al.* (1968) similarly found the affinity of the alkaline earth cations to be in the order Mg < Ca < Sr < Ba. Sposito (1981, pp. 128–132) gave a clear explanation of these effects. The hydrated Li^+ ion more readily loses a proton than the hydrated Cs^+ ion because, being small, its positive nucleus is closer to the water molecules. It is therefore more negative and less attracted to the negatively charged siloxane surfaces of the clay minerals. In contrast to the siloxane surfaces of montmorillonite, the positive surfaces of hydrous oxides tend to show selectivity in the order Li > Na > K > Rb > Cs.

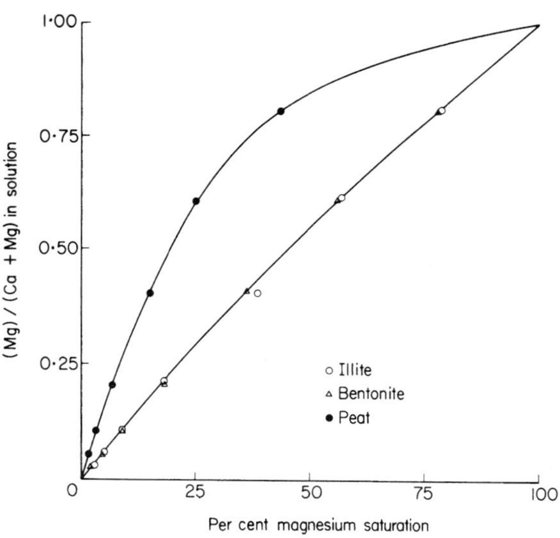

Figure 3.2 Relation between magnesium saturation and the activity ratio (Mg)/(Ca + Mg) in solution for illite, montmorillonite and peat (after Salmon 1964).

The edges of illitic 2:1 type clays can readily accommodate large monovalent cations such as K^+, Rb^+ and NH_4^+ within the lattice (Bruggenwert & Kamphorst 1979, pp. 187–190). Hence, for a given activity ratio $(K)/(Ca^+ Mg)^{1/2}$, it is usually found that the proportion of potassium on the exchanger is in the order: illite > montmorillonite > humus as in figure 3.3 (Salmon 1964).

In real soils, the exchangers are often different from the 'type' clays on which so much work has been done, because of poor crystallinity, mixed lattice forms and amorphous complexes. Hence, it is rarely possible to deduce the exchange behaviour of a soil from its mineral composition, except in broad outline (Sposito 1984).

3.1.6 Acidic Cations

As we have seen in section 3.1.5, the hydrated cations of the alkali and alkaline earth cations tend to lose a proton. They are therefore acids, though only very weak acids.[1] On the other hand, the hydrated heavy metal cations are more acidic, for example

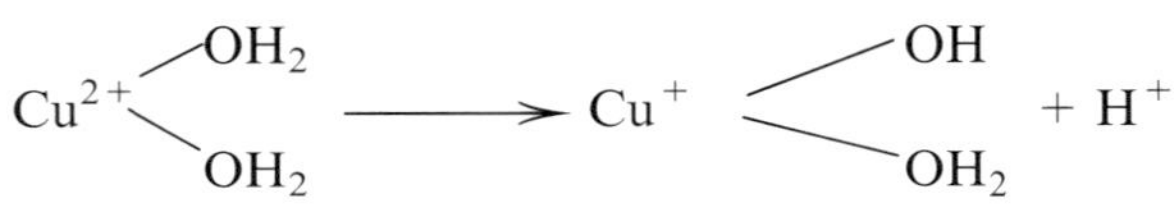

Here, the balance between the states of Cu will depend on the pH, which greatly affects the adsorption of these metal ions. For example, Jopony & Young (1994)

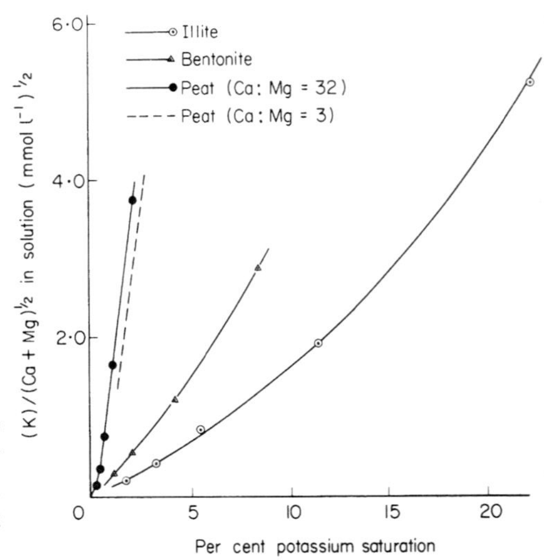

Figure 3.3 Relation between percentage potassium saturation and the activity ratio $(K)/(Ca + Mg)^{1/2}$ for illite, bentonite (montmorillonite) and peat (after Salmon 1964).

[1]These cations are still described in some of the current literature as exchangeable bases or basic cations. They were originally so called because their oxides were basic. Since a base is a proton acceptor, today this nomenclature is wrong and confusing, and should have been abandoned many years ago.

found the concentration of Cd on the soil to be related to the concentration in solution and the pH by the empirical equation

$$\log (C/C_L)_{\mathrm{Cd}} = 1 + 0.305\,\mathrm{pH} \tag{3.5}$$

These metals also readily form complexes with other ligands besides OH^- that are in the solution, such as acetate, citrate, and other organic ions and molecules of higher molecular weight. These ligands are especially important in determining the solution concentration of heavy metals in the rhizosphere and in soils to which manures and sludges have been applied. They may prevent adsorption, that is 'solubilize' the metal; but if the metal–ligand complex is itself adsorbed they may increase adsorption. Lindsay (1974, pp. 239–265) surveyed the complexes that may exist in solution, and Sposito (1981, p. 133) gave an outline of the many possibilities for adsorption that exist for a particular metal–ligand complex. Real soils are even more complex. Ligands may simultaneously exist in true solution, colloidal solution, or on the soil solid. Their effectiveness depends on the balance between these forms. When the variety of adsorption sites available in soils is added to these complications, it becomes difficult to generalize about the adsorption behaviour of the heavy metals. Bruggenwert & Kamphorst (1979, pp. 178–181) have tabulated much of the available information.

As mentioned above, these ions have a strong affinity for higher molecular weight organic molecules, which may be in true solution, as 'fulvic acid' and 'humic acid', or as humus in colloidal solution or insoluble form. The exact composition of these polymers is uncertain and variable. However, Mountney & Williams (1992) describe a promising method of predicting their affinity for metals. The program RANDOM aims to predict the statistical distribution and average concentration of the main C–H–O ligands (e.g. the hydroxycarboxyl arrangement as in hydroxybenzoic acid, $C_6H_4(OH)(CO_2H)$) in a sample of organic matter from its proportions of C, H and O, its molecular mass and its degree of aromaticity. Since the individual metal–ligand formation constants are available, these can be used in a metal speciation program (similar to GEOCHEM (section 3.1.2)) to predict the adsorption affinity for a defined metal–organic material combination.

3.2 Buffer Power

The relation between the concentration of an exchangeable ion in the soil and its concentration in solution is very important in determining its mobility (chapter 4), and the ability of the soil to maintain the solution concentration as the ion is removed by root uptake or leaching. In considering these processes, we are rarely concerned with the complete isotherm, but only with a section of it, in the range from zero to a few percent of saturation. This is true even for a nutrient like potassium, required in large amounts by plants. For elements required in trace amounts, such as zinc, or for heavy metals where problems of toxicity arise, the range is even smaller.

Figure 3.4 shows a typical relation between the reduced activity ratio $(\mathrm{K})/(\mathrm{Ca} + \mathrm{Mg})^{1/2}$, which is nearly equal to the reduced concentration ratio (section

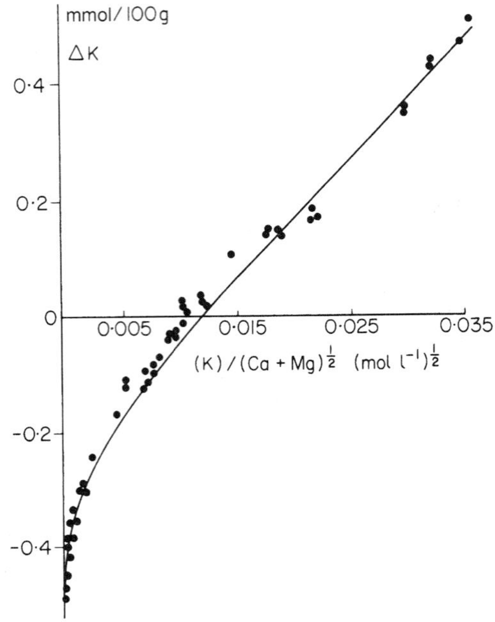

Figure 3.4 Relation between the reduced activity ratio $(K)/(Ca+Mg)^{1/2}$ and potassium adsorbed or desorbed from Lower Greensand soil (after Beckett 1964).

3.1.1) in dilute solution, and the potassium adsorbed by or desorbed from the exchange complex of a field soil (Beckett 1964). In many soils, it is very difficult to replace all the exchangeable ion by calcium, owing to release from slowly exchangeable sites. The amount of potassium replaced also varies with the replacing ion. Thus, the zero point of an isotherm such as that illustrated in figure 3.4 is often indefinite.

From the slope at any point on the curve in figure 3.4, the buffer power, dC/dC_L, may be derived, where C is the concentration of labile ions in the soil, including the soil solution, and C_L is their solution concentration. The increase in buffer power at low concentrations of potassium shows that, relative to calcium and magnesium ions, the potassium is adsorbed with greater affinity, probably by sites that have a special affinity for the potassium ion at the edges of the clay mineral lattices (Beckett & Nafady 1967a). Such curved isotherms for exchangeable ions present in low proportion have also been observed for sodium (Tinker & Bolton 1966; Bolton 1971), aluminium (Nye *et al.* 1961) and rubidium (Deist & Talibudeen 1967). Though not usually presented in this context, it is also true of the hydrogen ion, as shown in figure 3.5 (Farr *et al.* 1970). Here, the sorbed hydrogen is covalently bonded to oxygen. As the concentration of hydrogen ion in solution decreases, the buffer power (dC/dC_L) increases. For hydrogen ion, the plot of C_{H+} against pH is more linear and the pH buffer power, the slope (dC_H/dpH), is more nearly constant and is more useful in practice.

Given the diversity of clay minerals and other exchange materials in soils, and the corresponding variety of types of exchange site they offer, it seems very likely that most soils will have a small proportion of sites with a strong affinity for any particular cation. Thus, trace amounts of micronutrients (e.g. copper), or toxic or

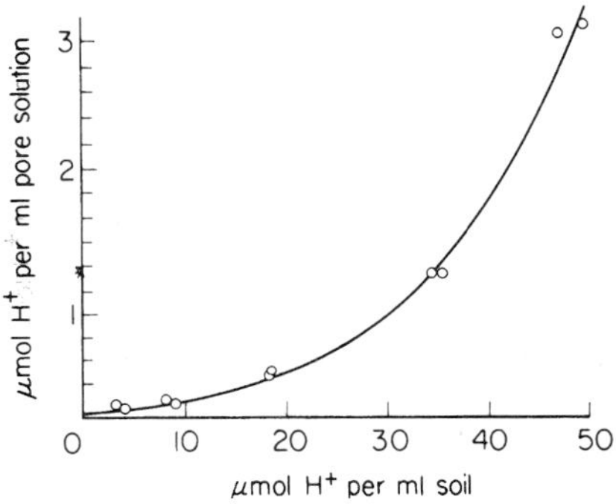

Figure 3.5 The relation between the exchangeable hydrogen ion on a soil and the concentration of hydrogen ion in an equilibrium solution of calcium chloride (from Farr *et al.* 1970).

unusual cations arising from industrial processes (e.g. mercury), or radionuclides of elements not already present in quantity (e.g. caesium), are likely to be relatively strongly adsorbed (Poelstra *et al.* 1974). Their effect on plant uptake and growth will be less than if they were adsorbed with low affinity, but their rate of leaching will also be less.

3.3 Poorly Soluble Compounds

The concentration in the soil solution of cations normally absorbed in trace quantities by the plant (e.g. iron, aluminium, manganese and zinc) may be controlled by almost insoluble salts (e.g. $ZnSiO_3$), oxides or hydroxides (e.g. $Al(OH)_3$), or ill-defined complexes, which may exist as separate crystals or adsorbed on crystal surfaces (Hodgson 1963). Usually, the concentration of these elements in the soil solution can be calculated from the solubility products of the solid forms, if these can be identified; but in practice the formation of mixed crystals, slow equilibration (section 3.6), and the presence of poorly ordered adsorbed and colloidal material makes accurate calculation impossible. The fact that the experimental sorption data can be fitted to an adsorption isotherm equation or a constant activity product (section 3.5.2) is no evidence of the actual sorption mechanism (Sposito 1981, p. 122). Understanding of these ions' behaviour is further complicated because they form soluble complexes with organic matter, and the proportion between free and complexed ion in the soil solution varies widely between different soils (Cottenie & Kiekens 1972). If a single compound with low solubility product does control the concentration of an ion, then in principle the solution is infinitely buffered until all the solid phase has dissolved. Lindsay (1974) gives a full account of true equilibrium solubility data for compounds that may exist in soil.

3.4 Cations with Multiple Valency

The prediction of solid–solution equilibria is particularly difficult in poorly drained soils where cations such as Fe^{III} and Mn^{IV}, which form very insoluble oxides, may be reduced to Fe^{II} and Mn^{II} ions whose oxides are more soluble. Thus the reduced forms occur in much higher relative concentrations in the soil solution, where they are subject to the same controls as other divalent cations. Following submergence, a wealth of redox and microbiological reactions occur that are very important both for understanding plant nutrition in badly drained and rice paddy soils, and in studying soil development. These have been reviewed by Ponnamperuma (1972) and Rowell (1981). Successful recent attempts to model these simultaneous processes in the rice rhizosphere are described in chapter 11.

3.5 Adsorption of Anions

3.5.1 Non-specific Anion Bonding

As we have already emphasized at the beginning of this chapter, since soil colloids usually carry a net negative charge, there is a general tendency for anions to be repelled from solid surfaces (negative adsorption) and retained in the soil solution, where their amounts together with the soil water content control the overall salt concentration of the soil solution. Anions may, however, be adsorbed on the soil solid both by non-specific and specific bonding.

Soils contain free hydrous oxides of iron(III) and aluminium that may become positively charged when one of their oxygen atoms accepts a proton. Similarly, at the edges of aluminosilicate clay mineral lattices, there are oxygen atoms whose electron orbitals are not fully coordinated with aluminium or silicon atoms. These oxygen atoms readily accept protons. The amount of positive charge developed depends on the concentration of hydrogen ion, which is thus a surface potential-determining ion, that is, the hydrogen ions held at the surface alter and control the potential difference between the surface and the surrounding solution. These localized positive charges attract anions by electrostatic forces. Such anions are freely exchangeable with anions in the soil solution; and since this is dominated by chloride, nitrate and sulphate, it is these anions that predominate on the non-specific sites. Since there are few positive sites at pH greater than 6, and carbonic acid is too weak an acid to provide a high concentration of bicarbonate ions at pH less that 6, the amount of bicarbonate held in this way is small.

3.5.2 Specific Anion Bonding

Many anions form strong bonds with the aluminium, iron(III) and other cations in hydrous oxides and aluminosilicate clay lattices by replacing the surface ligands attached to them: thus,

$$> \text{Al-OH} + H_2PO_4^- = > \text{Al-}H_2PO_4 + OH^-$$

Examples of these anions are F^-, $H_2PO_4^-$, $HMnO_4^-$, $HSeO_3^-$, $H_2BO_3^-$, $H_2SiO_3^-$, and HCO_3^-. Specific anion adsorption has been systematically studied on pure mineral hydrous oxides, such as goethite, FeO(OH), by Posner, Quirk, and co-workers in Western Australia, and a brief account of their findings and those of others has been given by Mott (1988). At a given pH, the relation between amount adsorbed and solution concentration follows a Langmuir adsorption isotherm (see figure 3.12), modified to take account of the increasing negative charge on the surface as the number of anions adsorbed increases. The maximum adsorbed is very sensitive to the pH (figure 3.6) (Hingston *et al.* 1968). The curve relating the maximum that can be adsorbed to the pH shows a sharp break in slope at about the p*K* value of the acid formed by the anion when protonated. Unlike cation adsorption, the maximum amount of anion adsorbed specifically differs markedly from one anion to another, as shown in figure 3.6. A specifically adsorbed anion may be displaced by another specifically adsorbed anion, but not by a non-specifically adsorbed one. When two anions are in competition, the ion that increases the negative charge on the surface to the greater extent is adsorbed preferentially.

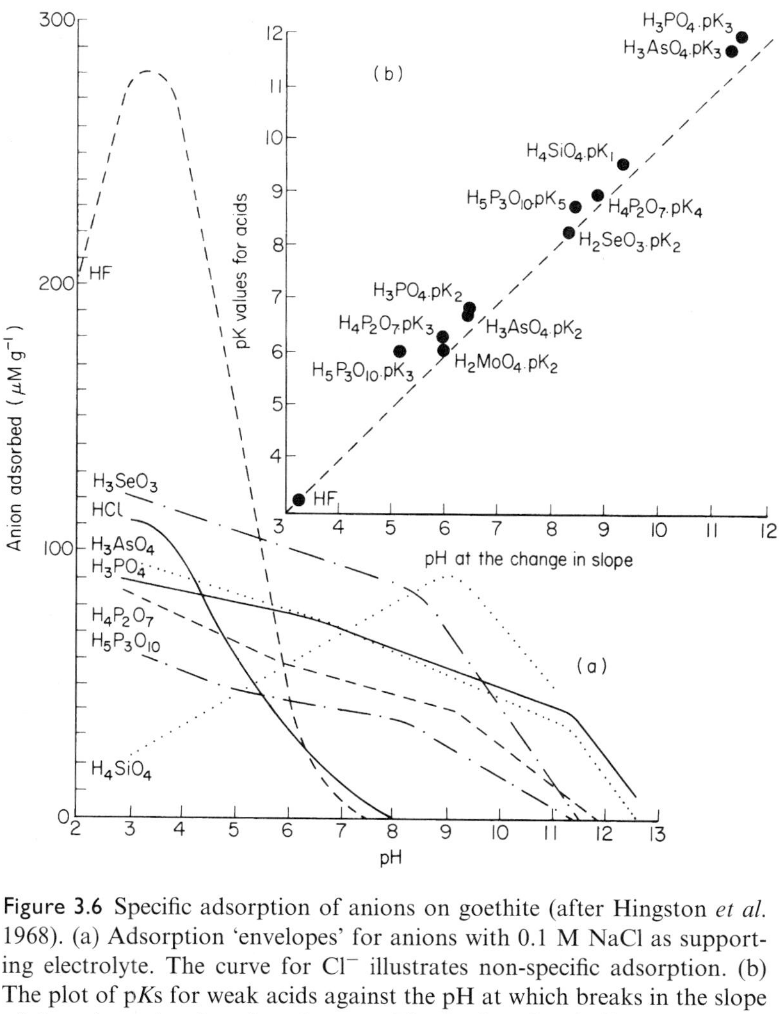

Figure 3.6 Specific adsorption of anions on goethite (after Hingston *et al.* 1968). (a) Adsorption 'envelopes' for anions with 0.1 M NaCl as supporting electrolyte. The curve for Cl^- illustrates non-specific adsorption. (b) The plot of p*K*s for weak acids against the pH at which breaks in the slope of the adsorption 'envelopes' occur. The broken line indicates the ideal relationship.

Barrow (1987, pp. 16–29), following a model first proposed by Bowden *et al.* (1977), has been able to explain the effects of anion competition, pH and solution salt concentration on the adsorption on goethite of a comprehensive range of anions — phosphate, sulphate, selenite, silicate, citrate, fluoride, and arsenite, as well as the variable-charge cations — copper, zinc, and lead. To do this, it was necessary to split the single Stern adsorption layer at the solid surface into three parallel adsorption planes associated successively with surface hydroxyl groups, inner-sphere complexes, and outer-sphere complexes. The many parameters of this model, such as the electrical potentials of the planes, have to be found by curve fitting, so in this sense the model has not been verified, and may need further refinement. Models with fewer planes would not simulate satisfactorily results obtained with the very many combinations of the experimental variables that have been studied. Though this model is too idealized for use with soils, which offer such a variety of sorption sites, it provides a good indication of the processes that are involved in adsorption on some of them.

In soils, the relation between the amount of an anion adsorbed and its concentration in solution is usually well described for practical purposes by the empirical Freundlich equation (see figure 3.12):

$$C_s = aC_L^n \qquad n < 1 \tag{3.6}$$

Sposito (1982) and Barrow (1987, p. 83) have found that an isotherm based on a model in which there is a normal (Gaussian) distribution of site affinities fits the experimental data rather better.

The method (section 3.1.2) by which we showed that any pair of exchangeable cations, with charges zA and zB, tend to buffer the reduced activity ratio $(A)^{1/zA}/(B)^{1/zB}$ in the soil solution, can be applied to anions adsorbed by the soil in a definite phase, simply by giving z_A and z_B negative signs. Likewise, an adsorbed cation A with charge z_A and anion X with charge $-z_X$ will buffer in the equilibrium solution the activity ratio $(A)^{1/zA}/(X)^{-1/zX}$, which equals the activity product $(A)^{1/zA}(X)^{1/zX}$. Such a constant activity product does not require that an insoluble salt $A_{zX}X_{zA}$ should exist in the soil. As an example, Aslyng (1954) showed that the activity product $(Ca)^{1/2}$ (H_2PO_4), though unequal to that of crystalline $Ca(H_2PO_4)_2$, remained constant in the soil solution when the calcium concentration changed 10-fold.

Among the specifically bonded anions, phosphate has received by far the most detailed study in soil, from the point of view both of solid–solution equilibrium (section 3.5.3) and transport characteristics (section 4.2.1) (Goldberg & Sposito 1985).

3.5.3 Phosphate Equilibria in Soil

The elements of soil phosphate chemistry are well described by Wild (1988, chapter 21). A great deal of effort has been expended on trying to relate the concentration of phosphate in the soil solution to specific insoluble phosphate compounds or surface complexes. This complicated story has been reviewed by Larsen (1967) and by Lindsay (1974, pp. 163–215). The difficulties arise because the rates of equilibration of phosphate in solution with the crystalline forms are very slow; and because the pure crystalline forms, particularly hydroxyapatite, may have variable surface compositions. After prolonged equilibration of samples, or in field soils to which no phosphate has been added for many years, the concentration of phosphate in the soil solution is probably controlled by a surface of hydroxyapatite, not only in alkaline and neutral soils, but also in slightly acid soils. In very acid soils, below pH 3, variscite, $Al(OH)_2(H_2PO_4)$, may control the solution concentration, but this dissolves incongruently above pH 3.1 to form a more basic solid phase of aluminium hydroxyphosphate, possibly as a surface complex of variscite. Bache (1963) concluded that strengite, $Fe(OH)_2(H_2PO_4)$, is never likely to be in equilibrium with any soil solution. In soils that contain abundant iron and aluminium oxides, most phosphate will exist as complexes either on their surfaces or at sites within the solid that are accessible by diffusion. As Barrow points out, the meaning of the term calcium phosphate can be questioned:

> Imagine that phosphate may react with an aluminous goethite, and has formed one link to an iron atom and another to an aluminium atom in the goethite. The charge on the phosphate molecule will be balanced by ions in solution, and it could well be balanced by calcium ion. This phosphate molecule is then simultaneously an iron phosphate, an aluminium phosphate and a calcium phosphate. (1987, p. 56)

It is useful to be able to characterize the adsorption relations of a soil by the minimum number of parameters, for example for computer programs. Although some workers have found that phosphate adsorption curves may be fitted to a Langmuir isotherm (see figure 3.12) (Olsen & Watanabe 1957), generally no maximum of adsorption is attained as the phosphate concentration is increased (Gunary 1970). Bache & Williams (1971) found that a plot of the sorbed P against the logarithm of the concentration of P in the soil solution was linear for their adsorption data on Scottish soils. We have found a similar equation to describe many desorption isotherms in the form $\Delta C_s = k \ln C_L/C_{Li}$, where ΔC_s is the change in concentration of sorbed P and C_{Li} is the initial concentration of P in the soil solution. This equation corresponds to the middle range of surface coverage of the Temkin isotherm, which may be derived theoretically if it is assumed that the energy with which each phosphate ion is bonded decreases linearly with the number already adsorbed. The Langmuir isotherm assumes that the bonding energy is unaffected by the number of ions already adsorbed, which is inconsistent with the idea that each phosphate ion adds negative charge to the surface. For soils very low in native phosphate, the Freundlich isotherm fits the data in the form of a linear plot of log C_s versus log C_L (Mead 1981; Raven & Horsner 1993). But in richer soils this plot is inexact because the amount of native phosphate equilibrating is uncertain, that is, the value of C_s is uncertain. The characteristics and assumptions of various types of isotherm are discussed by Hayward & Trapnell (1964).

3.6 Rates of Ionic Interchange between Solid and Solution

3.6.1 Cations

Although we have seen that a fairly stable equilibrium is usually established within minutes between the exchangeable cations on the soil solid and in the solution, we have to consider slower processes that may occur over longer periods comparable to rainfall events, or the life of a root, crops or a stand of natural vegetation.

It is convenient to divide these reactions into rapid, with half-times of the order of an hour or so; intermediate, with half-times of the order of a day; and slow, with half-times of a week or more.

3.6.1.1 Rapid Reactions

The exchangeable cations are held on the external surfaces of clay or humus particles, or in the interlayers of expanded clay minerals in which the aluminosilicate sheets are separated by a distance of at least two, and usually three, molecular diameters of water molecules. The half-time of a reaction whose rate is controlled by diffusion is approx. L^2/D, where L is the average distance that a particle travels. If the diameter of a clay particle is 1 μm, and the apparent diffusion coefficient of an interlayer cation is 10^{-8} cm^2 s^{-1}, such as in vermiculite, then the half-time for exchange with the pore solution would be of the order of 1 s. In the pores themselves, the diffusion coefficient of cations is about 10^{-6}–10^{-8} cm^2 s^{-1} (allowing for adsorption on the sides of the pore), and the diameter of a very large pore might be 1 mm, so the half-time for equilibration across such a pore with solution just outside it could be of the order of 10^4–10^6 s. Thus, it would seem that attainment of rapid exchange equilibrium in a structurally intact soil with no water movement is often limited by diffusion across pores rather than release from the solid phase. Equilibrium when a soil is shaken with solution will be more rapid. Malcolm & Kennedy (1969) found that 75% of the exchange between potassium and barium on kaolinite, illite and montmorillonite occurred in less than 3 s, and Ikeda *et al.* (1984) found the exchange between alkali metals on zeolites to be complete in the order of seconds. The rates of rapid reaction are needed to decide what should be described as a ‘diffusible’ ion — a matter to be discussed in the next chapter.

3.6.1.2 Intermediate Reactions

For the major cations, the fraction showing an intermediate rate of exchange is small or negligible. However, for potassium and other ions of similar size and charge, like ammonium, rubidium and caesium, the amount exchanging at an intermediate rate is significant. The fraction of potassium that shows an intermediate rate is from $\frac{1}{4}$ to 5 times that exchangeable to ammonium acetate. The rate of release appears, as a rule, to obey first-order kinetics, though the rate-controlling processes are diffusion coupled with lattice deformation. This intermediate potassium is probably held within the interlayers of weathered illitic clays (Arnold 1970). In unweathered illites, the aluminosilicate sheets do not readily expand to allow interchange of potassium with ions in solution. However, weathering of illites is accompanied by a lowering of the charge density on the lattice, which consequently tends to expand more easily. Stanford & Pierre (1947) have found that ammonium was fixed in a similar way to potassium on eight Iowa soils. Such reactions are described in greater detail by Black (1968). The rates quoted above have been obtained in laboratory experiments under controlled conditions. The rates of release or fixation are accelerated by drying, the effect of which is only likely to be large if the matric suction exceeds 1 MPa, so that, as a rule, only the topsoil will be affected. Beckett (1969) has summarized knowledge of the release of potassium from soils under field conditions (as in figure 3.7).

3.6.1.3 Slow Reactions

We do not propose to review the subject of mineral weathering, which is an essential part of soil formation, but refer readers to the symposium edited by Hallsworth & Crawford (1965), the book by Loughnan (1969), the chapter by Northcliffe (1988), and the extensive compendiums of Dixon & Weed (1989) and Bryant & Arnold (1994). According to figure 3.7, the amount of potassium released at the slow rate would not satisfy crop requirements. The rate approximates to zero order, that is it is nearly constant with time, though there are some reports that it decreases with time. Though such slow reactions may not be significant during one growing season, under natural vegetation they are the main means by which exchangeable potassium, which is slowly lost to drainage, is replaced in the profile. Heavy metals also react slowly with soil after a period of rapid sorption. In this respect, they behave much the same as anions, which will now be considered.

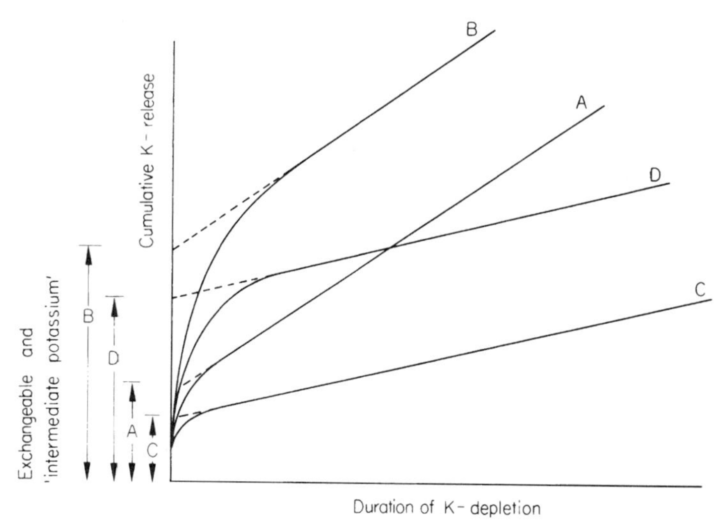

Figure 3.7 Representation of release of potassium from soils under continuous cropping. (A) Unfertilized and (B) fertilized plots on a virgin soil; (C) unfertilized and (D) newly fertilized plots on the same soil following prolonged exhaustion (after Beckett 1969).

3.6.2 Anions

A fast, followed by a slow, continuing adsorption by the soil has been observed for many specifically adsorbed anions, for example molybdate, fluoride, sulphate, arsenite, and, most notably, phosphate (Barrow 1983, 1987, pp. 57–77).

Although a great deal of work has been done on the rate of exchange of soil phosphate with the radioisotope ^{32}P, we must emphasize that for practical application we are concerned with the rate of change of the phosphate concentration in the soil pores as it is disturbed by plant uptake or percolating solutions. The ^{32}P exchanges more rapidly, and finally to a much greater extent, with phosphorus in the soil than will the other adsorbed anions that predominate in solution.

A typical rate of exchange with ^{32}P is shown in figure 3.8 (McAuliffe *et al.* 1947). To explain the exchange rate, the exchange is usually decomposed into two or more reactions; for example, Arambarri & Talibudeen (1959) distinguished a fast reaction with half-times between 0.3 and 1.6 h, which involves exchange with phosphate in pore solutions and on readily accessible surfaces; an intermediate reaction with half-times of 1.8–8.6 h; and a slower reaction with half-times of 25.8–46.1 h. These slower reactions represent exchange with phosphate in micropores and in the interior of crystals and complexes which are accessible to exchange through microcracks and crystal defects. In fact, there is no definite limit to the amount exchanged, and given sufficient time most of the soil phosphate would exchange. Strauss *et al.* (1997) have shown that on samples of goethite (FeO·OH) of differing crystallinity, phosphate is initially sorbed on the surface and then diffuses into finer and finer pores more slowly as the degree of crystallinity increases.

Studies of the kinetics of phosphate adsorption have usually been carried out by shaking the soil with a larger volume of solution. Staunton & Nye (1989a) found the reaction was greatly accelerated by shaking. A better method, which causes minimal mechanical disturbance, is to incubate the moist soil with added phosphate and extract the soil solution by centrifugation with a heavy immiscible liquid (Kinniburgh & Miles 1983).

Evidence from experiments with shaken soils suggests that, in general, a quasi equilibrium is established after about 2 h, and that this changes slowly thereafter because of further reaction with less accessible sites, and disturbances caused by

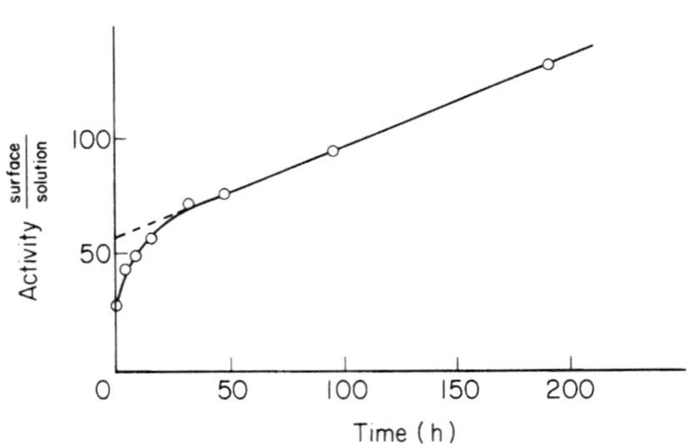

Figure 3.8 Exchange of ^{32}P in solution with 'surface' phosphate (^{31}P) of Caribou soil (after McAuliffe *et al.* 1947).

microbial activity, pH changes and the like. For example, Larsen *et al.* (1965) found that the half-time for soluble phosphate added to mineral soils to become non-exchangeable is between 1 and 6 years. The rate of diffusion of phosphate to the sites with which it will equilibrate will clearly depend on their accessibility, which will vary greatly within any one soil. Hence, no simple kinetic expression is to be expected to describe the whole sorption process. Nevertheless, an approximate equation is useful in models, and three kinetic equations have been fitted to phosphate sorption data (Aharoni & Sparks 1991).

(a) One or more first-order reactions

$$Q = q_{1\infty}(1 - a_1 e^{-k_1 t}) + q_{2\infty}(1 - a_2 e^{-k_2 t}) + \ldots + \ldots \tag{3.7}$$

where

Q = total sorbed,
$q_{1\infty}$, etc. = maximum amount sorbed by site 1, etc.,
a_1, etc. = constant for site 1, etc.,
k_1, etc. = kinetic constant for site 1, etc.

This equation might be expected ıı adsorption were controlled by a simple first-order reaction with one or more types of adsorption sites. It implies that the electrical potential of the sites is independent of the amounts adsorbed and it takes no account of transport limitation to the sites. It should therefore be regarded as a convenient empirical fit.

(b) the Elovich equation

$$Q = A + (1/b)\ln(t + t_0) \tag{3.8}$$

where A, b and t_0 are constants. This equation can be derived theoretically if it is assumed that there is a heterogeneous distribution of activation energies on the surface (Atkinson *et al.* 1971).

(c) a Freundlich-type equation

$$Q = kt^n \qquad n < 1 \tag{3.9}$$

where k and n are constants. This equation was used by Kuo & Lotse (1974). It represents well the adsorption of phosphate, molybdate, and fluoride over a 10 000 range of time (Barrow & Shaw 1975a, b, c, 1977), and on a range of world soils (Barrow 1980).

Each of these equations has merit, since Aharoni & Sparks (1991) show theoretically that when a solute at constant concentration diffuses into an initially empty, limited heterogeneous medium of any surface geometry, such as soil, the uptake should initially vary as (time)^n ($n < 1$), then as the Elovich equation, and finally as an exponential equation. If the amount that is transferred from the surface to internal sites varies as $t^{1/2}$, this is often taken as evidence of a diffusion-controlled process; however, this relation strictly applies only to the special conditions of a planar surface, a constant surface concentration, and a constant surface electrical potential. Adsorption of anions will clearly change both the surface concentration and its electrical potential.

Barrow (1987, pp. 81–100) describes a comprehensive model that includes the following effects on adsorption: anion concentration in solution, reaction time, temperature, effect of repeated additions of solute, pH, and salt concentration in solution. By suitable adjustment of its parameters, the model also describes the behaviour of the hydrolysable heavy metals, for example $ZnOH^+$ and $CuOH^+$. The many parameters required for the model have to be found by fitting curves to the experimental data for each soil and sorbate. The model is valuable not only for the insight it provides into the mechanisms involved, but also because it can be used as a submodel in wider models of solute leaching.

3.6.3 Relaxation and Hysteresis in Sorption Isotherms

Adsorption and desorption isotherms do not always follow the same curve. Figure 3.9 shows an example for potassium (Arnold 1970). Similar observations have been made for phosphate by Muljadi *et al.* (1966) and for sulphate by Sanders (1971), when only a short time — less than 24 h — has been allowed for equilibration. Laudelout *et al.* (1968) have noted that to determine accurately the exchange equilibrium between two cations, it is essential to approach the single equilibrium point, if it exists, from both directions.

If the difference between the two curves is due to insufficient time being allowed for equilibration, the phenomenon is called relaxation (Everett & Whitton 1952). In many instances, it seems likely that the difference would persist however long a time was left for equilibration, in which case it is known as hysteresis (section 2.3.1). In most instances, the distinction is not made and may not matter in practice. If the isotherm is being used to interpret a desorption process, for example removal of solute by plant uptake, then the desorption limb must be used, and likewise the adsorption limb to follow an adsorption process. Newman (1970) instances the sodium–potassium exchange in vermiculite as an example of true hysteresis in solid–solution reactions.

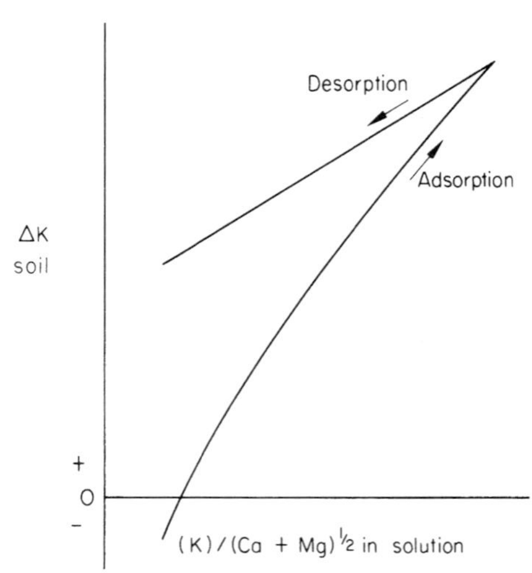

Figure 3.9 Hysteresis in the sorption isotherm for potassium on a soil (after Arnold 1970).

Barrow (1987, pp. 65–71) has treated relaxation in phosphate adsorption–desorption quantitatively. Based on the Freundlich rate equation, he generated sorption–desorption curves from the expression

$$S_d = k_1 S_0[(t - t_1)/t]^{b_2} - k_1 a c^{b_1}(t - t_1)^{b_2} \tag{3.10}$$

where S_d is the amount desorbed, S_0 is the amount initially adsorbed at $t = 0$, t is the time after adsorption begins and t_1 is the time after desorption begins, c is concentration, and k_1, a, b_1 and b_2 are constants.

3.7 Mineralization and Immobilization in Organic Forms

These are changes in covalent bonding effected by microorganisms, and differ from the ionic sorption–desorption and exchange reactions just described. They are particularly important in determining the level of nitrate in the soil solution (section 3.8). This controls not merely the nitrogen supply to plants, but in fertile soils also the overall strength of the soil solution. It therefore partly determines the concentration of the cations in solution, and hence their availability and ease of leaching. Supply of phosphate, sulphate, and organically bound trace elements by mineralization is also of some importance.

3.7.1 Mineralization

Unfortunately, it is possible to make only the most general statements about rates of mineralization since large fluctuations in average rates are caused by such factors as temperature, wetting and drying, freezing and thawing, and cultivation techniques — these are discussed at length in standard textbooks on soil science, such as Wild (1988).

There is a 'flush' of mineralization in the spring in temperate arable soils, and after the early rains, following a dry season, in tropical soils. Thereafter, the amount of nitrogen released by mineralization over the growing period may be small compared with the amount already present plus that added as fertilizer, and a rough indication of mineralization rate may be sufficient.

Over sufficiently long periods, the simplest equation for representing the organic carbon balance is

$$\mathrm{d}C/\mathrm{d}t = -kC + A \tag{3.11}$$

where k is a measure of the rate of mineralization, C is the concentration of carbon in the soil, and A is the annual addition of humus carbon. Though there may be wide variations in the ratios of C:N:S:P in soil humus, they tend to be of the order of 100:10:1:1. Equation (3.11) may be used for each element.

The decomposition constant k is an average over many organic compounds and ages of humus, and equation (3.11) has been refined by many workers (Jenkinson 1990). Jenkinson (1966), for example, has shown that the humus that resulted from addition of ^{14}C-labelled ryegrass was, 4 years after the addition, still decomposing at four times the average rate for the soil humus present initially.

Values of k for nitrogen, derived from long-term arable systems in North America, lie in the range 0.02–0.10 $year^{-1}$ (Bartholomew & Kirkham 1960). After clearing tropical forest and cropping continuously, k for organic C ranges from 0.018 to 0.09 $year^{-1}$ (Nye & Greenland 1960) and in savannas from 0 to 0.068 $year^{-1}$. At Rothamsted, a field under continuous barley for over 100 years gave $k = 0.025$ for nitrogen.

Under undisturbed vegetation, decomposition rates are slower. In South Australia, under pasture, $k_N = 0.029$ and 0.013 (Greenland 1971). Under tropical forest, $k_C = 0.02$–0.05 and, under savanna, $k_C = 0.005$–0.012 (Greenland & Nye 1959). In uncultivated Iowa topsoils under grass, the average age of the organic matter was determined by ^{14}C dating as 210–440 years ($k_C = 0.005$–0.002). In a woodland, the sodium hydroxide-soluble fraction had an average age of between 50 and 250 years (Broecker & Olson 1960), though residual material was about 2000 years old. In peat bogs of the northern English uplands, aerobic peat (0–12 cm) yields $k_C = 0.07$, while anaerobic peat yielded $k_C = 0.0014$ (Gore & Olson 1967). Stevenson (1982) has summarized much of the available data on long-term organic nitrogen changes under arable and perennial crops.

Levels of nitrate under naturally vegetated sites are usually low, and hence the nitrate levels in drainage water from these sites are also low. In regions remote from airborne sources of chloride and sulphate, and in the absence of nitrate, the bicarbonate ion becomes the major anion in the soil solution, except at low pH. For example, McColl & Cole (1968) found that the total cation concentration in the leachates under Douglas fir was nearly balanced by the bicarbonate ion concentration.

3.7.2 Immobilization and Denitrification

Nitrate (and phosphate) may be removed from the soil solution by microorganisms stimulated by additions of energy-rich material, when this contains less than about 2% nitrogen. Reduction of nitrate may also occur under anaerobic conditions. Under normal free drainage conditions, losses as nitrous oxide may still occur, because of very local and temporary anaerobic pockets. Long-term soil nitrogen balance sheets usually show a small unaccountable loss of nitrogen (Allison 1965), but an increase overall, due to N fixation. Greenland (1971) concluded from experiments in South Australia that under 3 years of wheat and pasture, the annual losses would not have exceeded 23 kg ha^{-1}.

3.8 Applications to Whole Crop and Drainage Models

In chapters 10 and 11, we discuss the modelling of the growth of a crop or drainage losses. If this depends on the concentration of nutrients in the root zone, it is essential to understand how the soil solution is controlled. This chapter has been largely concerned with analysing the factors that determine this. We have seen that the total strength of the soil solution is determined by the concentration of unadsorbed anions, mainly nitrate, chloride, sulphate and bicarbonate. These anions are balanced by the cations calcium, magnesium, potassium and sodium; traces of heavy metal cations and their complexes; and, increasingly as the pH falls, hydrogen and aluminium ions.

In practice, fluctuations in the concentration of nitrate and bicarbonate caused by temperature and rainfall are very difficult to predict so period-average concentrations have to be used.

3.9 Sorption Reactions of Organic Materials

An already great and ever-increasing variety of herbicides, insecticides, fungicides, stimulants and repressants, pollutants and wastes are added to the soil; and their uptake by plants, and their movements within the soil and out of it in drainage, urgently need to be rationalized.

We have less detailed knowledge about adsorption of organic compounds than we do of the inorganic ions, but in principle the same ideas govern their behaviour and the same approaches are appropriate. Many of them are, indeed, ions. The sorption reactions of the remainder, which are uncharged, are more straightforward than those of the ions, though they may have the complication that they are volatile, whereas only a few inorganic ions in solution are in acid–base equilibrium with uncharged volatile forms: for example, NH_3–NH_4^+, (H_2O + CO_2)–H_2CO_3–HCO_3^-, H_2S–HS^-. The organic solutes are also decomposed by soil organisms, and the amounts of agrochemicals and their decomposition products that enter drainage water depend on the balance between their rates of addition, leaching, and metabolism.

3.9.1 Solution–Gas Equilibria

Solution of gases in liquids is controlled by Henry's law: 'the mass of gas dissolved by a given volume of solvent, at constant temperature, is proportional to the pressure of gas with which it is in equilibrium'. Since the pressure of gas is, ideally, proportional to its concentration in the gas phase, the solubility is most conveniently expressed by the ratio C_L:C_g, which is thus the distribution coefficient between liquid and gas phases at a given temperature. This distribution coefficient was proposed by Ostwald and is sometimes listed in tables as Ostwald's solubility coefficient, b.

3.9.2 Solid–Solution Equilibria

Organic materials exhibit a tremendous range of sorption behaviour. The main properties controlling their sorption are their charge, their molecular weight and the possibility of hydrogen bonding. We will discuss them in order of decreasing affinity with the soil. A more complete account has been given by Hamaker & Thompson (1972) and Hartley & Graham-Bryce (1980). Weber & Miller (1989) have surveyed the adsorption by soils and clays of a great range of herbicides and pesticides, classified according to the shapes of their isotherms, and have given extensive lists of references.

3.9.2.1 Cations

The bipyridylium (quaternary ammonium) herbicides (e.g. paraquat) have received most attention. The flat pyridyl ring, containing a positively charged N atom, is very strongly held on the flat surface of clay minerals by normal ionic and also Van der Waals forces. Figure 3.10 shows the sorption isotherm of paraquat, and the extremely low concentration maintained in solution at normal applications is evident (Knight & Tomlinson 1967). The effect of Van der Waals forces is apparent in the adsorption of methyl-substituted ammonium cations on calcium montmorillonite. The strength of adsorption increased regularly from $MeNH_3^+$ to Me_4N^+ (figure 3.11) (Theng *et al.* 1967).

3.9.2.2 Uncharged Bases

Such compounds can accept a proton to become cations ($B + H^+ = BH^+$). The extent to which this occurs depends on the base strength of the organic compound and the activity of hydrogen ion, which is considerably greater near the negatively charged clay surface than in the free solution. As for any other cation pair, at equilibrium the activity ratio $(H^+)/(BH^+)$ will be constant throughout the system. Consequently, BH^+ is concentrated on and near the clay surfaces (Bailey *et al.* 1968). The triazine herbicides and amitrole are important examples in this group, and also the systemic pyrimidine fungicides, such as ethyrimol. As would be expected, titratable soil acidity is well correlated with the adsorption of simazine and atrazine, and amitrole, over a series of soils (Nearpass 1965).

3.9.2.3 Uncharged Compounds

Dry Soils Adsorption of uncharged organic vapour on the surfaces of dry soils usually follows the theoretical BET (Brunauer–Emmett–Teller) isotherm, up to a relative vapour pressure, p/p_{sat}, of about 0.3 (figure 3.12):

$$A/A_m = xc/[(1-x)(1-x+xc)] \tag{3.12}$$

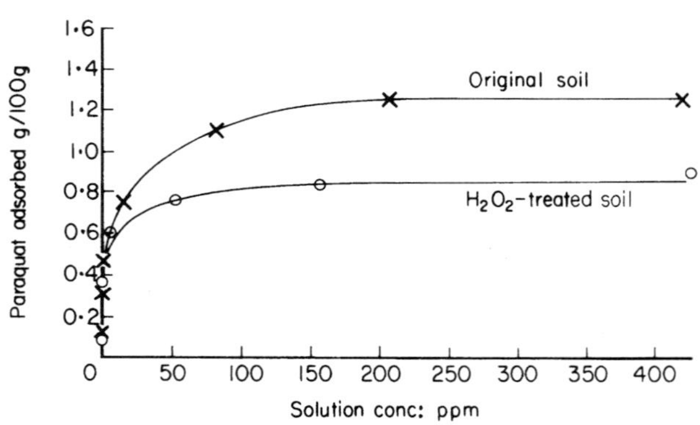

Figure 3.10 Paraquat adsorption isotherm for Jeallot's Hill soil (after Knight & Tomlinson 1967).

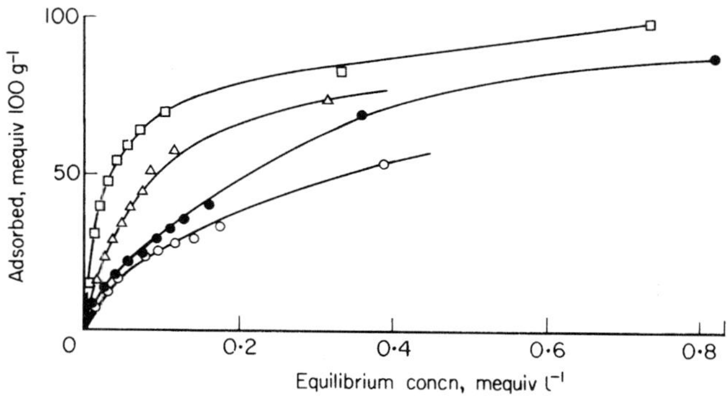

Figure 3.11 Adsorption isotherms of methyl-substituted ammonium cations on Ca-montmorillonite (after Theng *et al.* 1967). ○, $MeNH_3^+$; ●, $Me_2NH_2^+$; △, Me_3NH^+; □, Me_4N^+.

where A is the amount of gas adsorbed per unit of soil; A_m is the amount of adsorbed gas forming a monolayer; c is a dimensionless parameter related to the difference in heat of adsorption per mole of vapour on condensed vapour and the bare soil surface; and x is the relative vapour pressure, p/p_{sat}. When p/p_{sat} exceeds *c*. 0.3, the amount adsorbed is less than that predicted by the simple BET equation. The lower portion of the isotherm, to the point of inflection, corresponds to the formation of a monolayer on the surfaces of the dry soil pores. At higher relative vapour pressures, multilayers are formed, until near the saturation vapour pressure the multilayers merge into liquid in capillary pores. The adsorbed molecules are mainly held by Van der Waals forces, and both mineral and organic surfaces are involved. Such isotherms are required for predicting the rate of volatilization of spilled petroleum contaminants of groundwater, through arid soils (Nye *et al.* 1994b).

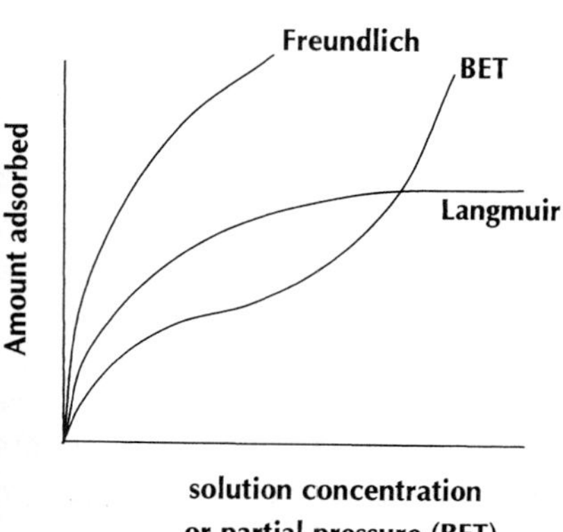

Figure 3.12 Three types of adsorption isotherm.

Moist Soils Addition of water greatly depresses the adsorption because mineral surfaces are very hydrophilic, the polar water molecules competing strongly with the organic molecules for sites on the surface. For a given concentration of organic solute in water, there is usually a linear relation between the organic matter content of the soil and the amount of solute sorbed. Since the molecules of solute appear to penetrate the whole of the soil organic matter, their distribution between water and soil should be described as partition, rather than adsorption (Chiou 1989). The partition is expressed by the equation

$$K_{OM} = S_{OM}/S_w \tag{3.13}$$

where S_{OM} and S_w are the concentrations in organic matter and water, respectively.

In soils with more than 6% organic matter, the value of K_{OM} remains constant over a wide range of solute concentration. Its value tends to be inversely related to the solubility in water, S_w, and so to the tendency of water to expel the molecules of solute. Whereas S_w varies greatly with the nature of the solute, the solubility of a variety of solutes in organic matter, S_{OM}, is relatively constant. Consequently, the value of S_w is a rough guide to the value of K_{OM}.

In soils low in organic matter, the mineral surfaces also contribute to the sorption, and the apparent value of K_{OM} rises (Walker & Crawford 1968; Hassett & Banwart 1989). In surface soils, it is not clear to what extent the preferential sorption on organic matter compared with clay surfaces is due to direct preference or to the fact that the soil organic matter is shielding the clay surfaces from the herbicide.

Adsorption on the siloxane surfaces of aluminosilicate clay minerals is better understood. The following interactions that lead to adsorption have been identified (Greenland 1970):

(a) Hydrogen bonding between water in clays and added organic molecules, and between the organic matter and added organic molecules.
(b) Non-specific Van der Waals bonding. Though the molecules are uncharged, they are often highly polarizable in the aromatic ring.
(c) When a single organic molecule displaces several water molecules from the surface of a clay, there is a net increase in entropy. The effect increases with the size of the organic molecules.

In general, at comparable concentrations in solution there is a much lower degree of adsorption for the physically bound than the ionically bound molecules.

Organic Anions Species with a net negative charge are unlikely to be adsorbed to an appreciable extent, though there is some evidence (Greenland 1965) that they may be attached as ligands to polyvalent exchangeable cations. They may also be electrostatically attracted to positive sites on iron and aluminium hydrous oxides, which occur in acid soils, by protonation.

Many anions are derived from weak acids, and the un-ionized form may be adsorbed by non-specific forces already discussed. An example is picloram ($pK_A = 3.7$), which was adsorbed more strongly as the pH of the soil was changed from 10 to 2 (Hamaker *et al.* 1966).

3.9.3 Rates of Equilibration

In experiments in which pesticides are added to stirred soil slurries, most of the sorption occurs within minutes and equilibrium is attained within a few hours (Graham-Bryce & Briggs 1970). Hamaker & Thompson (1972) and Pignatello (1989), however, record many examples of a further slow reaction over weeks or months. The process is largely reversible, though desorption may be considerably slower than adsorption, particularly on soils with much organic matter, and a fraction of the adsorbed material may be very slowly removed, particularly if the soil has been dried and rewetted. Chen & Wagenet (1997) have evidence that the rate of adsorption of atrazine is controlled by a multiplicity of sites whose individual sorption rates are statistically distributed as the bell-shaped gamma curve.

There is no general agreement on whether the slow adsorption–desorption reactions are caused by slow diffusion or chemically rate-limited sorption reaction processes. Probably, both play a part. The question is fully discussed by Brusseau & Rao (1989).

3.9.4 Decomposition Rates of Organic Materials Sorbed by Soil

Adsorption by the soil may be expected to decrease the accessibility of a compound to microbes; on the other hand, there is a greater concentration of microorganisms on surfaces than in the pore solution. Hance (1970) concludes that it is not possible to predict the net effect of adsorption on decomposition rates. Nevertheless, the overall kinetics for disappearance of a biocide are often approximately first order at the usual rates of application in the field. If the decomposition is largely due to microorganisms, there is usually a lag phase while the microorganisms multiply, followed by rapid decay. When decay is slow, it is difficult to distinguish between direct chemical and microbial decomposition. Losses of material are not due only to decomposition; volatilization, leaching,

Table 3.3 Persistence of materials in soils: approximate times for at least 75% of added dose to disappear.

Compound	Use	Persistence	Normal rate of use per application (kg ha^{-1})
Methyl bromide	Fumigant for soil sterilization	< 1 week	1000
Ammonia	Fertilizer	3 weeks	up to 100
2,4-D	Foliar herbicide	1 month	1
Disulfoton	Systemic organophosphorus insecticide	6 weeks	1–2
Chlorfenvinphos	Organophosphorus soil insecticide	6 months	1–4
Diuron	Soil-applied substituted urea herbicide	8 months	0.5–2
Simazine	Soil-applied triazine herbicide	1 year	0.5–1.5
Dieldrin	Soil and crop organochlorine insecticide	> 3 years	1–5
DDT	Soil and crop organochlorine insecticide	> 5 years	1–5

Source: Graham-Bryce & Briggs (1970).

and uptake and removal in crops are also involved. Detailed information on the persistence of pesticides has been given by Kaufmann *et al.* (1976), and later studies have been surveyed by Rache & Lichtenstein (1985). The great range in persistence of biocides is illustrated in table 3.4 (Graham-Bryce & Briggs 1970).

Alexander & Scow (1989) discuss various proposed models that are based on the kinetics of microbial growth and decay. Nelson *et al.* (1982) give an example in which the growth of organisms and the hydrolysis of parathion in homogenized soil were simultaneously studied. The model was verified by predicting the rate of hydrolysis of successive additions of parathion. Unfortunately, models that work in controlled laboratory conditions rarely work in so complex a medium as soil in the field, though they may throw light on field performance.

4

Local Movement of Solutes in Soil

In the previous chapter, we dealt with the distribution of solutes between gas, liquid and solid phases in the soil at equilibrium; and with the rates of redistribution between these phases within soil pores. In this chapter, we consider movement of the order of 1–1000 mm from one volume of soil to another. Such movements occur largely by diffusion and mass flow of the soil solution or soil air, and by mass movement of the body of the soil. Major movements that involve the balance and amount of solutes in the whole soil profile, including plant uptake and drainage losses, are treated in chapter 11.

4.1 Diffusion

The process of diffusion results from the random thermal motion of ions, atoms or molecules. Consider a long column of unit cross-section orientated along the x axis, and containing a mixture of components in a single phase at constant temperature and external pressure. If the concentration of an uncharged component is greater at section A than at section B, then on average more of its molecules will move from A to B than from B to A. The net amount crossing a unit section in unit time, which is the flux, is given by the empirical relation known as Fick's first law:

$$F = -D\, \mathrm{d}C/\mathrm{d}x \tag{4.1}$$

where F is the flux, and $\mathrm{d}C/\mathrm{d}x$ is the concentration gradient across the section. The minus sign arises because movement is from high to low concentration in the direction of increasing x. The diffusion coefficient, D, is thus *defined* by the equation as a coefficient between two quantities, F and dC/dx, which can be measured experimentally. It is not necessarily a constant.

The diffusion coefficient of the molecules in a phase is directly proportional to their absolute mobility, u, which is the limiting velocity they attain under unit force. Terms D and u are related by the Nernst–Einstein equation:

$$D = ukT \tag{4.2}$$

where k is the Boltzmann constant and T is the temperature on the Kelvin scale.

The Nernst–Einstein equation is derived as follows (Atkins 1986, p. 675). For a system with concentration gradients in the x direction only, the force on a molecule is the gradient of its chemical potential, $d\mu/dx$. Therefore, the flux across a plane normal to the x axis is

$$F = -C(d\mu/dx)u \tag{4.3}$$

Now for a single molecule

$$\mu = \mu_0 + kT \ln Cf \tag{4.4}$$

where f is the activity coefficient. Therefore, differentiating equation (4.4) with respect to x, and using $d \ln C/dx = 1/C\, dC/dx$, then substituting for $d\mu/dx$ in equation (4.3), gives

$$F = -ukT(dC/dx + C d \ln f/dx) \tag{4.5}$$

The term in $d \ln f/dx$ is usually very small for uncharged particles, and can be neglected. So, comparing the definition of D in equation (4.1) with equation (4.5), we find $D = ukT$ (equation (4.2)).

Since ions are charged, their movements depend not only on their concentration gradients, but also on the electrical forces set up in the solution by movements of other ions or by external electric fields (section 4.2.1). Thus, the interpretation of D for ions may not be simple, but once it has been defined in any given situation, equation (4.1) can be applied. The interactions between ions in the diffusion of salt solutions are described by Robinson & Stokes (1959, chapter 11) and their application in soils by Nye (1966a).

Table 4.1 gives an idea of the enormous range of diffusion coefficients that are of interest to us. Molecules in the vapour phase are approximately 10 000 times as mobile as they are in the aqueous phase: for example, compare O_2 in air and water. Adsorbed ions show a wide variation in mobility. In pure clays, for example, the mobility of sodium ion in montmorillonite ($D = 4 \times 10^{-6}\ cm^2\ s^{-1}$) is one third of its mobility in free solution. In this clay, the hydrated sodium ion lies within a layer of water a few molecules thick between the aluminosilicate sheets of the clay mineral lattice. On the other hand, where there is no water between the sheets, as in the unhydrated mineral illite, the self-diffusion (diffusion of an isotope) coefficient of the potassium ion is only $10^{-23}\ cm^2\ s^{-1}$.

The physical significance of these diffusion coefficients may be appreciated more clearly if we fix our attention on one of the particles that is moving randomly. Then, its average random movement in a linear system, in the x direction, in time t, will be a distance of $(2Dt)^{1/2}$ from its starting point. Thus, a molecule of

Table 4.1 Some typical diffusion coefficients.

	T(°C)	D(cm^2 s^{-1})	Reference
Gas phase			
O_2 into air	25	0.209	
CO_2 into air	25	0.163	
Chloropicrin into air	25	0.088	
Liquid phase			
O_2 in water	25	2.26×10^{-5}	
CO_2 in water	25	1.66×10^{-5}	
NaCl in water	25	1.61×10^{-5}	
Glucose in water	25	0.67×10^{-5}	
Solid phase			
Na in montmorillonite gel	25	4×10^{-6} [a]	Lai & Mortland (1962)
Na in vermiculite	20	6×10^{-9} [a]	Lai & Mortland (1968)
K in illite	lab.	10^{-23} [a]	De Haan *et al.* (1965)
Soil			
Cl in sandy clay loam (40% H_2O by vol.)	lab.	9×10^{-6} [a]	Rowell *et al.* (1967)
Cl in sandy clay loam (20% H_2O by vol.)	lab.	2.4×10^{-6} [a]	Rowell *et al.* (1967)
Na in sandy clay loam (40% H_2O by vol.)	lab.	2.2×10^{-6} [a]	Rowell *et al.* (1967)
Na in sandy clay loam (20% H_2O by vol.)	lab.	0.5×10^{-6} [a]	Rowell *et al.* (1967)
PO_4 in sandy clay loam (40% H_2O by vol.)	lab.	3.3×10^{-9} [a]	Rowell *et al.* (1967)
PO_4 in sandy clay loam (20% H_2O by vol.)	lab.	0.3×10^{-9} [a]	Rowell *et al.* (1967)

[a]Indicates self-diffusion coefficient.

gas will move about 1 m in 10^5 s (1.2 days). In solution, it will move about 1 cm in the same time. The potassium ion in the illite will move only 1 nm in 5×10^8 s (16 years). The long-lived radioactive ^{237}Np, which is a particularly hazardous component of nuclear waste, has a diffusion coefficient in montmorillonite of about 10^{-12} m^2 s^{-1} (Staunton *et al.* 1990). If the waste is surrounded by a barrier of unfissured clay, the neptunium can be expected to diffuse through the clay an average of 11 m during its half-life of 2×10^6 years. We will discuss here, in turn, the factors affecting diffusion of particles in gas, liquid, and solid phases.

4.1.1 Gases

Diffusion of gases has an important role in respiration of soil organisms and roots, and in predicting the fate of volatile soil pollutants. When, at a constant overall external pressure, molecules of gas A diffuse into those of gas B, there is a countermovement of molecules of gas B, and the net diffusion rate is expressed in terms of an interdiffusion coefficient D_{AB}. The interdiffusion coefficient can be predicted approximately from the kinetic theory of gases (Bird *et al.* 1960, p. 508).

The simplest case arises in self-diffusion when gas molecules diffuse into others of the same kind. This is nearly true of the interdiffusion of isotopes. Here,

$$D_{AA^*} = 2/(3\pi^{3/2})[(kT)^{3/2}/(Pd^2m^{1/2})] \tag{4.6}$$

where d is the molecular diameter, m is the mass of a molecule and P is the pressure. This equation is derived below (equation (4.12)). It will be noted that the self-diffusion coefficient varies inversely not only as the square root of the molecular weight, but also as the square of the diameter. A similar, but more complex, expression may be derived for the interdiffusion coefficient D_{AB} of molecules A and B:

$$D_{AB} = \frac{2}{3\pi^{3/2}}\left[(kT)^{3/2}/P\left(\frac{d_A + d_B}{2}\right)^2\right]\left(\frac{1}{2m_A} + \frac{1}{2m_B}\right)^{1/2} \tag{4.7}$$

Notice that the diffusion coefficient even for a component at very low concentration relative to other components is modified by the other component, since D_{AB} depends only on the total pressure and is independent of the concentrations of the separate components. Thus, the diffusion coefficient in air rather than the self-diffusion coefficient should be used for components such as CO_2 or NO, when at concentrations of only a few percent of the total.

From the kinetic theory for pure rigid spheres at low density,

$$D_{AA^*} = \tfrac{1}{3}u\lambda \tag{4.8}$$

where u is the average molecular speed and λ is the mean free path; also,

$$u = (8kT/\pi m)^{1/2} \tag{4.9}$$

where m is the mass of molecule A and

$$\lambda = 1/(\sqrt{2}\pi d^2 n) \tag{4.10}$$

where d is the molecular diameter and n is the molecular concentration. Hence,

$$D_{AA^*} = \tfrac{1}{3}(8kT/\pi m)^{1/2}[1/(\sqrt{2}\pi d^2 n)] \tag{4.11}$$

Using the ideal gas equation, $P = CRT = nkT$, to substitute for n, we obtain

$$D_{AA^*} = 2/(3\pi^{3/2})[(kT)^{3/2}/(Pd^2m^{1/2})] \tag{4.12}$$

A more accurate prediction of D_{AB}, within 5%, may be made from the Chapman–Enskog kinetic theory of gases (Chapman & Cowling 1951), which takes account of intermolecular forces.

For multicomponent systems, such as occur when mixed petroleum products are evaporating through soil (Baehr & Bruell 1990), prediction of interdiffusion coefficients is complex. Curtiss & Hirschfelder (1949) derive the formula

$$D'_{12} = D_{12}\{1 + x_3[(M_3/M_2)D_{13} - D_{12}]/[x_1D_{23} + x_2D_{13} + x_3D_{12}]\} \tag{4.13}$$

where D'_{12} = the diffusivity of gases 1 and 2 in the 3 component system,
D_{12}, etc. = the diffusivities of gases 1 and 2, etc., in a binary mixture,
x_1, etc. = the mol fractions (not distances),
M_2, etc. = the molecular weights.

This formula has been applied to gaseous diffusion in soils by Wood & Greenwood (1971).

4.1.2 Solutions

Diffusion is one of the main means by which nutrients reach plant roots and pollutants enter the soil drainage.

4.1.2.1 The Relation between the Diffusion Coefficient and Particle Size

In solution, the force opposing the motion of a rigid sphere is, by Stokes law, $6\pi\eta rv$, where η is the coefficient of viscosity, r is the radius of the sphere and v is its velocity. The force opposing motion is balanced by the force impelling the sphere. If this force is unity, by definition of mobility, $v = u$, the mobility. Hence,

$$6\pi\eta ru = 1 \tag{4.14}$$

Now the mobility is related to the diffusion coefficient by the Nernst–Einstein equation:

$$D = ukT \tag{4.15}$$

Hence, substituting for u in equation (4.14) gives

$$D = kT/(6\pi\eta r) \tag{4.16}$$

and D is inversely proportional to the radius.

The simple nutrient cations and anions have self-diffusion coefficients in solution in the range $(0.5\text{–}2.0) \times 10^{-5}$ cm^2 s^{-1} at 25°C, as shown in table 4.2. It will be noted that the mobility increases in the order Li < Na < K < Rb < Cs, and Mg < Ca < Sr < Ba, because the cations of higher atomic weight have the smaller hydrated radius. The hydrogen ion, with $D = 9.4 \times 10^{-5}$ cm^2 s^{-1}, has an exceptionally high diffusion coefficient, because it is carried along by a chain mechanism from one water molecule to the next (Glasstone *et al.* 1941, p. 559). A

Table 4.2 Self-diffusion coefficients of ions in aqueous solution at 25°C.

Ion	Li	Na	K	Rb	Cs
D(cm^2 s^{-1} × 10^{-5})	1.0	1.3	2.0	2.1	2.1
Ion	Mg	Ca	Sr	Ba	
D(cm^2 s^{-1} × 10^{-5})	0.7	0.8	0.8	0.8	
Ion	F	Cl	Br	I	
D(cm^2 s^{-1} × 10^{-5})	1.5	2.1	2.2	2.2	
Ion	H	OH			
D(cm^2 s^{-1} × 10^{-5})	9.4	5.3			

Source: From Robinson & Stokes 1959, Appendix 6.1 'Limiting equivalent conductivities of ions at 25°C in water', and the relation

$$D^\circ = \frac{RT}{F^2}\frac{\lambda^\circ}{z} = 2.66 \times 10^{-7}\frac{\lambda^\circ}{z}$$

similar explanation accounts for the abnormal diffusion coefficient of the hydroxyl ion, $D = 5.3 \times 10^{-5}$ cm^2 s^{-1}.

The effective radius of organic molecules in solution depends on their molecular weight and shape. For spherical molecules, the molecular weight $M = 4\pi r^3 \rho/3$, where ρ is the density. Hence, the diffusion coefficient is approximately inversely proportional to the cube root of the molecular weight; for example, for a range of lignosulphonates Goring (1968) found $D = 4 \times 10^{-5}/M^{1/3}$ cm^2 s^{-1}. The globular proteins also fall into this category. The diameter of spherical soil bacteria (cocci) is *c*. 0.5 μm. This is about 10^3 times the effective diameter for diffusion of simple cations, so that the diffusion coefficient of bacteria free in the soil solution should be of the order 10^{-8} cm^2 s^{-1}.

Large long-chain molecules commonly assume in solution a random coiled shape that is roughly spherical. Solvent molecules are trapped in the coil and move with it. The radius of the equivalent sphere is $2/3R_G$, where R_G is the radius of gyration of the random coil. Since $R_G \propto M^{1/2}$, for such molecules $D \propto 1/M^{1/2}$ (Meyerhoff & Schultz 1952). It may be calculated that for thin rod-shaped molecules of constant diameter, $D \propto 1/M^{0.81}$. Collagen is an example. A selection of diffusion coefficients of substances with high molecular weight, including enzymes, is given in table 4.3, extracted from Tanford (1961, chapter 6), who deals with this topic in detail.

4.1.3 Solid Surfaces

In 1:1 clay minerals, such as kaolinite, all the exchangeable cations are held on the external surfaces of the crystals, and can move relatively freely over them. In 2:1 clay minerals, such as montmorillonite, some of the exchangeable cations are on the external surfaces, but the majority occupy interlayer positions between the aluminosilicate sheets. Here, their mobility depends mainly on the expansion of the clay mineral lattice, and the consequent thickness of the water layer between the sheets. Some typical values for surface diffusion coefficients are given in table 4.4. Staunton *et al.* (1990) found no evidence for surface diffusion of the large 5-valent cation ^{237}Np on a range of clays, even though its size restricted interlayer adsorption. Diffusion of unhydrated ions through crystal lattices is an extremely slow process ($D < 10^{-20}$ cm^2 s^{-1}). It occurs mainly along defects in the lattice.

Table 4.3 Diffusion coefficients of macromolecules in water.

Shape	MW	D(cm^2 s^{-1} $\times 10^{-7}$)
Random coil		
Ribonuclease	13 683	11.9
Haemoglobin	68 000	6.9
Urease	480 000	3.5
Rod		
Collagen	345 000	0.7

Source: after Tanford (1961).

Table 4.4 Self-diffusion coefficients of interlayer ions in aluminosilicate clays.

Ion	T(°C)	System	Interlayer water molecules thick	$D(\mathrm{cm^2\,s^{-1}})$	Reference
K	lab.	Illite	nil	1×10^{-23}	De Haan *et al.* (1965)
Mg	25	Vermiculite	2	7×10^{-13}	Graf *et al.* (1968)
Ca	25	Montmorillonite	3–4	1×10^{-6}	Van Olphen (1957)
Na	25	Montmorillonite	3	4×10^{-6}	Mott (1967)

It seems likely that anions, such as $H_2PO_4^-$, specifically adsorbed on clays or oxides, have a negligible surface mobility since they are covalently bonded. Non-specifically adsorbed anions held at positively charged sites (section 3.5.1) may be mobile, but there is no experimental information available.

It seems likely that non-polar organic molecules may have some surface mobility when adsorbed on clay or other organic particles. In moist soils, such surfaces usually have a monolayer of very strongly adsorbed water molecules, and a substantial fraction of the organic molecules will be in outer water layers. However, polar organic molecules are more strongly adsorbed (Goss 1993). Barraclough & Nye (1979) satisfactorily accounted for the diffusion of the polar polyethylene glycol (MW 4000) and polyvinyl pyrollidone (MW 40 000) molecules by assuming that their surface mobility was negligible. However, Gerstl *et al.* (1979) have found that the diffusion coefficient of parathion over a water content ranging from $\theta = 0.18$ to $\theta = 0.34$ indicated that 10% of the adsorbed parathion was as mobile as in solution.

4.2 Diffusion in Soils

Following this brief review of diffusion in gas, liquid, and adsorbed phases, we may now turn to soils, where a substance may be present in one, two, or all three phases. More details of diffusion of solutes in soils will be found in reviews by Nye (1979) and Staunton (1995).

4.2.1 Diffusion of Non-volatile Solutes

Here, we are concerned only with diffusion in solid and solution phases, conditions that obtain for most plant nutrients. For porous systems with well-defined simple geometry — for example close-packed spheres — it is possible to express the total flux of solutes in terms of their concentration gradients and diffusion coefficients in the solution and solid phases, together with pathway shape factors. The flux is not simply the sum of the solution and solid pathways because the diffusate, as we have seen, passes rapidly between the surface and solution. The resulting equations are complex and may be consulted in the reviews of Goring & Churchill (1961), Meredith & Tobias (1962), and in Crank (1975, chapter 12). Nye (1968a) has discussed their application to soils.

In a medium so heterogeneous as soil, a more fruitful approach is to treat the soil as a quasi-homogeneous body in which Fick's first law, $F = -D\,\mathrm{d}C/\mathrm{d}x$, may be applied. This is legitimate as long as we are concerned with diffusion between volumes large enough to average over microscale variations of particle and pore size; that is, to include a representative sample of gas- and liquid-filled pores, and the adjacent adsorbed phases. Then, C is the concentration of diffusate in the whole soil system, that is those ions, atoms or molecules that are in, or pass through, a mobile phase during a time that is short in comparison with the time of the overall process to which the diffusion coefficient is to be applied. Thus, we have seen in section 3.6.1 that equilibrium between exchangeable cations and the adjacent pore solution may be limited by release from the exchange site and diffusion through the interlayers of a clay mineral, giving rise to time-dependent exchange. In stirred suspension, cation exchange is usually rapid, so that release from the solid is not normally rate limiting. In undisturbed soil, equilibrium by diffusion across the adjacent pore liquid may be the slowest process. A large pore 1 mm wide may require 10^4–10^6 s to achieve 95% equilibration for a local difference in concentration across it. Hence, it seems unwise to apply the diffusion coefficient to processes that take much less than a few hours. Solutes that do not exchange nearly completely within the chosen period are not defined as diffusible. In the solution of practical problems, such solutes are treated as having a rate of reaction and they are covered by the source/sink term, $S(C, x, t)$ in equation (1.7).

In practice, the solute may move in the direction of diffusion both through the pore solution, and in the adsorbed form by surface diffusion, and combined surface and solution phase diffusion in series; each is a continuous pathway. Consider now the diffusive flux of solute through a cross-section of soil normal to the x direction. It is given by

$$F = -D_L f_L \theta\,\mathrm{d}C_L/\mathrm{d}x - D_L f_s \theta_s\,\mathrm{d}C_s/\mathrm{d}x \qquad (4.17)$$

where the first term on the right-hand side (RHS) represents exclusively solution phase diffusion and the second term all diffusion that involves the surface. Here

D_L = the diffusion coefficient of the solute in free solution;
θ = the fraction of the soil volume occupied by solution;
θ_s = the fraction of the soil volume occupied by soil solid, and gives the cross-section for surface diffusion;
f_L and f_s = liquid- and solid-phase impedance factors, respectively (see below);
C_L and C_s = amounts of solute per unit volume of liquid and solid phase, respectively.

The use of D_L in the second term is explained in section 4.2.1.4.

Now $\theta_s C_s = \rho S_s$, where ρ is the weight of dry solid per unit volume of whole soil and S_s is the amount of solute adsorbed on unit weight of solid. So ρS_s is the amount of sorbed solute associated with unit volume of soil and now θC_L is the amount in solution. Hence,

$$C = \theta C_L + \rho S_s \qquad (4.18)$$

Since, by definition, $F = -\mathrm{d}C/\mathrm{d}x$, substituting for F in equation (4.17) gives

$$D = D_L f_L \theta\,\mathrm{d}C_L/\mathrm{d}C + D_L f_s \rho\,\mathrm{d}S_s/\mathrm{d}C \qquad (4.19)$$

The terms in equation (4.19) may now be discussed.

4.2.1.1 The Diffusion Coefficient in Solution — D_L

We have already discussed in section 4.1.2 the self-diffusion coefficients of ions in solution. Also, as mentioned in section 4.1, ions, in contrast to neutral molecules, do not diffuse independently, because every microvolume must remain electrically neutral. This is achieved in counterdiffusion by ions of like charge moving in opposite directions, and in salt diffusion by cations and anions moving together. Hence, ions with greater mobility tend to be slowed down by ions with less mobility. The resulting expression for the ionic flux is complex (Nye 1966a). However, it can be shown that the flux calculated using the self-diffusion coefficient is nearly true (a) if the ion of interest is a small proportion of the total solution concentration, which is often the case for phosphate, potassium and ions of trace elements, as well as for most organic ions; (b) if all the ions have similar mobilities. As has been seen, the self-diffusion coefficients of the main cations and anions in solution range from $0.7 \times 10^{-5}\,\mathrm{cm^2\,s^{-1}}$ for Mg to $2.0 \times 10^{-5}\,\mathrm{cm^2\,s^{-1}}$ for Cs, so the modifying effect is usually fairly small in practice. But, if the concentration of the very mobile hydrogen ion is high the modifying effect may be greater — for instance, the counterdiffusion between H and Ca ions in solution decreased from $8.2 \times 10^{-5}\,\mathrm{cm^2\,s^{-1}}$ to $4.4 \times 10^{-5}\,\mathrm{cm^2\,s^{-1}}$ as the pH in M/100 $CaCl_2$ changed from 4.5 to 2.5 (Farr *et al.* 1970). When the proportion of Ca was very small and that of H very high, the H–Ca interdiffusion coefficient would approach that of the Ca self-diffusion coefficient, $0.8 \times 10^{-5}\,\mathrm{cm^2\,s^{-1}}$.

4.2.1.2 The Moisture Content — θ

The concentration of the soil solution is that of an equilibrium dialysate (i.e. a solution in equilibrium with, but separated from, the soil by a permeable membrane), and θ refers to the water associated with this solute; thus, it excludes water hydrating exchangeable cations, or water in the anion exclusion layer (negative adsorption; section 3.1.2). Thus, θC_L is the amount of solute in solution per unit volume of whole soil.

4.2.1.3 The Liquid-Phase Impedance Factor — f_L

(a) Simple Ions and Molecules The f_L is readily calculated by measuring the diffusion coefficient of a non-adsorbed ion, such as Cl^-, since then $C = \theta C_L$ and, by equation (4.19), $D = D_L f_L$. It is similar for all simple ions and molecules. The liquid-phase impedance factor takes account primarily of the tortuous pathway followed by the solute through the pores. This has the effect both of increasing the pathlength to be traversed and of reducing the concentration gradient along the pathlength. It may also include the effect of the increase in the viscosity of water near charged surfaces, which will affect the mobility of all solutes, though it is unlikely to be significant except in dry soils. Kemper *et al.* (1964) found the viscosity of the first three molecular layers of water on charged surfaces to be, respectively, 10, 1.5, and 1.1 times that of free water. Another effect that may be included is the negative adsorption of anions. These may be excluded from very

narrow pores and thin water films, which may thus cut off connections between larger pores. In a soil, all these effects are difficult to separate experimentally.

Clearly, as soil dries, the pathway for diffusion will become more tortuous and f_L will decrease. The relation between the impedance factor and soil moisture for chloride in a sandy loam is well illustrated by figure 4.1. Further illustrations from different experiments were collected by Nye (1979) (figure 4.2). It will be seen (figure 4.1) that in very dry soil f_L is very low and approaches zero when the water content of the soil is a few percent, which probably corresponds to water adsorbed as a monolayer on external surfaces, or in clay interlayers or water in isolated water bodies. Over the field moisture range, -10 to -1000 kPa, the product θf_L, and thus the solute flux at a given concentration gradient (equations (4.1) and (4.19)), may change by a factor of as much as 100. Figure 4.2 shows that at a given moisture content, clay soils have a lower value of f_L than sandy soils, probably because a greater proportion of their water is held closer to surfaces and thus has a higher viscosity. At a given water potential, clay soils usually have a higher value of f_L because they hold more water. Figure 4.3 (So & Nye 1989) shows that increasing the bulk density of soils at a given moisture content slightly reduces the value of f_L.

(b) Macromolecules Two effects will reduce the mobility of a molecule in pores of diameter less than 10 times the molecular diameter. The cross-section of the pore available to the molecule is only (pore radius)^2 – (molecular radius)^2; and the viscous drag on a moving particle increases near the wall of the pore by a factor of $(1 - 2.09x + 2.14x^3 - 0.95x^5)$, where x is here the ratio of the molecular to the pore radius (Faxen 1922). When the medium is finite, the drag factor modifies Stoke's law (section 4.1.2), which assumes an infinite medium. When $x = 1/10$, the reduction in the diffusion coefficient is nearly 40%. This hydrodynamic effect is not to be confused with any effect due to an increase in viscosity in the first few molecular layers of water near charged surfaces, as mentioned in section 4.2.1.3(a).

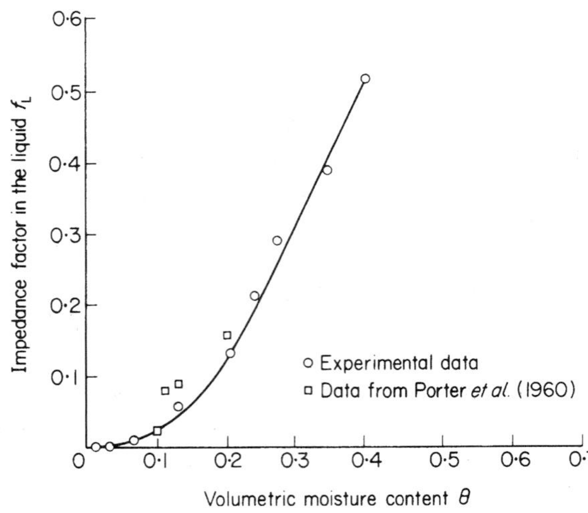

Figure 4.1 Relation between diffusion impedance factor, f_L, and moisture content, θ, for chloride in a sandy loam soil (after Rowell *et al.* 1967).

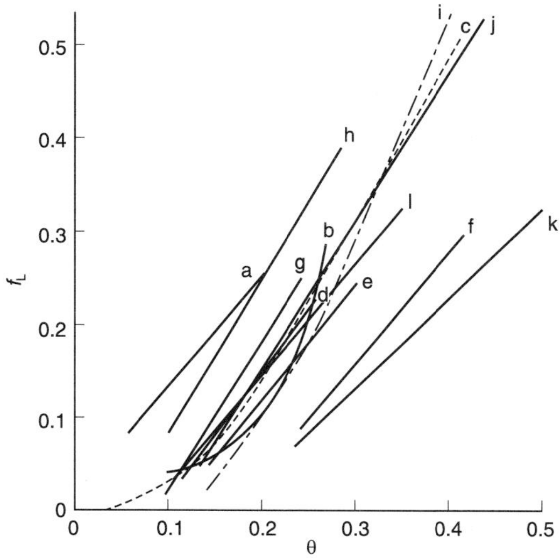

Figure 4.2 Relation between the impedance factor, f_L, and moisture content, θ, for soils of increasing clay content:

Soil	% Clay	Reference
(g) Sand	4	Nielsen (1972)
(a) Wambi sand	5	Paul (1965)
(h) Sandy loam	15	Nielsen (1972)
(l) Hottenrode series loess clayey silt	16	Bhadoria *et al.* (1991)
(b) Urbrae loam	19	Clarke & Barley (1968)
(j) Hooke series silty clay loam (sieved)	23	Barraclough & Tinker (1981)
(c) Sandy loam	24	Rowell *et al.* (1967)
(d) Ft. Collins loam	26	Porter *et al.* (1960)
(i) 6 silt loams (average)		Warncke & Barber (1972)
(k) Woburn sandy loam	33	Barraclough & Tinker (1981)
(e) Apishapa silty clay loam	37	Porter *et al.* (1960)
(f) Pierre clay	53	Porter *et al.* (1960)

Experiments have been carried out with molecules of different size and shape. Williams *et al.* (1966, 1967) found that polyvinyl alcohol (PVA) penetrated aggregates with pores of maximum diameter 6 nm more slowly as its molecular weight increased from 25 000 to 100 000. There was little penetration of pores less than 3 nm across by PVA of MW 70 000. Barraclough & Nye (1979) measured the self-diffusion coefficients of Cl^-, polyethylene glycol (PEG) 4000, and polyvinyl pyrollidone 40 000 in a sandy loam over a wide range of water content (figure 4.4). These solutes have effective radii of 0.18, 1.9, and 18.3 nm, respectively. The impedance factors of PEG 4000 and Cl^- were similar. In moist soil it appeared that the PEG did not diffuse rapidly into 8.5% of the soil volume whereas there was no evidence of any exclusion of Cl^-. The PVP 40 000 did not diffuse into 28% of the soil volume, which corresponded to the intra-aggregate pore space. In dry soil its f_L value was correspondingly small, but in moist soil its f_L value exceeded

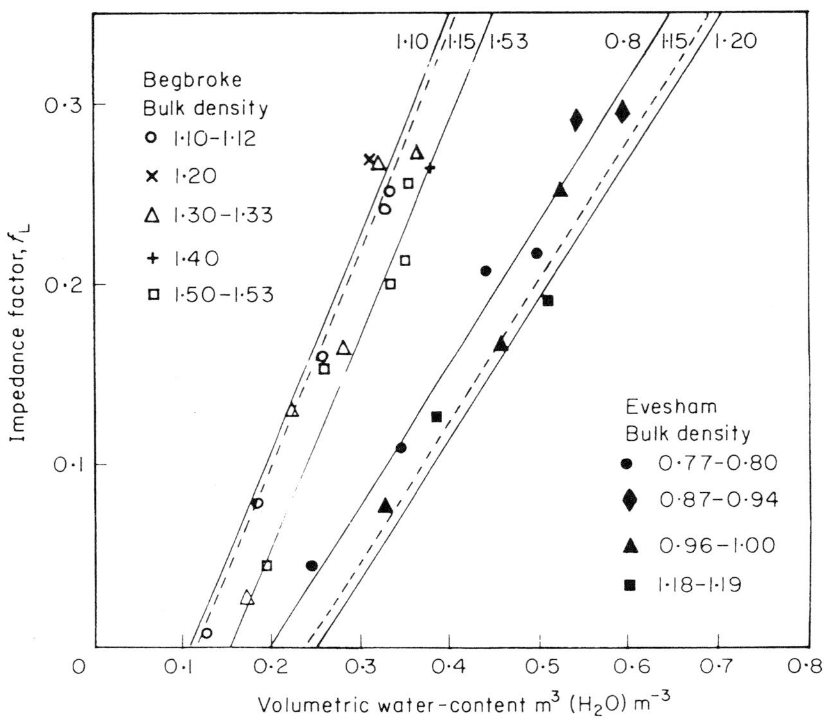

Figure 4.3 The effect on the impedance factor, f_L, of increasing the bulk density of a soil at a given moisture content (from So & Nye 1989).

that of Cl^-, probably because the inter-aggregate pores to which it was confined offered a more direct pathway.

These experiments suggest that the common herbicides and pesticides, whose molecular weights are usually less than 5000, should diffuse with impedance factors similar to simple ions. But, the small proportions that are in the soil solution of humus compounds, synthetic polymers, and plant and microbial exudates and enzymes of high molecular weight or linear shape — and living bacteria — will diffuse largely in pores more than 10 times their size; they will penetrate smaller pores very slowly, and may so reduce the pore's size by adsorption on their walls as to block them completely.

4.2.1.4 The Solid Phase Impedance Factor — f_s

This takes account of all those processes that decrease the mobility of the adsorbed solute from the mobility it would have in free solution. They include the restricted number of water layers available for surface diffusion, orientation of the particles, and close association with the surface by electrostatic forces or by covalent or hydrogen bonding.

As we have seen, as a solute diffuses it moves between solution and solid surfaces, so that the liquid and solid pathways cannot be treated as independent.

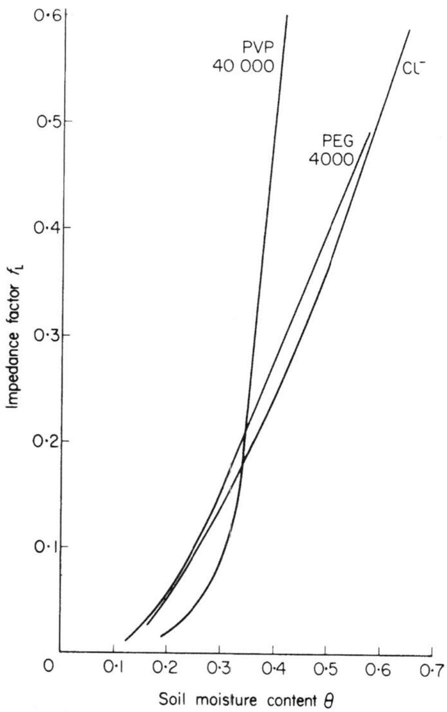

Figure 4.4. The effect of soil moisture level on the impedance factor, f_L, of PVP (MW 40 000), PEG (MW 4000) and chloride ion (after Barraclough 1976).

However, if it is known that a solute is confined to the soil solution, f_L can be determined unequivocally. For example, for Cl^-, which is not usually adsorbed, $C = \theta C_L$. Hence, in equation (4.19), $D = D_L f_L$. Likewise, if surface diffusion is significant and the fraction of a solute that is in the soil solution is negligible, as it will be if θ is small and the solution is very dilute, $C = \rho S_s$. Hence, in equation (4.19), $D = D_L f_s$. When surface diffusion is negligible, as with strongly adsorbed anions, such as phosphate, $f_s \approx 0$ and $D \approx D_L \theta f_L \, \mathrm{d}C_L/\mathrm{d}C$.

In general, when both terms in equation (4.19) contribute to D, independent values of f_L and f_s cannot be assigned. It is convenient to assume that the solution pathway is the same as that for a non-adsorbed ion, and it is assumed that the mobility of an adsorbed ion when following the solution pathway is the same, D_L, as it is in free solution, save that it is reduced by a factor, f_s, which takes care of all the retarding effects due to its adsorption *and* all the complexity arising from the interdependence of the solution and solid pathways (see Staunton 1986 for further theoretical discussion).

With these assumptions, Staunton (1990) has compared the surface impedance factors of Na, Ca, Rb, and Cs in soils of varying mineralogy, as shown in table 4.5. All the f_s values in these moist soils are considerably lower than the f_L values. Table 4.5 further shows that the higher the solid–solution partition coefficient, S_s/C_L, and therefore the more strongly the cation is adsorbed, the smaller is the value of f_s. However, in equation (4.19), an increase in S_s/C_L, and therefore in $\mathrm{d}S_s/\mathrm{d}C$ in a particular experiment, tends to increase the contribution of the surface diffusion to the overall diffusion coefficient. Because the effects of f_s and

Table 4.5 Solid-phase impedance factors, f_s, and solid/liquid concentration ratios for exchangeable cations in a range of soils.

	Banbury[a]		Batcombe[b]		Denchworth[c]		Tedburn[d]		Teigngrace[e]	
	top	sub	top	sub	top	sub	top	sub	top	sub
Surface impedance factor, f_s										
Na ($\times 10^2$)	6.1	4.4	3.5	10.5	—	9.5	5.3	12.9	5.3	18.9
Ca ($\times 10^2$)	1.8	1.1	3.8	1.6	—	2.6	2.3	2.0	2.9	4.1
Rb ($\times 10^3$)	2.0	0.3	4.1	0.9	—	1.1	1.8	0.6	1.9	0.8
Cs ($\times 10^5$)	6.4	7.1	21.0	4.9	—	8.2	11.0	6.1	5.2	4.1
Solid/liquid concentration ratio, S_s/L										
Na	0.9	2.2	0.5	0.4	—	1.6	0.7	1.7	0.7	0.8
Ca	7.1	42.0	5.2	5.4	—	39.0	7.1	87.0	15.1	23.0
Rb	41.5	172.0	22.0	80.0	—	927.0	174.0	625.0	105.0	160.0
Cs	553	25 823	267	1101	—	35 235	2927	6885	1725	1817

[a]Banbury (micaceous very ferruginous); clay: top 35.7%, sub 34.5%.
[b]Batcombe (complex vermiculite); clay: top 18.9%, sub 24.5%.
[c]Denchworth (montmorillonite); clay: sub 51.4%.
[d]Tedburn (micaceous); clay: top 43.4%, sub 41.5%.
[e]Teigngrace (kaolinite); clay: top 38.0%, sub 41.5%.

S_s/C_L work in opposite directions, the net effect on the contribution of surface diffusion to the overall diffusion coefficient is variable, as Staunton (1990) shows. For these exchangeable cations, surface diffusion contributed between 27 and 97% to the overall diffusion coefficient. No correlation between f_s and the mineralogy or composition of the soils was found.

4.2.1.5 *The Reciprocal of the Buffer Power — the Derivative* $\mathrm{d}C_L/\mathrm{d}C$

The great range in values of the slope of the sorption isotherms for various ions and uncharged molecules and for different soils is discussed in chapter 3. A correspondingly wide range of diffusion coefficients has been measured. For a solute that is not usually adsorbed by the soil solid, such as the nitrate ion, $\theta\,\mathrm{d}C_L/\mathrm{d}C = 1$. Hence, in equation (4.19), $D = D_L f_L$, and in a moist soil the diffusion coefficient is about $2 \times 10^{-5} \times 0.5 = 10^{-5}\,\mathrm{cm}^2\,\mathrm{s}^{-1}$. In contrast, $\mathrm{d}C/\mathrm{d}C_L$ for phosphate ion may be 1000, giving a diffusion coefficient $D_L \theta f_L\,\mathrm{d}C_L/\mathrm{d}C$ of $10^{-5} \times 0.3 \times 0.5 \times 10^{-3} \approx 10^{-9}\,\mathrm{cm}^2\,\mathrm{s}^{-1}$ in a similarly moist soil. Thus, in 1 day the root-mean-square displacement of nitrate in a moist soil may be as much as 1 cm, while that of phosphate may be as low as 10^{-2} cm. Since the relation between C and C_L is often non-linear, the diffusion coefficient may vary with concentration. Figure 4.5 shows the slope of the adsorption isotherm of a clay soil at different levels of potassium and the corresponding diffusion coefficient measured (Vaidyanathan *et al.* 1968).

(a) Uncharged Solutes At low concentrations, the adsorption isotherms of a great range of herbicides and pesticides are approximately linear (Hamaker 1972). Since the surface diffusion coefficient of these compounds is likely to be negligible, their diffusion coefficients, by equation (4.19), should be independent

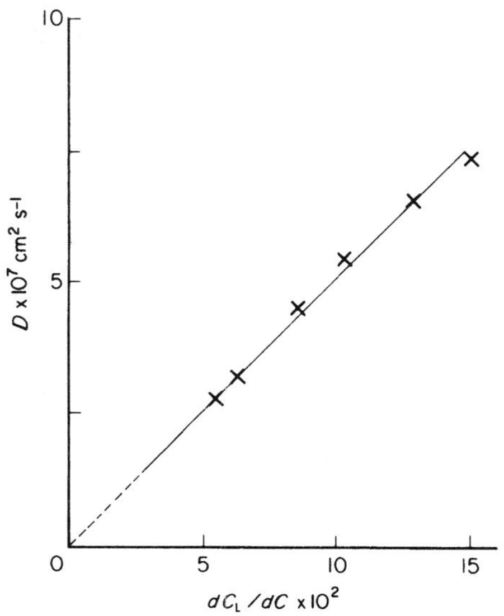

Figure 4.5 The relation between the diffusion coefficient of potassium in Coral Rag clay soil and the reciprocal buffer power $\mathrm{d}C_L/\mathrm{d}C$ (after Vaidyanathan *et al.* 1968).

of concentration. Often the differences between the diffusion coefficients of non-volatile herbicides can be accounted for solely by variation in their solid–liquid distribution coefficients. However, as discussed in section 3.9.2, adsorption isotherms are often non-linear and in accurate work the variation in $\mathrm{d}C_L/\mathrm{d}C$ over the concentration range of interest must be taken into account. Nevertheless, for many practical predictions of the fate of organic compounds, it is sufficient to use the average value of $\mathrm{d}C_L/\mathrm{d}C$ over the concentration range.

(b) Ions In those instances when surface diffusion is negligible, such as for phosphate, the diffusion coefficient is inversely proportional to the buffer power. However, the correct determination of the relation between C and C_L is not as easy as it might appear, since it must reproduce exactly the same conditions as occur in the diffusion process. The following conditions arise.

(i) The True Pore Solution Concentration Methods of determining this have been described in section 3.1.1. The value will be influenced by the soil water content and by the concentration of other ions in the soil solution. Hence, concentrations measured in saturation extracts are not usually sufficient. Nor is it possible to prepare unsterilized soils with electrolyte-free pore solutions by washing with distilled water, and even sterilized soils will have HCO_3^- in solution.

(ii) The Choice of the Exchanging Ion It is particularly important that the exchanging ion should be correctly chosen. For example, in a diffusion process in which phosphate is being desorbed from the soil, the relation between C_L and C is very different in a solution that contains an indifferent anion, such as Cl^- or NO_3^-, from one that contains a specifically adsorbed ion, such as bicarbonate or citrate. If the exchanging ion is an isotopic form of the ion of interest,

$dC/dC_L = C/C_L$ which is a constant provided that the soil is at equilibrium except for the movement of the isotopes; hence, D_{self} is constant. But in normal non-isotopic diffusion, it is clear from figure 3.12 that dC_L/dC may differ greatly from C_L/C. Hence, the self-diffusion coefficient cannot normally be used as the effective coefficient over a given range of concentration. Further discussion of the effects of the derivative dC_L/dC has been given by Olsen & Kemper (1968), Nye (1968a) and Tinker (1970).

In practical applications, it has not proved necessary to measure the absolute amounts of diffusible ions. This would be a difficult task if C and C_L are related as shown for potassium in figure 3.4, since at very low concentration the amount that will desorb is indefinite, and is often affected by release of ions that are only slowly exchangeable (see (iii) below). In practice, one is always concerned with diffusion between certain concentration limits, and hence with a difference ΔC. If, as is frequently the case, the concentration limits are expressed in terms of the solution concentration, then ΔC becomes the change in the amount of diffusible ions between the specified limits of the solution concentration.

(iii) The Rate of Equilibration between the Solution and the Solid In equation (4.18), dC_L/dC and dS_s/dC, and hence D, will be independent of time only if there is virtually instantaneous equilibrium between the ions on exchange sites and the adjoining solution. If, for a small change, δC_L, C changes with time, then D will be time dependent. Fortunately, as we have seen in chapter 3, cation exchange is usually rapid; but if there is a slow further exchange the system must be treated as one of diffusion with reaction, in which the source term in equation (1.7) is significant. Many pesticides equilibrate only slowly with the soil (Walker *et al.* 1995). For example, the distribution coefficients for cyanazine and metribuzin were two or three times greater after 56 days than they were initially. There is a useful rough method of deciding whether slow release into solution will limit the diffusive supply of a solute to a sink (Crank 1975). If the half-time for the release is the same as or less than the diffusion time, the amount diffusing is approximately the same as it would be if the release were infinitely rapid. If the release is slower, it will significantly decrease the amount reaching the sink in the given time.

Many of the distribution coefficients in the literature have been determined by equilibrating a suspension of the soil in an agitated solution. This may give a false impression of the rate of equilibration of the undisturbed soil. Staunton & Nye (1989a) found the rate of phosphate adsorption to be greater in a stirred suspension than in either a soil column percolated by a solution, or in an undisturbed soil, as measured by extracting the soil pore solution by centrifuging with a dense immiscible liquid (figure 4.6). The reason for slow equilibration may lie in slow adsorption on the solid surfaces, or it may be due to slow access to the adsorption sites through fine and tortuous aqueous diffusion paths or subsequent diffusion into the solid. For example, Staunton & Nye (1989b) interpreted the concentration gradient of ^{32}P diffusing through a soil column in terms of a rapid diffusion path through macropores and a slow one through micropores. Barrow (1983) has shown that phosphate is rapidly adsorbed on the surface of goethite (FeOOH), an important soil constituent, and then slowly diffuses into the goethite lattice.

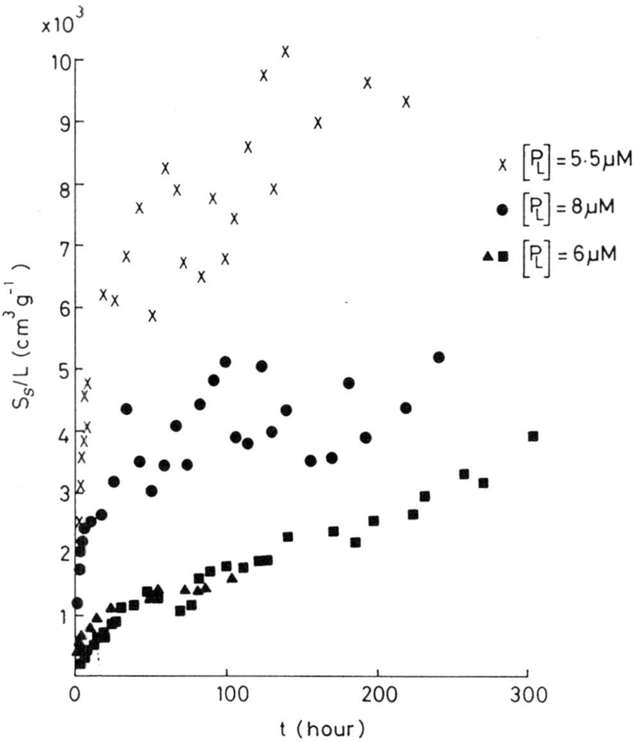

Figure 4.6 Effect of method of equilibration on rate of phosphate sorption (after Staunton & Nye 1989a). Plots of concentration ratio, S_s/L, in the sorbed and liquid phases obtained from the following: ×, 1:10 suspension; ●, 1:1 suspension; ▲, circulating solution; ■, extraction of solution from moist soil.

Relaxation (section 3.6.3) is an aspect of slow reaction. In calculating the diffusion coefficient for a desorption process, it is clearly important to use the desorption limb of the isotherm; and for an adsorption process, to use the corresponding adsorption limb.

(iv) The Self-Diffusion Coefficient Much has been learned about diffusion in soils from self-diffusion experiments, in which one isotope is allowed to counterdiffuse against another of the same species. For instance, as in the classic experiment of Schofield & Graham-Bryce (1960), a block of soil, labelled with a radioactive isotope, may be placed in contact with an unlabelled but otherwise identical block and the movement of the labelled isotope into the unlabelled block followed. In such an experiment, assuming complete equilibrium initially in the labelled block, we have $C'_L/C' = C''_L/C'' = C_L/C$, where the primed terms refer to the two diffusing isotopes and the unprimed to the total. Since the distribution of the total number of ions between the soil and the solution in the blocks remains constant, C'_L/C' also remains constant, and we may write

$$dC'_L/dC' = C'_L/C' \quad (4.20)$$

By a similar argument dC'/dS'_s remains constant. Hence from equation (4.19) we obtain

$$D_{self} = D_{Lself}\theta f_L C_L/C + D_{Lself} f_s \rho S_s/C \quad (4.21)$$

which is a constant, independent of C'_L and C''_L for the experiment.

4.2.2 Diffusion of Volatile Solutes

We have noted that the mobility of molecules in the gaseous phase is about 10 000 times greater than in aqueous solution. Hence, in a moist soil in which volumes of water and air are about equal, gaseous diffusion may be expected to dominate if the distribution coefficient $C_L/C_g \ll 10^4$. (This distribution coefficient is often tabulated as Ostwald's 'coefficient of solubility', β.) For example, in water at 0°C, β for $O_2 = 0.048$, $CO_2 = 1.71$ and $SO_2 = 80$. When the distribution coefficient is about 10^4, the gaseous phase may replace the solution without greatly affecting the overall steady-state diffusion — a conclusion confirmed experimentally by Graham-Bryce (1969) for disulfoton.

We may estimate the diffusive flux of a volatile solute if the gaseous and liquid pathways are assumed to be independent and any solute adsorbed on the solid to be immobile. We may then write

$$F = -(D_g\theta_g f_g + D_L\theta f_L\beta)\,dC_g/dx \quad (4.22)$$

where we assume Henry's law (C_L/C_g is constant) is obeyed. In reality, the gaseous and solution pathways, while usually continuous, are not independent, as we found for the surface and solution pathways, and the true flux will be greater than the formula indicates because there will be an extra term for the interaction between them. Hartley & Graham-Bryce (1980, chapter 5) cite many examples of the true flux being greater than the formula indicates (equation (4.20)).

4.2.2.1 Gaseous Diffusion — the Impedance Factor f_g

In dry materials, the Bruggeman equation, $f_g = \theta_g^n$, satisfactorily accounts for the relation between f_g and air-filled porosity. The value of n, which can be derived theoretically, depends on particle shape; it is 0.5 for spheres and larger for more complicated shapes. Figure 4.7 shows values calculated from Currie (1960) for diffusion of hydrogen and air through a range of materials, including dry soil crumbs for which $n = 2$.

When water is added to a dry material with a unimodal distribution of pore sizes, such as sand, it has a greater effect in lowering f_g than would consolidation of the dry material to the same θ_g. Figure 4.8 shows this striking difference. The reason is that the water is held at the narrowest constrictions of the channels between the pores, just where it exerts the greatest blocking effect.

For several types of material the behaviour on wetting is described by the equation derived from Currie (1961):

$$f_g = \theta_g^n[\theta_g/(\theta_g + \theta)]^{2.5} \quad (4.23)$$

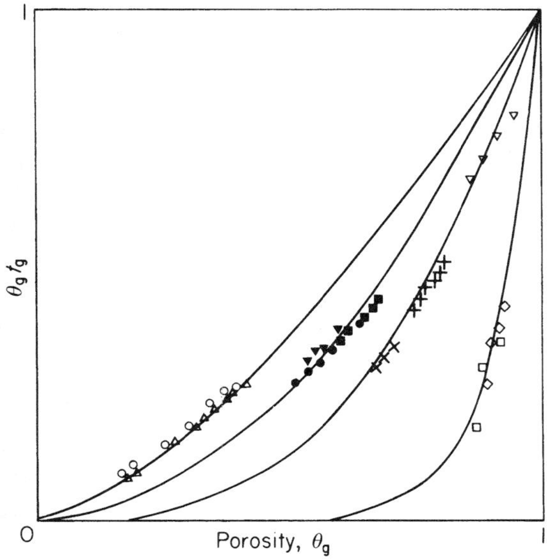

Figure 4.7 Diffusion of gas through a range of porous materials—relation between $\theta_g f_g$ and porosity, θ_g. Curves are plotted according to the Bruggeman equation, $f_g = (\theta_g)^n$ (after Currie 1960). $n = 0.5$: ○, glass beads; Δ, sand. $n = 1$: ● ▼ ■, soil crumbs. $n = 2$: ×, pumice; ∇, diatomaceous earth; +, kaolin. $n = 9$: ◇, vermiculite; □, mica.

In soils having well-marked structural aggregates, the pore size distribution is bimodal, corresponding to the micropores within the crumbs and the macropores between them. When water is added to these soils and the air-filled porosity decreases, the value of f_g may actually increase until the micropores are saturated (see figure 4.8). On further wetting, f_g then follows the same course as in sand. The reason is that the diffusion path is simpler when the air-filled pore space contains only macropores than when it contains micropores as well.

Millington (1959), following a model of Childs & Collis-George (1950) originally used for hydraulic conductivity, has calculated the impedance factor by

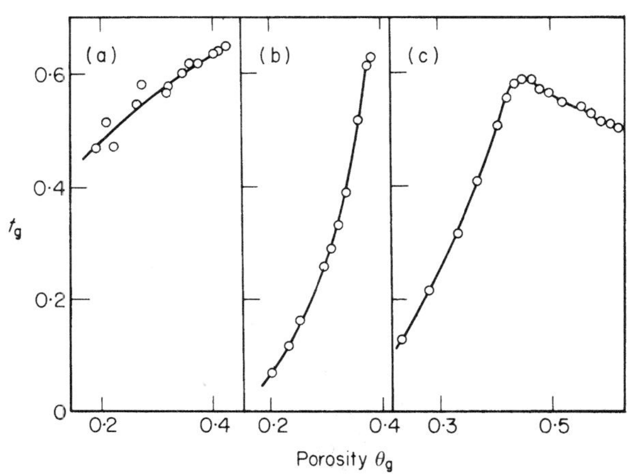

Figure 4.8 Effect of (a) consolidation, (b) adding water on f_g for a sand. (c) Effect of adding water on f_g for an aggregated soil (after Currie 1961).

considering the probability of two pores being opposite each other when 'adjacent' pore-thickness layers of soil are superposed. In effect, he calculates the continuous pore space across the two layers. Millington & Shearer (1971) have used this approach to include soil aggregates. Figure 4.9 shows that the effect of aggregation is to increase the value of f_g, but to decrease that of f_L.

Sallam *et al.* (1984) have reviewed the various models that have been proposed. They point out that while all models agree fairly well when the air-filled porosity exceeds 0.3, they differ when it is less. For an aggregated silt loam with θ_g between 0.05 and 0.15, they find

$$f_g = \theta_g^{2.1}/(\theta_g + \theta)^2 \tag{4.24}$$

4.3 Mass Flow and Dispersion in Solution

In practice, solutes are moved both by diffusion and by convection (mass flow) of the soil solution caused by such processes as transpiration, evaporation or drainage; and the relative importance of diffusion and convection to root surfaces will be an important theme in chapter 6. Here, we are concerned with the effect of water movement through the soil on the dispersion of the solute, since this augments and interacts with the molecular diffusion discussed in section 4.1.

When water flows steadily through a tube by laminar (in contrast to turbulent) flow, its velocity profile across the tube is given by $v_r = K(a^2 - r^2)$ where K is a constant, a is the radius of the tube and r is the radial distance from the centre.

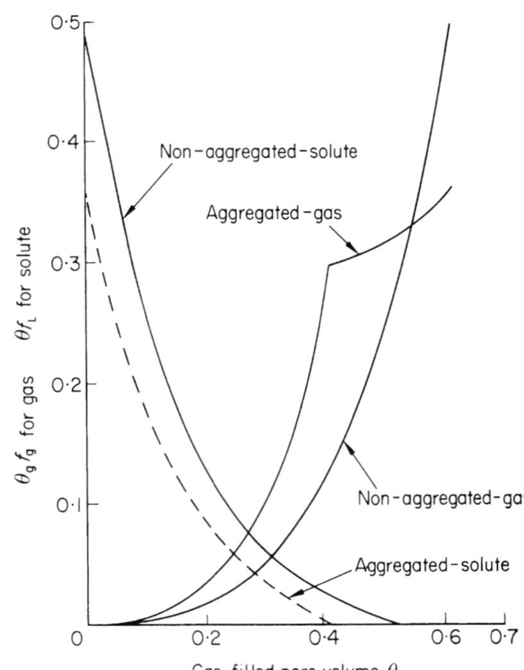

Figure 4.9 The theoretical effect of aggregation on the relation $\theta_g - \theta_g f_g$ for gases and $\theta_g - \theta f_L$ for solutes (after Millington & Shearer 1971).

The velocity is zero at the wall and greatest at the centre. If at time zero, t_0, a thin layer of dye were injected across the tube, it would after successive times t_1, t_2 and t_3 have assumed the parabolic forms shown in figure 4.10. Thus, even in a simple tube convection spreads a solute. In a porous medium the microscopic flow pattern of the water is far more complex. Philip (1969) has pointed out that in a spherical pore there are even zones where eddies create a flow in the opposite direction to the main stream.

A complete theory for dispersion is lacking, though the many approximate theories have been reviewed by Dullien (1979, chapter 7) from a hydrodynamic viewpoint; and by Rose (1977) and Nielsen *et al.* (1986) from a soil viewpoint. In these approaches, the molecular diffusion coefficient D_L, due to random thermal motion of the solute molecules, is replaced in equation (4.17) by an effective solution dispersion coefficient D_L^*. Frissel & Poelstra (1967) found that the equation

$$D_L^* = D_L + v\lambda d/\theta f_L \tag{4.25}$$

in which the effective solution dispersion coefficient, D_L^* increases linearly with the average flow rate in the pores v/θ, described the dispersion of strontium in saturated columns of sand–resin and sand–clay mixtures, over a range of v from 2×10^{-6} to 20×10^{-6} cm s^{-1}. The average particle diameter is d and λ is a packing factor that is 1 for spheres, but may increase to 10 for irregularly shaped particles.

In soil, it is impossible to separate λd, and the combined effective size and packing of the aggregates must be determined experimentally. In natural soils, the dispersion coefficient may be expected to exceed that measured in columns of uniformly packed solids or aggregates because of cracks due to shrinkage, and channels created by growth and decay of roots and the passage of worms and other fauna. For example, Fried & Ungemach (1971) found λd to be as high as 1100 cm in water seeping laterally from a well, compared with an average laboratory value in soil of 0.1 cm. Frissel *et al.* (1970a, b) found the product λd had mean values of 0.7, 0.8, and 6.0 cm in columns of *undisturbed* sandy, clay, and loess loam soils, respectively, over a range of v from 0.6×10^{-6} to 230×10^{-6} cm s^{-1}.

Porous aggregates tend to increase dispersion, since there is not instantaneous equilibrium between the intra- and inter-aggregate pore solution. This last effect

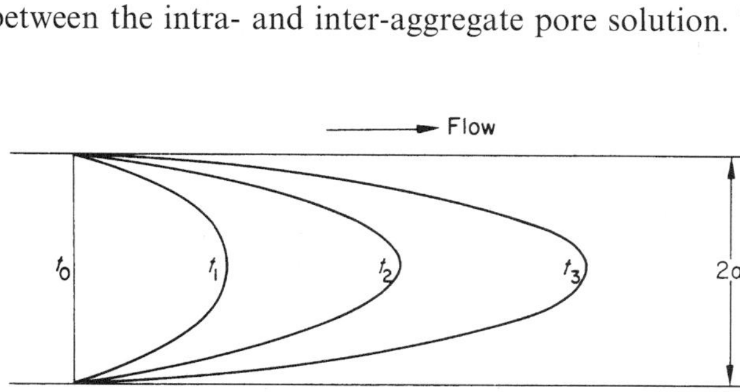

Figure 4.10 Sketch of movement of a thin layer of dye by water flowing in a tube.

has been shown theoretically by Gluekauf (1955) and Passioura (1971) to increase the dispersion coefficient D_L^* by an amount

$$1/15\,(\varepsilon_A/\varepsilon_T)(v^2/\varepsilon_M)(a^2/D_A f_{LM})$$

where ε_A, ε_M, and ε_T are the porosities of the aggregates, the interaggregate spaces and the total porosity, respectively; a is the aggregate radius; and D_A is an intra-aggregate molecular diffusion coefficient $= (D_L f_L)_A$. This expression has been tested by Passioura & Rose (1971). For values of va/ε_M of less than $2 \times 10^{-5}\ \mathrm{cm^2\,s^{-1}}$, the expression is negligible, but it becomes comparable with the linear velocity term $v\lambda d/\theta f_L$ in equation (4.25), when va/ε_m is *c.* $10^{-4}\ \mathrm{cm^2\,s^{-1}}$. This will only be attained with large aggregates and rapid drainage: for example, for crumbs of 2 mm diameter and $\varepsilon_M = 0.2$, v would be $10^{-4}\ \mathrm{cm\,s^{-1}}$ or *c.* 10 cm $\mathrm{day^{-1}}$. It leads to an increasing slope in the plot of dispersion coefficient against flow velocity.

All the work covered so far has been done with saturated or fairly moist materials. Kirsch (1992) has studied dispersion of chloride ion near the surface of a column of sandy loam soil that was evaporating freely at a steady rate into the atmosphere. His experiments include water contents as low as 0.06. Figure 4.11 summarizes the relation between the effective solute dispersion coefficient and the water flux in these experiments. In agreement with equation (4.25), the relation is linear and the slope gives a value for $\lambda d = 0.04$ cm. It is interesting that

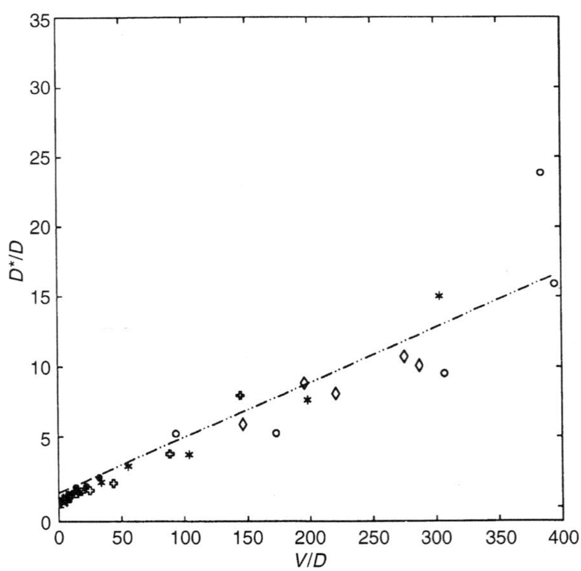

Figure 4.11 The relation between v/D, and D^*/D for chloride in very dry soil in seven experiments (from Kirsch 1992). The theoretical relation (equation (4.25)) is $D^*/D = 1 + d\lambda v/D$. The slope of the regression line gives $d\lambda = 0.04$.

equation (4.25) seems to describe the effect of water flow on the dispersion coefficient even in very dry soil.

In many cases in the field, the influence of soil structure on the movement of the solute is so strong that a clear distinction can be made between rapid movement by convection in the water flowing between structural aggregates, and the much slower movement by diffusion and convection within them. This process is often described as bypass flow (see section 11.2.2).

In natural soils, at flow rates that occur during infiltration and drainage, it seems clear that the dispersion coefficient will usually greatly exceed the molecular diffusion coefficient. Whether D_L^* is appreciably greater than D_L for solute convected to root surfaces by transpiration depends on the aggregate size and packing, the transpiration rate and the moisture content. To take a typical instance, if v at the root surface $= 10^{-6}\ \text{cm}^2\ \text{s}^{-1}$, $f_L = 0.1$ and $\lambda d = 1.0$ cm, then in equation (4.25), $v\lambda d/\theta f_L = 10^{-5}\ \text{cm}^2\ \text{s}^{-1}$, which is comparable with D_L. However, since the water flow is radial, v decreases away from the root. Also, the value of λd will be lower over short distances around roots than in columns in which long channels and cracks are influential. Unfortunately, quantitative experiments with roots, described in chapter 6, have been made with carefully packed aggregates, so there is no direct evidence about the situation in the field.

4.4 Gaseous Convection and Diffusion

Transfer of gases within the soil is mainly by diffusion, and convection resulting from hydraulic pressure gradients is of minor importance. The hydraulic conductivity of aerobic soil for gaseous movement is so large that the overall pressure within the soil adjusts rapidly to changes in the external barometric pressure. An increase in atmospheric pressure forces an amount of gas into the air-filled pore space in the soil that can be predicted from Boyle's law. Buckingham (1904) estimated that, because of fluctuations in atmospheric pressure, air would enter and later be expelled from only 3–5 mm depth of a soil 3 m deep.

Pressure fluctuations caused by wind have a more important effect because they mix the air in surface layers, and hence aerate soils that have a surface crust that hinders the escape of carbon dioxide. Farrell *et al.* (1966), in a theoretical study, treated the turbulent air at the soil surface as a train of sinusoidal pressure waves. They predicted that a wind speed of 15 mph would cause mixing of air in a coarse soil or a surface mulch to a depth of several centimetres. The effective diffusion coefficient of air in coarse mulches could be as much as 100 times the molecular diffusion coefficient. Currie (1970) has verified that passage of turbulent air does decrease the accumulation of carbon dioxide beneath a surface crust.

Temperature gradients within the soil, and temperature differences between the soil surface and the atmosphere, also cause mass movement of soil air, but in comparison with movement by gaseous diffusion their effects are small (Romell 1922). Water infiltration, following rainfall or irrigation, also results in displacement of soil air.

4.5 Mechanical Movement

In most of the examples we treat, the soil matrix is regarded for simplicity as fixed; but in any applications of theory to field situations due weight must be given to mechanical disturbances that arise from a wide range of sources that are difficult to quantify. These are fully discussed in standard textbooks such as Wild (1988). Here, we note without elaborating that shrinkage and swelling caused by moisture changes create cracks and channels through which air and water pass readily. An extreme situation arises in so-called self-mulching soils (vertisols), in which the whole solum gradually turns over as the topsoil at the edges of the cracks falls to their base.

During the growth of a crop, solutes may be redistributed by cultivation treatments, such as earthing up of potatoes. Erosion by wind or water may redistribute the immediate surface layers. Fauna, such as worms and termites, carry soil from subsoil to surface, and create channels for the passage of roots, gases, and water. In spite of these effects, it is noteworthy that sharp boundaries are maintained between adjacent plots under different treatments in long-term trials such as Parkgrass and Broadbalk, at Rothamsted, which suggests that no major surface transfer has taken place with these treatments.

5

The Uptake Properties of the Root System

The uptake of nutrient and other ions into the root from the surrounding soil is the main topic of this book. To understand it, we need to know how the nutrient uptake and demand of the plant is expressed at the root surface. The main interest is on how the demand at the root surface can be quantitatively defined in terms of its uptake characteristics. For this reason, our explanation of the ion uptake mechanism of the root itself is brief, and is intended mainly for readers who have not studied the subject deeply. The subject has become considerably more complex since 1977, but this detailed knowledge has not yet coalesced into a full model of how ions are absorbed, such as ultimately will allow root uptake properties to be predicted. There have been many good reviews in the recent past, and the following may be consulted: Clarkson & Hanson 1980; Glass 1983; Luttge 1983; Clarkson 1985; Sanders 1990; Clarkson & Luttge 1991; Marschner 1995.

5.1 Root Morphology

5.1.1 General

We will describe the structure of a single root only briefly here, since this information can be found in standard texts (Troughton 1957; Esau 1965; Cutter 1978; Fahn 1982). Figures 5.1–5.5 show the general structure, but here we stress points that have a special bearing on the process of ion uptake or root behaviour in soil. Byrne (1974) noted that the anatomy of soil-grown roots may differ somewhat

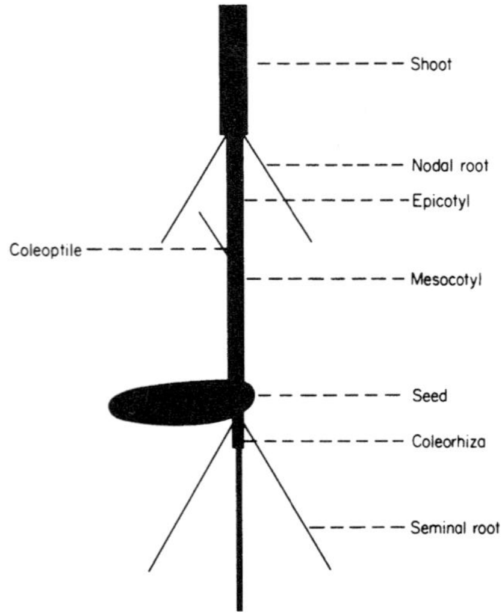

Figure 5.1 Schematic drawing of the basic structure of a young grass seedling, showing development of both seminal and nodal roots (after Troughton 1957).

from that of solution-grown roots. The architecture of whole root systems in soil is dealt with in chapter 9.

5.1.2 Root Tip Structure

The root tip is a highly important part of the root. The apical meristem (the 'quiescent centre') is a fraction of a millimetre behind the visible root tip; cells that form behind the centre of this develop into the root, whereas those in front of the centre form the root cap (figure 5.2). These cells gradually reach the surface of the cap, and there are rubbed off and lost into the soil at a rate of several thousand per day in maize. Often, these cells are visible in the mucigel that forms from the base of the root cap and covers the young root (section 8.1.3), and can remain alive in the gel for a period.

The root cap has long been thought to be the site of the geotropic response of the root, but this is not yet settled. Salisbury (1993) concluded that the amyloplasts (specialized bodies in the cell that store starch) of the root tip cells do, indeed, have a function in relation to gravitropism. These 'statoliths' have a density of around 1.4, well above that of the cytoplasm in which they lie, so they respond quite rapidly to gravity by settling out. This behaviour normally, but not always, accompanies a gravitropic response, so the process is not fully understood (Sievers & Braun 1996). Auxin, and an auxin inhibitor, are probably involved in the actual response of the root. With such uncertainty about a fundamental aspect of root architecture development, it is hardly surprising that root distribution cannot be predicted in detail.

The roots of many species also show a hydrotropic response, growing towards high relative humidities. Oyanagi *et al.* (1995) showed that these tropisms may interact, so possibly the site of this effect is also in the root tip.

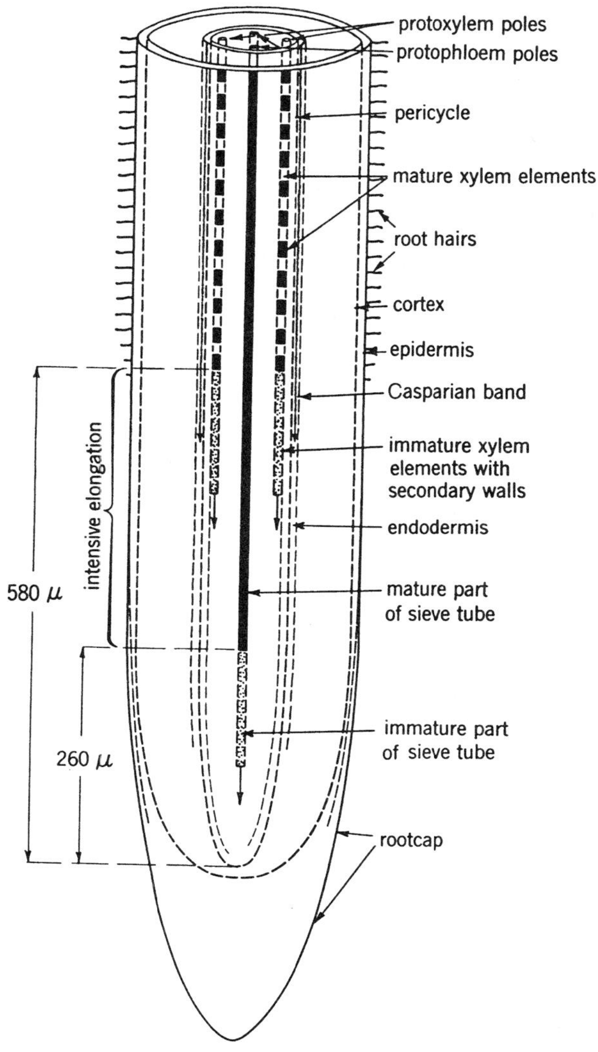

Figure 5.2 Diagram of the internal structure of the root tip of pea (after Esau 1965).

5.1.3 Structure of Epidermis and Root Hair

The outer surface of the root is formed by the epidermis, a layer of thin-walled small cells (figure 5.2). As the root ages, their cell walls can become impregnated with the hydrophobic substance called suberin, which may affect the ability of water and solutes to enter the root (Canny & Huang 1994). Later the whole cortex may be lost.

Root hairs are elongated outgrowths from some of the epidermal cells; typically, they are 100–1500 μm long and 5–17 μm in diameter (Dittmer 1949; Hofer 1996; Peterson & Farquhar 1996). Numbers range from 2 per mm^2 on the surface of roots of loblolly pine to 50–100 per mm root length in the *Gramineae* (Drew & Nye 1970; Newman 1974). In some species, they appear only on specialized epidermal cells, the trichoblasts. The root hair zone on roots is short because of their short life, and Fusseder (1987) observed that in maize roots the internal cytoplasmic structures started to break down by the second day. The root hair walls can remain for quite a long time after the root hair cell has ceased to function (Clarkson 1996), so even though Sparling (1976) found that up to one third of all roots in natural grassland in northern Britain still carried root hairs, some of which must have been several months old, they were probably not active in uptake. It is, however, possible that they are still useful to the plant by ensuring close contact between root and soil and by giving the plant a firmer anchorage. When measurements are made, the physical condition of root hairs should be noted.

The frequency and size of root hairs depend upon many environmental factors (Jungk 1996): water potential (Mackay & Barber 1987), phosphate (Bhat & Nye 1974b), calcium (Cormack *et al.* 1963), bicarbonate concentration, microorganisms (Bowen & Rovira 1961), pH, and soil impedance. High concentrations of P_i and nitrate can diminish root hair length sharply (Bates & Lynch 1996). It is likely that root hairs add little to the ability of a root to take up the necessary nutrients from solution culture, but they have an important function in uptake from soil (section 6.3.2). This has been measured directly, and recently ammonium and nitrate transporters have been isolated from the root hairs of tomato (Lauter *et al.* 1996).

5.1.4 Structure of Cortex, Endodermis, and Stele

The cortex is the outer part of the root (figures 5.2 and 5.3). After being differentiated in the root tip, the cells elongate and become highly vacuolated. There are conspicuous spaces between the cells running longitudinally along the root, that are air filled in living roots. They remain air filled even when under a positive hydrostatic pressure, as for example in rice roots, which seems surprising. In some species and conditions, cells break down to form larger air-filled cavities (aerenchyma) that can keep the root aerated even when growing below the water table (Armstrong 1979) (section 9.3.6).

The cortex is bounded on its inner surface by a single annular layer of cells called the endodermis, whose radial walls are impregnated with suberin (the Casparian band) (Petersen 1988). This, in effect, blocks and waterproofs the intercellular spaces and the cell wall pores so that passage of water and solutes across this layer can occur only through the plasmalemma membranes of the cells. The latter are strongly attached to the Casparian band on the inside of the cell walls, so that water and solutes must pass through the symplasm (section 5.2.1) at this point in moving from the root surface to the stele (Clarkson 1993). The endodermis is therefore semipermeable, and the root is a somewhat imperfect osmometer. This is clearly important in the development of pressure or suction in the vessels of the stele, and in the transport of water and solutes into the plant (section 2.3.3). The Casparian band forms within a few millimeters of the root tip,

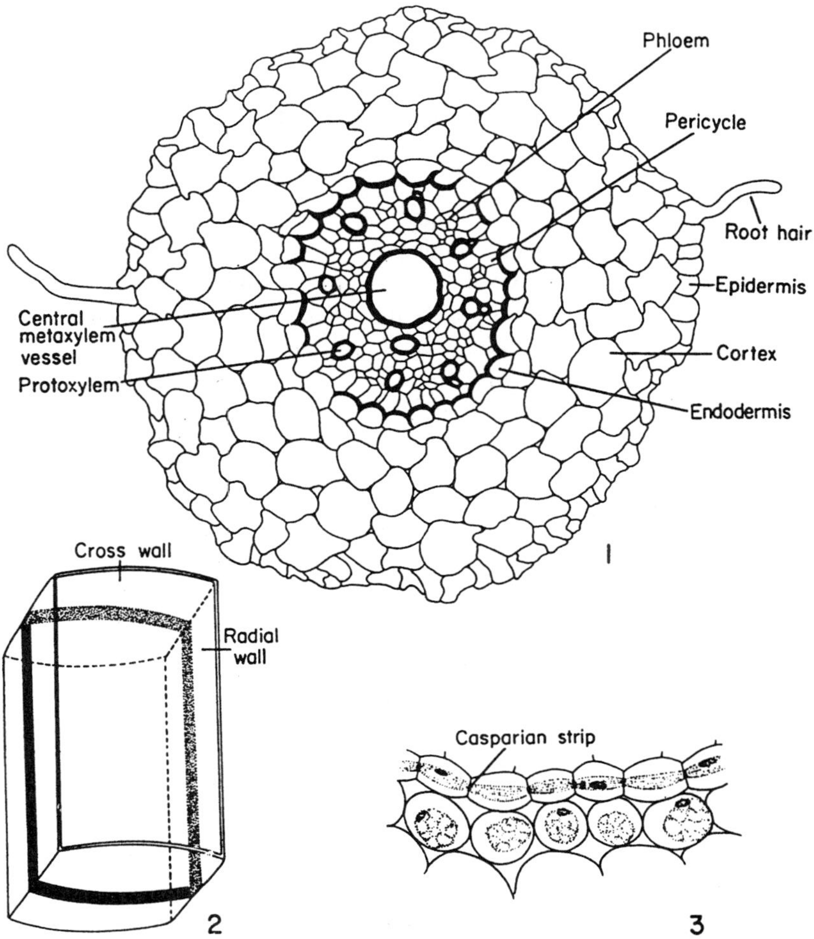

Figure 5.3 (1) Cross-section of a fully developed root of wheat. (2) Three-dimensional diagram of an endodermal cell showing the Casparian strip. (3) Enlarged representation of the endodermis, with cortical cells, all in a state of plasmolysis (shrinkage of the cell contents within the plasmalemma away from the cell wall). Note the continued attachment of the plasmalemma to the Casparian strip in endodermal cells (after Fahn 1982).

so it is present in nearly all parts of the root that are active in ion uptake (Esau 1965).

Within the endodermis lies the stele. This contains unspecialized cells, the pericycle, and two systems of connecting vessels. These gradually form as the root matures, and are probably effective from a short distance behind the root tip. The phloem system transports sugars, carboxylic acids, potassium, nitrogen and phosphate from the leaves to the rest of the plant (Peel 1974; Canny 1991). The xylem vessels consist of functionally dead cells, the ends of which have broken down to give a continuous tube of cell wall material, though it may still contain a plasmalemma for a considerable distance from the tip, and it is surrounded by living pericycle cells. This system transports water and mineral nutrients from the roots to the rest of the plant. There is appreciable transfer of water and solute between these vascular systems (Peel 1974).

5.1.5 Changes with Age

As the root ages, branching will usually occur (section 9.2.1). Lateral initials develop in the pericycle inside the stele, depending upon signals that are caused by environmental factors (section 9.3.3), and these meristems gradually expand. They then break out through the endodermis and cortex of the main root (Charlton 1996) to form a new root tip, which is very similar in organization to the main root tip (Malamy & Benfey 1997). The endodermis and the stele of the branch root gradually join up to that of the main root, and vascular connections are established in the xylem and phloem. Branching is promoted by excision or impedance of the root tip, by the presence of auxin from the upper parts of the plant, or by various environmental stimuli. The process of lateral root initiation is now understood much more clearly (Malamy & Benfey 1997), but the exact way in which it is integrated into the root system in its response to the environment is not yet understood.

Many changes occur in the structure of the main root as it ages, and these vary among species. Suberin continues to be deposited in the endodermis and sometimes in the epidermis, so that the thickness of the endodermis increases (Martin & Juniper 1970). In some grasses, the whole epidermis and cortex may slough off, though the stelar cylinder that remains may still be functional. Dicotyledons may also lose the cortex, but as part of a more complex process of 'secondary thickening' in which a secondary cambium (a cylinder of meristem cells around the root) forms inside the endodermis. This continues to grow, producing more xylem, which finally forms the woody part of the root. It is often assumed that such roots, which commonly have corky deposits on their surfaces, cannot absorb water and nutrients, but water — at least — seems able to enter (Kramer & Boyer 1995), probably via lenticels and other breaks.

5.1.6 Root Classification

Roots can be described in terms of their origin from the plant (nodal roots, which spring from the nodes of the stem, or seminal, which form directly from the embryo) (figure 5.1), their position with respect to the root system as a whole (e.g. main axis, tap root or lateral), or the type of plant to which they belong (monocotyledon, dicotyledon, or gymnosperm), but the general structure differs little from that described here, and there are no clear generalizations about differences in function (section 9.2.1). The most important difference is developmental, in that gymnosperm and dicotyledon roots show secondary thickening, which produces a root architecture that is characteristically different from that of the monocotyledons (section 9.2.2). The ways in which the vessels are arranged in the stele are used to classify roots anatomically, but these seem of no interest for our subject.

5.2 The Ion Uptake Process

5.2.1 The Physical System

The uptake into the mature root requires movement of ions through the cortex, through the endodermis, through the xylem parenchyma cells (the pericycle) and into the xylem vessels, in which they are transported to the upper root and the shoot (figures 5.2 and 5.3) (Clarkson 1993). There are two general routes across the cortex, with different implications for different types of solute. The apoplasmic route lies through the root 'free space': intercellular spaces, the pores of cell walls and other water-filled voids that are not within a cell plasmalemma. The symplasmic route lies through the interior of cells, within the cytoplasm and possibly the vacuole (figure 5.4). This implies that the solute has moved through at least one membrane to enter this route. This is important because there may be energy changes on such a transfer, and the membranes have transport properties which influence solutes that move through this route. The route can be extensive because of the transfer connections between cells called plasmodesmata (Robards & Lucas 1990) (figure 5.5), by which nutrients can be passed on from cell to cell without moving back to the apoplast.

The Casparian band of the endodermis forms a barrier to radial movement of ions in the apoplast (section 5.1.4) (figure 5.3), and water and solutes generally must pass through the symplasm at this stage. However, some transport occurs wholly through the apoplast, by exploiting breaks in the Casparian band, such as those where young lateral roots emerge (Petersen *et al.* 1981). Apart from this, a single ion may use both routes during its full journey to the xylem vessels (figure 5.4). As explained for the 'composite membrane' model of the root (section 2.3.3), the apoplasmic and symplasmic routes for water and ion movement form a series/parallel connection network. The 'free space' of the cortex contains the apoplasm outside the Casparian band. About 5% of the root volume is occupied by free space (Grignon & Sentenac 1991). Much of this consists of the fine pores, with a diameter of 3–8 nm, lying between the cellulose microfibrils, which are the main wall constituent. The wall contains 10–20% of polysaccharide to wet weight, which is in gel form, and uncharged low-molecular-weight compounds have access to all of the water-filled part of the wall (Michel 1971). Larger molecules, above about 3 nm diameter, and negatively charged ions may be excluded (Clarkson 1993; Marschner 1995, p. 8). The intercellular spaces are normally air filled, so they contribute little to the apoplasmic space.

The movements of a low-molecular-weight water-soluble dye in the cortical cells suggest that the real situation is more complicated (Canny & Huang 1994). In parts of the root, the dye could not enter through the exodermis, so there was no diffusion through the cortex there. In contrast, even in thickened endodermal walls, there were still very fine pathways along which the dye could slowly diffuse. Only in fine roots did solutes have general and rapid access to all the cortical plasmalemma surface.

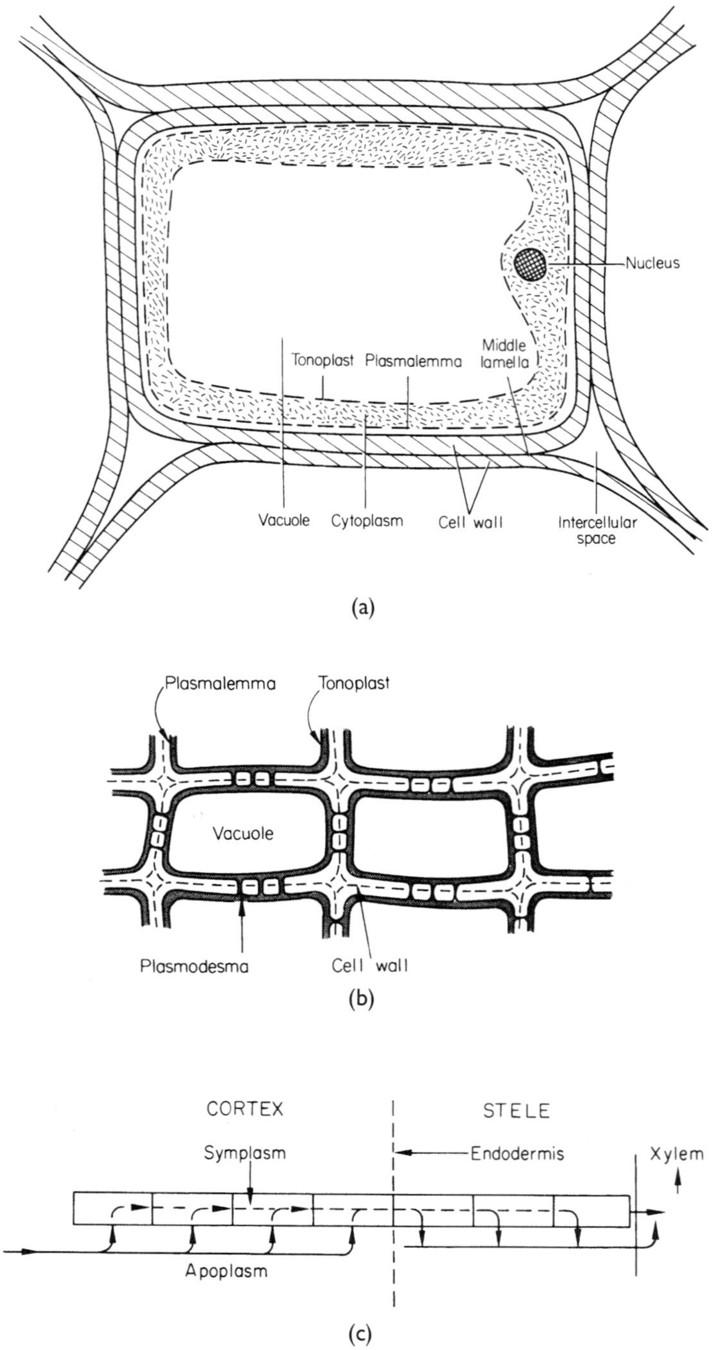

Figure 5.4 (a) Schematic diagram of a single vacuolated cortical cell (not to scale). (b) Outline diagram of cells in mature cortex tissue (not to scale). (c) Schematic diagram of the interaction of symplasm and apoplasm pathways of movement in cortex and stele (also see figure 2.8) (a–c after Nye & Tinker 1977).

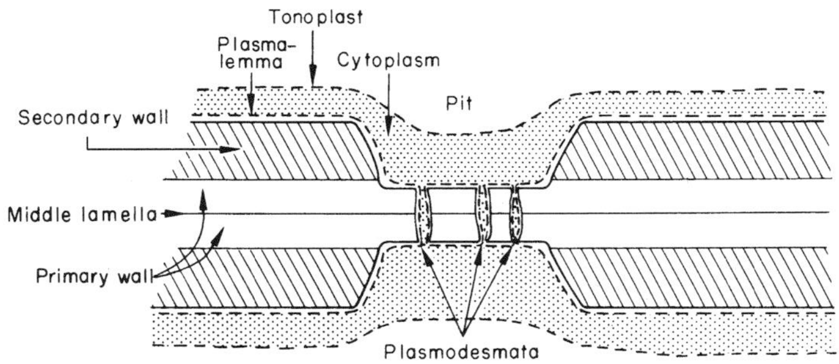

Figure 5.5 Detailed diagram of a pit field in the cell wall containing the plasmodesmata (after Nye & Tinker 1977). For a detailed representation of the internal plasmodesmata structure, see Clarkson (1996).

5.2.2 Uptake of Water and Uncharged Solutes

The properties of the membranes around cells, vacuoles and the cell organelles are fundamental to the uptake and control of ions in the plant. The structures of the typical lipid bilayer membrane have been well described by many authors (e.g. Nobel 1991) (figure 5.6). The membrane is hydrophilic on the outside because of the carboxyl or other charged groups on the ends of the alkyl chains, and hydrophobic inside where the alkyl chains interact. The thickness ranges from 6 to 10 nm, because the lipid composition in different membranes varies. Membranes are semipermeable to various solutes, and contain, packed into their structures, a range of protein molecules which give the membranes their varied properties in ion transfer.

The relative ease of movement of substances through a membrane is best expressed by the permeability coefficient, which is the ratio of the flux of a chemical species to the concentration difference (not gradient) across the membrane (Nobel 1991). Water moves freely through cell membranes, with a large permeability coefficient of around 10^{-2} cm s^{-1} (Stadelmann 1977). The coefficient for small uncharged and moderately lipophilic compounds is around 10^{-4} cm s^{-1} (Stein 1986), whereas that for a charged ion such as K^+ may be of the order of

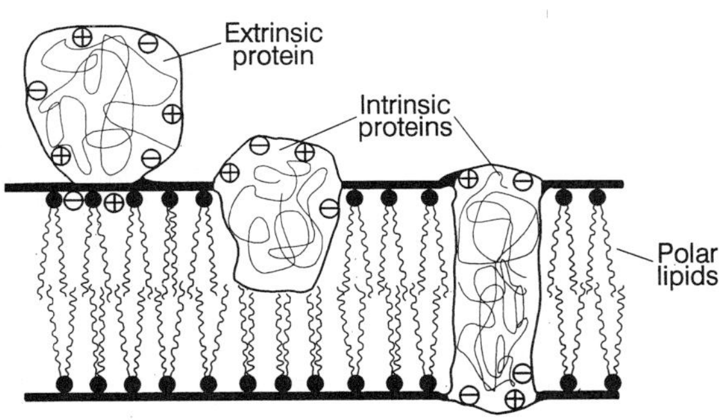

Figure 5.6 Schematic diagram of the structure of a bilayer lipid membrane, showing hydrophilic heads and lipophilic hydrocarbon tails and associated proteins within the membrane or attached to the outside (after Marschner 1995).

10^{-7} cm s^{-1}. Uncharged solutes of small molecular weight therefore appear to move readily through the root with the transpiration stream, normally without much dilution or concentration from its original state; the behaviour of silicic acid is a good example of this (Jones & Handreck 1967). This is an important property for uncharged and water-soluble biocides that are active within the plant (section 4.2.1).

The driving forces that move water through semipermeable membranes into cells are osmotic pressures that are caused by the solutes taken up by the cells of the cortex. These provide the turgor pressure within the cell membranes, which are prevented from expanding by the rigid cell wall. The thermodynamic treatment for osmotic pressures is given in section 2.1.2 and that in relation to water flow in section 2.2.3.

5.2.3 Uptake of Nutrient Ions across Membranes

In contrast to uncharged solutes, the most striking characteristic of the uptake of ions is that their concentrations, both in absolute value and relative to other solutes, are virtually always altered from those in the external solution as a consequence of uptake (table 5.1). The cation exchange properties of the cell walls of the cortex may affect the flow of charged species into the root (Robson & Pitman 1983), and we consider this question first. Afterwards, we consider two complementary ways of looking at the uptake process through the plasmalemma into the cell: the mechanistic way and the thermodynamic way (figure 5.5).

5.2.3.1 The Ion Exchange Properties of the Cell Wall

The cell wall contains pectic material that has free carboxyl groups. When these ionize, the cell walls become a cation exchange material (Haynes 1980). Earlier discussions of the free space used the term 'Donnan free space', but the concept of exchange capacity is much simpler (section 3.1.4). The cation exchange capacity usually lies between 10 and 70 meq per 100 g dry tissue, and dicotyledonous plants

Table 5.1 Differences in the initial and final (4 days) ion concentrations of the external solution supplied to maize and bean plants and that of the sap expressed from the roots.

	External concentration (mM)			Concentration in the root press sap (mM)	
		After 4 days			
Ion	Initial	Maize	Bean	Maize	Bean
Potassium	2.00	0.14	0.67	160	84
Calcium	1.00	0.94	0.59	3	10
Sodium	0.32	0.51	0.58	0.6	6
Phosphate	0.25	0.06	0.09	6	12
Nitrate	2.00	0.13	0.07	38	35
Sulphate	0.67	0.61	0.81	14	6

Source: after Marschner (1995).

usually have a value roughly twice as large as monocotyledons. There is occasionally a small anion exchange capacity, ascribed to amine groups. Polyvalent cations are bound much more strongly than monovalent cations, as is also found with soil organic matter (section 3.1.6), and plants with high root exchange capacity have relatively large calcium contents in their tissues.

There is no evidence that adsorption on exchange sites is necessary before cations are absorbed. The exchange capacity can directly affect the ion composition of the roots by holding exchangeable ions, but it is important to distinguish between ions that have been absorbed by the cells or adsorbed on the cell walls (i.e. that have or have not passed through a membrane).

However, there is the possibility that the exchange capacity of the cell walls, and the ions that may be attracted to them (section 3.1.4), can affect the uptake step, by changing the mobility of ions in the cortex apoplasm, or by perturbing the plasmalemma with electrical fields (Grignon & Sentenac 1991). The diffusion coefficient for cations could be enhanced by the mobility of exchangeably held cations, and that for anions could be reduced by their exclusion from a diffuse double layer (Shone 1966), which would reduce the already small volume of the free space. Pitman (1965) reported diffusion coefficients for sodium and potassium, with 0.01 M iodide solutions, of $3\text{–}4 \times 10^{-7}$ cm^2 s^{-1}, which implies a value for f_L of around 0.6 (section 4.2.1) if the water-filled free space is 5% of cortex volume. Several reported effects, such as that added calcium causes increased uptake of phosphate, have been explained by the collapse of the electrical double layer on root cell wall surfaces by calcium, so that the volume available for anion diffusion and the average phosphate concentration in the pore solution both increase (Haynes 1980).

The importance of the diffusion coefficients of ions in the root cortex depends upon how much of the total ion flux goes through the apoplast, and whether there is an important mass flow contribution (section 6.1.2). If the whole flux occurred by diffusion through the apoplasm, with a root of diameter of 0.05 cm and endodermal diameter of 0.03 cm, and with an inflow of 10^{-13} mol cm^{-1} s^{-1}, the decline of concentration from the external solution to the endodermis would be around 6×10^{-5} mol L^{-1} assuming Db to be 10^{-7} cm^2 s^{-1}. This is a significant concentration difference (Nye 1973), and with thick cortex layers this could well restrict uptake. In very dilute solutions, almost all the uptake may be in the epidermal layer of cells.

Grignon & Sentenac (1991) noted that even the unstirred layer in the cell wall pores and outside them would introduce an impedance to solute fluxes that could cause errors in measuring K_m of up to an order of magnitude. They also pointed out that the pH at the membrane surface is almost impossible to measure, yet it will affect and be affected by the proton extrusion rate of the cell, the membrane potential and hence its uptake properties, and it may be very different from the pH of the surrounding liquid. It is very difficult to make precise measurements on roots at low solution concentrations that are free of errors. Thus, it is easy to show that if the unstirred layer thickness around a root is d, then α (section 5.3.2) tends to D_L/d as the solution concentration declines, and cannot be any larger than this (Nye & Tinker 1977, p. 112).

5.2.3.2 Ion Uptake Thermodynamics

Thermodynamics is, in theory, applicable only to equilibrium states (Nobel 1991), but, despite this, it has always been applied to dynamic systems — such as growing plants, by assuming that the system departs from equilibrium by only small amounts (see chapters 2, 3 and 4). Non-equilibrium thermodynamics are used to address the problem of simultaneous fluxes of different components.

There was doubt whether thermodynamics added anything to the understanding of ion uptake into higher plants, when all ion uptake was believed to occur by independently energized 'uptake sites'. However, it is now clear that there is effectively one main energy supply for ion transfers. This is the electrical potential across the plasmalemma created by the expulsion of H^+ ions by an electrogenic H^+-ATPase. This electrical potential difference across the plasmalemma in part determines the electrochemical potential difference, and thus if entry of an ion to the cell is passive or active, and if an active pump is necessary for each ion. These factors make thermodynamics more valuable than before in understanding the total process.

The detailed selectivity of uptake arises from the structure of the entry sites (section 5.2.3.3), but the thermodynamics explains the equilibrium positions that determine how the sites operate (Clarkson & Luttge 1991; Nobel 1991; Marschner 1995). An expression for the work done in transferring ions across a membrane is given by the electrochemical potential difference (section 2.1.1):

$$\begin{aligned} \Delta\mu_e &= RT \ln a_i - RT \ln a_o + zF'E_i - zF'E_o \\ &= RT \ln a_i/a_o + zF'(E_i - E_o) \end{aligned} \tag{5.1}$$

where E refers to electrical potential, subscripts i and o refer to inside and outside the membrane and other terms are as in chapter 2, with F' equal to Faraday's constant.

In this situation, neither the gravitational (mgh) nor the macroscopic pressure (VP) terms in equation (2.1) are likely to be important (Nobel 1991, p. 111). The electrochemical potential then depends only upon the chemical activity of the ion ($RT \ln a$), and the electrical force acting upon it ($zF''E$). In real roots, the flow of solution will also affect the equilibrium position and the flux of solute (section 5.4.1).

If the electrochemical potential is the same on both sides of the membrane, then $\Delta\mu$ is zero, and the ions are in equilibrium, even if the concentrations are very different. The Nernst electrical potential difference E_N is that at which there is equilibrium across the membrane; if the concentrations (really activities (section 3.1.3)) differ by a factor of 10, for example, this potential difference will be 59.2 mV, with z equal to 1:

$$E_N = 59.2/z \log_{10}(a_o/a_i) \tag{5.2}$$

In general, the flux of a chemical species is proportional to the gradient of its electrochemical potential, that is, $F = u\,\mathrm{d}\mu_e/\mathrm{d}x$; hence, if we can replace a by c, assuming that the activity coefficient remains constant over the transfer path, then

$$F = RTu\,\mathrm{d}c/\mathrm{d}x + uczF'\,\mathrm{d}E/\mathrm{d}x \qquad (5.3)$$

where u is the mobility of the species, or the velocity of movement under unit potential gradient. The first term reduces to Fick's first law (section 1.4.1), and the second is the movement under an electrical gradient (section 4.1)

By using these equations and measurements of the electrical potential difference across membranes, it can be shown that within the cell all major anions are more concentrated than they would be at equilibrium, potassium is often near equilibrium, and sodium and calcium are almost always less concentrated. With the exception of potassium in some cases, the cell must continually expend energy to maintain the concentration of the major nutrient at the optimum level for the cell to function. The energy demand associated with ion uptake is considerable; Lambers *et al.* (1996) concluded that, averaged over 24 species, the fraction of root energy budgets used for uptake ranged from 50 to 70%, depending on the relative growth rate of the plant, which implies that around 30% of the total respiration of the plant is devoted to maintaining ion concentrations at the required levels (table 5.2) (section 9.1.2).

5.2.3.3 Membrane Structure and Ion Uptake Processes

The cytoplasm is a layer lying just inside the plasmalemma; it contains a variety of organelles, membranes and compounds, and is the metabolically active part of the cell. It surrounds the nucleus, which contains the genetic information for the cell, but is normally only a few percent of the total cell volume. The vacuole is an internal compartment bounded by the tonoplast membrane, surrounded by the cytoplasm and containing a dilute solution of low-molecular-weight solutes, including inorganic salts, carboxylic acids and their salts, and sugars (figure 5.4). Despite its small volume and large surface area, the cytoplasm has a rather rigidly maintained concentration of important ions and pH (Leigh & Wyn Jones 1986). The vacuole concentration, on the other hand, can change over a wide range without affecting the function of the cell (subject to osmotic pressure considerations), and therefore offers the possibility of storage of ions, with later release if the plant requires them. The maintenance of the constant composition of the cytoplasm of the cortex, through which all absorbed ions have to pass, therefore requires a remarkable balance between influx and efflux rate through the plasmalemma, loss or gain by the vacuole, and the rate of onward transport via

Table 5.2 Respiratory energy costs in roots of *Carex* for ion uptake.

Proportion of total ATP demand required for	Plant age (days)		
	40	60	80
Ion uptake	36	17	10
Growth	39	43	38
Maintenance of biomass	25	40	52

Source: after Marschner (1995).

the symplasm to the rest of the plant. The ion transport activities of the various membranes must be controlled with great accuracy.

A fundamental property of the cell is the H^+-extrusion system, powered by the hydrolysis of ATP and catalysed by a plasmalemma-located ATPase (Briskin & Hanson 1992); the yield is one proton transported per ATP molecule hydrolysed. There is consequently a fairly steady electrical potential difference across the plasmalemma of around 100 mV or more, with the inside negative with respect to the outside. The electrical potential difference decreases the energy expenditure on uptake of cations, and increases it for anions (section 5.2.3.2). The electrochemical difference for H^+ across the plasmalemma also acts as the immediate energy source for the transport of other ions across the membrane against an electrochemical gradient.

This energy source is provided by the simultaneous and coupled movement of hydrogen ions into the cell, which is down the proton electrochemical potential gradient, and another ion into or out of the cell. The uptake site where a proton and another ion can move into the cell together is a 'symport'; where the ion moves out of the cell as the proton moves in, it is an 'antiport'. The original idea of ATPases as 'uptake sites' (Epstein 1972) has therefore been greatly elaborated over the last few years, with a variety of structures with different properties (section 5.2.3.4).

Whereas 'transporters' move ions against the electrochemical gradient, there is also passive leakage by many constituents of the cell through the membrane by diffusion down an electrochemical gradient (Luttge 1983; Elliott *et al.* 1984). Some of the passive leakage is general across the plasmalemma, but some can occur in a specific and controlled way through 'ion channels' (Sanders 1990; Tester 1990). These allow rapid diffusion, and they are strongly ion-specific, so that they provide a selective leak. They also have the important property of 'rectification', in that the change of potassium ion flow with voltage across the membrane has a different value above and below the potassium equilibrium point, that is, for influx and efflux (figure 5.7). The full explanation of the functions of such channels is extremely complicated (Sanders 1990). Because the ion flux through a single channel (ion turnover per second of *c.* 10^7–10^8) is very large by comparison with the flux through a symport or antiport (turnover 10^3–10^4), it is possible to measure the flux through a single channel by the technique of 'patch clamping'. At millimolar concentrations of the major cation nutrients, much of the influx is through ion channels, though the numbers per cell are quite small.

Nutrient ions, in general, leak out at a rate which is positively related to the internal concentration of the ion, but uptake increases proportionately much more rapidly than efflux with external concentration. There is consequently an equilibrium concentration of the external solution (C_{min}) at which the efflux and influx are equal and opposite (Bieleski & Ferguson 1983; Devienne *et al.* 1994). Efflux is an important control mechanism for plant ion concentrations at low and moderate levels, but at high concentrations efflux can probably be ignored.

Calcium occupies a rather unusual position (Evans *et al.* 1991). It has a very low concentration in the cytoplasm, *c.* 0.1 μM, but this has to be closely controlled because of the signalling functions of this ion. With the need for extreme constancy in the cytoplasm, the absence of symplasmic transport, and the need to

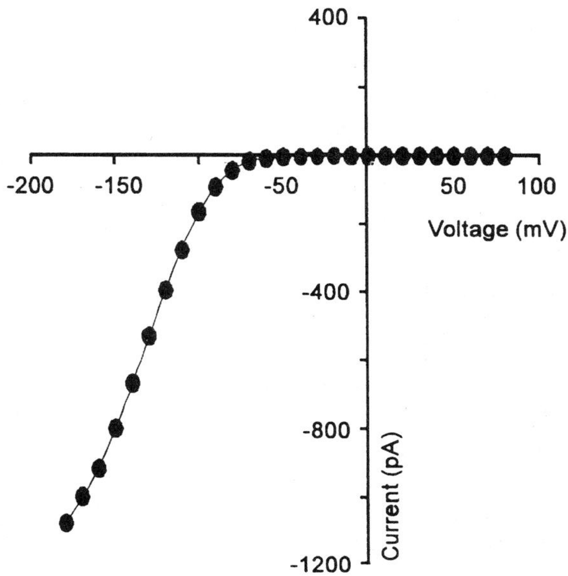

Figure 5.7 Non-linear relation of the current and voltage across a plasmalemma membrane of a maize cell caused by inwardly rectifying K^+-selective channels (after Tester 1997).

constantly expel calcium ions, the mechanisms needed for this ion must be highly sensitive, with an outward Ca transporter in both the plasmalemma and the tonoplast.

Sodium is also unusual (Bradley & Morris 1991), as it has to be maintained at low concentrations in the cytoplasm, though it can in some species be accumulated in the vacuole. Apart from the general need to expel sodium from the cell, so that it is well below the equilibrium concentration, a number of whole-plant adaptations exist in different plants to exclude sodium. For example, it may be removed from the xylem stream before reaching the shoot (Wolf *et al.* 1991), it may be accumulated in non-damaging ways, or it may be secreted into non-functional parts of the plant (Serrano & Gaxiola 1994) (section 9.3.3).

5.2.3.4 The Molecular Mechanisms of Ion Uptake

For some years, work has been proceeding on the molecular mechanisms that must underlie the ion uptake processes described above, but it is only recently that substantial progress has become apparent (Clarkson & Hawkesford 1993; Smart *et al.* 1996). The integration of this membrane-level information into our understanding of the behaviour of whole plants is still only beginning. All these transport sites are proteins embedded in the membrane structure (figure 5.6) and providing a 'pore' through it, but they are different to the ATPase enzyme that transports H^+ and thereby establishes the general electrical and proton gradients. Each transporter is specific to a single ion, with very few exceptions, and contains a binding site for that ion. A rearrangement of peptide chains then exposes the ion to the opposite side of the membrane, so that it can be released. The polypeptide then rearranges back to its initial conformation, ready for the next ion. A further insight into the mechanisms of the uptake of ions is coming from the use of microelectrodes, which allow ion activities in small intracellular compartments — such as the cytoplasm — to be measured (Miller & Smith 1996).

The most basic process in ion uptake is the 'high-affinity mechanism', originally identified by Epstein (1972) with 'uptake sites' that allow plant roots to absorb from very dilute solution in the micromolar range. Uptake proteins have now been identified that do precisely this for potassium, and their corresponding gene *KAT1* has now been cloned (Schachtmann & Schroeder 1994), after extraction from K-starved wheat roots. This 'transporter' has a K_m value of 29 μM K, which is well within the expected micromolar region, and it is well distributed throughout the cortex. Smart *et al.* (1996) list six different transporter proteins whose genes have been cloned. Meharg & Blatt (1995) showed that *Arabidopsis* roots contained a high-affinity transporter for nitrate that required two H^+ ions to move into the cell for each nitrate ion taken up. The high-affinity transporters for different ions are generally quite distinct and with different homology from each other. Thus, the sulphate transporter in onions is quite similar to those found in other eukaryotes, but it is different from all phosphate transporters. This specificity indicates that very few ions can be carried by the same transporter, selenate and sulphate being exceptions to this rule. Uptake competition between ions cannot therefore, in general, be explained by this mechanism.

Sometimes, transcription of the genes for these transporters, and hence their synthesis, may be induced or repressed (Clarkson & Luttge 1991). The nitrate high-affinity uptake system is induced by contact with external nitrate that starts the synthesis of the transporter proteins after a lag, which in the case of an algae studied by Watt *et al.* (1992) was of the order of 100 minutes. As an example of the very low concentrations at which this high-affinity system is induced and operates, Watt *et al.* (1992) could detect a high-affinity uptake system for N in the algae *Chlamydomonas reinhardtii* only when it was in an N-deficient condition.

The outward-directed transporters for calcium and for sodium are analogous to the inward-directed 'high-affinity' uptake transporters. Calcium-pumping ATPases have been identified in the plasmalemma and the endoplasmic reticulum (the internal membranes within the cytoplasm), and a Ca/H antiporter has been found in the tonoplast.

The 'low-affinity mechanism' for potassium or ammonium, which is important in the millimolar range, is now firmly associated with K^+ or NH_4^+ ion channels (Smart *et al.* 1996; Tester 1997) that allow passive movement under an electrochemical gradient. These also are complex proteins that span across a membrane, but must be distinguished from transporters that cause active transport. They can be open or closed ('gated'), depending upon the transmembrane potential and other factors; and ion flow rate due to an electrical field differs in the two directions (inward and outward rectifying channels) (figure 5.7). Cation transport through these channels does not require energy directly, because it is driven by a favourable electropotential gradient. Divalent cations will have an even more favourable electrochemical potential gradient (equation (5.2)), and probably enter by channels even at low external concentrations. Indirectly, transport through channels may require energy, because a cation that moves through the channel into the cell tends to discharge the normal membrane potential, and requires that a similar number of positive charges on protons shall be transported outwards to maintain it. The net result is that the low-affinity process usually requires one ATP molecule per K ion transported inwards, whereas the high-affinity process requires two molecules (Maathuis & Sanders 1997).

The low-affinity pathway observed with anions cannot, however, be explained by ion channels. Because of the normal membrane potential of 100 mV or more, anions such as nitrate and single-charged phosphate ions can very rarely move down an electrochemical gradient into the cell, and divalent ions such as sulphate are even less able to do so. An open ion-selective channel for nitrate into the cytoplasm would therefore be of no use. For anions, there appears to be a set of different transporters, operating at a higher concentration (i.e. with a higher K_m) than those for the high-affinity pathway (Miller & Smith 1996), and these constitute the low-affinity pathway. There are consequently two quite distinct sets of transporters for anions, and these high- and low-affinity transporters are very different in their structures (Smith *et al.* 1995; Trueman *et al.* 1996; Leggewie *et al.* 1997).

It is still not possible to explain fully all root uptake properties, particularly that of regulation (Smart *et al.* 1996; Maathuis & Sanders 1997) (section 5.3.3). However, the interaction of the two main mechanisms described above can parallel the classical type of uptake isotherm (section 5.3.2) with mechanisms I and II. Maathuis & Sanders (1997) estimated that the potassium symport high-affinity system transports almost all potassium inwards up to 0.1 mM, but the contribution declines progressively to zero at 10 mM. The amount moving through inward rectifying channels increases until it is dominant at 10 mM, while outward rectifying channels make a small contribution at millimolar levels.

5.2.4 Transfer within the Symplasm and into the Xylem

We have dealt with the processes that load the ions into the symplasm of the cells of the cortex. The next step, their passage onwards within the symplasm, occurs through the plasmodesmata (Robards & Lucas 1990). These are fine pores full of cytoplasm and around 5 nm in diameter that penetrate the cell walls and join the cytoplasm of one cell to that of the next. They are grouped in 'pit fields' where the cell walls are very thin (figure 5.5). These are extremely important components of the plant, in that they bind the whole plant together into a single organism; the plasmalemma lines each plasmodesma, so that the cytoplasm is effectively continuous from cell to cell. There is considerable internal structure in the plasmodesmata (Robards & Lucas 1990), so that the cross-section for transport is very small, the individual spaces being around 3 nm across. If much of the flux of water and solutes into the plant has to pass through the plasmodesmata in the cells of the Casparian band, it raises questions of how this flux is driven, but it seems unlikely to be by simple diffusion. Transport through the symplasm occurs to different extents for different ions, and calcium and magnesium are not transported to a significant extent in the symplasm (Luttge 1983).

There is surprisingly little certainty about the processes that transfer these ions into the xylem, which is part of the apoplasm, and what controls the rate of this transfer (Luttge 1983; Clarkson 1993). Water must flow into the xylem at the same time, if the plant is transpiring, but mass flow is not adequate as an explanation of the transfer. The concentration of ions in the xylem stream is inversely related to the flow rate of water, suggesting that the ion-loading process is not influenced by transpiration rate in the short term (Marschner 1995).

This transfer step into the xylem may turn out to be as complex as the uptake into the symplasm. It is clear that the xylem loading process is separate from the cortex uptake step (Marschner 1995). The usual view has been that active transport of ions is necessary to explain movement into the xylem, and there is evidence for an active transporter which transfers protons into the xylem. However, Wegner & Raschke (1994) used patch-clamping techniques to show that there were three different cation-specific 'rectifiers' (ion channels) in the xylem parenchyma of barley roots, through which ions could move to the xylem vessels. They therefore visualized release into the xylem of cations and anions as being essentially passive flow through ion channels.

There is a recirculatory process in plants in which nitrogen, potassium, phosphate and sulphate are brought down to the root in the phloem, transferred into the xylem and transported back up to the shoot (section 5.3.3). This ensures that nutrients can be redistributed in the plant (Cooper & Clarkson 1989). Simpson *et al.* (1982) found that in wheat 56% of the nitrogen going into the shoot was translocated down again, but the mechanism whereby it is transferred from phloem to xylem is not known. However, the need to load most of this circulating flow of ions into the xylem means that the total loading process there is considerably greater than the flow into the cortex from the external solution.

5.3 Ion Uptake Kinetics and Plant Demand

5.3.1 Experimental Systems for Plant Uptake Rate Measurement

The main question in kinetics is the rate at which roots absorb ions, and how this is affected by the external and internal concentrations of the ions. However, by using different systems for investigating problems of ion uptake, plant physiologists have, in effect, asked different questions, sometimes without defining them exactly. Most of the early work on the free space and ion uptake was done with tissue slices, giant algal cells or excised low-salt barley roots (Epstein 1972). All these had a low capacity for ions and were disconnected from a growing plant shoot, so they were useful only for equilibrium or short-term transient state experiments. An attached shoot is necessary to allow nutrient recirculation, to give a larger sink for ions and to supply photosynthate to the root. Thus, White *et al.* (1992) found fluxes of Ca into intact plants to be 4–40 times larger than in excised roots.

A normal plant can be maintained at close to a steady state, at least during growth in daylight hours. An agricultural plant such as a cereal may grow at 10% per day, and may have a root/shoot ratio as low as 10% (Welbank *et al.* 1974) (see table 9.8). If 30% of the root wet weight consists of cortical cells, these will have to absorb and pass on to the xylem three times the amount of nutrients they contain every day. Given variable uptake in different parts of the cortex and at different times, some cortical cells must have a throughflow rate considerably greater than this. This steady-state high-throughflow type of root is very different from an excised root.

The ideal experiment would measure uptake rates on a short defined piece of root still attached to an intact growing plant, but rather few experiments of this type have been performed (Brouwer 1965; Clarkson *et al.* 1968; Clarkson 1996), and, to our knowledge, in no case has a complete isotherm for a defined length of root been measured. Such an experiment would need the nutrient status of the plants to be precisely defined, because uptake rates at given external concentrations are not constant, but vary with the nutrient status of the plant (regulation) (section 5.3.3).

There are two main types of uptake experiments. In the first type, plants are grown at a single concentration, so that they are all uniformly pre-treated and have the same nutrient status. They are transferred into a range of different concentrations, uptake rates are measured, and the ion uptake isotherm is constructed (figure 5.8). The plants are not, however, in equilibrium with the new concentration, and the results are essentially records of a transient state, as plants placed in high concentrations will accumulate ions, and plants placed in low concentrations will dilute their ions by growth that is more rapid than uptake.

The second main type is flowing solution culture, as described by Asher *et al.* (1965) and Wild *et al.* (1987). In this system, high rates of flow allow low concentrations to be used, with some confidence that these values are maintained close to the root. The plants are grown under these conditions for a considerable period before final measurements are made, so that the plants are in balance with those solution concentrations and other environmental factors. From such work, it is clear that plants can meet their full nutrient needs by uptake from a wide

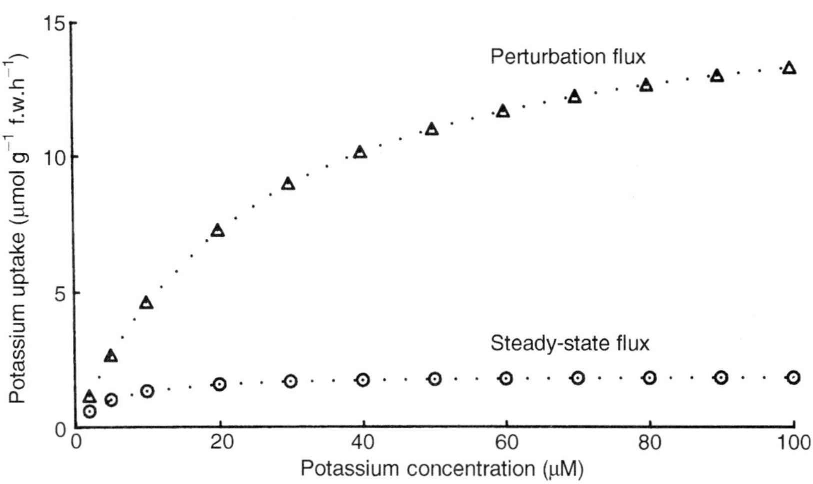

Figure 5.8 Relationship between the uptake rate and the external concentration for short-term uptake for plants all pre-treated in the same way (perturbation flux) and for uptake by plants in flowing solution culture pre-treated at the same concentrations as in the experiment (equilibrium flux) (after Glass & Siddiqui 1984).

range of solution concentrations, including low concentrations in the micromolar range, so long as these are maintained (figure 5.8).

It is therefore essential to define the preceding conditions, and the consequent state of the plants in ion uptake experiments (Devienne *et al.* 1994). The results with these two systems can differ greatly (figure 5.8). The isotherms measured in flowing solution culture show a more rapid initial rise in rate with concentration, and a maximum and constant value at a much lower concentration than with the constant concentration pre-treatment method, because the plants' regulatory systems have been able to adjust to the extreme concentrations.

A third general method is the 'depletion' method (Barber 1995, p. 64), in which a plant absorbs from a nutrient solution, which is analysed at intervals, and from which the uptake is calculated. In this case, the absorbing plant may not be in a state of balance with the solution, because the latter is constantly changing. Mullins & Edwards (1989) compared influx measurements with maize in a depletion system and a 'replenishment' (roughly equivalent to flowing solution culture) system, and found that the K_m values differed by a factor of 2–3.

A fourth type is discussed in section 2.5.2, in which plants are grown with exponentially increasing quantities of added nutrients, without reference to the concentration in the solution.

Many of the differences in results are due to the initial states of the plants in the different systems, but some may arise from different errors in defining the actual concentration at the root surfaces in stirred systems (section 5.2.3.1). An unstirred layer of liquid of around 10–100 μm thick (d) surrounds individual roots even in stirred cultures (Helfferich 1962, p. 253; Grignon & Sentenac 1991). Where dense masses of root are present, solution concentrations may be very different from those intended. If the root surface flux of an ion such as NO_3^- is 5×10^{-12} mol cm^{-2} s^{-1}, D_L is 10^{-5} cm^2 s^{-1} and d is 50 μm, then the concentration difference across the unstirred layer needed to drive the flux solely by diffusion is 2.5×10^{-6} M. Uptake measurements at these levels must therefore be made with great care. Robinson (1986) has given a set of curves from which the error can be estimated.

Several methods can generate curvilinear relationships of solution concentration and uptake rate. These are usually interpreted by equation (5.4) or (5.5) and give results as shown in tables 5.3 and 5.4. However, the physiological implications may differ greatly:

(1) Long-term flowing solution culture systems ensure that all plants are in equilibrium, but the isotherm (figure 5.8, equilibrium or steady-state flux) is not equivalent to a 'Michaelis–Menten' curve, because plants at each concentration are physiologically different. The parameters simply describe the curve, and the two-straight-line model (equation (5.9)) is probably most appropriate. Such results show how plants react to long-term changes in external concentration.
(2) Very short-term uptake experiments of a few hours on plants all given the same pre-treatment (figure 5.8, perturbation flux; see also figure 5.9) are most appropriate for deriving Michaelis–Menten parameters. These results represent the way in which plants react to very short-term fluctuations in the external concentrations.
(3) All systems in which the isotherm is measured over longer periods using plants that are not in equilibrium with the solutions have no precise interpretation,

Table 5.3 Examples of parameters from uptake isotherms obtained with equilibrium systems in flowing solution culture. This isotherm is fitted to the straight-lines model (equation 5.5; figure 5.9b).

Ion	Root radius a (cm)	I_{crit} (mol/cm s $\times 10^{-14}$)	C_{Lcrit} (μM)	αa (cm/s $\times$ 10^{-5})	Species	Reference
K	0.008	15	10	23	*Dactylis glomerata*	Wild *et al.* (1974)
K	0.015 (est.)	37	2	3	Fodder radish (32 days)	Wild *et al.* (1987)
P	—	7	2	0.6	Maize	Jungk *et al.* (1990)

I_{crit} and C_{Lcrit} define the critical point on the curve. αa is the average over the rising line interval, during which it is relatively constant. These parameters define the uptake properties of plant roots when in stable equilibria with the external solution.

because the plant nutrient status and its uptake system must vary. Each plant is exposed to solutions of varying concentration, and as in nature the external concentration and the plant nutrient status will be continuously varying.

(4) The best experiment is one in which sets of plants are brought to equilibrium at different external concentrations, and then each set is used for a short-term isotherm determination (Jungk *et al.* 1990).

5.3.2 Nutrient Uptake Rate in Relation to External Concentration

At the root surface there is a relationship between the concentration of a nutrient in the external solution and the uptake rate, or inflow, into the root. This is usually called an isothern (figure 5.9a,b). The shape of the single concentration

Table 5.4 Examples of parameters from perturbation isotherms (equation 5.4; figure 5.9a).

Ion	I_{max} (mol/cm s $\times 10^{-14}$)	K_m (μM)	C_{Lmin} (μM)	I_{equil} (mol/cm s $\times 10^{-14}$)	C_{Lequil} (μM)	Species	Reference
P	50	10				Onion	Brewster *et al.* (1976)
P	17	5				Rape	Brewster *et al.* (1976)
P	36	5	0.4			Barley	Drew *et al.* (1984)
K	600	11	1			Barley	Drew *et al.* (1984)
P	17	1.7	0.02	6	0.3	Soybean	Jungk *et al.* (1990)
P	37	6	0.01	3	0.1	Maize	Jungk *et al.* (1990)

I_{max}, K_m and C_{Lmin} have their usual meaning, resulting from fitting the data to equation (5.8). I_{equil} is the I at the pretreatment concentration C_{equil}, where these are known. An approximate value for αa at low C_L can be found from the equation

$$\alpha a = I_{max}/(2\pi\{K_m + C_L - C_{Lmin}\})$$

pre-treatment isotherm (figure 5.9a) often approximates to a rectangular hyperbola at low and moderate concentrations, when the high-affinity uptake mechanism is dominant. This is expressed by equation (5.4), where I is the measured inflow, C_{La} is the solution concentration at the root surface, I_{max} is the highest value of I, and K_m is a constant. I_{max} and K_m are calculated by applying equation (5.4), often by graphical methods such as the Lineweaver or Hofstee plots (Marschner 1995). At higher concentrations (figure 5.4a) the inflow rises slowly

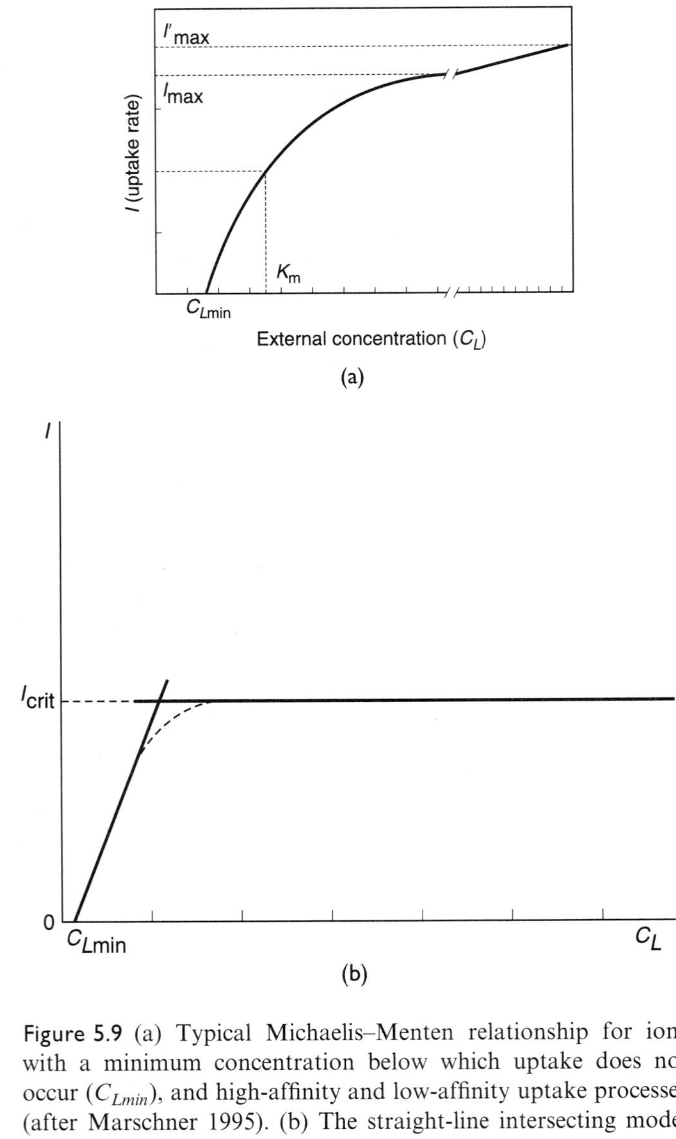

Figure 5.9 (a) Typical Michaelis–Menten relationship for ions with a minimum concentration below which uptake does not occur (C_{Lmin}), and high-affinity and low-affinity uptake processes (after Marschner 1995). (b) The straight-line intersecting model especially for nutrient uptake by plants in flowing solution culture systems with constant concentrations with which plants are in equilibrium. The slope is $2\pi\alpha a$. The final equilibrium inflow in such cases is called I_{equil}.

to an ill-defined maximum (I'_{max}) due to the low-affinity mechanism, before declining again:

$$\bar{I} = I_{max} C_{La}/(K_m + C_{La}) \tag{5.4}$$

The initial justification for using this Michaelis–Menten type of equation to express the isotherms was that the uptake process was enzyme-catalysed, but it is clear that the whole uptake process for an entire plant is too complex for such a simple view. There are many different types of transporters, each of which could be regarded as a different enzyme; there are ion channels that operate in a quite different way; the impedances to flow within the root system will modify the uptake rate; and different parts of the root system will differ in their properties. The justification is therefore empirical. The relationship can be stated in terms of the uptake rate of the plant, U, or of the mean inflow, $\bar{I}$, using the relationship $U = L\bar{I}$. Uptake is still sometimes expressed in terms of root weight, but the reason for preferring root length is that the calculation of inflow is central in theoretical work (equation (5.4)) (chapter 6).

An alternative way of expressing the relationship is as two intersecting lines (figure 5.9b), which seems suitable for the isotherm obtained from flowing solution culture experiments. The important variable is the position of the crossover point, which defines the horizontal part of the isotherm (I_{crit}) and the slope of the rising line. Obviously, this point will be subject to regulation as are I_{max} and K_m. The slope of the rising line may be derived from equation (5.5):

$$\begin{aligned} I &= 2\pi\alpha a C_{La} \qquad & C_{La} &< C_{Lcrit} \\ I &= I_{crit} \qquad & C_{La} &> C_{Lcrit} \end{aligned} \tag{5.5}$$

where C_{Lcrit} varies with internal concentration.

The term αa or the 'root-demand coefficient' was defined by Nye & Tinker (1969), where a is the root radius and α is given by the following relation: flux into root $F = \alpha C_{La}$. At low values of C_{La}, equation (5.4) can reduce to

$$I = I_{max}\, C_{La}/K_m \qquad C_{La} \ll K_m \tag{5.6}$$

hence, from equation (5.5),

$$\alpha a = I_{max}/2\pi K_m \tag{5.7}$$

The root-demand coefficient is empirically based, and is used in this book in several ways. First, it is a coefficient that expresses in a general way the strength of root uptake for a particular ion in a particular situation relative to the external concentration C_{La}, and which can be regarded as an expression of plant demand at the root surface. Second, it allows approximate solutions to be obtained to the equations that define the concentrations around absorbing roots in soil by assuming αa to be constant (chapter 6). In the Michaelis–Menten model, αa decreases across the whole curve, but it is often reasonable to regard it as constant at the lower values of C_{La} (figure 5.8) (equation (5.7)). Third, in many cases αa is, in fact, constant over particular ranges, and, in particular, it expresses well the rising limb of the intersecting line model (figure 5.8). Beyond the intersection point at C_{Lcrit}, αa varies inversely with C_{La}.

These two basic relationships are taken to express the 'high-affinity' uptake system (Epstein 1972). However, they need two modifications to be closer to the behaviour of real plants. First, there is normally a gradual but continuing rise in the uptake rate as the external concentration increases into the millimolar range (figure 5.9a), which is usually accompanied by a steady increase in internal ion concentration of the plant (luxury consumption). This indicates the low-affinity uptake system, in which ion selectivity is less, and ion competition is therefore greater. This may be due to the operation of ion channels for some ions. It has been suggested that there are a number of uptake systems that operate at increasing concentrations (multiphasic uptake) (Nissen 1974, 1996). However, the high- and low-affinity pathways can be related to separate physical uptake mechanisms, so the model described here appears the best in practice. Another linear term can be added to equations (5.8) and (5.9) to take account of this.

Second, we have assumed that the origin for the isotherms is a C_{La} of zero, but, because of efflux, there is a finite concentration C_{Lmin} at which net I is zero (figure 5.9a). This will vary (section 5.3.3), because it depends upon both the rate of efflux and the regulation of the high-affinity uptake system. For phosphate, C_{Lmin} has been given by various authors as 0.12 μM for tomato, 0.04 μM for soybean and 0.01 μM for ryegrass, and for potassium 2.0 μM for maize and 1.0 μM for barley (Marschner 1995). For nitrate, very variable values have been given. The high-affinity uptake system for nitrate has to be 'induced' by the presence of nitrate, and without this, C_{Lmin} would clearly be large. Therefore, C_{Lmin} will vary with conditions both inside and outside the plant. The resulting equations from adding in these terms are then

$$I = I_{max}(C_{La} - C_{Lmin})/(K_m + C_{La} - C_{Lmin}) \tag{5.8}$$

$$\begin{aligned} I &= 2\pi\alpha a(C_{La} - C_{Lmin}) \qquad && C_{La} < C_{crit} \\ I &= I_{crit} && C_{La} > C_{crit} \end{aligned} \tag{5.9}$$

The discussion of uptake kinetics should, in principle, relate to a defined short piece of uniform root, along which the various parameters can be taken as constant. These parameters are certainly not constant over longer distances, and will vary between different types of root (section 5.3.3). Uptake by a whole plant is the sum of the uptakes through all the root segments. If, for example, the Michaelis–Menten equation given here (equation (5.4)) applied precisely to individual root segments, but K_m varied widely, then the equation would not be obeyed by the whole plant, and it is perhaps surprising that the equation is approximated to at all closely. This may be because most test plants will have been in rapid growth before the experiment, so that most roots will be young, and the root system may therefore be reasonably uniform.

Of the systems discussed, the long-term flowing culture type seems most important, as the conditions are well defined and constant. With the single concentration pre-treatment system, the value of this concentration is arbitrary, and the plants in test solutions with concentrations different from the pre-treatment concentration will usually undergo an unknown degree of regulation during the experiment.

The main essential elements for plant nutrition can be divided into those that depend upon active uptake (K, P, NO_3 , S), those which enter passively, but are controlled by excretion (Na, Ca, Mg), and those which enter as uncharged molecules (B). Marschner (1995) has grouped these in terms of their typical uptake isotherms (figure 5.10).

Heavy metals also seem to follow this general pattern, though the chemistry of the heavy metals is complicated by complex formation with ligands in the soil solution and in the plant, and different valency and hydrolysis states. Rengel & Wheal (1997) measured the uptake parameters for Zn, in a factorial arrangement with three genotypes of wheat, with and without Zn and chlorsulfuron pre-treatments. Chlorsulfuron is a herbicide known to decrease Zn uptake. They used the solution-depletion experimental system, which may give unreliable values (section 5.3.1), but the comparisons between measurements should be dependable. Table 5.5 shows the consistent increase in C_{min} and decrease of I_{max} with chlorsulfuron, and the very marked difference between two genotypes. Regulation was also evident with decrease in I_{max}, and a small increase in C_{min} after Zn pre-treatment. This regulation was strongest in the Zn-efficient genotype.

5.3.3 Plant Nutrient Status and the Regulation of Uptake Rate

When the internal nutrient concentration declines, plants develop a stronger ability to absorb the deficient nutrient (but normally not others) at high rates from dilute solutions (Clarkson *et al.* 1978; Cogliatti & Clarkson 1983; Glass & Siddiqui 1984), and vice versa. Thus, the classical 'low-salt' roots or plants will absorb by the high-affinity mechanism at high rates that are sensitive to the external solution concentration. Gradually, such a plant will tend to increase its internal concentration, and the uptake rate will lessen. Plants thus have a strong tendency to homeostasis for their internal nutrient concentration, and the process in the roots which produces this is called 'regulation' (figure 5.8) (Glass 1983; Clarkson & Luttge 1991). This effect is of course the scientific basis for bioassays in which the nutrient intake rate of roots is used as an indication of nutrient status (Harrison & Helliwell 1979; Keith 1998).

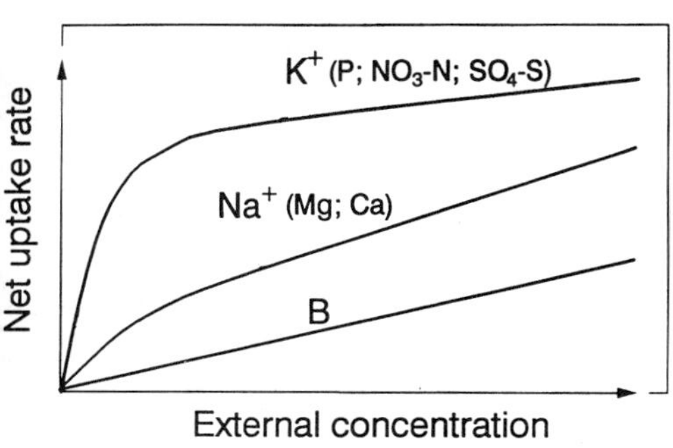

Figure 5.10 Schematic diagram of the relationship of net uptake rate of various ions into roots and their concentration in the external solution. Top line, ions absorbed against an electrochemical concentration gradient; middle line, ions excreted against a gradient; lower line, passive absorption (after Marschner 1995).

Table 5.5 Effects of Zn nutrient and chlorsulfuron (CS) pre-treatments and plant genotype on Zn uptake characteristics of wheat.

Genotype	Zn pre-treatment	CS pre-treatment	I_{max} (μg Zn g^{-1} root DW h^{-1})	K_m (μM Zn)	C_{Lmin} (μM Zn)
Excalibur	0 Zn	0 CS	0.36	0.86	0.11
		+ CS	0.19	0.83	0.21
	+ Zn	0 CS	0.23	0.76	0.13
		+ CS	0.16	0.71	0.45
Durati	0 Zn	0 CS	0.18	0.95	0.29
		+ CS	0.10	0.85	0.51
	+ Zn	0 CS	0.16	0.93	0.39
		+ CS	0.10	1.01	1.05

Source: after Rengel & Wheal (1997).

The concentrations of K and P in the plant differ in their effects on the respective uptake parameters (Drew *et al.* 1984) (figure 5.11). For potassium, K_m declined by 75% in a single day, when potassium supplies to plants were removed, but there was no change in I_{max} during that time. Over the subsequent 3 days, I_{max} increased sharply. The same effect on I_{max} was found for phosphate, but there was no change in K_m, and, indeed, the results of Jungk *et al.* (1990) showed that a 1000-fold decrease of external P concentration increased K_m from 1.0 to 1.6 μM (figure 5.12).

Both efflux and influx of phosphate and nitrate into cortical cells are changed during regulation, as measured by dual labelling. There have been many contradictory results in this subject, and it has proved difficult to fit them into a single working hypothesis. Siddiqui & Glass (1982) have suggested that for potassium, the parameters K_m and I_{max} could be expressed empirically in terms of the ionic concentration c in the tissue, as

$$I_{max(s)} = (\text{Max}I_{max})\,e^{(-a_1 \times c)} \tag{5.10}$$

$$K_m(s) = (\text{Min}K_m)\,e^{(a_2 \times c)} \tag{5.11}$$

where $(\text{Max}I_{max})$ and $(\text{Min}K_m)$ are, respectively, I_{max} and K_m for an extremely deficient plant, and a_1 and a_2 are constants.

The mechanisms whereby the internal condition of the plant regulates uptake rate are not understood. It is certainly specific to each ion, because the mechanism works separately for each ion; a single ion deficiency does not produce a general increase in nutrient uptake rates, such as would be produced by an increase in the plant's relative growth rate (Glass & Siddiqui 1984). Bearing in mind that different ions may be either absorbed or ejected against a potential gradient, or be in approximate equilibrium across the plasmalemma, the mechanism of regulation must be elaborate, and probably unique for each ion. However, there are some interactions; for example, nitrate and chloride uptake rates both seem to react to the combined internal concentration of the two ions, so regulation for one regulates the other (Cram 1973).

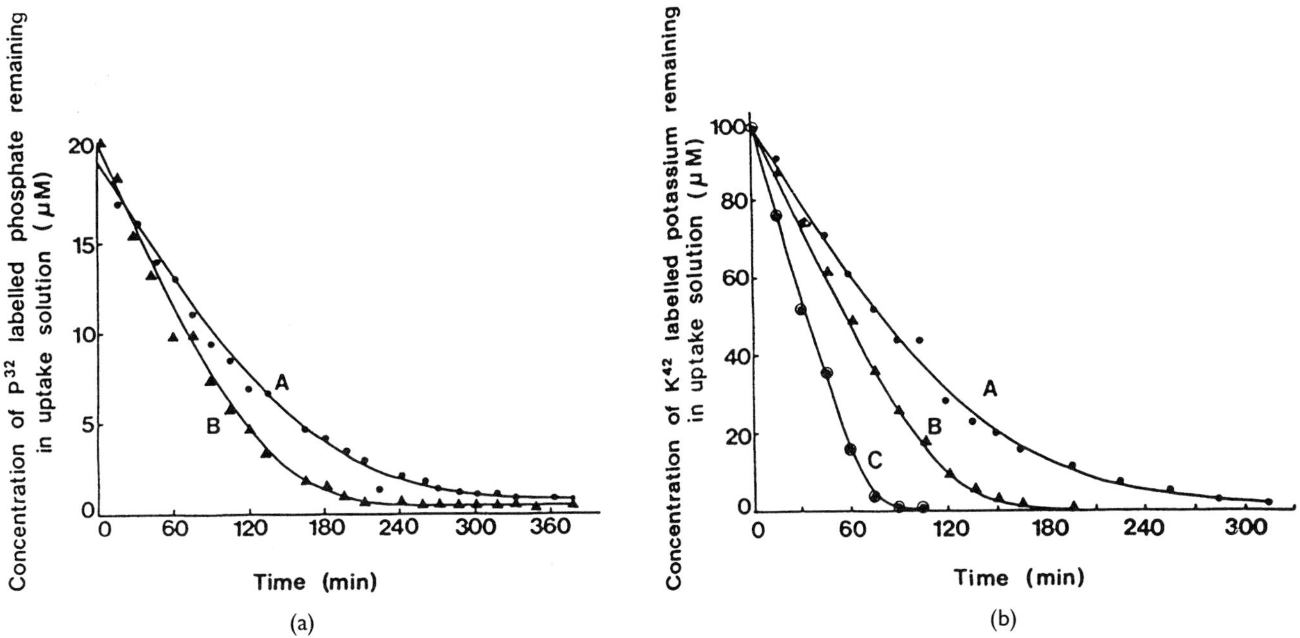

Figure 5.11 Short-term depletion experiments to give values of V_{max} (μmol g DW root^{-1} day^{-1}) and K_m (μM) for uptake of (a) phosphate (^{32}P) and (b) potassium (^{42}K) by barley after different periods of nutrient deprivation as indicated by the letters on the curves. (P) For previous phosphate deprivation of, respectively, 0 (A) or 4 (B) days, the uptake parameters were V_{max}, 153 and 208; K_m, 5.5 and 5.1. (K) For potassium deprivation of 0 (A), 1 (B), or 4 (C) days, the uptake parameters were, respectively, V_{max}, 1780, 2240, and 3350; K_m 44.6, 10.2, and 9.8 (after Drew *et al.* 1984).

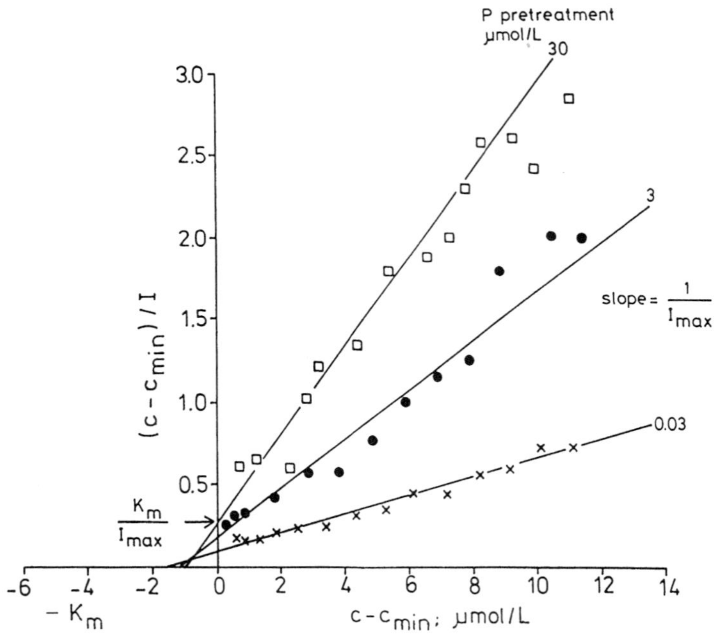

Figure 5.12 Hanes plot of phosphate uptake data for soybean in a set of short-term P-depletion experiments after pre-treatment of the plants at the phosphate concentrations marked on the lines. Note marked change in I_{max} but little change in K_m (after Jungk *et al.* 1990).

Thus, regulation affects both K_m and I_{max} over different timescales (Smart *et al.* 1996). The changes in I_{max} can be accounted for by the production of more transporter proteins, but a K_m change seems to call for a physical change in the transporter structure, such as the allosteric changes suggested by Glass (1976) or the activation of a gene that produces a different type of protein. More generally, possible mechanisms are changes in the rate of efflux; changes in the membrane potential; changes in the numbers of high-affinity uptake proteins and ion channels synthesized; changes in the relative numbers of different sorts of uptake protein; and allosteric changes in a single protein. Possibly, several of these operate at the same time. Dunlop & Gardener (1993) measured phosphate uptake, proton extrusion and membrane potential simultaneously in clover supplied with varying levels of phosphate. They concluded that at different levels of P regulation, the uptake mechanisms were qualitatively different, which suggests a different, or a changed, transporter rather than simply the acceleration of the same mechanism, or a general process such as the membrane potential. For example, one P_i transporter in potatoes is only synthesized when these are P-starved (Leggewie *et al.* 1997) (section 5.2.3.4).

The cytoplasm ion concentrations must remain within very narrow concentration ranges for it to function (Leigh & Wyn Jones 1986). Ion concentrations in the cytoplasm are therefore unlikely to be the direct signal for change in uptake rate, and Lee & Ratcliffe (1983) showed by the use of NMR that the concentration of P_i in the cytoplasm of pea root tips remained constant whilst the level of supply of P_i in solution was varied, and the P_i in the vacuole changed sharply. Miller & Smith (1996) have discussed possible mechanisms for the maintenance of constant cytosolic nitrate concentrations. The vacuolar concentrations are variable over a wide range, and simple ions of the major nutrients (nitrate, phosphate, chloride, potassium) start to accumulate rapidly in the vacuole at a sharply defined concentration of that nutrient in the plant (Zhen & Leigh 1989). Their state could therefore act as a trigger for ion uptake, though nothing is known of how a signal might be transferred across the cytoplasm. Regulation does not occur wholly within the single root, because split-root experiments have shown that the uptake parameters of a given root depend upon the state of the whole plant rather than upon its own immediate environment or nutrient condition (Drew & Saker 1984), so there must be a long-distance signal emanating from the shoot which controls local root uptake properties. This could be the nutrient concentration in the phloem–xylem recirculatory system which operates for some ions, particularly potassium, nitrogen and phosphate (Cooper & Clarkson 1989). However, a number of the immobile elements in plants do not circulate in this way (e.g. Fe, B, Ca), but nevertheless the siderophore system for iron uptake does respond to low availability of iron (section 7.3.4).

Regulation of nitrogen is particularly interesting, because of the complex chemical transformations it undergoes in the plant, its two main mineral forms, and uptake from a symbiotic system. Touraine *et al.* (1994) discussed the possible signals that result from nitrate assimilation (figure 5.13). Nitrate moves rapidly

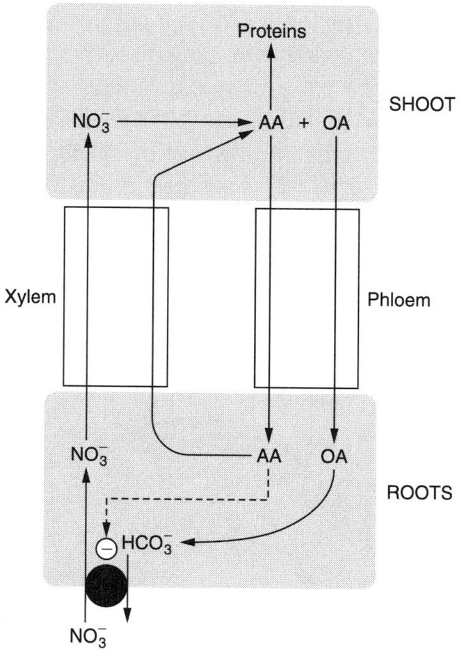

Figure 5.13 Model of regulation of nitrate uptake by circulation of amino acids and negative charge in the phloem and xylem. Export of organic acids (OA) to roots provides bicarbonate ions, and enhances nitrate uptake. Amino acids (AA) circulate, and depress nitrate uptake (after Touraine *et al.* 1994).

up to the leaves, and is there assimilated, with the simultaneous formation of organic acids (to maintain pH balance) and amino acids. Both of these will be translocated to the root in the phloem, and it is proposed that either of these may act as the signal; the organic acids tend to increase nitrate uptake, and the amino acids repress it. If so, the cycling of ions in the phloem and xylem may integrate the nutritional state of the whole plant with regard to translocatable nutrient elements, and thereby tend to maintain homeostasis in nutrient concentration (Touraine *et al.* 1994).

5.3.4 Interactions and Competition between Elements

Interactions can be positive (synergisms) or negative (competition) between different nutrient ions. Low pH decreases cation uptake, probably because this lowers the membrane potential (figure 5.14). There is little effect of pH on anion uptake, except for phosphate, where the pH controls the relative amounts of the differently charged phosphate ions.

Some of the cations compete in uptake, largely depending upon the similarity of their ionic charge and size. The alkali metals compete, and K and Rb are not distinguished by the uptake system. Ammonium competes against potassium, but the apparent lack of a reciprocal effect is explained by uptake of uncharged NH_3 through the plasmalemma. In terms of whole plant composition, there is very often a generalized competition, in that increase of K, Na, Mg, or Ca will tend to decrease the other cations. Magnesium suffers strong competition by K^+, NH_4^+ and H^+, and magnesium deficiency in the field is often increased by one of these. The competitive interaction between potassium and magnesium seems to arise from the external concentrations of the ions, and to involve interference at the uptake mechanism stage.

Calcium (Marschner 1995, p. 43) can increase uptake of potassium and of anions (figure 5.14). This may be due to the collapse of double layers in the cell walls, or to the improved membrane integrity caused by calcium. Calcium is also important in regulating the K/Na balance, where it tends to shift uptake in favour of K. High pH tends to increase cation uptake, but the presence of calcium can modify the effects on other cations (figure 5.14). At the present time, most of these observations on ion interaction can be explained only tentatively, because of the many possible mechanisms that can be involved, and the complicated boundary between regulation and competition.

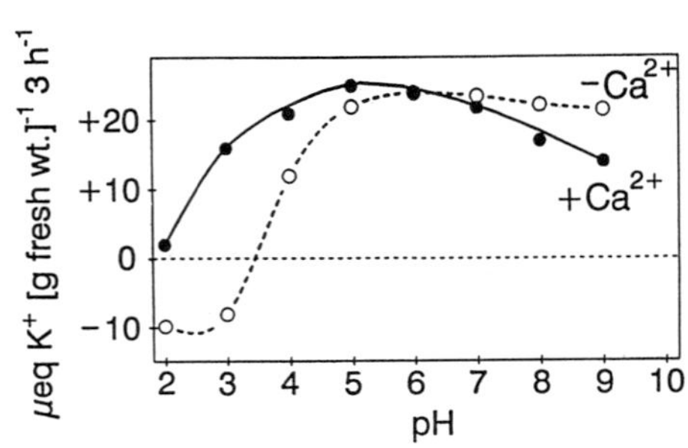

Figure 5.14 Effect of pH value and presence of calcium (5 mM) on uptake rate of potassium. Note the small change between pH 4 and 7 in the presence of calcium and the small change between pH 5 and 9 in the absence of calcium, giving balanced regulation (after Marschner 1995).

Aluminium also has multiple effects, and Al and Ca may tend to counteract each other (Brady *et al.* 1993a). In general, Al does not appear to cause serious damage to the plasmalemma, as the membrane potentials remain reasonably normal (Kochian 1995). It inhibits the uptake of cations, but may increase that of anions, and Nichols *et al.* (1993) suggested that the primary process is that Al is bound to the phospholipids of the plasma membrane and forms a layer of positive charge there. Within the cell, low levels of Al also have complex effects, for example it greatly increased the complexation of phosphate within the vacuole (Macklon & Sim 1992).

The main macroscopic effect of aluminium is in the root apex, where Al may inhibit root growth rate by inhibition of DNA synthesis and a rapid reduction in cell elongation rate (Kochian 1995). Genotypes that are Al-resistant respond to Al^{3+} at the root tip by secreting citrate and/or malate with the root mucilage (table 5.6), which complexes the aluminium or increases the pH. The mechanism may be the opening of an organic ion-permeable channel (Degenhardt *et al.* 1998).

5.3.5 Plant Growth and Plant Nutrient Demand

Nutrient demand arises from changes in the plant's growth rate and nutrient concentration. A plant is a sink for two different types of materials, carbon and mineral nutrients, which are supplied from different sources, and the complexity of its growth arises because the sources are not simple, and because both interact with the same sink and hence with each other. The plant has feedback regulatory systems for the acquisition for both of these, such that they shall remain in balance. This overall balance may be expressed simply by a nutrient ion conservation equation (Nye & Tinker 1969) that is described in section 10.2.6.

5.4 Plant Factors that Affect Uptake Rates

5.4.1 Effects of Water Relations

Water relations might affect nutrient uptake in many ways, such as a change in growth rate, or production of roots with different properties. A change in tran-

Table 5.6 Effect of removing mucilage three times a day (– mucilage) from cowpea roots growing in solution culture on the aluminium content of the roots (0–5 mm) and of the mucilage. Experiments done with and without 5 mg/L Al in solution.

Treatment	Mucilage	Root growth (cm day^{-1})	Al content of root tips (0–5 mm): Roots (μg Al (25 tips)$^{-1}$)	Mucilage (μg Al (25 tips)$^{-1}$)	Roots (mg Al g^{-1} dry wt)	Mucilage (mg Al g^{-1} dry wt)
–Al	+	6.3	—	—	—	—
	–	5.9	—	—	—	—
+Al	+	4.8	12.4	16.6	2.1	16.6
	–	2.1	20.6	3.6	3.2	14.5

Source: after Marschner (1995).

spiration rate, which may or may not be associated with an important change in plant water potential, will produce a larger flow rate through the soil and the root. If the plant grows in soil, drying out can slow the flow of nutrients to the roots, as the rate of transpiration decreases. In solution culture, uptake of ions is increased by greater transpiration if the solution is concentrated, but has little effect if it is dilute (Hsiao 1973). Thus the rate of loading of nutrients into the xylem from dilute solutions is not affected by the rate of transpiration, so that the concentration of the xylem exudate (after detopping the plant) is inversely proportional to the transpiration rate (Marschner 1995). Work with sodium by Yadav *et al.* (1996) makes this effect of water inflow clearer. In two lines of rice with different sodium uptake properties the only correlation with the sodium uptake rate was with that of a chemical tracer that is taken up via the apoplasmic pathway. This suggests that the sodium enters by an apoplasmic route, which has been estimated to carry up to 5% of the transpiration in rice, so the effect is noticeable if the external solution is concentrated. If not, the uptake of nutrients by this route is small compared with that by any high-affinity uptake process. In saline soils, the sodium concentration is high, and a large apoplasmic flow could therefore cause sodium toxicity. For the effect of water potential on the uptake characteristics, see section 6.5.

5.4.2 Effects of Plant Species and Variety

The majority of higher plants are surprisingly similar in their nutrition. There is a major difference between monocots and dicots, in that the former have higher K/Ca ratios in their tissues. The C3 and C4 species are also generally similar, but the former are more sensitive to phosphorus deficiency.

Some species, such as buckwheat, have an inherently high cation/anion uptake ratio, and maintain electrical neutrality by losing protons. They therefore acidify their rhizospheres, and are traditionally good at utilizing rock phosphates (Van Raij & Van Diest 1979) (section 7.2.5). Rye is particularly efficient at absorbing copper, zinc, manganese and phosphate amongst the small grains (Graham 1984). Species vary widely in their ability to tolerate acidity. Other generalizations on species differences can be found in Clark (1982) and El Bassam *et al.* (1990).

Apart from such differences of degree in widespread species, there are some species that are effectively restricted to mineralized soils with high concentrations of metals such as nickel, zinc and cadmium. Such plants are accumulators, and contain quite large amounts of these heavy metals. Their uptake physiology is necessarily specialized. They show in extreme form an ability to tolerate very high levels of sodium or heavy metals, which varies widely among many species (Petersen 1993). Some plants deal with excess of an ion in the soil by excluding the potentially toxic element, some by excreting it, and some by biochemical adaptations that allow it to accumulate without damaging the plant, such as in complexes in the vacuole. Hyper-accumulators use the last mechanism; accumulation of nickel is most frequently found, even up to 3% dry matter. This accumulation is associated with the formation of histidine in proportion to the metal concentration (Kramer *et al.* 1996). Supplying histidine to non-accumulating species also greatly increases nickel uptake and tolerance. This property is used in soil remediation (McGrath 1998).

Below the species level, there can be very large differences in the ability of different cultivars or varieties to resist deficiency or toxicity. Better uptake and use efficiency can be attained in a breeding programme by various mechanisms (Graham 1984), such as better root architecture (chapter 9), faster uptake rate (chapter 5), rhizosphere modification (chapters 7 and 8), and improved internal allocation or a lower requirement for the element per unit weight of tissue (chapter 10). With the trace elements, a single major gene may control the susceptibility to an element deficiency. Table 5.7 (Graham 1984) shows the variation in resistance to manganese deficiency in cultivars of wheat, and among wheat, triticale and rye.

5.4.3 Effects of Age, Position on Root, and Root Radius

The effect of plant age shows mainly in drastic changes in the allocation of carbon to the root, and the consequent slowing of root growth, and increase in mean root age (section 9.1.1). For the major elements, the total uptake rate does not vary greatly over a distance of 44 cm from the root apex for barley (Clarkson *et al.* 1968, 1978; Marschner 1995; Clarkson 1996) (table 5.8). For the three macronutrients, it is therefore not unreasonable to assume uniform uptake over the surface of a rapidly growing root system (Brady *et al.* 1993b).

By contrast, the uptake of calcium, magnesium, and iron is much more restricted to near the apex; for the last, this is associated with the special processes involved in its uptake (section 7.3.4). More measurements of the distribution of I values along roots for various elements would be desirable.

It is assumed that if uptake of ions occurs throughout the cortex, the uptake rate per unit surface of root (αC_{La}) will increase with root radius. However, the real concentration close to the plasmalemmas of the cortical cells will decrease with distance from the surface, and in the limit uptake will only occur in a layer close to the surface, as the concentration closer to the stele will be C_{Lmin} or less. If

Table 5.7 Cultivar and species effects on wheat, triticale, and rye grown in a manganese-deficient soil in chlorosis rating, shoot Mn concentration, and yield relative to the same cultivar grown in the same soil supplied with manganese.

Genotype	Chlorosis rating (0–3)	Shoot Mn concentration ($\mu g\ g^{-1}$)	Shoot Mn content (μg/plant)	Relative yield (%)
RAC 311 wheat	2.5	8.3	0.84	58
Oxley wheat	2.5	9.5	0.96	65
Halberd wheat	2.5	10.3	1.09	80
Gatcher wheat	0	12.7	1.50	83
Bodallin wheat	0	18.3	2.94	96
Venus triticale	0	15.0	3.39	90
S.A. Commercial rye	0	17.9	2.01	112

Source: after Graham (1984).

Table 5.8 Uptake rates of P and Sr (analogue of Ca) by portions of barley roots at different distances from the apex.

Content of phosphorus (mole per plant per day × 10^{-12})				
Concentration of solution (M)	3×10^{-6}		10^{-4}	
Distance of treated part from apex (cm)	1	44	1	44
In treated part	395	335	2070	770
In rest of plant	53	241	330	560
Total	448	576	2400	1330
Content of strontium (mole per plant per day × 10^{-12})				
Concentration of solution (M)	1.2×10^{-6}		10^{-4}	
Distance of treated part from apex (cm)	1	44	1	44
In treated part	89	58	2910	2240
In rest of plant	51	2	1100	200
Total	140	60	4010	2440

Source: after Clarkson *et al.* (1968).

we assume that uptake by unit cortex volume is given by kC_L, Nye (1973) showed that when $a(k/D)^{1/2} > 1.6$ (where D is the diffusion coefficient of an ion through the cortex), then α for the whole root is effectively independent of root radius a. The uptake rate will then depend on root surface area rather than volume. These conditions are likely to apply to nutrients in the deficiency range of external concentration, when k will be at its largest value.

5.5 Environmental Variables that Affect Uptake Rate

5.5.1 Temperature

Temperature affects the physiological characteristics of roots also (Clarkson *et al.* 1988). Uptake rate increases for almost all ions up to about 30°C, followed by reduction. At low temperatures, the permeability of the roots may be greatly changed. Effects arise both from the temperature at which an experiment is performed and from the temperature at which the roots are grown (Gur & Shulman 1971). The uptake rate of phosphate, in particular, is decreased by low temperatures, which also changes the K/Ca ratio. Moorby & Nye (1984) exposed the individual roots of a single rape plant to different temperatures and found phosphate inflows to be independent of temperature between 10°C and 23°C, but halved at 5°C. Inflows to a single root were little affected by the temperature of the rest of the root system.

Cumbus & Nye (1982) found that root temperature had two distinct effects on the growth of rape that relate to nitrate absorption. First, temperature extremes of 10°C and 35°C led to high concentrations of free carbohydrate in the shoot and low ones in the root, and high shoot/root ratios. Second, nitrate inflow through the available root was determined by the shoot demand, demonstrated by the very high inflow at 35°C required to satisfy a rapidly growing shoot on a small root.

5.5.2 Oxygen Supply

The uptake process depends upon the metabolism of the root and consumes a large fraction of the total root respiration (Lambers *et al.* 1996). A lack of oxygen therefore decreases uptake rates, and continued anaerobiosis leads to leakage of ions from the roots. In soil, the lack of oxygen causes many other effects, such as the production of toxic organic compounds and toxic reduced forms of Mn and Fe. For species that grow in waterlogged habitats, the diffusion of oxygen down through the root is of critical importance, and the roots are often modified with a large air-filled pore space to increase this rate. The growth of roots in anaerobic soils is well described by Drew (1988) and Drew & Stolzy (1996).

5.5.3 Allelopathy

The still controversial subject of allelopathy is dealt with elsewhere (section 9.3.3.4). There is, however, considerable evidence that ion uptake may be implicated (Balke 1985). A direct test on whole cucumber plants (Booker *et al.* 1992) in short-term solution culture showed that K uptake was strongly decreased by ferulic acid, a cinnamic acid derivative found in soils and considered to be an allelochemical. Treatment with ferulic acid caused net efflux of K from the roots, so that net uptake only occurred over 1.5 mM K in solution, and the high-affinity uptake process appeared to have been strongly inhibited.

5.6 Conclusion

The insights into the mechanisms of ion uptake have increased rapidly. However, this knowledge about membrane and cell behaviour has not yet been built into a full understanding of the way the whole plant works. Single processes are becoming well understood, with some notable exceptions, but the interactions of these processes, and the control of the resulting systems, has not yet made similar progress.

6

Solute Transport in the Soil near Root Surfaces

6.1 Transport Processes

6.1.1 The Relative Importance of Diffusion and Mass Flow

We discussed in chapter 4 the movement of solute between small volumes of soil, and in chapter 5 some properties of plant roots and associated hairs, particularly the relation between the rate of uptake at the root surface and the concentration of solute in the ambient solution. In the chapters to follow, we consider the plant root in contact with the soil, and deal with their association in increasingly complex situations; first, when the root acts merely as a sink and, second, when it modifies its relations with the surrounding soil by changing its pH, excreting ions, stimulating microorganisms, or developing mycorrhizas. In this chapter, we take the simplest situation that can be studied in detail, namely, a single intact root alone in a volume of soil so large that it can be considered infinite.

The essential transport processes occurring near the root surface are illustrated in figure 6.1. We have examined in chapter 3 the rapid dynamic equilibrium between solutes in the soil pore solution and those sorbed on the immediately adjacent solid surfaces. These sorbed solutes tend to buffer the soil solution against changes in concentration induced by root uptake. At the root surface, solutes are absorbed at a rate related to their concentration in the soil solution at the boundary (section 5.3.2); and the root demand coefficient, αa, is defined by the equation

$$I = 2\pi\alpha a C_{La} \tag{6.1}$$

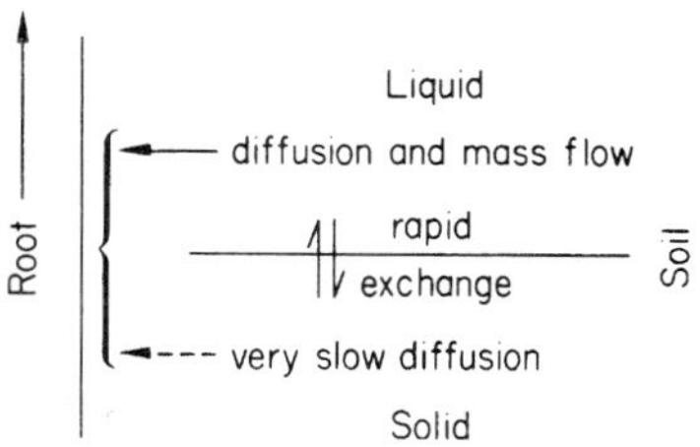

Figure 6.1 Solute transport processes near an absorbing root.

where I = inflow (rate of uptake per unit length),
a = root radius,
C_{La} = concentration in solution at the root surface.

To calculate the inflow, we have to know C_{La}, and the main topic of this chapter is the relation between C_{La} and the soil pore solution concentration C_L. The root also absorbs water at its surface due to transpiration (chapter 2) so that the soil solution flows through the soil pores, thus carrying solutes to the root surface by mass flow (convection). Barber *et al.* (1962) calculated whether the nutrients in maize could be acquired solely by this process, by multiplying the composition of the soil solution by the amount of water the maize had transpired (table 6.1). The authors did not give figures for nitrogen, but assuming there is 20 000 ppm in the plant (2%) and the transpiration ratio is 500, typical of a warm temperate region, the soil solution would need to contain 40 ppm N to satisfy the requirement. This is a high value in the absence of recent fertilizer application. Such calculations give only a rough idea of the importance of mass flow since the concentration of major nutrients in a plant is usually greater in the early stages of growth than later, and the transpiration ratio will often be lower. In general, mass flow will transport more than sufficient sulphur, calcium, sodium and magnesium (except in sodic, very acid, or Mg-deficient soils) to the root surface, but insufficient potassium and nitrogen, and quite insufficient phosphorus. Similar calculations made for micronutrient elements in one soil suggested that mass flow can account for only a small part of the uptake of copper, iron, manganese, and zinc, but more than half the uptake of boron (Oliver & Barber 1966). Mass flow should be adequate to supply molybdenum to an 'average' plant on most soils, but not on those with less than about 4 ppb molybdenum in the soil solution (Lavy & Barber

Table 6.1 Supply of elements to maize by mass flow.

Element	Composition of maize at harvest (ppm)	Concentration of soil solution needed if transpiration ratio = 500 (ppm)	Soil solution of 145 mid-Western USA topsoils modal values (ppm)
Ca	2 200	4.4	33
Mg	1800	3.6	28
K	20 000	40	4
P	2 000	4	0.05

Source: after Barber *et al.* (1962).

1964). Since the concentration of these elements in the soil solution is so sensitive to pH (section 3.5.2), generalization is unsafe. Barber (1995) gives more details of the importance of mass flow in supplying these elements, with calculations based on average values. Weather variations will lead to wide changes in mass flow.

Some comparatively small organic solutes in the soil solution, such as many herbicides, are absorbed at nearly the same rate as the solution, probably by passive uptake (section 5.2.2). Figure 6.2 shows the relation between transpiration of wheat and turnip seedlings and the total uptake of atrazine from two soils (Walker 1971). Variable amounts of linuron, atrazine and simazine are retained in the root depending on plant species, so that the concentration of the xylem stream is less than that of the soil solution (Shone & Wood 1972; Walker & Featherstone 1973).

If the solute is absorbed at a relatively greater rate than the water, as with phosphate and potassium, then its concentration at the root surface must fall, but less than if water were not being transpired. In response, the solid releases these ions, tending to buffer the concentration. Nevertheless, the solution concentration at the root surface will still fall somewhat, and this will induce ions to diffuse towards the root. Consequently, a zone of depletion develops (figure 6.3), as revealed by an autoradiograph of a root growing in soil labelled with ^{33}P (figure 6.4a).

If, on the other hand, water is taken up at a relatively greater rate than solute, the solute must accumulate at the root surface, and tends to diffuse away from the root (figure 6.3). Figure 6.4b shows this increase in concentration at the surface of two roots of a rapidly transpiring plant growing in $^{35}SO_4$-labelled soil.

Thus, there are only two processes by which solutes move in response to the disturbance created by the active root: mass flow and diffusion. Very few ions will be so close, less than 1 nm, to the root or root hair surface that they can be directly absorbed without these two transport processes. Roots do not 'pick up' or 'intercept' or otherwise ingest solutes as roots grow through the soil, since the root cap is not an absorbing organ. Roots are rather to be pictured as following pores and channels, pushing aside soil solution and, to some extent, soil particles, whose solutes then flow and diffuse to the new surfaces created.

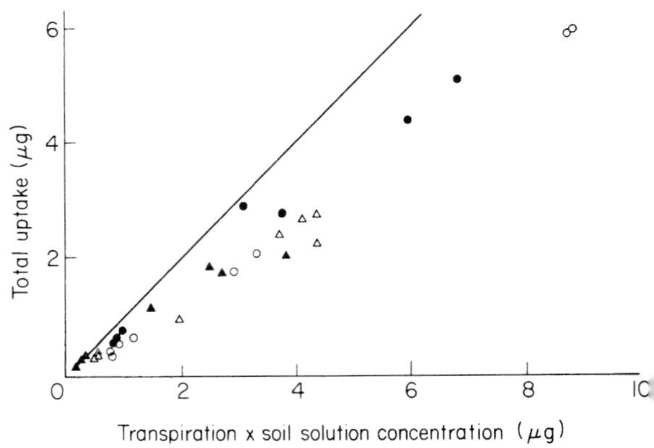

Figure 6.2 Relation between transpiration and uptake of atrazine by wheat and turnip seedlings from two soils (after Walker 1971). The straight line represents the theoretical uptake if all the material supplied by mass flow were taken up. ●, wheat, Soakwaters; ○, turnip, Soakwaters; ▲, wheat, Little Cherry; △, turnip, Little Cherry.

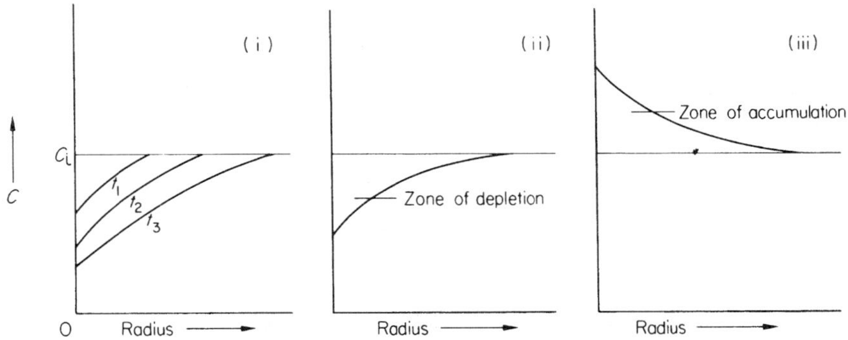

Figure 6.3 Concentration of solute near a root surface: (i) diffusion alone—increasing time t_1, t_2, and t_3; (ii) water absorbed relatively slower than solute ($v_a < \alpha$); (iii) water absorbed relatively faster than solute ($v_a > \alpha$).

However, the boundary between the root surface and the soil cannot be precisely defined. Here, epidermal cells are continually being sloughed off, a polysaccharide mucilage is being excreted and a high concentration of bacteria develop. So, the boundary is really a narrow zone, some micrometers thick, containing a mixture of soil particles, living and dead root cells, mucilage and bacteria (section 8.2.1). Very immobile elements, such as iron, present in high concentration in the soil, but often at very low concentration in the soil solution, may be solubilized by chelates excreted at the root surface (section 7.3.4). In all such cases, the solute has to move to reach an absorbing root surface even if the distance is only a few micrometers.

6.1.2 Quantitative Treatment of Mass Flow and Diffusion near a Single Root

Although diffusion and mass flow occur simultaneously, quantitative experiments that investigate them together are few and difficult to devise. It is therefore convenient to consider first the theory of diffusion in the root zone; and to describe experiments that test it, in which mass flow is negligible. Experiments that are not complicated by the presence of root hairs will be discussed first.

We may assume that movement of solute is normal to the root surface if the spread of the diffusion zone, equal to $(4Dt)^{1/2}$, is small compared with the root's elongation in the same period, which varies roughly linearly with time. Hackett & Rose (1972a, b), for example, report that barley root axes extend 20 mm day^{-1}, first-order laterals 4 mm day^{-1}, and second-order laterals 2 mm day^{-1}. Meanwhile, if $D = 10^{-7}$ cm^2 s^{-1}, the disturbance zone spreads only about 4 mm in 4 days. On the other hand, the short stubby but long-lived mycorrhizal roots of pine elongate much more slowly, and it may often be more correct to regard these as being in the centre of a spherical zone of depletion. Anderssen *et al.* (1969) and Darrah (1991) have published numerical solutions of the transport equations that take elongation into account.

(a)

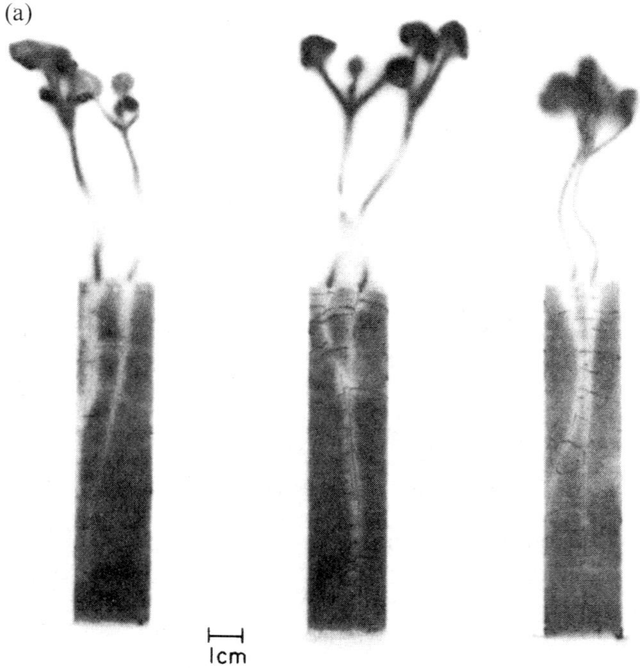

Figure 6.4a Autoradiograph of roots of rape in a soil labelled with ^{33}P showing zone of depletion around roots, and accumulation within the axis and laterals (after Bhat & Nye 1974a).

(b)

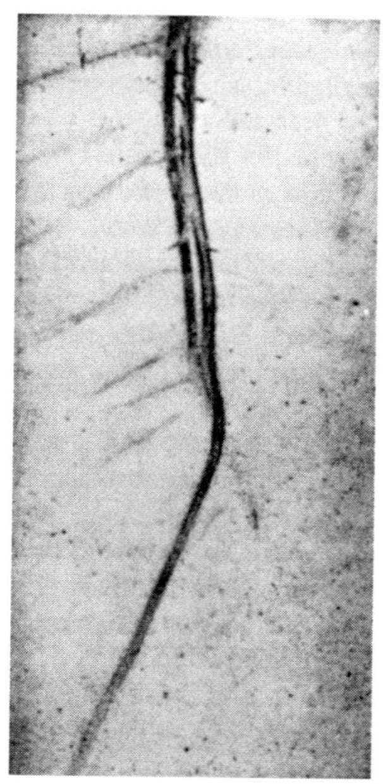

Figure 6.4b Autoradiograph of two maize roots in a soil labelled with $^{35}SO_4$ showing accumulation near surfaces (after Barber *et al.*, 1963).

The appropriate continuity equation for cylindrical symmetry around the root has been given in section 1.4.1:

$$\partial C/\partial t = 1/r\,\partial/\partial r(rD\,\partial C/\partial r) \tag{6.2}$$

This must be solved for the boundary conditions that will now be discussed. The exact analytical solution is complicated, but the following simplified account gives the essence of what happens when a root initiates a diffusion zone in a large volume of soil. There are three situations to be considered (section 5.3.2).

(1) $C_{La} > C_{Lcrit}$ $\quad I = I_{max}$ or I_{crit}

Most plants growing with a sufficiency of nutrients maintain a roughly steady inflow controlled by their growth requirements. This inflow is I_{max} in figure 5.9a. Its maintenance depends on the diffusive supply providing a concentration at the root's surface that is adequate to give I_{max}. Thus, plant uptake is the rate-limiting process, and the concentration at the root's surface, C_{La}, remains above the critical concentration, C_{Lcrit}, at which the inflow drops below I_{max} in figure 5.9a,b. As long as $C_{La} > C_{Lcrit}$, the concentration gradient at the root surface must remain constant as the depletion zone spreads (see figure 6.5), since

$$I_{max} = 2\pi a D_L \theta f_L (\mathrm{d}C_L/\mathrm{d}r)_{r=a} \tag{6.3}$$

or

$$(\mathrm{d}C_L/\mathrm{d}r)_{r=a} = I_{max}/(2\pi a D_L \theta f_L) \tag{6.4}$$

which is constant in this situation. It is clear from figure 6.5 that as the depletion zone spreads, the concentration at the root surface must fall, until it reaches C_{Lcrit}. We now have the second situation.

(2) $C_{La} < C_{Lcrit}$ $\quad \alpha a$ constant

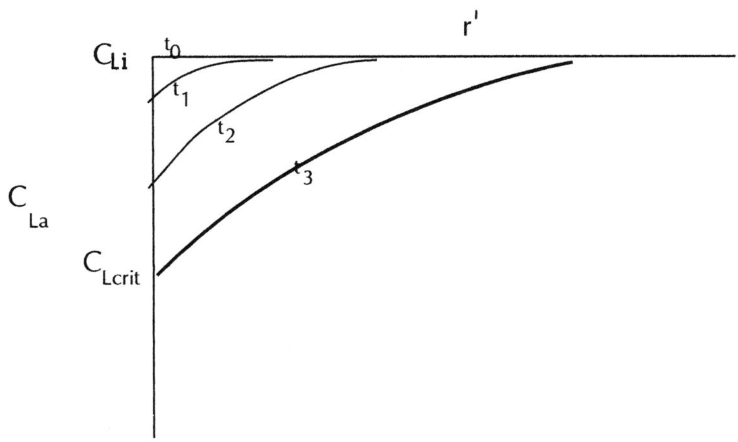

Figure 6.5 Concentration profiles in steady state at successive times in a soil with $C_{Li} = C_{Lcrit}$ so that the root demand coefficient αa is effectively constant.

As shown in figure 5.9a,b, when $C_{La} < C_{Lcrit}$ the root demand coefficient is effectively constant. If reduced uptake continues for a long period, the nutrient status of the plant will decline and αa will be affected (section 5.3.2).

Now, since the inflow at the root surface must equal the diffusive supply from the soil,

$$I = 2\pi\alpha a C_{La} = 2\pi a D_L\theta f_L(\mathrm{d}C_L/\mathrm{d}r)_{r=a} \tag{6.5}$$

At this stage, growth may also decrease. As C_{La} continues to fall both the inflow and the concentration gradient at the root surface also decrease. The implications of the analytical solution of equation (6.2) (Carslaw & Jaeger 1959) are not readily discerned since the solution is complicated by Bessel functions; it is illustrated in figure 6.6, which shows that in this condition (αa constant), uptake is affected by both the root demand coefficient and the diffusion term (i.e. $D_L\theta f_L$).

An approximate quantitative idea of the influences of αa and $D_L\theta f_L$ on the inflow may be obtained if we assume that the initial concentration in the soil solution is C_{Lcrit}; and that as the expanding depletion zone spreads, the shape of the curve relating C_L to r is the same as it would be if nutrient were diffusing to the root in a steady state from the edge of the depletion zone, r', as in figure 6.5. We then have the steady-state equation (9.6) derived in chapter 9.

$$C_{Lr'} = C_{Lcrit} = C_{La}[1 + (\alpha a/D_L\theta f_L)\ln(r'/a)] \tag{6.6}$$

Therefore,

$$I = 2\pi\alpha a C_{La} = 2\pi\alpha a C_{Lcrit}/[1 + (\alpha a/D_L\theta f_L)\ln(r'/a)] \tag{6.7}$$

in which I depends on αa and $D_L\theta f_L$. Initially, $\ln(r'/a) = 0$ and

$$I = I_{max} = 2\pi a\alpha C_{Lcrit} \tag{6.8}$$

and the inflow depends only on the root demand coefficient. As the depletion zone spreads and r' increases, $\ln(r'/a)$ in equation (6.7) becomes increasingly important and I depends on αa and $D_L\theta f_L$. Eventually, when r' is large and $(\alpha a/D_L\theta f_L)\ln(r'/a) \gg 1$, we obtain from equation (6.7),

$$I = 2\pi D_L\theta f_L C_{Lcrit}/\ln(r'/a) \tag{6.9}$$

and since, in equation (6.8),

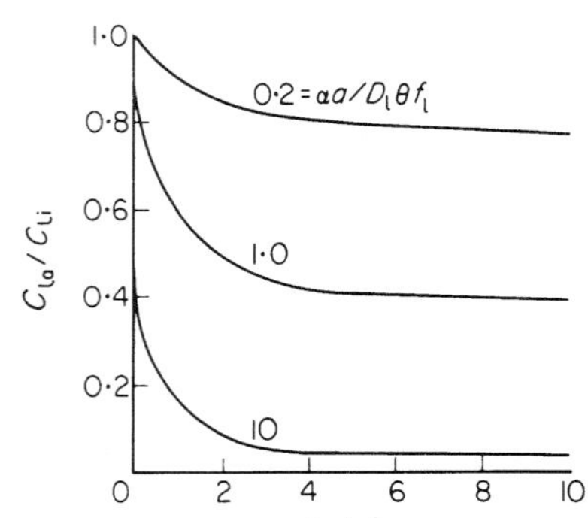

Figure 6.6 Change in the root surface concentration ratio C_{La}/C_{Li} with Dt/a^2 for different values of $\alpha a/D_L\theta f_L$ (after Carslaw & Jaeger 1959).

$$I_{max} = 2\pi\alpha a C_{Lcrit}$$

equation (6.9) gives

$$I = I_{max}(D_L\theta f_L/\alpha a)/\ln(r'/a) \quad (6.10)$$

Equation (6.10) shows in simple terms the important parameters that control the inflow when C_{La} falls below C_{Lcrit}: I is lower than I_{max} as the zone of depletion spreads, according to the term $\ln(r'/a)$; I tends to be increased by easier diffusion and reduced by a higher root demand coefficient (though I_{max} is increased by this, as equation (6.8) shows) and αa does not appear in equation (6.9).

In practice, if the boundary conditions change in the course of depletion from a constant inflow to a constant root demand coefficient condition, or, alternatively, if the relation between inflow and root surface concentration is described by a Michaelis–Menten condition (for practical purposes, these are often equivalent), the continuity equation has to be solved by numerical methods. Barber & Cushman (1981) have made available a computer program for this purpose. A numerical solution is also needed if the soil diffusion coefficient depends on concentration.

With continued uptake, C_{La} finally falls to a value approaching C_{Lmin}, the threshold concentration for uptake. We now enter the third situation.

(3) $C_{La} = C_{Lmin}$ zero sink condition

In this condition, in which C_{La} is very small, even if the root demand were to increase, there would be no increase in inflow because $(\mathrm{d}C_L/\mathrm{d}r)_{r=a}$ in equation (6.5) cannot increase because C_{La} can fall no lower. Inflow is now entirely controlled by diffusion to the root surface, and it continues to decrease as the depletion zone spreads and the concentration gradient decreases further.

When a root exploits and exhausts a fresh zone of fertile soil, ion buffering influences the importance of these three situations. The contrast between nitrate and phosphate exemplifies this fact, as sketched in figure 6.7. In a soil fertilized with nitrate, $C_{Lnitrate}$ greatly exceeds C_{Lcrit}. Because nitrate is unbuffered, when it is depleted to C_{Lcrit} there is little left to exploit between C_{Lcrit} and C_{Lmin}, so that uptake declines rapidly. In the case of phosphate, the shape of the isotherm reveals that very much more phosphate remains to be taken up when the concentration has been lowered to C_{Lcrit}. Consequently, whereas nitrate inflow tends to be either I_{max} or nearly zero, for phosphate the intermediate period where I depends on αa is much more important. This makes it easier to model nitrate uptake than phosphate uptake (chapter 11).

6.2 Experimental Evidence for Theory of Diffusion near Roots with Restricted Mass Flow

6.2.1 Indirect Methods

The model outlined in the previous section has been tested in several experiments. In the earlier work, which is more fully described in Nye & Tinker (1977), the root surface concentration, and hence the root demand coefficient, was not measured

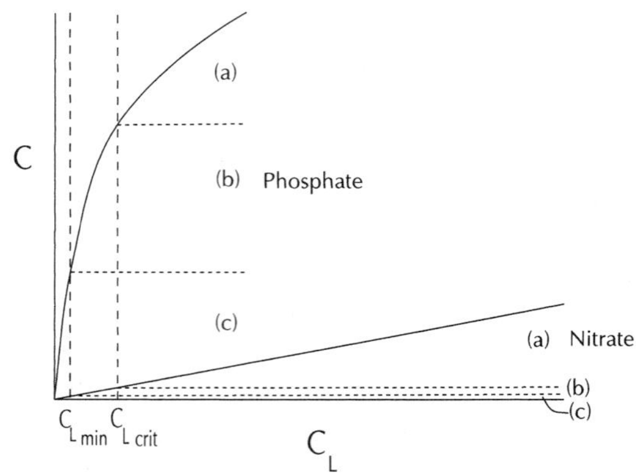

Figure 6.7 The effect of the low buffer power of nitrate and the high buffer power of phosphate on the progress of nitrate and phosphate depletion from a typical soil. In region (a) ($C_L > C_{Lcrit}, I = I_{max}$), the amount, C, of nitrate remaining in the soil is low, and the amount of phosphate is high. In the region (b) (C_L between C_{Lcrit} and C_{Lmin}), over which αa is constant, the amount of phosphate exceeds the amount of nitrate. In the region (c) ($C_L < C_{Lmin}$), there is little nitrate left in the soil.

directly, but only indirectly by measuring the inflow through a portion of root and the diffusion characteristics of the soil. The model was therefore tested only to the extent that 'reasonable' values of uptake were obtained. For example: (i) The observed uptake should not greatly exceed the amount that could diffuse to a zero sink. (ii) Values of the root demand coefficient derived from the experiments should be consistent with values obtained in experiments in stirred or flowing nutrient culture solutions. (iii) These values should change with the concentration of nutrient or moisture status of the soil in a way consistent with theory.

6.2.2 Direct Methods

The concentration of solute in the soil around roots has been measured directly by three main methods.

(a) Autoradiography This technique was initiated by Walker & Barber (1961), who grew plants in a box of soil which was uniformly labelled with radioisotope. One side of the box consisted of mylar film, 1 μm thick, against which some of the roots grew. Autoradiographs of their contact with the soil were taken with X-ray film. The earlier work with this method was qualitative since (a) visual assessment of concentration from density on the film is very deceptive, and the contrast is greatly influenced by the time of exposure and development; (b) the isotopes used, such as ^{32}P, ^{86}Rb, and ^{90}Sr, have high maximum β energy, and the resolution was correspondingly poor. For example, ^{32}P emits β particles of maximum energy 1.71 MeV, which can emerge from a depth of about 5 mm of soil.

More quantitative results have been achieved by strict calibration of film density against isotope concentration, scanning with a microdensitometer, and use of low-energy emitters. For example, ^{33}P, with maximum β energy of 0.25 MeV, has a maximum depth of emergence of 0.33 mm, giving a resolution (defined as the distance at which the grain density is one half that observed directly over the source) of 0.25 mm. Figure 6.8 shows a typical autoradiograph and figure 6.9 shows the solute concentration 'contours' that may be derived by this method.

(b) The Root Plane Technique Farr *et al.* (1969) placed single seedling onion roots side by side to form a 'plane' of roots, which was then sandwiched between two blocks of soil, and the uptake measured over several days. The blocks were then separated, frozen in liquid nitrogen, and sectioned parallel to the root plane in a freeze microtometer technique developed by Brown *et al.* (1964) for measuring diffusion coefficients in soil. The sections were then assayed. Since sections of 20 μm can be cut in a fine-textured soil, concentrations very close to the root surface can be measured.

Kuchenbuch & Jungk (1982) introduced another useful planar technique. They grow roots in a cylinder of soil which is divided in two by a membrane that is permeable to soil solution, but not to roots or root hairs. The roots form a mat against the membrane surface, and the soil can then be sliced on the other side of the membrane for assay.

The disadvantage of these planar methods is that they do not develop the same concentration profiles in the soil as a root cylinder would.

(c) The Wrapper Method Zriek (1987) solved the problem of measuring concentration profiles around roots by making a paper impregnated with soil clay from pulped chromatography paper, wrapping it around a needle of the same diameter as a root, withdrawing the needle, and introducing the tip of a growing root into the hole. After a period of uptake, he unwrapped the paper and cut it into strips for analysis. Since the paper was less than 0.25 mm thick he obtained good resolution (figure 6.10).

With these methods, the influence of a number of factors (represented in the theoretical treatment given earlier) have been measured and compared with pre-

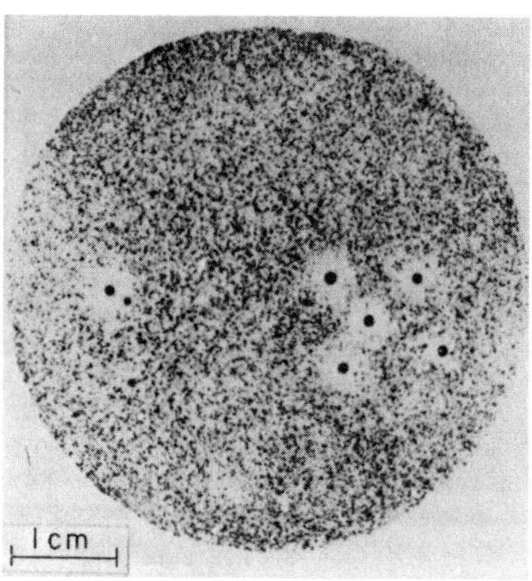

Figure 6.8 Autoradiograph of section across anion roots in a soil labelled with $^{35}SO_4$ (after Sanders 1971).

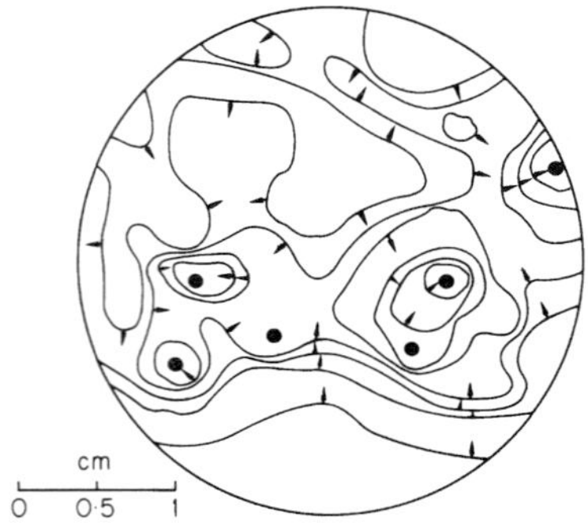

Figure 6.9 Concentration contour diagram of $^{35}SO_4$. Contours are at equal intervals on the densitometer scale (after Sanders 1971).

dictions. It is well to recall that they measure the total concentration of diffusible solute in the soil. The concentration in the soil solution has to be deduced from an appropriate desorption isotherm.

6.2.3 Diffusion of Phosphate and Potassium

Figure 6.11 shows the concentration profiles obtained by Bagshaw *et al.* (1972) near a plane of onion roots in a poorly buffered sandy soil. The concentration of exchangeable phosphorus was reduced by nearly 50% at the root surface. The spread of the depletion zone was greater than that predicted from a simple isotherm. The pH gradients were also measured and pH was found to be reduced by at least 0.5 units near the root surface. In the soil used, the concentration of

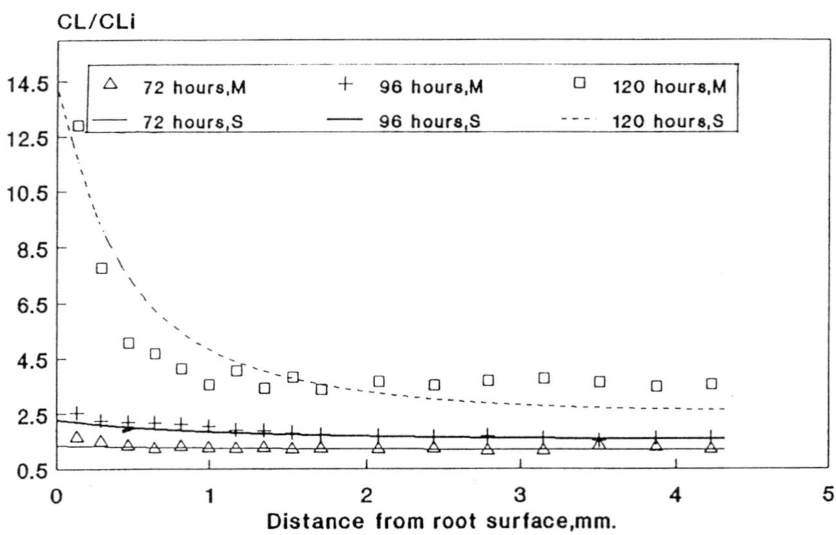

Figure 6.10 Comparison with theoretical predictions of the concentration gradients of sodium ions around the primary root (0.8 mm diameter) of a rapidly transpiring maize seedling measured by Zriek's wrapper method (from Zriek 1987). Points are measured data; lines are theory predictions. M, measured; S, simulated.

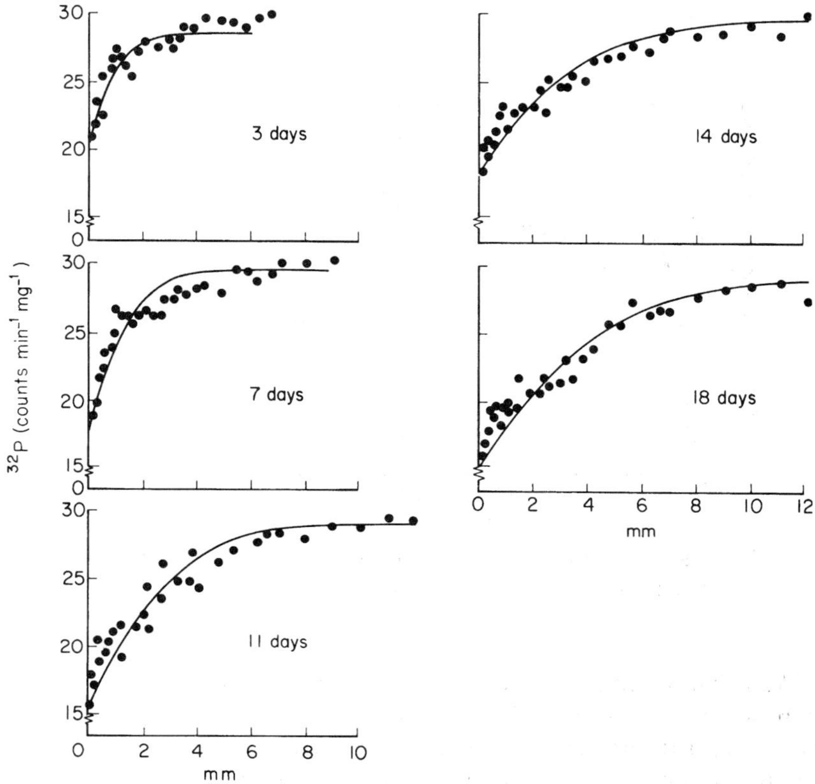

Figure 6.11 Concentration profiles of ^{32}P exchangeable phosphate in a sandy soil in contact with an onion root plane (after Bagshaw *et al.* 1972). The experimental points are shown. The curves are those expected for a concentration-independent diffusion coefficient and a constant flux across the root plane.

phosphate in the soil solution was increased as the pH was lowered, and Bagshaw *et al.* suggested that this was the reason for the greater spread.

The concentrations of potassium in the same soil are shown in figure 6.12 (Farr *et al.* 1969). No comparison with predicted values was made, but the figure illustrates that the spread of the potassium depletion zone is greater than that of phosphorus at comparable times and on a similar soil, as expected.

In predicting concentration profiles for comparison with those observed experimentally, it is necessary to know the diffusion characteristics of the soil, and the value of the root demand coefficient. The spread of the disturbance zone depends mainly on D and t rather than αa, and the comparison largely tests whether the diffusion coefficient measured in bulk soil is changed close to the root. The value of the root demand coefficient can, in principle, be determined independently in culture solution, though the extent to which solution and soil-grown roots behave in the same way, even in the absence of root hairs, has not been critically tested. Exudation is certainly larger in soil-grown roots (Meharg & Killham 1995). It

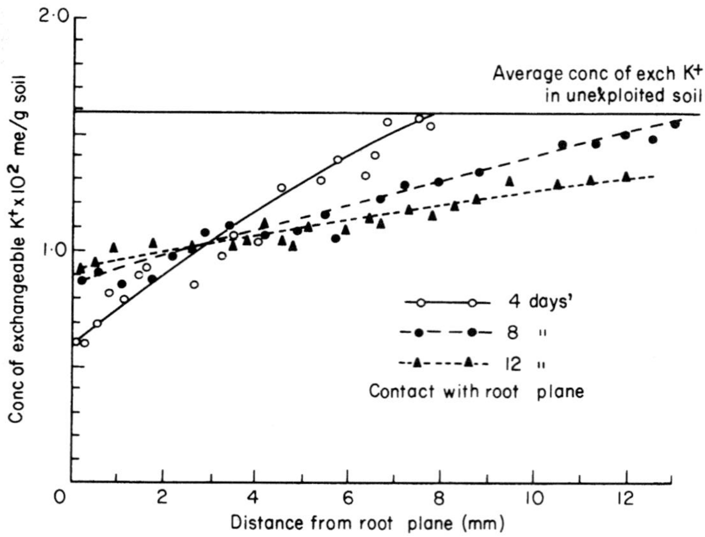

Figure 6.12 Concentration of exchangeable potassium in a block of soil in contact with a plane of onion roots (after Farr *et al.* 1969).

seems likely that the epidermal cells of soil roots will be more leaky than solution-grown roots because of bacterial damage, or handling, and this will affect the value of C_{Lmin}. It is also difficult to reproduce in solution culture the concentrations of all the other ions in solution at the root surface.

When the concentration of phosphorus in the soil solution at the root surface is less than about 10^{-6} M, the root demand coefficient of young roots is high, approaching the 'zero sink' condition, and their uptake rate depends very largely on the rate of diffusion through the soil. Using the autoradiograph scanning method, Bhat & Nye (1974b) found very good agreement between observed and predicted phosphorus uptake along the entire length (13 cm) of a single onion root. The concentration profiles of individual scans also agreed well with the predicted profiles.

6.3 Roots with Root Hairs

Most roots in soil bear a dense cluster of root hairs, numbering 100–1000 per centimetre of root and up to several millimetres long (section 5.1.3). The hairs appear a few millimetres behind the root apex, and in the larger roots the hair zone usually extends for several centimetres towards the base.

Since the hairs are protuberances of 'piliferous' epidermal cells, we may anticipate that they absorb nutrients through their walls like any epidermal cell (section 5.1.3). As a demonstration of this, Barley & Rovira (1970) grew pea radicles through channels in a clay soil into which root hairs penetrated, and in the same soil compacted so that no hairs penetrated it. Uptake of phosphorus was 78% greater from the uncompacted soil.

Attention has often been called to the greatly increased absorbing area provided by root hairs (Dittmer 1940). This must not, however, be taken to imply that absorption by the root axes should increase in the same proportion, since the hairs are normally so crowded that they rapidly interfere with each other's uptake. Hence, the simple theoretical treatment of a root as a cylinder, given in section 6.1, must be modified for roots with hairs.

6.3.1 Theoretical Treatment of Roots with Hairs

Since the hairs usually appear to be clustered within a fairly well-defined cylinder, this has been treated as an effective volume of soil being exploited by roots for water and nutrients (Kramer & Coile 1940; Wiersum 1961) and an effective surface to which nutrients flow and diffuse (Passioura 1963). While this picture is roughly correct, a more exact model (Bhat *et al.* 1976) shows that there is a gradient of solute concentration from the bulk soil to the root axis across the root hair zone (figure 6.13). The boundary of the cylinder is not sharp, partly because the hairs are of variable length, and partly because the distance between adjacent hairs increases with radial distance from the axis. The model is based on the solution of the continuity equation for diffusion around the root axis cylinder (equation (6.1)), modified to include uptake by the root hairs in successive radial zones around the central axis. Competition between the hairs in each radial zone is treated by the method developed by Baldwin *et al.* (1973) for competition between whole roots in a soil volume (section 10.4.3).

Because they provide a soil zone that is potentially intensely exploited, hairs are more important for nutrients that diffuse only slowly from the bulk soil to the axis

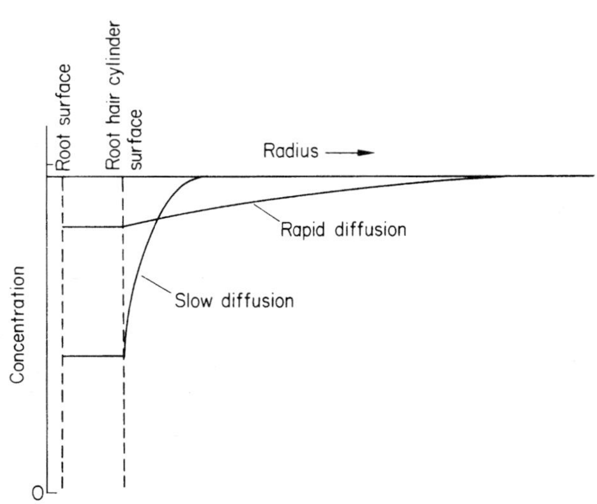

Figure 6.13 Predicted effect of densely clustered root hairs on the concentration profile of a root axis for a nutrient with a low- or high-diffusion coefficient (after Nye 1966b).

than for those that diffuse rapidly. For this reason, most experiments have been carried out with phosphorus.

6.3.2 Experiments on Roots with Root Hairs

Drew & Nye (1969, 1970) measured uptake of potassium and phosphorus by a 1-cm length of ryegrass root. Uptake of potassium was up to 77% and phosphorus 200–300% greater than that predicted from a hair-free axis. However, Bole (1973), using strains of wheat that developed different root hair densities, found a slight but not significant increase in phosphorus uptake per centimetre root length as the root hair density increased to 50 per millimetre length of axis in a soil at both low and high phosphorus level. There was no further increase at greater hair densities. The plants were harvested after 4 weeks, and the data presented are presumably an average for roots of all ages and orders. Competition between individual roots, and a long period of uptake, will tend to reduce the importance of the density of root hairs. Itoh & Barber (1983a, b) obtained a good correlation between uptake of phosphorus and root hair length in six crop species (figure 10.12). Using a modified version of the Bhat & Nye model, they also observed a good correlation with the predicted uptake if absorption by root hairs was taken into account. Jungk *et al.* (1982) found the same for potassium, while Caradus (1982) noticed that P uptake by different genotypes of white clover increased with their root hair length. Fohse *et al.* (1991) calculated that in soils low in phosphorus, root hairs contributed up to 90% of the phosphorus uptake by seven species.

Autoradiography has given more direct evidence. Lewis & Quirk (1967) found there was a pronounced diminution in ^{32}P intensity in a zone around wheat roots that corresponded with the extent of the root hairs. They calculated that this zone should extend only 0.2 cm from the axis if the hairs were inactive. Bhat & Nye (1973, 1974a, b), using ^{33}P and roots of rape seedlings, obtained a typical densitometer scan across the primary root axis, as shown in figure 6.14. It shows how the zone of intense depletion in the hair zone merges into a zone of less intense depletion. At both low and high levels of phosphate, the uptake was greater and the diffusion zone spread further than predicted beyond the root hairs. It appeared that rape had the ability to increase the concentration of phosphate in the soil solution, a conclusion supported by experiments with complete rape root systems in the same soil (section 7.2.1). There was a marked difference between the primary root of rape and a single hair-free seedling root of onion in absorbing phosphate, particularly in the soil of low-phosphate status, from which the rape root absorbed 10.5×10^{-7} mol P in 5 days, while the onion root absorbed only 0.25×10^{-7} mol P in 12 days, when it had attained a similar length of 14 cm. Misra *et al.* (1988) compared cotton and rape. They found a depletion similar to that of onion around cotton roots, and around rape the same intense depletion in the hair zone as earlier workers had observed. Kraus *et al.* (1987b) also found that the depletion zone around maize roots did not extend beyond the root hair zone. We may conclude that in soils low in phosphorus, there is usually a marked depletion within the root hair zone, but that autoradiographs may or may not show depletion outside this zone for reasons that will be discussed in chapters 7 and 8.

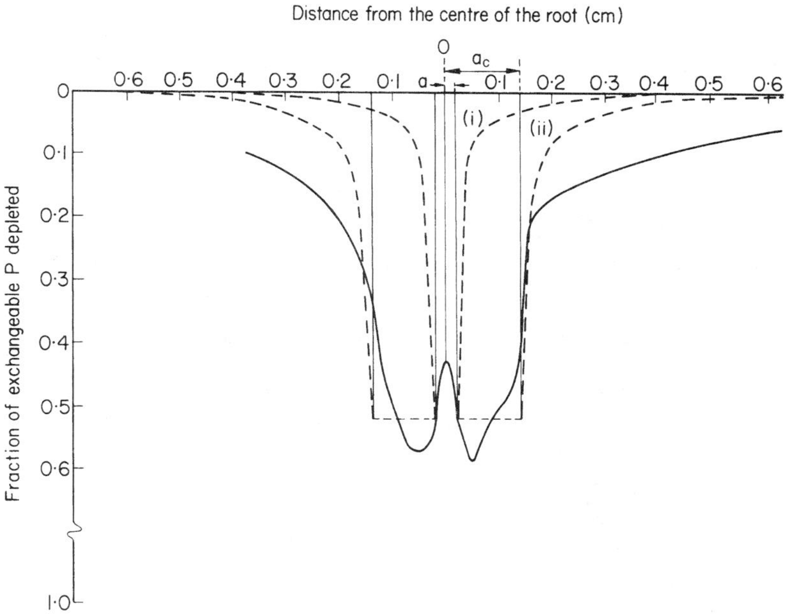

Figure 6.14 Measured phosphate concentration profiles around a rape root with dense root hairs (after Bhat & Nye 1973); *a*, radius of root axis; a_c, radius of root hair cylinder; – – – – (i), calculated assuming the root hairs are inactive; – – – – (ii), calculated assuming intense root hair activity and uniform depletion from the root hair cylinder; ———, observed in the experiment.

Kraus *et al.* (1987a, b), Hubel & Beck (1995) and others, by scanning soil sections cut horizontally across the primary root of maize, have detected a narrow zone of phosphorus accumulation very close to the root axis. We suggest this may be caused by accumulation of P in the dense rhizosphere population that forms close to the root surface (chapter 8).

We conclude that when root hairs are taken into account, reasonably good predictions of phosphorus uptake from phosphorus-deficient soils can often be made from models of phosphorus uptake, using independently measured plant and soil parameters. In some instances, the spread of the depletion zone and the uptake is considerably greater than that predicted, because the root is probably solubilizing phosphate by changing the soil pH or excreting anions such as citrate, as discussed in chapter 7.

6.4 Simultaneous Diffusion and Convection

Since roots normally absorb water as well as solutes, these solutes move in the neighbourhood of an absorbing root surface both by diffusion and by mass flow; and we have to consider how these two processes interact.

As we have seen, if diffusion acts alone, a zone of depletion develops. If at the same time solute is drawn into this zone of depletion, its extent will be reduced. If the solute is drawn towards the root faster than it is absorbed, a zone of accumulation will develop (figure 6.3).

6.4.1 Theoretical Treatment

The continuity equation applicable to a cylinder of soil with a root at its centre, is, from equation (1.9),

$$\partial C/\partial t = 1/r\, \partial[rD^* \,\partial C/\partial r + rvC_L]/\partial r \tag{6.11}$$

where v is the water flux *towards* the root. This equation cannot readily be solved analytically except for special cases, such as when $Db = rv$ (Geering 1967). Cushman (1982) has reviewed these cases. Numerical solutions for different conditions at the root surface have been given by Passioura & Frere (1967) for $\alpha a = 0$ and by Nye & Marriott (1969) for the Michaelis–Menten condition. Nye and Marriott assumed, in the first instance, that the constants of the Michaelis–Menten equation (5.4) remained steady as the root aged and that the diffusion coefficient was independent of the solute concentration and the water flux. The numerical solution can readily accommodate variations in the parameters (Cushman 1984), provided they are known — which involves a lot of experimental work. Amijee *et al.* (1991) have listed the history of the various phosphorus uptake models that have been proposed (table 10.1). These take account of complications such as age-dependent root demand coefficient, root hairs and concentration-dependent diffusion coefficients.

The effect, in theory, of an increasing water flux on concentration in the disturbance zone is shown in figure 6.15.

6.4.1.1 The Effect of Mass Flow on Solute Uptake

The processes of mass flow and diffusion occur together and the resulting concentration profile is not the result of two processes acting independently. Hence, it is not possible to divide the total uptake into separate contributions by mass flow and diffusion. It is, however, nearly correct to state that a given amount of solute has been drawn into the zone of solute disturbance by mass flow, provided nearly all the water absorbed by the root has been drawn from outside this zone, which will generally be true in a moist soil. Nye & Marriott (1969) considered by how much transpiration might increase the solute inflow above that which would occur by diffusion in any case. They showed that the formula

$$F = \alpha C_{Li}[1 + v_a a/(bD\gamma)]/[1 + \alpha a/(bD\gamma)] \tag{6.12}$$

was correct within 10% when $v_a < \alpha$ (γ is a term tabulated by Jaeger & Clark 1942). Provided Dt/a^2 is greater than 1 and less than 1000, $\gamma = 0.5$ within a factor of 2. The important point is that the solute inflow increases linearly with the water inflow, which is proportional to $v_a a$; and, further, that transpiration should increase the inflow of any solute by $100 v_a a/bD\gamma\%$ independently of any change in the root demand coefficient.

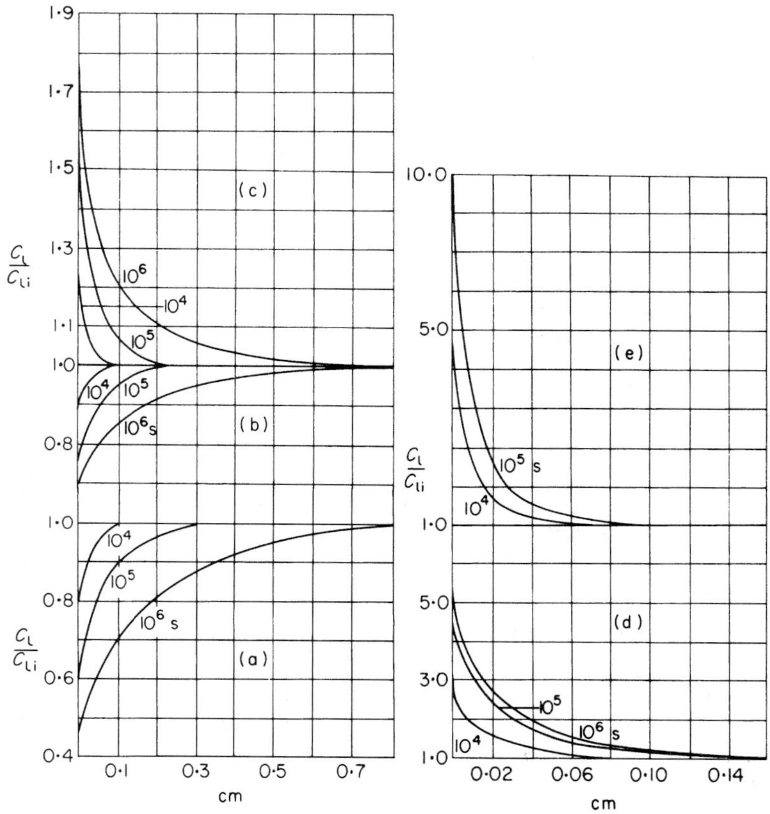

Figure 6.15 The effect of the rate of soil solution flow on the relative concentration near a root surface ($\alpha = 2 \times 10^{-7}\,\mathrm{cm\,s^{-1}}$; $D = 10^{-7}\,\mathrm{cm^2\,s^{-1}}$; $b = 0.2$; $a = 0.05\,\mathrm{cm}$; F_{max} high).
(a) $v_a = 0\,\mathrm{cm\,s^{-1}}$; diffusion alone. C_L/C_{Li} at the root surface continues to drop and the zone of depletion continues to spread outwards at 10^6 s.
(b) $v_a = 10^{-7}\,\mathrm{cm\,s^{-1}}$; C_{La}/C_{Li} has nearly reached the limiting value $v_a/\alpha = 0.5$ after 10^6 s.
(c) $v_a = 4 \times 10^{-7}\,\mathrm{cm\,s^{-1}}$; C_{La}/C_{Li} has nearly reached the limiting value $v_a/\alpha = 2$ after 10^6 s.
(d) $v_a = 10^{-6}\,\mathrm{cm\,s^{-1}}$; $\alpha v_a/Db = 2.5$. After 10^6 s, C_{La}/C_{Li} has almost reached the steady-state value $v_a/\alpha = 5$, and the zone of accumulation has ceased to spread outwards.
(e) $v_a = 2 \times 10^{-6}\,\mathrm{cm\,s^{-1}}$; $\alpha v_a/Db = 5$. Both C_{La}/C_{Li} and the zone of accumulation have nearly attained the steady state after only 10^5 s. C_{La}/C_{Li} has increased to $v_a/\alpha = 10$ but the zone of accumulation is compressed (after Nye & Marriott 1969).

Since, by equation (4.17), $bD \approx D_L f_L \theta$ and $D_L \approx 10^{-5}\,\mathrm{cm^2\,s^{-1}}$, for simple olutes the increase for a given value of $v_a a$ will depend greatly on the moisture ontent, θ. For example, at a high transpiration rate ($v_a a = 10^{-7}\,\mathrm{cm^2\,s^{-1}}$), ne increase in inflow is calculated as only 12% in a very moist soil $\theta = 0.4, f_L = 0.52$); but in dry soil ($\theta = 0.16, f_L = 0.09$), the increase is 120%

provided the same transpiration rate is maintained, which in practice is unlikely (see chapter 2). These conclusions assume that αa is independent of water flow rate. Accordingly, they do not apply to solutes passively swept into the root in the transpiration stream, for example many herbicides, in which case there is no depletion around the root, but there could be accumulation if the water enters faster than the solute.

Multi-ion uptake We noted in section 4.2.1.1 that ions with different mobilities did not, as a rule, move independently of each other because of the need to maintain overall electrical neutrality in the soil solution, resulting in a diffusion potential being set up. Further, roots do not take up nutrient cations and anions in equal amounts of charge. Electrical neutrality is maintained across the root–soil interface by a compensating export of H^+ or HCO_3^- (section 7.2.1). Consequently, the diffusion–convection equation (6.11), developed for a neutral solute, has to be modified for ions to take account of the mobilities and charge on all the ions in the system and the interaction between them. Bouldin (1989) has provided a computer program for this. Essentially, it ensures that the ions move in such a way that the root, the soil solution and the soil solid remain electrically neutral at all points. Fortunately, when an ion species of interest is present in fairly low proportion it can be considered to move independently of the other ions, so that calculations of the diffusion and convection of phosphate, potassium and micronutrients are little affected by these refinements. However, the effect on the major ions can be important, as the following simple example will show. Accumulation of cations, especially of calcium, which is the cation usually dominant in the soil solution near the root, is not necessarily caused by mass flow. As we have seen, roots that take up more anions than cations excrete bicarbonate ions. These will react with the soil adjoining the root to make it less acid. The degree of cation saturation of the soil exchange complex necessarily increases, and more calcium initially accompanied by anions moves by salt diffusion to this part of the soil, without any mass flow.

6.4.2 Experimental Evidence for Predictions from the Theory of Diffusion with Mass Flow

6.4.2.1 Qualitative Experiments

We have seen that the condition for an increase in concentration at the root surface due to convection is that $v_a a > \alpha a$ (figure 6.3b). Even if this is so, the increase in moist soil should be very small, because then $v_a a/bD\gamma$ in equation (6.12) is small. Conditions for a large increase are therefore high transpiration, low moisture, and low root demand coefficient. Such conditions are most likely to be found in saline soils under hot dry climates.

Direct measurements of accumulations have been given by Riley & Barber (1970), who found much greater salt concentrations in the 'rhizosphere' and 'rhizocylinder' volumes around soyabean roots than in bulk soil. Transpiration-induced salt accumulation in the rhizosphere of wheat and maize has been detected by Sinha & Singh (1974, 1976a, b) and around onion roots by Schlief

(1980). In all these experiments, 'rhizosphere' soil is the soil that adheres to the roots when they are shaken. Apart from being geometrically ill-defined, the analysis of this soil may be anomalous because nutrients may leak into it from roots damaged when they are separated from the bulk soil or when it is washed off the roots.

6.4.2.2 *Quantitative Experiments*

It is easier to devise an experimental system for measuring concentration profiles near the root surface when the geometry is planar rather than cylindrical. Such a system has the additional advantage that there is a convenient analytical solution of the continuity equation in one dimension (equation (1.6)) for the appropriate boundary conditions (Nye 1966c). At large times, the distribution of solute near the root tends to a steady state, expressed by the simple relation

$$C_L/C_{Li} = 1 + (v/\alpha - 1)\,\mathrm{e}^{-vx/(bD)} \tag{6.13}$$

The final concentration at the root surface ($x = 0$) thus depends only on v/α, and the distribution with distance from the root depends only on vx/bD.

To test the predictions of this equation, Wray & Tinker (1969) passed dry air over a single onion root that was growing through a rectangular block of very moist soil only 1 mm thick, so that movement of solutes to the root was effectively linear, with the results shown in figure 6.16. The approximately exponential form of the concentration–distance curve after 18 h is notable. When the transpiration rate was again severely restricted, the accumulation rapidly dispersed.

The time course of the accumulation at the surface followed approximately the rate predicted by theory, as shown in figure 6.17.

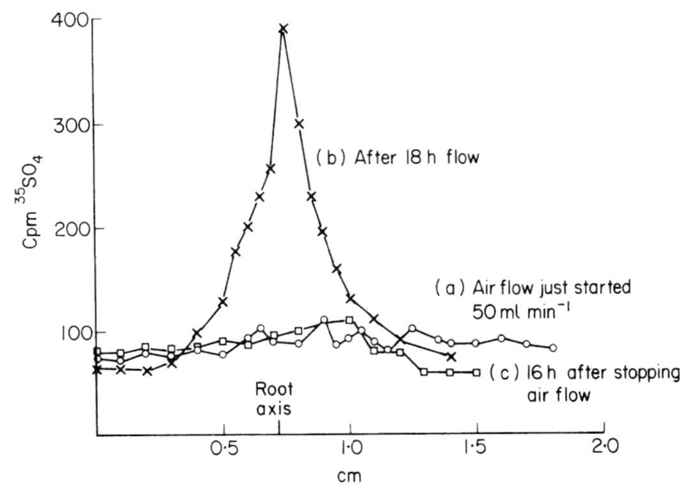

Figure 6.16 Concentration profile of $^{35}SO_4$ in moist soil near an onion root: (a) air flow over shoot just started; (b) after 18 h air flow; (c) 16 h after stopping air flow (after Wray 1971).

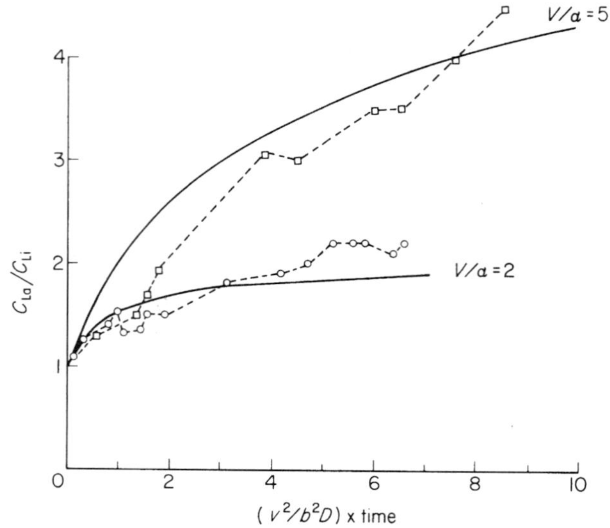

Figure 6.17 Change in root surface concentration ratio C_{La}/C_{Li} with time at two transpiration rates. Solid lines are the theoretical curves, points are measured data (after Wray 1971).

In the more realistic cylindrical geometry, Zriek (1987) used his wrapper method (6.2.2.3) to follow the concentration profiles of sodium and chloride around the primary root of maize seedlings. Figure 6.10 shows good agreement between his experimental results and the theoretical predictions of equation (6.11) using independently measured parameters (including the root demand coefficient, measured in solution culture).

6.5 The Effect of Soil Moisture Level on Solute Absorption by Single Roots

The soil moisture level has a marked effect on the absorption of solutes by whole plants. The detailed interpretation of these effects is extremely complicated because it involves both the transport of soil solutes, and complex plant physiological responses. Here, we consider the effect of reducing a favourable moisture level on solute absorption by a single root. Sections 6.5.1 and 6.5.2 describe some of the more obvious effects. We deal with the effect on a whole plant in chapter 9.

6.5.1 Plant Effects

(a) The root demand coefficient may be reduced by decreased water potential within the plant, affecting many features of plant growth. Transpiration rate will decline because of effects on stomata (section 2.3.3).
(b) Contact between root and soil may be reduced by shrinkage of the root away from the soil.

6.5.2 Effects within the Soil

(a) The diffusion coefficient will decrease because θ and f_L are reduced.
(b) Convection to a root may be reduced by decrease in the rate of transpiration.
(c) In drier soil θ decreases sharply near the surface, so the diffusion coefficient near a root may decrease sharply, if the plant still transpires.
(d) The solution concentration of non-adsorbed solutes, such as nitrate and chloride, will increase.
(e) The concentration in solution of exchangeable cations will increase as the anion concentration rises. If calcium and magnesium are the dominant cations, their concentration will increase approximately directly with the total anion concentration. The monovalent cations, such as potassium, will increase so that the reduced activity ratio is maintained; that is,

$$(\mathrm{K})^2_{\mathrm{dry}}/(\mathrm{K})^2_{\mathrm{wet}} = (\mathrm{Ca})_{\mathrm{dry}}/(\mathrm{Ca})_{\mathrm{wet}}$$

(f) The concentration of adsorbed anions, such as phosphate, will tend to decrease since the activity product $(\mathrm{Ca})(\mathrm{H_2PO_4})^2$ tends to be constant and (Ca) increases according to (e).

Early work with rubidium and chloride, reviewed in Nye & Tinker (1977), showed that decreased moisture reduced uptake in spite of increased concentration of ions in the bulk soil solution. It indicated that the decrease in the diffusion coefficient was responsible.

Dunham & Nye (1974) attempted to identify all the factors involved using the root plane technique. They measured the uptake of $^{36}\mathrm{Cl}$ by a plane of onion roots from a block of wet, moist or dry soil, the remainder of the roots being in unlabelled soil of similar moisture content. In the same experiments they measured the transpiration rate, the chloride concentrations near the root surfaces, and the soil moisture profiles. Since the chloride concentration at the root surfaces was measured they were able to measure the effect of water potential on the root demand coefficient for chloride. This was also measured in solution culture at various chloride concentrations.

The profiles of soil moisture have been shown in figure 2.10. Figure 6.18 shows the profiles of chloride concentration at the different soil moistures. The solid lines are those predicted from a numerical solution of equation (1.6) with values of α determined from the ratio of the chloride flux to the measured solution concentration at the root surface.

The main findings are shown in table 6.2. In the wet soil, the value of α agreed fairly well with that determined in solution culture. In the moist soil, the value of α was somewhat lower than that of the solution culture and in the dry soil it was much lower than was measured in solution culture. As expected, there was a marked drop in v in the moist soil from that of the wet soil, and a further drop in the dry soil, in which the water potential at the root surface fell to −2.5 MPa. As predicted from equation (6.6), the root surface concentration ratio, C_L/C_{Li}, at the end of the experiment approached v/α; there was a marked drop in the dry and moist soils; but in the wet soil C_L/C_{Li}, although reduced to less than 1 at 2.5 days, rose to 1.4 at 6 days because v became much greater than α.

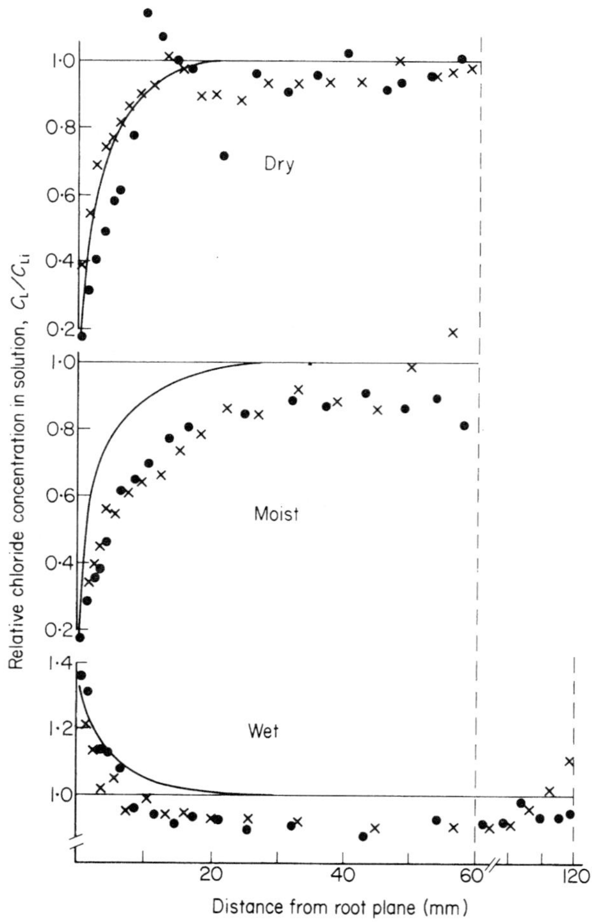

Figure 6.18 Relative chloride concentration after 6 days uptake by an onion root plane from a dry, a moist and a wet soil. Curves are derived from equation (1.6) with α selected to give the best fit at the root surface (after Dunham & Nye 1974). × •, symbols indicate measurements on opposite soil blocks.

In the dry soil, lower v, lower α, and lower D contributed to the lowered uptake, but since they are not additive, it is meaningless to say how much each contributed. The importance of the experiment is rather that using values of α determined from the chloride flux and the surface concentration (or in the wet soil, from solution culture), but otherwise independently determined diffusion parameters, it was possible to predict satisfactorily the concentration profiles of a non-adsorbed ion from a diffusion–convection model. It also shows that the root demand coefficient is sensitive to the water potential at the root surface.

Table 6.2 Effect of soil water potential at the root surface on uptake, root absorbing power α and surface concentration ratio of a plane of onion roots.

Initial soil moisture (θ)	Duration (days)	Average water flux ($cm^3\ cm^{-2}\ s^{-1} \times 10^{-6}$)	C_L/C_{Li} at root surface	α ($cm\ s^{-1} \times 10^{-6}$)		Final water potential at root surface (bar)	Uptake (μmol)	
				Soil experiment	Stirred[a] solution		Measured	Predicted
Chloride								
0.20	6.0	0.25	0.20	1.0	7.0	−25	0.84	0.51
0.27	6.0	0.63	0.20	2.0	7.0	−6.5	2.7	1.7
0.45	6.0	2.09	1.40	1.2	1.0	> -0.05	5.7	4.3
Phosphate								
0.14	4.5	0.12	1.0	0.4		−3.3	0.004	0.003
0.18	4.5	0.32	0.06	10	> 100	−0.15	0.015	0.020
0.36	4.5	0.70	0.06	20	> 100	−0.10	0.043	0.050
Potassium								
0.14	4.5	0.12	0.004	(100)		−3.3	5.7	1.3
0.18	4.5	0.32	0.077	10		−0.15	3.7	1.4
0.36	4.5	0.70	0.036	50		−0.10	7.3	4.5

Source: after Dunham & Nye (1974, 1976).

[a]The root absorbing power for uptake from a free solution that has the same concentration as the final concentration at the root plane surface.

The geometry of this system is linear, and the results do not represent, in detail, the effects to be expected around an isolated root — for example, the moisture gradients near an isolated root would not be so steep. However, now that the plane model has been verified experimentally, one can have confidence in predictions about behaviour around an isolated root, based on a cylindrical model that uses the same ideas and parameters.

In a similar single experiment, Dunham & Nye (1976) measured uptake of the adsorbed ions, phosphate, potassium, calcium and magnesium from blocks of wet, moist and dry soils. The rest of the roots were not in soil, but in contact with moist filter paper. In interpreting the experimental data, we recall that in each case the exchangeable ion was measured. The concentration in pore solution, which was required for the estimate of α (column 5 of table 6.2), was inferred from a desorption isotherm determined in a separate experiment. The main effects are described below.

Phosphate In the wet and moist soils, the root surface concentration ratio C_L/C_{Li} was low, showing there was considerable diffusive resistance. In the dry soil at a water potential of −330 kPa, the value of α was markedly reduced. This was the dominant effect in limiting uptake; the supply by diffusion, even with very high resistance, and mass flow were well able to satisfy the greatly reduced plant demand.

Potassium There was considerable diffusive resistance at all moisture levels and the value of α remained high in the dry soil. The roots had been cultured in calcium nitrate and hence were low in phosphate and potassium. It is clear that the uptake of phosphate and potassium are differently affected by moisture stress in the root. The high uptake of potassium from the dry soil proves that the low uptake of phosphate was not due to lack of contact between the roots and the soil.

Calcium and Magnesium Under all conditions, the relative solution concentration of these ions was between 1.0 and 1.5 — an indication that transport through the soil did not limit uptake.

6.5.3 The Prediction of Ion Uptake

The measured uptakes of chloride, phosphate, and potassium in these two experiments covered a range — in moles — of about 10 000-fold. The theoretical uptakes were calculated by numerical solutions of equation (1.6) that used independently measured parameters, with the exception that the values of α were not measured in roots under moisture stress, and were derived from the ion fluxes and the concentrations at the root surfaces. When the relative concentration at the root surface, C_L/C_{Li}, is below 0.2, large changes in α make little difference to the spread of the depletion zones or the total uptake, so that it is not a serious 'adjustment factor' in these experiments. The theoretical uptakes underestimated the measured uptakes of phosphate by a factor of 1.5, and systematically underestimated chloride uptakes by factors up to 2.0 and potassium by factors up to 4.5. These discrepancies could be accounted for by experimental difficulties, notably in the measurement of desorption isotherms and impedance factors, and by the possibility that the diffusion coefficient should be increased to allow for dispersion. Considering the very wide range in uptakes, the agreement between measured and theoretical values indicates that for simplified soil–root systems the theory presented here is substantially correct. We have now to consider some of the complications that arise in more natural root–soil situations.

7

Chemical and Physical Modification of the Rhizosphere

The term 'rhizosphere' tends to mean different things to different people. In discussing how a root affects the soil, it is well to bear in mind the spread of the zone being exploited for a particular solute: if this is wide, there may be no point in emphasizing effects close to the root; but if it is narrow, predictions based on the behaviour of the bulk soil may be wide of the mark. In a moist loam after 10 days, a simple non-adsorbed solute moves about 1 cm, but a strongly adsorbed one will move about 1 mm. In a dry soil, the spread may be an order of magnitude less. The modifications to the soil in the rhizosphere may be physical, chemical or microbiological. In this chapter, we discuss essentially non-living modifications, and in chapter 8 the modifications that involve living organisms and their effects.

7.1 Physical Effects

7.1.1 The Position of Roots in the Pore Structure

Roots tend to follow pores and channels that are not much less, and are often larger, in diameter than their own. If the channels are larger, the roots are not randomly arranged in the void (Kooistra *et al.* 1992), but tend to be held against a soil surface by surface tension, and to follow the channel geotropically on the down-side. If the channels are smaller, good contact is assured, but the roots do not grow freely unless some soil is displaced as the root advances. For example, in winter wheat, Low (1972) cites minimum pore sizes of 390–450 μm for primary seminal roots, 320–370 μm for primary laterals, 300–350 μm for secondary laterals, and 8–12 μm for root hairs, though some figures seem large (see figure 9.3). Whiteley & Dexter (1984) and Dexter (1986a, b, c) have studied the mechanics of root penetration in detail (section 9.3.5). It may compact and reorient the soil at the root surface. Greacen *et al.* (1968) found that wheat roots penetrating a uniform fine sand increased the density only from 1.4 to 1.5 close to the root; and a pea radicle, a comparatively large root, raised the density of a loam from 1.5 to 1.55. A pea radicle penetrating clay decreased the voids ratio progressively from 1.19 at the surface to 1.10 within 1 mm of the surface (Cockcroft *et al.* 1969). These changes in density should have little effect on the soil's diffusion characteristics (section 4.2.1.3). On the other hand, Guidi *et al.* (1985) found the porosity of aggregates adhering to maize roots was 0.141 $cm^3\ g^{-1}$ compared with 0.171 $cm^3\ g^{-1}$ in aggregates in the bulk soil (section 9.3.5).

A layer of clay orientated parallel to the root axis might well present a barrier to diffusion. To take an extreme case, diffusion of sodium ion across the well-oriented flakes of montmorillonite was 300 times less than diffusion in the plane of the flakes (Mott 1967). However, in rhizotrons, and in thin sections across roots in soil, orientation is rarely noted. Where oriented clay near roots has been observed, it is likely that it has been leached down channels of older roots after they have completed their life as absorbing organs.

In an aggregated soil, diffusion may well differ within the aggregates from that in the bulk soil. For small mobile solutes, such as chloride ion, Rowell *et al.* (1967) and Pinner & Nye (1982) could find no evidence for more rapid movement through favoured pathways, or for the existence of dead-end pores that do not form part of any pathway. Further, Green (1976) measured a chloride impedance factor, $f_L = 0.41$, similar to a value for a bulk soil, in saturated natural aggregates of porosity 0.38. However, for less mobile solutes, diffusion through aggregates may be slow. Gunary (1963) perfused natural and artificial aggregates, about 4 mm in diameter, with labelled phosphate solution. Autoradiographs showed the degree of penetration of ^{32}P increased with the concentration of the perfusing solution. When this was less than 0.1 ppm P, only a thin surface layer had been labelled after 40 days. This was probably caused by reduction of the diffusion coefficient associated with the increase in buffer power, dC/dC_L, as the solution concentration is lowered (section 4.2.1.5). Nye & Staunton (1994) explained the details of the diffusion of ^{32}P down a soil column in terms of a diminishing impedance factor, as the ^{32}P exchanged with ^{31}P within finer and finer intra-aggregate pores. Though diffusion of less mobile solutes to a root surface adjacent to inter-aggregate regions may differ from one near intra-aggregate regions, use of a single diffusion coefficient presupposes averaging over such microscale variations; and, in practice, longitudinal diffusion within the cortex will probably short-circuit irregularities to some degree. The fact that root hairs penetrate uncompacted crumbs contributes to this averaging process. Also, high resolution autoradiographs have not revealed significant irregularities in the phosphate depletion zones along roots. We conclude that, from a physical point of view, diffusion through the root–soil contact region will be nearly the same as through the bulk soil.

In drier soils, only the intra-aggregate pore space contains appreciable water, and a diffusing solute has to pass through these aggregates, so increasing their significance; but little critical work has been done under these conditions.

7.1.1.1 Effect of Root–Soil Contact

Purely visual assessment in moist, reasonably homogeneous soil suggests that most of the root surface is in contact with solid particles or separated from them by no more than the normal range of soil pore diameters; that is, the interface region is similar to an average cross-section through the soil. It may not be identical to it because the roots may avoid dense aggregates, and thus the proportion of their perimeter in contact with air may exceed the average soil air-filled porosity.

In pot trials with maize, Kooistra *et al.* (1992) found the average degree of root–soil contact in thin cross-sections to be 87%, 72%, and 60%, respectively, in an aggregated soil packed to give porosities of 44%, 51%, and 60%, after destroying the channels that were naturally present. Table 7.1 shows that very few roots had no contact with the soil. The majority of roots had more than 50% contact in all diameter classes and soil densities. Root–soil contact did not vary significantly with root diameter. In loose soil, roots had more contact with soil surfaces than would be expected if they were randomly located in macropores.

In the field, roots that follow existing macropores and cracks may have less contact with the soil. Van Noordwijk *et al.* (1993) obtained data on root–soil contact in thin cross-sections taken from two fields of winter wheat with different management histories. At 45 cm depth, in the subsoil, a rather high proportion of the root perimeters had no contact. At 15 cm depth, in the plough layer, the proportion with good contact was much greater, so the soil with least contact had a higher macroporosity. There was no relation between contact and root diameter.

Rogers (1939) observed apple tree roots growing against glass. He found that about 20–30% of the main roots followed worm holes. Contact was about 40% in

Table 7.1 Microporosity, root channels, and total macroporosity ($\geqslant 30\ \mu m$) as a percentage of the surface area of the thin sections. The number of roots observed (N_{roots}) per thin section and average percentage root–soil contact per sample are also given.

Treatment[a]	Microporosity (%)	Root channels (%)	Other macroporosity (%)	Total porosity (%)	N_{roots}	Average root–soil contact (%)	
C-v	39.1	1.4	3.2	43.7	109	90	87
C-h					94	83	
I-v	45.2	0.7	4.7	50.6	166	73	72
I-h					59	68	
L-v	43.8	1.2	14.6	59.6	266	58	60
L-h					59	68	

Source: from Kooistra *et al.* (1992).

[a]Treatment code: C = compacted, I = intermediate, L = loose soil, v = vertical and h = horizontal section.

these cases. Such roots can later fill their own holes by secondary thickening, or may stay loose for years. Older roots may be loosened by sloughing off their cortex. In the remainder, contact was good. Smaller roots of about 1.5 mm diameter tended to compress the soil 1.5 mm on either side of them, though the compression was less than 5%.

Lack of contact is reduced by the following effects: (a) Root hairs tend to proliferate in the humid air of large soil pores and channels. (b) Even though pores of diameter exceeding 10 μm are drained at -30 kPa water potential, there is always a film of water on the root and root hair surfaces, linking them to the soil moisture continuum. (c) Roots tend to excrete a layer of mucilage up to 5 μm thick — but, on average, more like 1 μm thick — which may bridge small gaps (Jenny & Grossenbacher 1963; Greaves & Darbyshire 1972). The mucilage consists of polysaccharide and polyuronic acid gels (Ray *et al.* 1988) across which ions and uncharged solutes should be able to pass freely. On root caps of most agricultural crops in solution, Samtsevich (1968) described a colourless gel of diameter 1.8–4.7 mm and length 2.2–14 times greater than the caps that produced them. Mucilage will almost certainly shrink at low water potentials at the root surface since the osmotic pressure exerted by the polysaccharides that oppose this will be low (section 2.3.4).

Roots shrink as their water potential falls (figure 7.1) and they thus lose all-round contact with the soil, as discussed in section 2.3.4. The extent to which lack of contact should reduce the rate of uptake of a solute, independently of any effect that low water potential may have on the root demand coefficient, has been examined theoretically by Sanders (1971) using an electrical resistance–capacitance network analogue (section 10.4.3). Figure 7.2 shows an example of his findings. The effect of poor contact is increased if the root cortex does not transmit the solute freely (cortical block). If, as is most likely, the cortex acts as a short-circuit, reduction of contact by 50% leads to a reduction in rate of uptake of water by a droughted root by about 25%. De Willigen & Van Noordwijk (1987) and Veen *et al.* (1992) have given an approximate analytical solution of this problem. We have seen in section 2.3.4 that poor contact may decrease transpiration, and, consequently, mass flow of nutrients. This should matter little if, as is usual, diffusion can satisfy the plant's needs.

De Willigen & Van Noordwijk (1987, 1989) calculated that insufficient oxygen may diffuse through the soil to roots with good contact with very moist soil. The roots may then depend on internal aeration through cortical air spaces (section 9.3.6).

7.2 Chemical Effects

The root may alter simple predictions of the diffusion of solutes by release of hydrogen or bicarbonate ions, by evolution of carbon dioxide from respiration, by changing the concentration of other ions and solutes that may affect the ion of interest, and by excretion of organic substances that may form complexes with it.

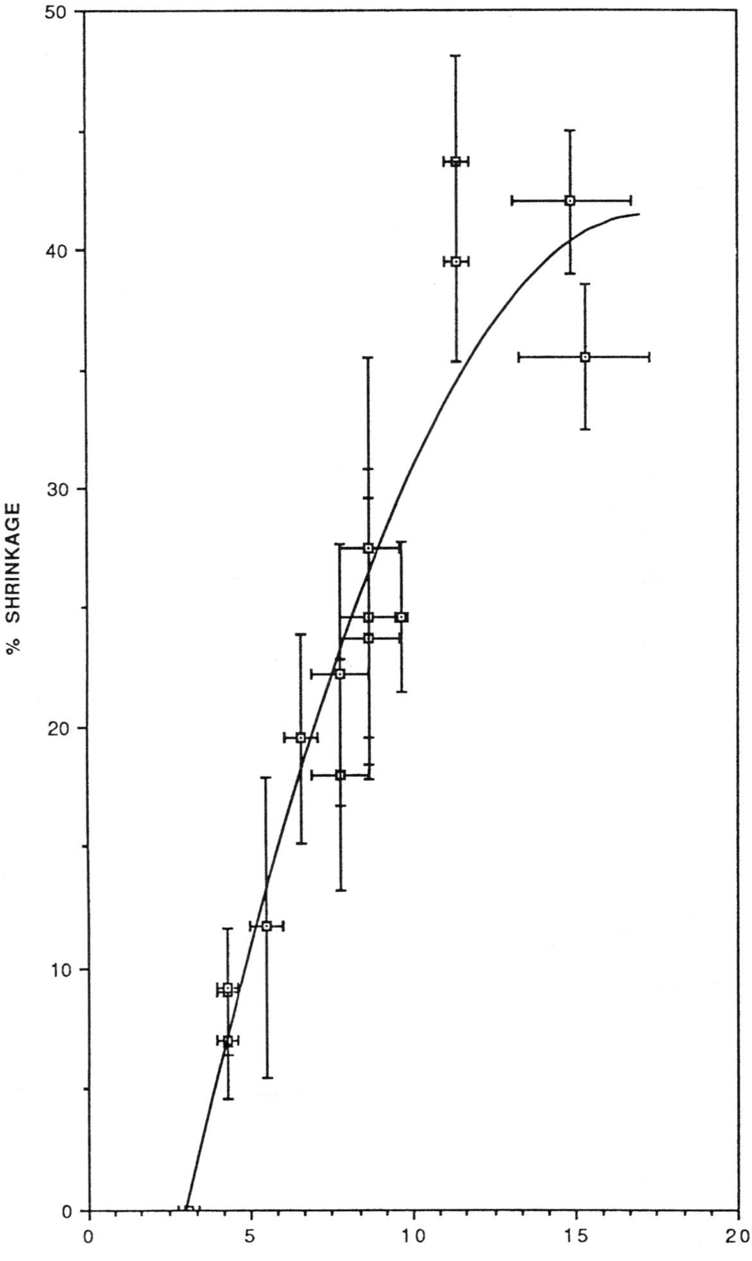

Figure 7.1 Relationship between root cell water potential (bars) and % shrinkage for series 2 and 3 (from Al-Najafi 1990).

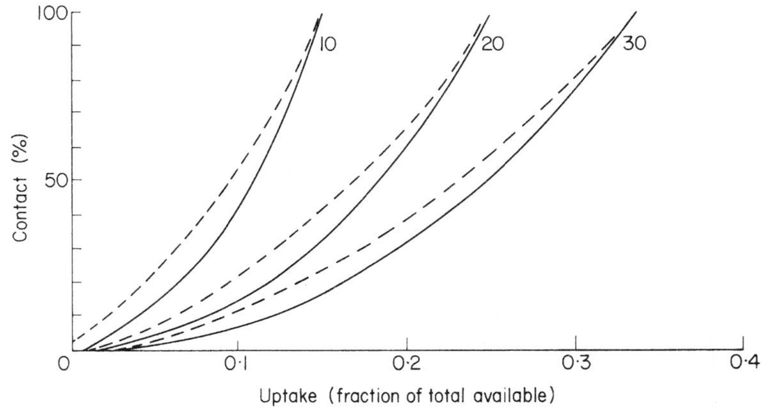

Figure 7.2 The theoretical effect of % contact with soil around the root circumference on uptake of potassium after 10, 20, and 30 days (after Sanders 1971). ——, Cortical short; – – –, cortical block.

7.2.1 Excretion of Hydrogen and Bicarbonate Ions

Cunningham (1964) collected analyses of 62 common temperate crops and weeds grown in soil and found that these contained a median average of 2.5 meq of absorbed cations ($K^+ + Na^+ + Mg^{2+} + Ca^{2+}$) and 3.6 meq of absorbed anions (as $NO_3^- + SO_4^{2-} + H_2PO_4^- + Cl^-$) per gram of dry tops, assuming all N to be absorbed as nitrate. To maintain electrical neutrality across the root–soil interface, the roots of these plants must release, on average, 1.1 meq of HCO_3^- ($OH^- + CO_2 = HCO_3^-$); thus, they must make the rhizosphere more alkaline (Nye 1968b). This deduction supposes that for plants exposed to rain, equal amounts of anions and cations are leached from the tops. Plant physiologists have long known (Sachs 1875) that nutrient culture solutions that supply nitrogen as nitrate become more alkaline as they are depleted. A few species, such as buckwheat (*Fagopyrum esculentum*), absorb more cations than anions and must excrete H^+ to maintain electrical neutrality.

Riley & Barber (1969) reported that the ‘rhizoplane’ soil (*c.* 0–2 mm from the root surface) of 3-week-old soybean had a pH of 6.9 compared with 6.2 in the pot soil. There was a corresponding increase in the bicarbonate ion concentration in the soil solution. The pH profiles near onion root planes were measured by Bagshaw *et al.* (1972). The pH fell from 6.0 in the bulk soil to 5.6 at the surface because the roots absorbed potassium rapidly. When the soil was treated with $Ca(NO_3)_2$ to reduce potassium and increase nitrate uptake, the pH rose from 6.4 in the bulk soil to 6.8 at the root surface. In general, potassium is absorbed more rapidly than calcium, and nitrate more rapidly than sulphate.

Nitrogen, if assessed as nitrate, normally constitutes more than half the total absorbed anions. When neutral soil is supplied with ammonium rather than nitrate ion, if nitrification is slow or is prevented with an inhibitor, then the root absorbs nitrogen as the NH_4^+ cation and the pH close to the root surface has been shown to fall by up to 2 units (Miller *et al.* 1970; Riley & Barber 1971; Smiley 1974; Soon & Miller 1977; Sarkar & Wyn Jones 1982).

In plants that fix nitrogen symbiotically, the uncharged di-nitrogen molecule crosses the root-nodule–soil boundary; if this is the major source of plant nitrogen, it is likely that the intake of cations will exceed the intake of anions (SO_4^{2-} + Cl^- + $H_2PO_4^-$ + a little NO_3^-) (Munns 1978). It is also likely that the root will lower the soil pH. Israel & Jackson (1978) found that nodulated soybean roots, growing in a culture solution free of inorganic nitrogen, released 1.08 meq of acid per gram total plant dry weight. Gillespie & Pope (1990a) found that black locust seedlings released up to 1.1 pmol of acid $cm^{-2}\ s^{-1}$ of root in N-free solution culture. It is well known that soils frequently or continuously cropped by nitrogen-fixing legumes become acid and need periodic liming. For instance, in parts of New South Wales, continuous legume culture has so acidified the soil that wheat can no longer be cultivated.

In some soils, such as acid podsols, the rate of nitrification is very low and nitrogen may well be taken up largely as NH_4^+. Cation uptake will then greatly exceed anion uptake, and acidification of the rhizosphere will result. Quantitative data are lacking. In many grassland soils, nitrate concentrations are low and nitrogen may be absorbed as NH_4^+ rather than NO_3^-. This has been extensively reviewed by Haynes & Goh (1978) for plants in general and by Lee & Stewart (1978) for natural communities.

These deductions from cation–anion balances rest on the fact that no continuing net charge flow, namely an electric current, can cross the root–soil interface. They do not depend on understanding how the plant controls its internal pH or equalizes the positive and negative charges it contains — a topic reviewed by Raven & Smith (1976) and Smith & Raven (1979). The main process within the plant tending to change its internal pH is the production of 1 mol of base for each mole of nitrate that is reduced to the amino form:

$$R\text{–}OH + NO_3^- + 8H = R\text{–}NH_2 + 3H_2O + OH^-$$

Since the internal pH of plants is controlled within fairly narrow limits, they must dispose of this base. If the reduction of nitrate takes place in the roots, they do so by exporting the base from their roots as HCO_3^-, or, if the reduction occurs in the shoots (as with buckwheat), by using it to form an organic anion — such as malate or citrate — that is generally retained within the cell vacuoles. The balance between these two processes varies widely among species and largely accounts for differences among them in the amount of acid or base they will release to the same soil.

7.2.1.1 Location of Acid or Base Release from the Root

Calculations of the rate of acid or base release from the root that are based on the balance of cation and anion intake by the plant yield the average rate of release over the whole root system. Electrical balance across the root–soil interface must be maintained not only for the whole root system, but also for each small element of it. Because some major nutrient ions, such as calcium, are absorbed much more rapidly near root apices than over the remainder of the root, it is to be expected that there will be corresponding differences in the rate of release of acid or base along the roots.

Marschner & Romheld (1983) have shown that at a high level of nitrate, the pH of the rhizosphere of maize seedling roots increased from 6.0 to 7.5 fairly uniformly along the root; but at a lower level of nitrate, that decreased the rate of anion absorption, the pH of the rhizosphere of the apical portions of the primary roots and the whole of the lateral roots became acid, although most of the primary root was alkaline compared with the bulk soil. Moorby *et al.* (1985), however, observed a uniform acidification around rape roots in nitrate-free solution. They showed that the pH change around roots is mainly a response to the level of nitrate in that zone and is affected only to a lesser extent by the treatment of the rest of the root system in the short term. Therefore, differences in nutrient level in the soil, for example between topsoil and subsoil, will lead to different release rates of acids and bases.

This discussion treats acid and base release as a necessary accompaniment of the balance of intake between the major cations and anions. When roots respond to deficiency within the plant of a micronutrient such as iron by acidifying the rhizosphere near their tips (section 7.3.4), or such as phosphorus by acidifying the whole rhizosphere (section 7.2.5), the effect is more likely to be the result of a direct effect on plant metabolism that leads to the export of hydrogen ions.

Metabolic differences may also underlie the acidification associated with differences in root morphology. Gardner & Parbery (1982) and Gardner *et al.* (1982) have noted that the soil zones occupied by the proteoid roots (section 7.3.2) of white lupin are more acid than the rest of the root rhizosphere.

7.2.1.2 The pH Profile across the Rhizosphere

The acids or bases released at the root surface diffuse into the rhizosphere soil. To understand their effects, it is necessary to describe how pH changes are propagated through the soil.

Changes in pH are caused by movement of protons. In every proton transfer affecting the pH change, there must be a proton donor and a proton acceptor, because free protons do not exist in aqueous solution. The movement of protons, therefore, involves movement of donors and acceptors of protons, namely acids and bases:

$$\text{Acid (proton donor)} = \text{Base (proton acceptor)} + H^+$$

An acid and base that are associated are called conjugate, and they form an acid–base pair. The two main *mobile* acid–base pairs in soil are H_3O^+ and H_2O, and H_2CO_3 and HCO_3^-. Water can act as an acid as well as a base; however, the system H_2O and OH^- is unimportant compared with H_2CO_3 and HCO_3^- when CO_2 is at or above its normal atmospheric concentration. The pH of a portion of soil in contact with a more acid zone will drop because of the arrival of an acid (H_3O^+ or H_2CO_3) and the simultaneous removal of a base (H_2O or HCO_3^-). As a result of the reaction, the proportion of hydrogen ions in the soil cation exchange complex will increase and the soil pH will decline.

The quantitative treatment of pH change is a problem of diffusion with simultaneous reaction, resulting in a concentration-dependent diffusion coefficient. In a way analogous to equation (1.4), a soil acidity diffusion coefficient may be defined by the continuity equation for soil acid:

$$\partial[\mathrm{HS}]/\partial t = \partial(D_{\mathrm{HS}}\,\partial[\mathrm{HS}]/\partial x)/\partial x \tag{7.1}$$

where D_{HS} is the soil acidity diffusion coefficient and [HS] is the concentration of all soil acid, namely all the titratable acid groups in the soil.

If the pH buffer capacity of the soil, defined by

$$b_{\mathrm{HS}} = -\mathrm{d}[\mathrm{HS}]/\mathrm{dpH} \tag{7.2}$$

is constant over the range of pH change, then, by substituting for d[HS], equation (7.1) may be written with pH as the acidity variable:

$$\partial\mathrm{pH}/\partial t = \partial(\mathrm{D}_{\mathrm{HS}}\,\partial\mathrm{pH}/\partial x)/\partial x \tag{7.3}$$

Nye (1972) and Nye & Ameloko (1986) showed that when H_3O^+ and H_2O, and H_2CO_3 and HCO_3^- are the only significant mobile acid–base pairs, the soil acidity diffusion coefficient is given by

$$D_{\mathrm{HS}} = 2.303\theta f/b_{\mathrm{HS}}(D_{\mathrm{LH}}[\mathrm{H_3O^+}] + D_{\mathrm{LC}}[\mathrm{HCO_3^-}]) \tag{7.4}$$

where D_{LH} and D_{LC} are the diffusion coefficients of H_3O^+ and HCO_3^-, respectively, in free solution. This equation shows that the relative importance of the two carriers H_3O^+ and HCO_3^- depends on their diffusion coefficients in free solution and their concentration, as illustrated in figure 7.3. The soil acidity diffusion coefficient is low, and pH gradients will be correspondingly steep, when the pH is in the range 5–6 (since $[H^+][OH^-] = K_W = 10^{-14}$, both $[H^+]$ and $[HCO_3^-]$ will be small, depending on the CO_2 pressure); but it will be increasingly larger as the pH moves outside this range. Figures 7.4a and 7.4b show the pH profiles that result from the release of H_3O^+ and HCO_3^- at the root surface, as obtained by a numerical solution of equation (7.3) expressed in radial coordinates (Nye 1981). It will be seen that for a soil with initial pH in the range 5–6, a change in pH of 1 unit at the root surface is likely, that the soil within a few millimetres of the root surface will be affected, but that soils with pH lower than 4 or higher than 7 will be much less changed.

7.2.2 CO_2 Production

Carbon dioxide is liberated at the root surface by respiration, and this has often been assumed to have a local acidifying effect. However, except in near water-saturated soils, CO_2 will readily diffuse away from the root through air-filled soil pores, and its acidifying effect will be distributed over the whole bulk soil. It has been calculated that the difference in pressure of CO_2 between the soil at the root surface and the neighbouring bulk soil under typical aeration conditions is only 2×10^{-6} atm (Nye 1981). The corresponding difference in the equilibrium pH is

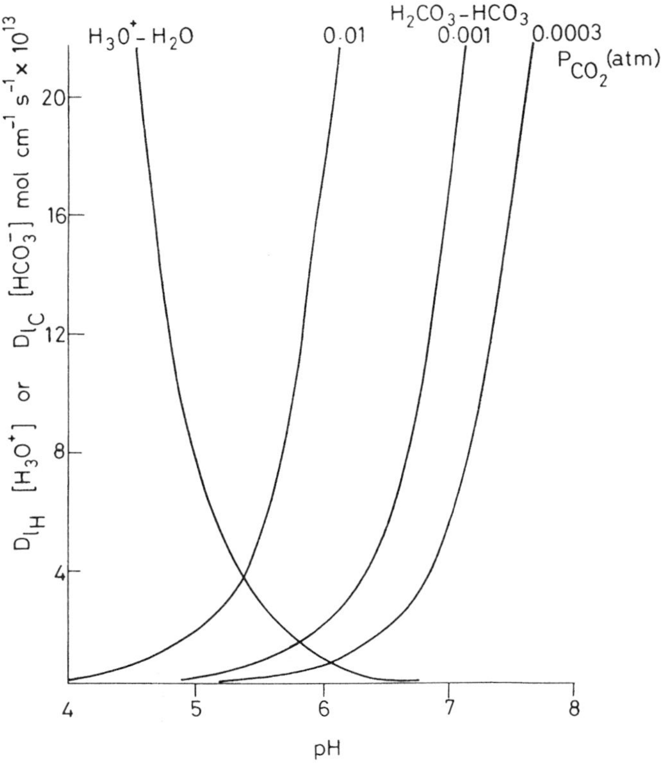

Figure 7.3 Relative importance of H_3O^+ or HCO_3^- in acid–base transport at different pHs (see equation (7.3)).

negligible. The situation is different in anaerobic soils, as described in section 7.2.7.

7.2.3 Excretion of Organic Acids

There have been reports of organic acids being produced by roots in axenic culture (Smith 1969, 1970; Gardner *et al.* 1982b) or under unnatural conditions (Moghimi *et al.* 1978). But significant quantities of low-molecular-weight organic acids have not been found in rhizosphere soil except in special cases (section 7.3.2), either because the amounts released are too small, or they are rapidly metabolized by microorganisms (Jones 1998). Root cortical cell vacuoles certainly contain ions such as citrate, oxalate, and malate, which are likely to be released by damage or necrosis. Even so, the pH of the cell cytoplasm is closely controlled within the range 6–7, well above the p*K* of organic acids (*c.* 2.5–3.5), so these substances will be released from the cytoplasm as the conjugate anion and should have no acidifying effect on the soil unless the free space of the roots is acid. The effect that these organic anions may have in solubilizing phosphate and micronutrients is considered in section 7.3.2.

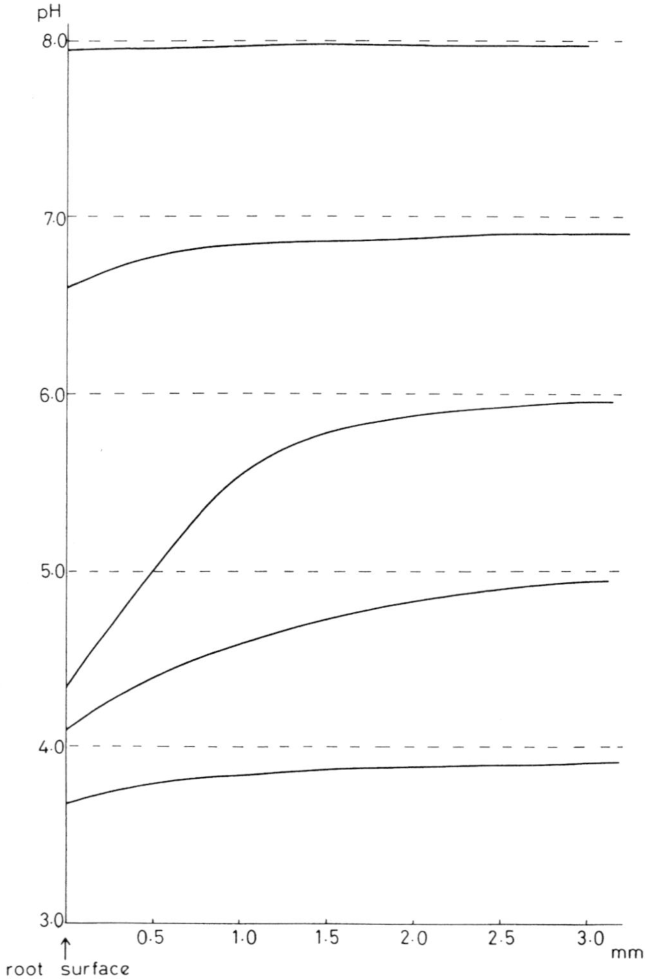

Figure 7.4a Effect of initial soil pH on pH profiles following H_3O^+ release at the root surface after 10^6 s.

7.2.4 Microbial Production of Acids from Root Carbon Release

Work surveyed in sections 8.1 and 9.1 suggests that up to 2–3% of plant dry weight may be released by roots into the rhizosphere. It includes soluble exudates, mucilage, and sloughed cells and lysates, all of which provide substrate for rhizobacteria that potentially are acid producers.

To generate 1 meq of acid per gram dry weight of plant growth, which is comparable to the amounts generated by cation–anion imbalance in nitrogen-fixing legumes, would require rhizosphere organisms to convert as much as one third of the total root carbon-release material to acids, if it is assumed that 60 mg

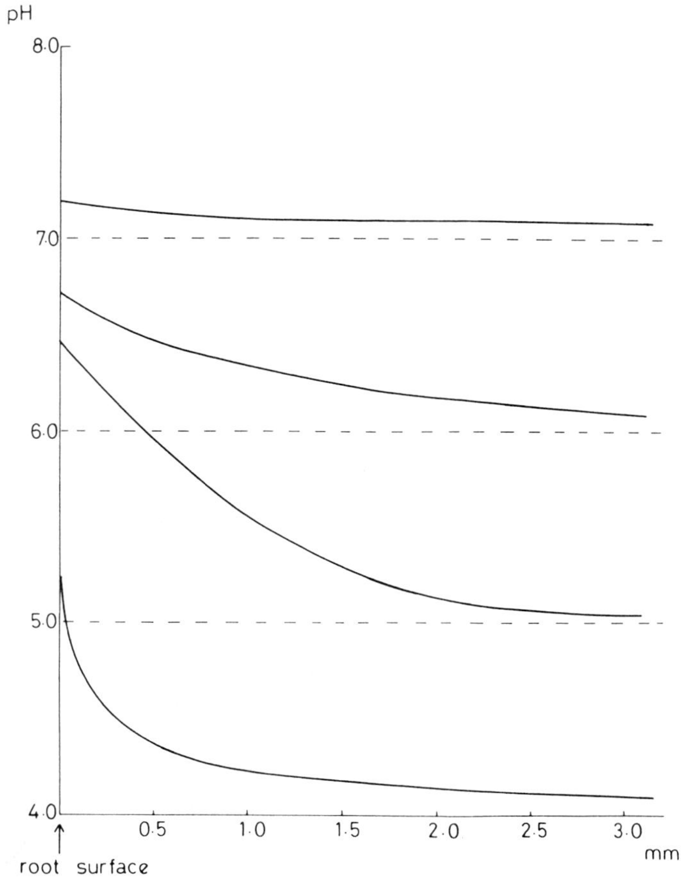

Figure 7.4b Effect of initial soil pH on pH profiles following HCO_3^- release at the root surface after 10^6 s.

acid yields 1 meq H_3O^+ (as for CH_3COOH) (Nye 1981). In practice, no differences in amounts of organic acids or their anions, which could have had significant effects on the rhizosphere pH, have been reported despite diligent research, except possibly in special conditions (Hale & Moore 1979). Table 7.2 shows an example of the amounts of organic acids found in rhizosphere soil (Hedley *et al.* 1982a). The evidence clearly indicates that cation–anion imbalance is by far the most important origin of pH changes in the rhizosphere.

7.2.5 Consequences of pH Changes

Clearly, pH changes of as much as 1 unit, representing a 10-fold change in concentration of H^+, may have far-reaching implication for a host of rhizosphere processes. We will mention here only some of the more direct effects on plant nutrition.

Table 7.2 Amounts of organic acids and their anions extracted from rhizosphere and non-rhizosphere Begbroke sandy loam.

	Extractable anion concentration (μeq g^{-1} soil)		
Method	Unincubated soil	Control soil	Rhizosphere soil
Resin extraction (for non-volatile acids	9.3 ± 0.9^a	12.0 ± 1.1^a	9.3 ± 0.5^a
excluding oxalic)	19.8 ± 1.4^b	17.3 ± 1.9^b	4.2 ± 0.7^b
Steam distillation (volatile acids + CO_2)	9.0 ± 0.8	9.3 ± 0.5	8.5 ± 0.5
Alkaline ethanol extraction (oxalate)	6.5 ± 1.8	9.3 ± 1.0	12.0 ± 0.5

Source: after Hedley *et al.* (1982a).

[a]Determined by titration to pH 6.1. [b]Determined by titration to pH 9.0. Difference measures weak acid groups.

7.2.5.1 Phosphorus

Miller *et al.* (1970), Riley & Barber (1971), Soon & Miller (1977) and Sarkar & Wyn Jones (1982) have noted that supply of ammonium rather than nitrate ion increased the phosphorus uptake from neutral soils and lowered the rhizosphere pH. In the soils studied, there was a corresponding increase in the phosphate concentration in the soil solution. By contrast, in soils rich in ferric and aluminium oxides, there is usually a rise in the concentration of phosphate in the soil solution when the pH is raised, and Gahoonia *et al.* (1992) even showed that in a slightly acid oxisol, ryegrass fed with NO_3^- took up more phosphate than that fed with NH_4^+. When roots mobilize P from rock phosphates the uptake of Ca^{2+} and other ions in the rock also plays a part in promoting dissolution. Hinsinger & Gilkes (1996) have discussed these effects in detail.

Legumes that derive their nitrogen by symbiotic fixation of N_2 tend to acidify their rhizospheres (section 7.2.1), and also to solubilize phosphorus (Aguilar & Van Diest 1981; Gillespie & Pope 1990b). The proteoid roots of white lupin solubilize soil phosphate and also lower the pH of the adjacent soil (Gardner *et al.* 1982) (section 7.3.2). Grinsted *et al.* (1982) have shown that phosphorus-deficient rape seedlings can solubilize phosphorus and increase their uptake even when their source of nitrogen is nitrate. The extra phosphate absorbed was directly related to the amount of acid released from the roots (Hedley *et al.* 1982b), and accorded with the phosphate solubilized by acid *in vitro*.

If a root is isolated, phosphate solubilized by the acid released from it has the possibility of diffusing away from the root into bulk soil as well as towards the root. Nye (1981, 1984) has deduced that the fraction of the P solubilized that can be taken up by the root is given approximately by the formula

$$\mathrm{P}_{\text{taken up}}/\mathrm{P}_{\text{solubilized}} = \{1/[1+\surd(D_{\text{HS}}/D_{\text{P}})]\} \times \{1/[1+(\surd(\pi D_{\text{HS}}t)/4a)]\} \quad (7.5)$$

The first term shows that if D_{HS} is small compared with D_{P}, that is the spread of the acidified zone is less than the phosphate-depleted zone, then most of the phosphate solubilized will be taken up by the root; however, if D_{HS} is greater than D_{P}, most will diffuse away from the root. The second term shows the effect of root radius. The larger the spread of the acidified zone compared with the root radius, the lower is the proportion of phosphate solubilized that will diffuse back to the root.

This analysis applies to a single root in a large volume of soil. If other roots are close by, they can take up P solubilized by H^+ released by, and diffusing away from, a neighbour. The arguments leading to equation (7.4) apply to other solutes that may be solubilized by acids or bases excreted by roots.

7.2.6 Effect of Concentration Changes of Other Ions in Addition to H^+ and HCO_3^-

Because ions are charged and the soil solution must remain neutral, uptake of one ionic species not only changes the concentration of that species, but also affects the concentration of other ions, so greatly complicating the simple solutions of transport equations that have been developed for single solutes. These effects need to be understood before it can be decided whether simple or complex solutions are required.

Some of the important considerations are as follows:

(a) At each point outside the root, the total strength of the soil solution is controlled by the concentration of unadsorbed anions, which will differ from point to point (section 3.1.2). To maintain neutrality, the total concentration of positive charge on cations in the soil solution must be the same as the negative charge on anions. The relative proportions among the cations will depend on their cation-exchange binding strengths (equation (3.4)).

(b) Electrical neutrality across the root–soil interface and within the plant must be maintained by excretion of H_3O^+ or HCO_3^- ions. The carbon dioxide concentration in the air-filled pore space is needed to assess the bicarbonate ion concentration.

Bouldin (1989) has developed a model that takes these factors into account. Here, we can only cite his conclusions:

> The important features of the model are (i) electrical neutrality of plant, soil and soil solution are maintained, (ii) plant uptake, transport and reaction are treated in successive steps in each time increment, this provides for flexibility in describing the processes, (iii) uptake, transport and reaction of seven ions is described; because of the structure of the programme, others could be added, and (iv) water and CO_2 are essential components of the reactions. The uptake of cations is a function of concentration of anions and the suite of exchangeable cations. The model emphasizes an essential role for CO_2 in soil chemistry and plant nutrition. The model requires over 20 soil and plant parameters; perhaps for the near future its usefulness will be mostly restricted to development of hypotheses. The most important hypothesis is that something about as complicated as the present model will be required to model uptake of several ions and crop yield.

To test the model, the uptake by wheat of Ca, Mg, and K at three levels of nitrate was determined (Bouldin *et al.* 1992). The model predicted the uptake reasonably well. Discrepancies were attributed to the effects of root ageing, and the mutual influences that Ca, Mg, and K have on their plant absorption parameters.

(c) Accurate knowledge of the diffusion coefficient of an ion requires a knowledge of the concentration and concentration gradients of all the other ions in the system (section 4.2.1.1). In a numerical solution, at each time step the concentrations of all the ions at each radius value are calculated, so the information needed to determine each ionic diffusion coefficient accurately should be available.

Some simplifications are possible. It is well to consider first the anions: chloride, nitrate, sulphate, and bicarbonate, since, as pointed out in chapter 3, their concentration controls the overall solution strength. When diffusing to the root they are accompanied by cations; hence, the appropriate approximate diffusion coefficient for these anions is often that of their calcium salt, since calcium is the dominant cation in most soils. The concentration gradient of the total anion concentration in solution is often negligible. We have seen that soil solution concentrations are usually greater than 10^{-3} M in agricultural topsoils, and they can be much less than 10^{-3} M in very poor natural soils. The total anion inflow rarely exceeds 10^{-11} mol cm^{-1} s^{-1} in a fertile soil. These values lead to a value of $\alpha a \ll 10^{-6}$ cm^2 s^{-1} for the anions as a whole, although individual anions like nitrate may well have much higher values. At the same time, we have seen that $v_a a$ lies in the range 10^{-8}–10^{-7} cm^2 s^{-1}. Since all the anions typically have diffusion coefficients in solution $\approx 10^{-5}$ cm^2 s^{-1}, they may, for this purpose, be treated as a common species. In moist soil, $Db(\approx D_L \theta f_L)$ is *c.* 10^{-6}. Hence, both $\alpha a/Db$ and $v_a a/Db$ will be much less than 1. Hence, whether there is a fall in concentration at the root surface or a rise, it will be shallow (Nye & Marriott 1969, figures 2 and 3, where $v_a a$ is 10^{-7}). An example of this behaviour is to be found in Brewster & Tinker's (1970) study of ion uptake from soil by leeks. In dry soils, the change in concentration may be greater; though both α and v_a are reduced, the product θf_L is reduced more. If there is a significant gradient in the overall concentration of the soil solution, then a numerical solution must be used. This involves calculating the solution strength at each time step and then using the appropriate value of b, adjusting (Ca + Mg).

For the cations, the calculation of b and hence D is simplified if the overall solution strength may be regarded as constant. Then, for an adsorbed cation present as a minority component, such as K^+, the value of b_K is obtained from the buffer curve of (K)/(Ca + Mg) in which (Ca + Mg) may be taken as constant.

(d) Another complication arises when, due to transpiration, the soil moisture level changes during the uptake period. If the hydraulic conductivity of the soil is sufficiently high compared with the rate of water inflow, there will not be an appreciable water gradient to the root, that is the whole soil will dry out uniformly (see chapter 2). In these circumstances, a numerical solution is readily achieved by adjusting θ and dependent parameters accordingly before calculating the ion concentration at each time step. If, however, there is an appreciable gradient in water content towards the root, then for each time step the water content at each radius value must be calculated first, and then the ionic concentration gradients calculated using the appropriate value of θ. Examples are provided by Dunham & Nye (1974, 1976). During a growing season, the topsoil is likely to be wet and dry by turns. It will then be very difficult to calculate water gradients, because of hysteresis.

7.2.7 Root Effects in Anaerobic Soil

The following additional effects are directly caused by a root growing in reduced soils.

(a) Oxygen, which has diffused down the root's internal gas channels, is released from its surface into the soil, which has a lower oxygen pressure.

(b) Near the root surface, Fe^{2+} is oxidized to $Fe(OH)_3$, causing the brown coating of soil commonly noted around the roots of marsh plants. Consequently, the concentration of Fe^{2+} near the root is lowered and more Fe^{2+} diffuses to the root from the bulk soil. This is then oxidized, resulting in an accumulation of $Fe(OH)_3$ near the root surface. The pH of anaerobic soil is usually about 6.

(c) The oxidation reaction generates H^+:

$$4Fe^{2+} + O_2 + 10H_2O = 4Fe(OH)_3 + 8H^+$$

and the pH close to the root falls.

(d) In the reduced soil, the concentration of the anion, NO_3^-, is low and the root takes up nitrogen as the cation, NH_4^+. Consequently, its intake of cations exceeds that of anions and H^+ is released by the root to maintain electrical neutrality, as we have seen. The amounts of H^+ formed by oxidation are comparable with the amounts released to maintain charge balance across the root surface.

(e) Because the air-filled pore space in anaerobic soils is low, diffusion of CO_2 is restricted to the liquid phase and its concentration in the soil is usually considerably higher than in the atmosphere. In spite of the CO_2 generated by root respiration, the carbon dioxide pressure outside the root may be greater than that inside it, resulting in a flow of CO_2 from the soil to the atmosphere through the root's aerenchyma (Higuchi *et al.* 1984). Any net uptake of CO_2 by the root decreases the concentration of the acid, H_2CO_3, near the root, and this, to some extent, may offset the acidity produced by oxidation and excess cation uptake. On the other hand, in the absence of much NO_3^-, the overall soil solution concentration depends largely on the concentration of HCO_3^-, which will fall if the concentration of H_2CO_3 falls. This will lower the concentration of NH_4^+ in solution, which, in turn, will lower the rate of diffusion of NH_4^+ to the root and limit the plant's nitrogen uptake rate.

(f) Lowering of the pH will usually tend to solubilize P and micronutrients such as Zn and Mn and produce hydrated ions of the latter. A pH decline will also alter the equilibrium between NH_3 molecules and ammonium ions NH_4^+:

$$NH_4^+ = NH_3 + H^+$$

All these processes have been studied experimentally, and models have been constructed that will predict their effects under specified conditions. Begg *et al.* (1994) followed the concentration profiles of pH, Fe^{II}, and Fe^{III} that developed near a plane of rice roots sandwiched between blocks of anaerobic soil. There was a steep fall towards the root plane in the concentration of Fe^{II} and a rise in pH and Fe^{III} that developed within 10 days. The H^+ formed by Fe^{II} oxidation was comparable with the H^+ exported to balance excess cation uptake. Since the overall lowering of pH was less than expected from the measured O_2 flux into the soil and the cation–anion balance across the roots, the authors suggest that CO_2 uptake by the roots may account for the discrepancy.

Saleque & Kirk (1995) measured the phosphate solubilized at varying distances from a plane of mature rice roots. Some 90% of this phosphate was drawn from acid-soluble pools. There was accumulation of $Fe(OH)_3$ in a narrow zone near the roots, and this adsorbed some phosphate. Kirk & Saleque (1995) developed a mathematical model of the diffusion of acid away from the roots, of the acid reaction to solubilize phosphate, and of the back-diffusion of the solubilized phosphate to the absorbing roots. The model was based on an analytical solution of the coupled continuity equations for pH and phosphate. There was good agreement between the measured and predicted concentration profiles of P and pH for different levels of P added to the soil. About 80% of the P taken up was solubilized by the acid produced. However, only about half of the total P solubilized was taken up by the roots, because the rest diffused away, as noted in section 7.2.5. In a similar experiment, Kirk & Bajita (1995) showed that much of the Zn solubilized by the acid released was adsorbed within 5 mm of the root plane by the $Fe(OH)_3$ and amorphous organic matter accumulated there.

7.3 Direct Effects of Soluble Exudates on Mineral Nutrition

Soluble root exudates include most of the soluble compounds found inside root cells (Uren & Reisenauer 1988), to which are added soluble products that arise from bacterial metabolism of mucilage and fine root debris (Lynch 1990b). The quantities released are discussed in chapter 8 because of the relevance to the support of rhizosphere organisms. The precise location of the release of exudates may be important for nutrients that are absorbed over a limited root length, such as near the root tip for trace elements, but less so for N, P, and K, which are absorbed over considerable distances along the root (section 5.4.3). Most soluble root exudates are exuded fairly near the tip, in the extension and root hair zones, and most mucilage comes from the root tip itself (Oades 1978; Curl & Truelove 1986, p. 77; Nielsen *et al.* 1994), but debris from root hairs and cortical cells are produced on older parts of the root.

A variety of environmental and physiological factors, especially 'stresses', affect the quantity and type of compounds released (Uren & Reisenauer 1988). Thus, root impedance by compacted growth media causes greater exudation, much of it being mucilage polysaccharides (Boeuf-Tremblay *et al.* 1995), which have been suggested to act as a lubricant. Strom *et al.* (1995) found that acetic acid predominated in the exudates of calcifuges, but citric and oxalic acids predominated in acidifuges, where chelate formation with calcium and iron would be beneficial. Soluble exudates can probably move well out from the root, because labelled carbon was detected up to 20 mm away from a root 25 days after feeding ^{14}C to the plant (Lynch & Whipps 1990).

7.3.1 Mucilage

Matar *et al.* (1967) postulated that the materials in the root mucigel can 'mobilize' phosphorus, micronutrients, aluminium and heavy metals from soil in immediate contact with the root (Marschner 1995). The concept that the mucigel solids directly mobilize materials from soil is doubtful, as they are insoluble long-chain polysaccharide polymers (Roy *et al.* 1988; Uren & Reisenauer 1988), and it seems unlikely that they could bring significant amounts of these elements into solution. However, if soluble exudates are held in solution within mucilage that touches soil colloid surfaces, their concentrations and contact times may be increased, so that they may react more than if they were solely in mobile water films on the same surfaces. Thus, Uren (1993) argues that mucigel permeates soil crumbs close to the root, and thereby provides the right conditions for reactions at the surfaces, including reduction. However, there is no real physical evidence for this yet.

Both aluminium and the heavy metals can be bound by the organic acid groups on mucigel, and thus can be detoxified for the plant (Horst *et al.* 1982; Marschner 1995). If phosphate is held on sesquioxide surfaces in the soil, and the metals in the oxides can be complexed by the mucigel in this way, some phosphate could be brought into solution.

7.3.2 Polycarboxylic Acids

The change of pH in the rhizosphere by the anion–cation balance of roots has been discussed in section 7.2.1, and happens with all plants to varying extents. However, some plant roots appear to release unusually large amounts of particular polycarboxylic acids in a specific way, such as oxalic acid from maize (Kraffczyk *et al.* 1984) and citric acid from white lupins (Gardner *et al.* 1982a, b; Dinkelaker *et al.* 1989). This seems to be associated with phosphorus deficiency in particular species, as the amount of carboxylic acids in exudates from most plants is usually quite small (Hale & Moore 1979). Whereas sugars appear to leak out of roots down a concentration gradient, the release of organic acids seems to be much more specific in response to the environmental and physiological conditions. Also, organic acids are not reabsorbed within the free space in the root as are other compounds (Jones & Darrah 1992), which may make it easier to attain high net exudation rates.

These acids can acidify the rhizosphere; sequester cations that would combine with phosphate ions, such as Al, Fe, or Ca; or they can displace phosphate ions previously held on the surface of insoluble sesquioxides or clays. The citrate and malate exuded from the roots of Al-tolerant genotypes of wheat and maize also protects them from injury by forming complexes with aluminium. Exudation of malate is 10 times or more larger from Al-tolerant than from Al-susceptible varieties (Ryan *et al.* 1995). However, their breakdown by microbes in the rhizosphere will counteract all these effects (Jones *et al.* 1996b).

Lupins (Gardner & Parbery 1982; Gardner *et al.* 1982, 1983a) seem to be exceptional in production of citric acid from their proteoid roots — both root formation and high exudation being a response to phosphate deficiency. The increased solubility of phosphate following acidification in the rhizosphere has been verified by many studies (section 7.2.5), but the direct effect of chelation by exuded organic acids on phosphorus nutrition is still much less well understood, and may be the exception rather than the rule. The distinction between release of phosphate due to general change in pH and that due to complexation is not easy to make.

Dinkelaker *et al.* (1989) found that, when lupins were grown in a low-phosphate soil containing 2.9% active $CaCO_3$ and 20% total $CaCO_3$, up to 50% of the root length became proteoid (section 9.2.1). Within these root clusters, the soil pH dropped from 7.5 to 4.8. A heavy precipitate of calcium citrate was found within the proteoid root clusters, reaching 50 μmol g^{-1} soil in the proteoid root mats after 8 weeks. The recovered calcium citrate corresponded to 23% of the plant dry weight at harvest.

The anion–cation balance was measured in the same experiment, and was only 44 meq acid excess per 100 g plant tissue, corresponding to only about 10% of the acidity in the citric acid that appeared as calcium citrate. Acidification by cation–anion balance does not seem to be an adequate explanation in all cases.

In the same experiment, the concentrations of iron, manganese and zinc increased greatly in the soil around the proteoid roots, but it was not clear if this was a direct result of the pH change or a more specific effect of the citrate (see also Gardner & Parbery 1982; Gardner *et al.* 1982, 1983a).

Braum & Helmke (1995) measured the specific activity of phosphorus absorbed from a labelled soil by soybeans and white lupin, and found that it was always lower in the latter, by a factor of as much as 3. The labile soil phosphorus level calculated from the uptake of the plants was therefore up to three times larger for white lupin. Braum & Helmke (1995) found no significant effect on phosphorus uptake by soybeans intercropped with the lupin, and deduced that the general soil solution level was not increased. However, the lupins took up large amounts of copper, iron and zinc, and especially manganese, and the intercropped soybeans also absorbed up to twice as much of these metals as when grown alone. This would be expected if the lupin effect for phosphorus occurs by displacement from soil surfaces of phosphorus that can readily be readsorbed, whereas the effect for trace elements is due to the formation of reasonably stable chelates that stay in the soil solution for a longer period, and thus may reach other plants.

The polycarboxylic acid mechanism for white lupins therefore has strong evidence in its favour. However, similar effects can also be produced by acidification; for example, Hedley *et al.* (1982) showed that the labile pool of phosphorus was increased by a factor of 2 around phosphorus-deficient rape roots that acidified the soil by excreting protons, but produced little polycarboxylic acid. Conclusive proof of the polycarboxylic acid chelating mechanism, as opposed to general acidification, is therefore still required in each case, and it remains to be seen whether many species other than white lupin show this particular ability, and whether all such plants form proteoid roots. It may be a special case rather than a general mechanism, but it is certainly important for some species, or perhaps even for some cultivars, as shown above for Al tolerance (Jones 1998).

7.3.3 Phosphatases

The function of phosphatases in the rhizosphere has been discussed for many years. These have repeatedly been postulated to break down the considerable amounts of organic phosphate esters that form part of the soil organic matter, and so make phosphate ions available to the root uptake sites (Anderson 1980).

The literature is not clear whether diffusion of soluble organic esters to membrane-bound enzymes, or of soluble enzymes to insoluble esters, or of hydrolysis of soluble esters by soluble phosphatases, is envisaged (Helal & Sauerbeck 1991b). These enzymes are acid phosphatases from the root, acid or alkaline phosphatases from fungi, or alkaline phosphatases from bacteria (Marschner 1995, p. 559). In total, there is a marked increase in acid and alkaline phosphatase concentrations near to and in the root, most of which is root-derived, though some will also be from microbes (Tarafdar & Jungk 1987). This does not prove that hydrolysis of esters increases proportionately and, indeed, the relative increase in phosphatases near the root to that in bulk soil is quite small compared with rhizosphere/bulk soil (RH/S) ratios for microbes from the same roots (Tarafdar & Jungk 1987).

Tarafdar & Claassen (1988) showed that the root-produced phosphatases from clover, barley, oats and wheat were sufficient to hydrolyse much more P_i than was needed for the plants' growth, so long as sufficient substrate — lecithin, glycerophosphate or phytin — was present. The principle of the effects thus seems well established, but the rate will depend upon the amount of soluble substrate present. The decrease in soil organic phosphates very close to the root surface (Tarafdar & Jungk 1987) supports the effects. However, even if microbes also produce phosphatases in the rhizosphere, their contribution is apparently not essential, and the role of 'phosphate-solubilizing bacteria' must therefore be questioned (Tinker 1980). Hedley *et al.* (1983) noted that root phosphatases increased following the onset of phosphorus deficiency, but could find no evidence that they solubilized soil organic phosphorus.

Obviously, phosphate is released in the general breakdown by microbes of soil organic matter, in which the C/P ratio is usually about 100, but the short-term importance of this process in plant nutrition is not yet established. Bennett-Leonard *et al.* (1993) noted that the root phosphatases of clover were able to hydrolyse various phosphate esters, including cyclic adenosine monophosphate (cAMP), but not inositol hexaphosphate (phytin), though Tarafdar & Claassen (1988) reported that the latter was hydrolysed. The phosphatases in the work of Bennett-Leonard *et al.* (1993) were largely located on the internal cell walls of the cortex, and these authors make the interesting suggestion that the main value to the plant of root phosphatases is to break down phosphate esters that exude from the root cells, rather than esters in the soil solution. The P_i produced is then readily reabsorbed (section 8.1.4). The synthesis of phosphatases is increased by both P and Ca deficiencies, and the latter will also make the plasmalemmas of the cells more leaky.

7.3.4 Phytosiderophores and Iron Uptake

Iron and, to a lesser extent, other trace metal oxides and hydroxides have extremely low solubility products (that for ferric hydroxide is about 10^{-36}), so that the metal ions are present in the soil solution at very low levels, and they are very poorly available to plants. There are, in principle, several ways in which the solubility of the metal can be enhanced. The most obvious one is to depress the pH to levels where the solubility product gives a reasonable concentration of free ionic metal, so that a pH of 4 would allow an iron concentration of about 10^{-6} M. The second mechanism is to form soluble complexes with organic ligands. It is well known that much of the trace elements in the soil are present as chemical complexes, and that some of these complexes are soluble and available to plants (Hill-Cottingham & Lloyd Jones 1965; Hodgson 1969; Tinker 1986). This applies to copper, zinc, and especially to iron. Some of the exudates of plant roots can act as ligands in this way, particularly the phytosiderophores. Third, a change in solubility can be effected by a change in valency, that is by reduction. Thus, ferrous hydroxide has a solubility product of 1.6×10^{-14}, so that reducing ferric iron will greatly increase the potential concentration of iron in the soil solution.

These basic chemical properties are relevant to the uptake systems for iron. There are two specific mechanisms, called strategy I and strategy II, for increasing the solubility of iron in the rhizosphere and the uptake rate of iron (Romheld & Marschner 1986; Marschner & Romheld 1994; Marschner 1995). Both are located in the apical part of the root, and the genes controlling both are expressed when a deficiency of iron occurs.

Strategy I is found in dicots and non-graminaceous monocots. The main responses are (i) increases in reducing capacity at the root surface by the production of membrane-bound reductases and, sometimes, soluble reductants; (ii) acidification by enhanced excretion of protons; and, often, (iii) the increased excretion of chelating compounds. With chelation and the lowered pH, a significant concentration of Fe^{III} chelate forms in solution. This is then reduced at the root surface by the iron deficiency-induced reductase, resulting in enhanced uptake of the ferrous iron released from the chelate, probably by ion-specific channels or carriers. The strategy I mechanism varies in detail among different species, and there is still disagreement about which processes must be regarded as essential, especially about the effective role of soluble reductants.

The strategy II system is found only in graminaceous plants, and is rather more specific. The main characteristic is the release of a small number of non-protein amino acids (phytosiderophores), such as mugineic acid, that form extremely stable complexes with ferric iron (Neilands & Leong 1986). These compounds diffuse from the root into the rhizosphere, form complexes with iron from sesquioxide surfaces or other insoluble compounds, and some of them eventually return to the root by diffusion and mass flow. The Fe–siderophore complexes are there absorbed without being broken down, possibly by a proton-anion co-transport system also produced in response to iron deficiency. Similar systems are well known in microbes, though these siderophores have a lower affinity for iron (Neilands 1984). It is to be expected that a considerable fraction of the siderophores, with or without complexed iron, will diffuse away and be lost to the plant, or be broken down by microorganisms in the soil (Von Wiren *et al.* 1993). It is therefore not surprising that the ability to resist iron deficiency in the plant is closely related to its quantitative capacity to produce phytosiderophores (Clark *et al.* 1988).

There has been some controversy about whether this description is adequate, because a number of exceptions and special cases have been discovered, for example C4 plants show little strategy II effect (Marschner & Romheld 1994).

Some authors claim that the siderophore–Fe complexes in strategy II, in fact, break up before uptake of iron, others that Fe complexes in strategy I are absorbed without splitting. A synthetic chelator of ferrous iron (BPDS) had no effect on uptake of Fe by oats and maize (strategy II), so apparently there is no stage with reduced, free ferrous iron at the root surface before uptake. The same compound strongly decreased uptake by a typical strategy I plant, cucumber (Romheld & Marschner 1986), thereby showing that, in this case, the iron was reduced before uptake.

Microorganisms are a major source of complications. In normal non-sterile soils, microorganisms will both excrete their own siderophores and break down those produced by the plant (Crowley *et al.* 1992). There seems no doubt that the phytosiderophores are readily metabolized by soil microorganisms, and, in that case, the iron they contain will presumably enter the microbial tissue. Fortunately, the siderophores are released in the root apical zones, whereas the main microbial populations build up in the rhizosphere further up the root, so metabolism is somewhat delayed.

It has also been suggested that microbial siderophores are important in allowing uptake by plants. The complicated and confusing effects in this area are clearly seen in the work of Crowley *et al.* (1992). They tested iron uptake in the presence of a phytosiderophore (hydroxy-mugineic acid, HMA) or a microbial siderophore (ferrioxamine B, FOB) on axenic and non-axenic plants, which were or were not Fe-deficient. The HMA was clearly the best source of iron and also caused large translocation to the shoot of the plant. The FOB also increased uptake, but only in non-sterile plants, and the iron was not translocated, so uptake was probably only into root bacteria. The effects varied with the source and type of bacteria, and it is difficult to draw clear conclusions, except that experiments with axenic plants are essential as part of any study of the subject. The outline of these processes is well defined, but it is extremely difficult to produce quantitative data for dynamic and natural conditions. The existence of fungal siderophores that are strong chelators may suggest effects of mycorrhizas, but they are not yet proven. There is some evidence that even if microbial siderophores do not help plant uptake in short experiments, they may help nutrition in the long term, by maintaining a larger pool of soluble or freshly precipitated iron, which may be on the root surface or even in the apoplast (Marschner & Romheld 1994).

Jones *et al.* (1996a) have made a study of the importance of organic acid excretion in strategy I uptake of iron. Their computer simulation included the condition that the soil was sterile, so no breakdown of the acids occurred. The dissolution of iron by complex formation with citrate or malate was followed by strategy I reduction at the root absorbing surface. Many factors affected the simulation results, but at pH values less than 6.8 the dissolution of amorphous Fe was rapid. Above pH 7, citrate-mediated dissolution was slow, because the complexes are unstable. These results suggest that the mechanism may contribute to iron nutrition, but not under the alkaline soil conditions where the most severe iron deficiency is found.

The transport processes of trace metals in the soil are not easy to treat quantitatively, but Bar-Yosef *et al.* (1980) produced a model for the transport of chelated zinc to a plant root that dealt with Zn^{2+}, $Zn(OH)^+$, H^+, L (ligand), and ZnL^+. All these species are sorbed, by various mechanisms. This is still relatively simpler than the situation for iron, because there is no valency change, and the ligand bonding is of moderate strength. Earlier, Hodgson (1969) produced a similar, but simpler, model, in which he ignored pH effects and adsorption by soil.

8

Microbiological Modification of the Rhizosphere

8.1 Microbial Substrates in the Rhizosphere

8.1.1 Carbon in Root Systems

The general questions of root/shoot ratio, allocation of carbon to the root system, and root system dynamics are discussed in chapter 9, and the detailed root structure in chapter 5. Root-derived carbon forms the substrate for rhizosphere and symbiotic organisms, and hence leads to the increase in their population densities close to or in the root. Some of the carbon compounds from the root have specific chemical effects also (see chapter 7). Both quantity and composition of these materials need to be known if their effects are to be understood, and we discuss this subject here.

The terminology of these materials is rather confused. The collective name for the injection of plant-derived carbon into the soil around living roots is 'rhizodeposition', but this has been used in different ways; for example, it may include root-respired carbon dioxide (Whipps 1990) (figure 8.1), but Darrah (1996) excludes carbon dioxide. The various forms include (Rovira *et al.* 1979; Lambers 1987; Whipps 1990) solid tissues lost from the root during growth; mucigel and debris from root surfaces and root cap; low-molecular-weight organic compounds in solution; carbon dioxide produced by root respiration for maintenance and for growth; faunal grazing of root tissues; and carbon transferred into symbionts, such as mycorrhizas and rhizobia. Some authors subdivide certain of these classes further. 'Rhizodeposition' is loss from a functioning root, but over a longer period the death and decomposition of whole roots

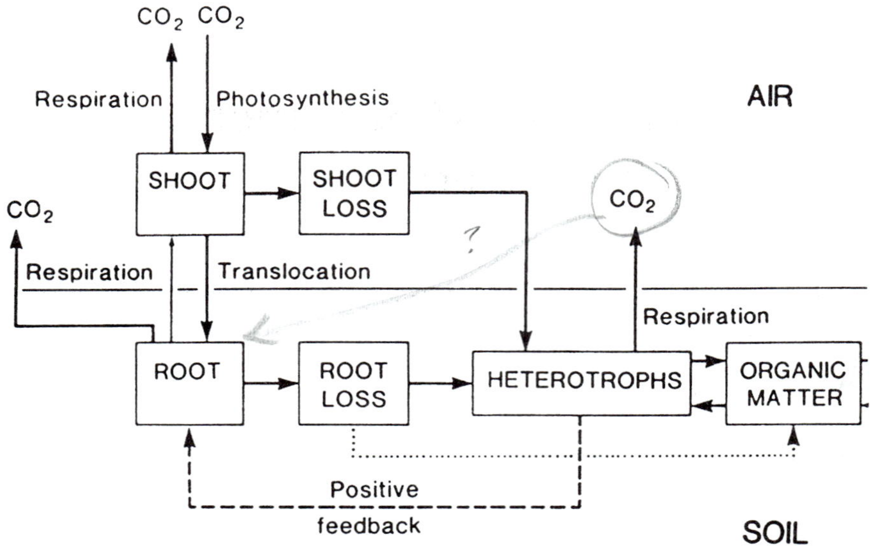

Figure 8.1 Carbon flows in the soil–root system associated with below-ground allocation of carbon and rhizodeposition (after Whipps 1990).

deposits large quantities of carbon into the soil, which continues to act as a more resistant microbial substrate (see chapter 9). All of these materials ultimately are converted to carbon dioxide (except for material formed into stable soil organic matter) and this is difficult to separate from carbon dioxide produced directly by root respiration.

The main issue here is how the various forms of deposition alter the ability of the living root system to absorb nutrients. We use the following terms for clarity, and because they relate to the practical means whereby these materials are quantified. As the rhizosphere situation is very dynamic, the results obtained will depend upon the timescale considered.

(a) Exudates: soluble low-molecular-weight material that comes directly from the living root (microbial metabolites may be similar, but are excluded).
(b) Rhizodeposition: this is (a) plus finely divided solids, root hairs, sloughed-off cells and mucigel that comes directly from the living root.
(c) Total rhizodeposition: this is (b) plus carbon dioxide respired directly by the living root.
(d) Root carbon: living and dead roots and their decomposition products.
(e) Below-ground carbon allocation: this is (c) plus (d) plus microbial tissue and all respiration that results from the presence of the plant and its photosynthesis.

8.1.2 Measurement of Carbon Allocation Components

Two main approaches to measuring rhizodeposition have been used. The first is to measure the lost material directly in non-soil media, such as sand or water culture. Measurement of exudation may be done in sterile media, to prevent rapid breakdown of the rhizodeposit, though the apparent exudation rate is increased by the presence of a solid rooting medium or of bacteria (Whipps 1990). Meharg & Killham (1995) found that the exudates from ryegrass increased from around 1% to over 3% of net photosynthesis in the presence of microorganisms, even when the latter were separated from the root by a Millipore filter. This suggests that microbial metabolites increase exudation, or that a local strong sink has the same effect (section 8.1.4). There is no known way of directly measuring rhizodeposition into a non-sterile soil system (Hodge *et al.* 1997). Extrapolation from work in sand or water culture is used, but cannot be relied upon for reasons given above, and also because faunal grazing occurs in field soils. Direct separation and weighing of visible dead root in the soil is simple in principle, though often difficult in practice (chapter 9).

The second general method is to label a plant by growing it in an atmosphere that contains $^{14}CO_2$ (Tinker *et al.* 1991; Meharg 1994; Swinnen *et al.* 1994a) or $^{13}CO_2$ (Simard *et al.* 1997), so that plant-derived materials can be identified. A rather different approach has used ^{13}C abundance to identify the carbon coming from the plant, by growing a C4 plant on soil that has previously carried only C3 plants, because C3 and C4 plants differ in their discrimination between the mass isotopes ^{13}C and ^{12}C (Balabane & Balesdent 1992).

This isotopic labelling has been used in two quite distinct ways, which give answers to different questions. First, measurements on plants grown continuously in a labelled atmosphere (Whipps 1990) distinguish all plant-derived carbon from that originally in the soil so that below-ground allocation can be measured directly. However, it does not give the allocation pattern for any specific part of the plant carbon. Second, the plant can be given a short pulse of labelled CO_2 and the distribution of this fixed carbon can be followed during a 'chase period'. From this, the allocation of one specific amount of fixed carbon can be determined in different carbon fractions and plant parts. The longer the chase period, the closer will the measured allocation approach the final distribution.

The units of measurement and conditions used have varied widely. The amounts may be expressed per unit weight of plant or of root, unit weight of soil, unit length of root, or per plant. The best way to understand plant function is to express the carbon allocation components as a percentage of the net carbon fixation over the labelling period.

8.1.3 Below-Ground Carbon Allocation Values

There is much data on exudation in nutrient or sand culture, some of it confusing (Hale & Moore 1979; Whipps & Lynch 1986; Whipps 1990). These results include sugars, amino acids, organic acids, and most other components of the cytoplasm (Uren & Reisenauer 1988; Marschner 1995, p. 549) but the extent to which solid materials are included is variable and often unknown. There is a broad consensus that losses of soluble exudate do not normally exceed 1–3% of plant weight (Lambers 1987). This value depends upon plant species, age, nutrition, light, temperature, drought, anoxia, microorganisms, mycorrhizal symbionts, the growth medium, and other variables (Hale & Moore 1979; Uren & Reisenauer 1988; Whipps 1990). However, recent ideas on recycling of exudate (see section 8.1.4) suggest that none of the values has any fundamental validity. Ranges of data for all below-ground carbon allocation components in the root are given by Lambers (1987) (table 8.1); in this case the 5% greater values for 'exudation' must include solid material.

The mucigel (section 7.3.1) is a polysaccharide gel that covers most parts of the young root system, where it is associated with 'soil sheaths' around the root (McCully 1987), and so helps to maintain root–soil contact. It is largely exuded by the root cap (section 5.1.2), but it may also contain bacterial mucilages, lysed cell contents and other components. It is difficult to determine the quantity of carbon represented by the mucigel, but it is probably well under 1% of the net photosynthesis, so rhizodeposition, as defined here, is probably of the order of 1–3%.

Overall, some 15–30% of the total carbon fixed appears as below-ground carbon dioxide production (Whipps 1990; Lambers *et al.* 1996) (section 9.1.2). It is very difficult to differentiate between the carbon dioxide produced by root metabolism, by symbionts that depend directly upon the root, and by microbial or faunal utilization of carbon compounds deposited outside the root. Lambers *et al.* (1987) estimated that the proportion of carbon dioxide from wheat root respiration to total below-ground respiration declined from 46% at tillering to only 5% at harvest. Helal & Sauerbeck (1991a) used antibiotics to suppress microbial respiration, and considered the remaining 20% of the initial respiration as being directly from the root. Swinnen *et al.* (1994a) injected labelled soluble compounds into the rhizosphere, and deduced that approximately between 85 and 95% of the carbon dioxide flux from below ground was from the wheat roots, decreasing towards ripening. The fraction of the CO_2 evolved directly from root respiration is thus extremely uncertain, and the methods used may be unreliable (Darrah 1996). It is therefore not presently possible to put precise values on total rhizodeposition as defined above, because the total CO_2 evolved is such a large fraction of the total below-ground carbon allocation.

The results of Swinnen *et al.* (1994b) (figure 8.2) show the variation with time of the allocations in wheat. Root growth and below-ground respiration start at high percentages of the net assimilation, but both decline to very small percentages near harvest, when nearly all the fixed carbon is going to the ear. By con-

Table 8.1 Allocation of carbon below ground, including rhizodeposition, in wheat.

		Swinnen *et al.* (1994b)		
	Lambers (1987)	130 days	150 days	183 days
Exudation	5	14	7	10
Growth of roots	11–35	28	16	4
Respiration of roots plus soil	12–29	28	15	8
N_2-fixation	5–23			
Mycorrhizas	7–10			

Averages from published data by Lambers (1987); specific values for different dates from Swinnen *et al.* (1994b), based on curves derived from the allocation model. 'Exudation' in Lambers is taken to be equivalent to 'Young photosynthate rhizodeposition plus root decay' in Swinnen *et al.* (1994b). Results in percent of net assimilation at time of measurement.

trast, carbon dioxide from plant-derived soil organic carbon and root decay increase gradually with time (Swinnen *et al.* 1994b) (figure 8.2). Other sets of similar data are now available (section 9.1.2).

8.1.4 Recycling of Exudates

The idea that nutrient elements can be simultaneously exuded and reabsorbed is familiar (section 5.3.2), and the suggestions that exuded phosphate esters are hydrolysed by exoenzymes in the apoplast before reabsorption of their P_i is men-

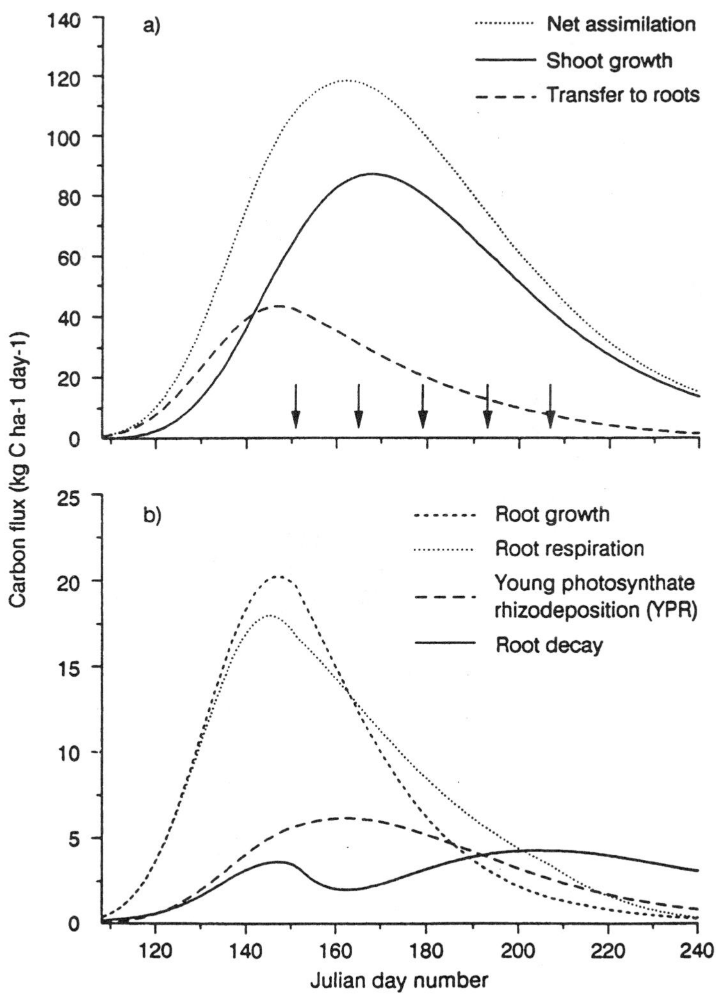

Figure 8.2 Carbon fluxes for a wheat crop, showing time changes in (a) root carbon allocation and (b) its components below ground, derived from ^{14}C distribution after pulse-labelling. Arrows indicate dates of labelling. Data before and after the arrows are extrapolations from the ^{14}C distribution pattern (after Swinnen *et al.* 1994b).

tioned earlier (section 7.3.3). There is evidence that some organic compounds can be exuded and re-absorbed in an analogous fashion (Schobert & Komor 1987; Jones & Darrah 1992, 1993; Darrah 1996), so that net exudation results from a two-way process. The measured net exudation rate therefore will depend critically upon the exact conditions of the experiment, and may only be a fraction of the gross exudation into the free space. This will be particularly true if exudation into a limited solution volume is measured, as the concentration there will soon attain an equilibrium value. This may explain why exudation is apparently greater in non-sterile media, where microbes form a competitive sink for the exudates.

This reabsorption depends upon there being a specific uptake transporter in the plasmalemma for the compound in question. These certainly exist for sugars and amino acids (Schobert & Komor 1987; Bush 1993), but probably not for organic acids (Jones & Darrah 1992). This may account for the large net output of organic acids from some plant roots. Where these transporters exist, the difference of net and gross exudation can be very large; thus, amino acid net efflux was less than 0.01% of gross efflux, and an equilibrium or zero net efflux condition was established at 10^{-6} M external concentration of the amino acid in one experiment (Jones & Darrah 1994). Maize roots normally absorb considerable amounts of glycine from external solution, but in presence of metabolic inhibitors, large effluxes were observed (figure 8.3). The uptake capacity was fairly constant up

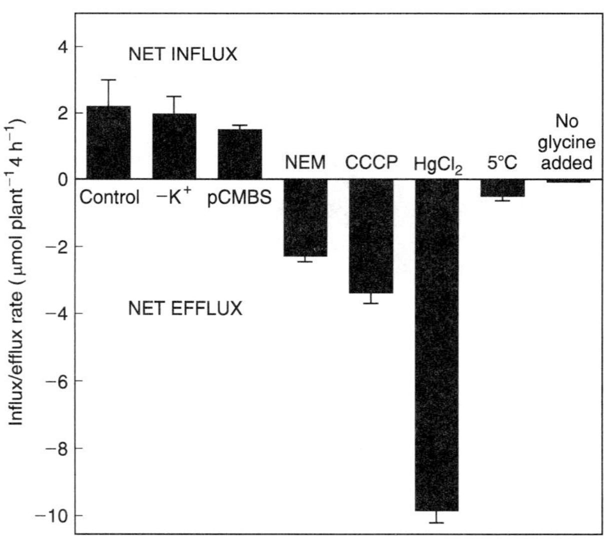

Figure 8.3 Net influx and efflux of glycine in the roots of intact maize plants supplied with external glycine, and effect of metabolic inhibitors. The net influx was not altered greatly by the absence of potassium or the presence of *p*-chloromercuriphenylsulphonic acid (pCMBS). *N*-ethyl maleimide (NEM), carbonyl cyanide *m*-chlorophenylhydrazone (CCCP) or mercuric chloride produced a net efflux (after Jones & Darrah 1994).

to 12 cm behind the root tip, whereas the internal concentration of amino acids, and their efflux, was greatest near the root tip.

The process of exudation is thus much more dynamic and responsive than had been thought before. Darrah (1996) has produced a model of this process, in which the exudation step is assumed to be simple diffusion, whereas uptake is active, with Michaelis–Menten kinetics:

$$AP(C_c - C_L) - V_{max}C_L/(C_L + K_m) = E \tag{8.1}$$

where A is the 'total internal and external area of the root exposed to soil solution' ($cm^2\,cm^{-2}$ of root cylinder area); P is the root membrane permeability; C_L and C_c are the exudate concentrations in the external solution and the cortical cytoplasm, respectively; and V_{max} and K_m are the maximum rate of uptake and the affinity constant in the Michaelis–Menten equation. The net flux of exudate E is linked to a diffusion equation for flow into the soil. If the V_{max} and K_m parameters depend upon the concentration of the organic compounds within the cell, the mechanism is readily responsive to the physiological condition of the plant.

8.2 The Microbiological Community and the Processes of the Rhizosphere

8.2.1 Rhizosphere Processes and Techniques

The 'rhizosphere' lacks a precise definition. In principle, it should include all that volume around a living root in which the physical, chemical and biological properties are perturbed by its presence. However, widely spreading perturbations, such as those of nitrate, water, or volatile compounds, would rapidly fill the whole soil volume, at least in the topsoil. We therefore include only that soil volume in which biological populations are significantly perturbed by a living root. The extreme range of variation of various properties in and along the rhizosphere is shown in Uren & Reisenauer (1988), as each stage of development of the root, in turn, passes through each volume of soil.

The carbon compounds delivered to the rhizosphere support chemical and biological processes that affect nutrient acquisition by the plant (Darrah 1993). The purely chemical processes have been dealt with in chapter 7. The microbial populations of the rhizosphere can certainly influence the growth of the host plants, but the mechanism is still often obscure. However, in some cases it is likely to affect plant nutrition, and for this reason the subject is appropriate to this book. The rhizosphere microbes may affect nutrient uptake by symbiotic and non-symbiotic nitrogen fixation, mycorrhizal uptake, root disease-causing organisms, hormonal effects and general faunal grazing of the roots. Many of these processes are still not well understood, but they are important, in principle, for this book, because they may modify the basic physicochemical uptake model developed in chapters 5, 6 and 7. The uptakes of immobile nutrient elements, such as phosphorus and the trace metals, are particularly dependent upon these rhizosphere processes, whereas others, such as potassium, are only marginally affected.

Other processes, such as N mineralization and denitrification, may proceed faster in the rhizosphere than in the bulk soil. However, it is unlikely that these have a specific effect on that root, rather than on the whole soil, so they are not further discussed here. This subject is complicated because the properties of the root and rhizosphere at a given location are changing progressively (chapters 5 and 9), and because of the constant interactions between the root and members of the microbial, protozoan and faunal populations. This complexity is well shown in figure 8.4. The fauna includes the microfauna, nematodes, worms, collembola, etc., and these will depend upon the vegetation type, the current condition of the soil, and its recent history. There is no space for a full discussion of all these factors here. Good reviews are given in Lynch (1990a), Box & Hammond (1990), Kapulnik (1996a, b) and others.

This complexity of the rhizosphere processes and populations is made more difficult by a lack of precise and quantitative techniques. The problems of defining root surface uptake characteristics are described in chapter 5. The traditional investigation of microorganisms uses metabolic behaviour, culturing or direct microscopic observation (Campbell & Greaves 1990), but none is ideal and the great majority of soil microorganisms have not been characterized (Paul & Clark 1996). New molecular biology techniques are becoming available that can identify

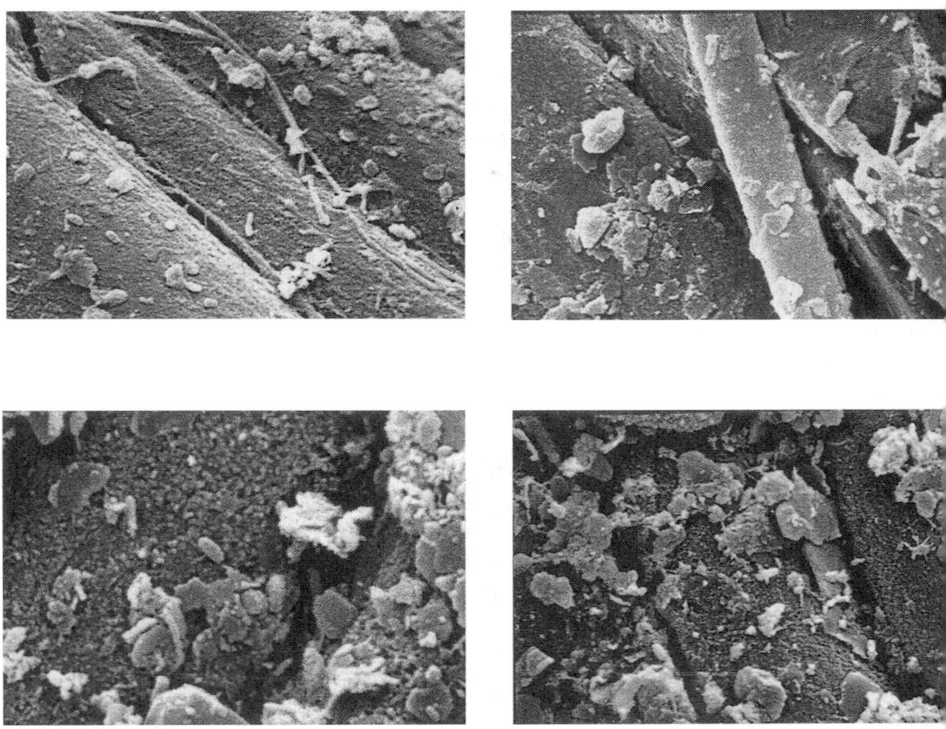

Figure 8.4 General views of root surface, showing cells, bacteria, and fungal hyphae un scanning electron microscope.

organisms more easily and certainly (Preston *et al.* 1998). Rattray *et al.* (1995) used a plasmid with a *luc* (luminescent) gene introduced into the bacterium *Enterocloaca* to follow the growth and behaviour of this organism in culture and in soil, where minimum detection limits were 90 and 445 cells, respectively. This system was used to determine colonization of the rhizosphere by the marked bacterium, and can be used in the environment generally (Prosser *et al.* 1996). Other methods can deliver information about the conditions in the rhizosphere. Thus, a *Pseudomonas* strain has been constructed that responds to phosphate deficiency by producing galactosidase, so that it can act as a phosphate-deficiency sensor within very small volumes (De Weger *et al.* 1994).

Other new techniques are aimed at advancing knowledge of the biological processes that are specific to the rhizosphere, and at determining the nature of the adaptive traits that enable bacteria to thrive there (Rainey 1999). The aim of this work is to identify genes that are expressed solely in the rhizosphere, and are therefore likely to encode for ecologically significant traits. Lethal mutants of *Pseudomonas fluorescens* that do not produce pantothenate (a vitamin that is essential for growth), and therefore cannot normally exist in the rhizosphere, are produced first. Their genomic DNA is broken into random fragments by restriction enzymes, and the fragments are fused to an effective gene for pantothenate production. These constructs are then reintroduced into the mutant *P. fluorescens*, which are tested for viability in rhizosphere conditions. If the fragment contains a promoter that switches on its attached genes in the rhizosphere, pantothenate will be produced, and that type of bacterium will be viable. From this study, a number of 'rhizosphere-specific genes' have been isolated and are in the process of being identified. New developments in molecular biology will clearly have a big impact in this subject.

8.2.2 The Population Dynamics of the Microbial Community

A general description of the rhizosphere community can be found in several excellent publications (Curl & Truelove 1986; Campbell & Greaves 1990; Marschner 1995; Kapulnik 1996a, b). Soil bacteria, in general, are adapted to a low supply of soluble substrate, and the mean metabolic rates are very low compared with those often found in laboratory experiments. The rhizosphere community appears to consist of the general soil community, but with a strong preponderance of those species able to grow rapidly when the substrate level is increased, such as various *Pseudomonas* species. Often the diversity of species is rather low in the rhizosphere. The organisms on the actual root surface, the rhizoplane, are basically similar, but may be embedded in the root surface mucigel (Campbell & Greaves 1990). Bacteria are also found inside the apparent surface of the root, in the endorhizosphere, living in intercellular spaces, or in the voids within the cortical tissues (Patriquin *et al.* 1983). Microbes in the rhizosphere will be exposed to widely differing conditions, with pH, ion concentrations, moisture content and substrate concentration likely to change considerably as the root ages, so the microbes cannot be too specialized (figure 8.4).

The microbial density can be determined by direct counting or by plate isolation, but the latter always gives lower values. The density of the bacteria declines with distance from the root surface (table 8.2), and the rhizosphere/soil (RH/S) ratio expresses the relative numbers per gram of soil in the rhizosphere soil close to the root, and the bulk soil beyond root influence. Typical RH/S (a term used here to prevent confusion with the root/shoot ratio or R/S) values are 5–20, but can reach over 100 (Bazin *et al.* 1990). Rouatt & Katznelson (1961) showed that six different crop species growing in the same soil had RH/S ratios ranging from 24 (red clover) to 3 (barley and maize). These comparisons are difficult, however, because Riviere (1960) showed that RH/S ratios were very dependent upon plant developmental stage, being 3.1 at germination, 27.7 at tillering, 16.8 at heading and 5.4 at maturity in wheat. This is not surprising, considering how carbon allocation in the root system changes with time (section 8.1.3). The RH/S ratio is very variable, and should therefore be used with caution. The change in density of fungi is less obvious, probably because they have the ability to translocate photosynthate and nutrients along their hyphae. The uncertainties attached to the various techniques used mean that the absolute numbers of microbes should not be stressed, but the relative values are probably more dependable.

The dynamics of this community are not easy to follow, as it is never in a steady state. Each segment of root goes through the development phases described in chapter 5 at the same time that rhizodeposition alters in both quantity and quality. Bazin *et al.* (1990) showed that the mathematical approaches used to describe variable populations in chemostat conditions are of little use in a situation as complex as the rhizosphere. Newman & Watson (1977) produced the first directly applicable mathematical model, based on diffusion of substrate from a central root, and leading to a wave of microbial growth both outwards into the soil and along the root as it grows into the soil. Numerical solutions of the equations show that (figure 8.5) the rhizosphere shrinks in diameter after its initial growth, because the rapidly increasing microbial population consumes the diffusing substrate progressively closer to the root. However, the model must be a great simplification on what actually happens in the rhizosphere.

Table 8.2 Microorganism populations at different distances from the root surface, for lupin seedlings, measured by plate count.

Distance from root[a] (mm)	Microorganisms (1000s per g oven-dried soil)			Fungi (per g oven-dried soil)		
	Bacteria	Streptomycetes	Fungi	*Aspergillus ustus*	*Cylindrocarpon radicicola*	*Paecilomyces marquandii*
0	159 000	46 700	355	5650	4940	9000
0–3	49 000	15 500	176	3360	0	2800
3–6	38 000	11 400	170	2920	0	1600
9–12	37 400	11 800	130	2880	0	1500
15–18	34 170	10 100	117	2270	0	0
80	27 300	9 100	91	1000	0	0

Source: after Papavizas & Davey (1961).

[a]Distance 0 is the rhizoplane; distance 80 mm was regarded as control soil.

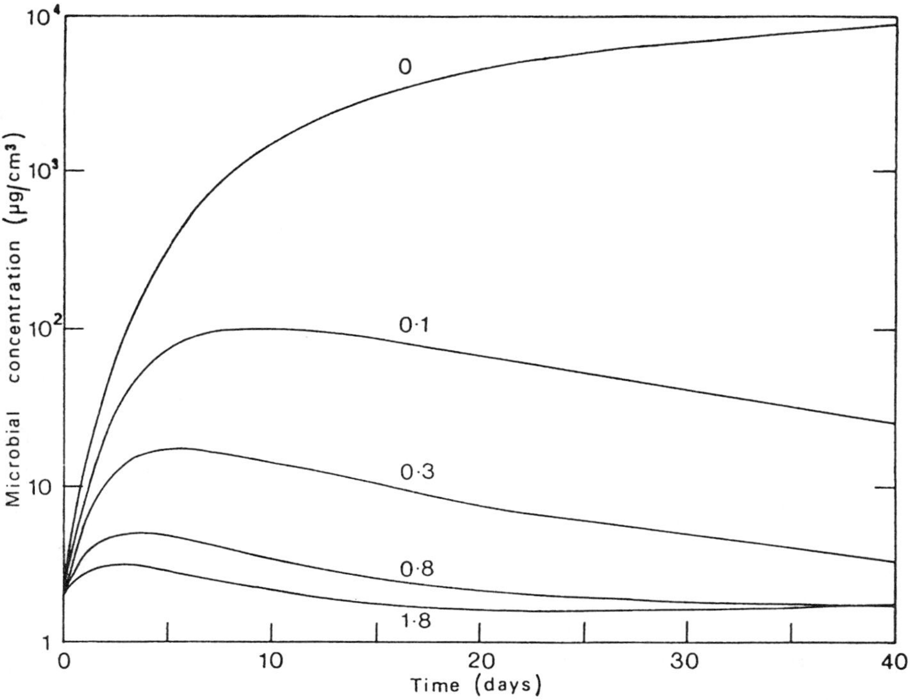

Figure 8.5 Variation with time of the growth of the simulated rhizosphere population along the length of a root at various distances from the root surface (after Newman & Watson 1977).

Darrah (1991) has developed a similar model with more complex carbon cycling and with varying distribution of exudation along the root. Biomass was predicted to reach a peak about 100–150 h after the arrival of the root. Darrah pointed out that the omission of predation on the microbial biomass in the model is serious, as this process has major effects on the microbial population (Clarholm 1981, 1985).

Yeates & Darrah (1991) modelled the conditions around an artificial 'root' of Millipore membrane. The same basic development as predicted by Newman & Watson (1977) was again found, with the zone of increased microbial biomass becoming progressively narrower. After 5 days, over 90% of the microbial biomass formed was within 2 mm of the 'root', and the soluble C that resulted from microbial metabolism was more concentrated than the original soluble carbon supply. Such models implicitly assume that the limiting nutrient for the microbes is carbon, but sometimes nitrogen deficiency can limit microbial growth (Marschner 1995, p. 561).

At present, it is not easy to identify the factors that determine which bacterial species thrive (have a high degree of fitness) in the rhizosphere, especially as many bacteria that live there have not been identified. This hinders the attempts to use artificially inoculated organisms to enhance plant growth (section 8.2.4). Hozore & Alexander (1991) found that the only property correlated with fitness was that of motility, because this ensures that the organism can follow the root as it grows through the soil.

Microbe–plant specificity is important, and there are many records of different levels of colonization of the roots of the same plant by different strains of the same bacterium, or by different bacterial species (Hornby 1990; Charney *et al.* 1991; Kloepper *et al.* 1991; Glandorf *et al.* 1993). However, it is very difficult to form generalizations from most of the reported work.

8.2.3 Effects on Plant Growth and Nutrient Uptake by Free-Living Microorganisms

Free-living microorganisms could alter the growth and nutrient uptake of plants in several ways.

(1) Change in carbon allocation. The effect of microbes on exudation and rhizodeposition has been noted in section 8.1.2, and if the exudates have direct effects on nutrient uptake, this may be, in part, an indirect effect of the microbes. However, so far no clear effects of this type appear to have been confirmed.

(2) Change in root morphology. Root morphology can certainly be altered by microbes, and this could be caused by a change in carbon allocation. Increase in root length after inoculation with bacteria (Martin *et al.* 1989) is not relevant if the size of the shoot has increased in proportion, but the changes in root hair numbers and length are important (Bowen & Rovira 1961; Kapulnik 1996a) (figure 8.6) (section 6.3.2). It has been suggested that bacteria on the root surface may block access to the apoplast for solutes in the soil solution. However, direct

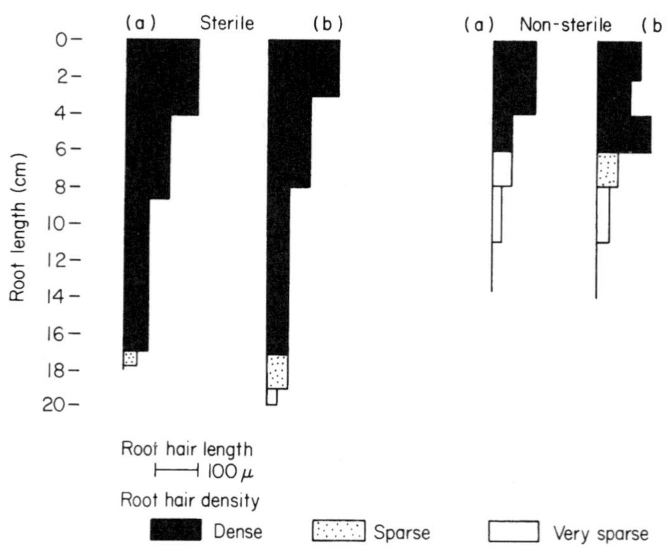

Figure 8.6 Effects of absence or presence of microorganisms on the root surface on root hair growth and development in subterranean clover: (a) in sand and (b) in agar (after Bowen & Rovira 1961).

observation showed that only about 10% (Rovira *et al.* 1974) or 1% (Schippers & Van Wurde 1978) or 8–20% (Suslow 1982) of the root surface carried microbes, so the effect is unlikely to be significant, even if the bacteria do live preferentially over the gaps between cells.

(3) Competition of microbe and plant. The idea that rhizobacteria can compete with roots for deficient nutrients has been put forward many times, but, according to the work of Tinker & Sanders (1975), the major nutrients that could be held in the bacteria are rarely sufficient to affect higher plants that are growing normally. There have been suggestions that competition for iron develops in the rhizosphere, both between different microbes and with the plant, but the evidence for plant effects is still unclear (Swinburne 1986).

(4) Alteration to nutrient. Some bacteria have the ability to oxidize divalent manganese ions to manganese dioxide, which will generally lower manganese availability. Microbial activity and plant exudates can also create reducing conditions, and so produce more divalent manganese, which favours uptake (Marschner 1995, p. 563).

(5) Change in uptake properties of roots. Bacteria have been found that apparently increase or reduce uptake of phosphate (Barber 1969; Barber & Rovira 1975). The mechanism is rarely certain, but the production of plant hormones could alter the root uptake characteristics.

(6) Causation or prevention of disease. Loss of active root through disease will decrease uptake, as shown by many root pruning experiments.

(7) Stimulation or inhibition of formation of ectomycorrhizas. Fitter & Garbaye (1994) have shown that some bacteria are important in determining the level of colonization of ectomycorrhizal fungi.

Some potential processes of enhanced nutrient supply have repeatedly been proposed for many years, and we will discuss these here in more detail.

(a) Phosphate solubilization can be regarded as a form of nutrient alteration. Over the years, a very large literature has built up on 'phosphate-solubilizing bacteria', in which 'solubilize' refers to dissolution of P_i (inorganic P) from less soluble Ca or Al phosphates, or from organic soil phosphate (Brown 1974) (section 3.6.2). Many of the experiments that apparently show this effect can be explained by bacterial pH changes or chelate accumulation in small volumes under artificial conditions, such as Petri dishes, or result in uptakes of 'solubilized' phosphorus by bacteria which would be too small to be important in crop nutrition (Tinker 1980). This is ascribed to bacterial organic complexing acids or phosphatase exoenzymes (May *et al.* 1993), but plant roots already have both mechanisms (sections 7.3.2 and 7.3.3) (Marschner 1995, p. 559). It seems unlikely that this is an important process, though minor effects cannot be discounted. It is likely that the usual rate-limiting factor in hydrolysis of organic phosphate is not the quantity of phosphatase, but that of soluble organic phosphate (Gahoonia & Nielsen 1992; Kapulnik 1996a).

There have been reports that phosphate-solubilizing bacteria and vesicular-arbuscular mycorrhizas interact positively, but the evidence that this is linked to phosphate-solubilization by the bacteria is slight (Reid 1990). Zhu & Ehrenfeld (1996) studied the rate of breakdown of organic litter due to the presence of mycorrhizal pine roots. The roots caused a larger mass loss and release of phosphate, but this appeared to be an increased general decomposition of litter rather than a specific release of phosphorus. These authors note the very contradictory results that have been reported in this subject.

It is possible that microbes can contribute to a flow of phosphorus to the root or in the soil profile generally, as discussed by Nye & Tinker (1977). Hannapel *et al.* (1964a, b) determined 1 $\mu g\ g^{-1}$ P in colloidal solution in the soil solution of a soil given soluble sugars to encourage microbial growth. There is much colloid-sized and dissolved organic phosphorus in some solutions, and Chapman *et al.* (1997) found some evidence that these concentrations are determined by changes in biological activity when soils are sieved or otherwise disturbed.

(b) Non-symbiotic nitrogen fixation. There are many reports (Zuberer 1990) of the ability of *Azospirillum* and other bacteria in the rhizospheres of grasses and other plants to fix nitrogen and thereby promote plant growth. Some of them are in very close association with the root surface, to which they adhere with their fimbriae (Korhonen *et al.* 1986), and may be specific for particular plant species. *Azorhizobium caulinodans* forms nitrogen-fixing nodules with a tropical legume, and Sabry *et al.* (1997) showed that it would also invade wheat roots. There are many indications that these microbes fix nitrogen, but the practical question is whether the amounts fixed are significant.

The energy demand for fixation is quite large (Whipps 1990; Zuberer 1990). An active rhizobium symbiosis in a legume usually consumes around 10% of the plant's photosynthate and the use of photosynthate within the nitrogen-fixing nodules must be more efficient than it can be in the rhizosphere. Rennie & Larson (1981) reported that lines of spring wheat which leaked most exudate also fixed most nitrogen, suggesting that this might be a limiting factor. The availability of energy for non-symbiotic fixation in the rhizosphere therefore needs to be considered. The average requirement is 1 g of carbon to fix 10–15 mg nitrogen (Kapulnik 1996b). If only 1–3% of plant net photosynthate, with 50% C, finds its way into soluble low-molecular-weight exudates, the maximum amount fixed would be 0.005–0.015% of the net photosynthesis, assuming it was all used for nitrogen fixation, which is improbable. This is a trivial contribution to a plant with at least 1% N in dry matter. The carbon allocation below ground is much greater than 1–2%, but much of this is as solid root debris. Some of this may become available to nitrogen fixers as it is degraded by microfauna and microorganisms (section 8.1.3), but this is not a rhizosphere process.

Hunt (1990) reviewed the situation and concluded that fixation by free-living bacteria had often been overestimated, and that observed growth responses were more often due to hormonal effects. In general, the current opinion is that non-symbiotic bacteria will rarely fix more than 20 kg N $ha^{-1}\ year^{-1}$ (Giller & Day 1985); the highest value found by Rennie & Thomas (1987) using a ^{15}N technique was 52 kg N ha^{-1} under optimum conditions. There are claims that sugar cane or other tropical grasses can yield for year after year without nitrogen fertilizer, with the implication that nitrogen is supplied by non-symbiotic fixation, but a set of long-term nitrogen balances (Kapulnik 1996b) gives most results well below 50 kg N $ha^{-1}\ year^{-1}$. We conclude that rhizosphere fixation of nitrogen is of marginal importance for high-yielding crops, though it might be useful for low-yielding crops or natural vegetation. The question of whether these small amounts can be increased by inoculation is discussed below.

8.2.4 Artificial Inoculation — 'Plant Growth-Promoting Rhizobacteria'

Groups of bacteria have been identified that either promote or that damage plant growth when inoculated on to their roots; those with a positive effect are the plant growth-promoting rhizobacteria (PGPR). Some of these bacteria are specific to certain host plants as found by Kloepper *et al.* (1991) for microorganisms that are antagonistic to nematodes. The mechanisms are difficult to identify (Lynch 1990b; Kapulnik 1996a), but they are probably among those listed in section 8.2.3. The most likely positive mechanisms include (Bowen & Rovira 1991) antagonism against detrimental or pathogenic organisms (possibly by competitive sequestration of iron), plant hormone production, or nitrogen fixation. Possible negative mechanisms include pathogenic effects, including subclinical infections, cyanide production and other toxin production. Experiments with PGPR bacteria inoculated on host plants roots in the field have usually given extremely variable and inconsistent results (Suslow 1982; Schippers *et al.* 1987; Kapulnik 1996a).

This inconsistency may be due to failure to colonize the root surface properly. Bacterial motility is believed to be important to maintain the root surface fully colonized with soil bacteria (section 8.2.2). If bacteria are inoculated only onto the seed, high motility must be essential for biological control agents (Suslow 1982; Kloepper *et al.* 1989; Garinin *et al.* 1991; Tan *et al.* 1991), whether by transport in flowing water, or by the motility of individual bacteria (Bowers & Parke 1993). Nijhuis *et al.* (1993) developed a set of techniques for determining the motility and fitness of introduced bacteria to survive in the rhizosphere, and De Weger *et al.* (1987) noted that the colonization of potato roots by *Pseudomonas* was slower if mutants without flagella were used as the inoculum. Other authors have come to the opposite conclusion (Kapulnik 1996a), but it seems unlikely that non-motile bacteria could effectively colonize a root system that may penetrate for a metre or more into the soil, if they are solely inoculated onto the seed.

Hormones are often implicated in the effects on higher plants, though they are difficult to investigate, and many experiments testing for them have given negative or conflicting results (Lynch 1990a). *Azospirillum brasiliense* inoculated onto wheat growing in soil increased root length, root hair length and number, and shoot weight (Martin *et al.* 1989), and these effects could be duplicated by supplying indole acetic acid (IAA) to the plant. However, Sanwar & Kremer (1995) reported an example of growth suppression by rhizobacteria that produced what they considered to be excessive amounts of IAA. Bacteria that produce strong siderophores (Neilands 1984) (section 7.3.4) could compete against pathogenic or cyanide-producing bacteria that produced weaker or less copious siderophores of iron (phytosiderophores are stronger ligands than siderophores, so the plant should not suffer from the competition). This is supported by the observation that added ferric iron prevented this bacterial competition in some cases.

It has been shown that many rhizosphere *Pseudomonads* produce hydrogen cyanide (Knowles & Bunch 1986), and some bacteria have specific cyanide resistance (Lynch 1990b). A *Pseudomonas fluorescens* bacterium that produces cyanide was toxic to *Thielaviopsis basicola*, which causes root rot of tobacco. A mutant without the cyanogenetic gene gave no control of the disease, but insertion of this gene restored the cyanide production and the control of the pathogen (Voisard *et al.* 1989). It is evident that bacteria on the roots of plants can affect their growth, and that artificial inoculation is sometimes successful. Over many years, the problem has been that the effects have uncertain mechanisms, and that the results are inconsistent.

8.3 Effects on Plant Growth and Mineral Nutrition by Mycorrhizal Fungi

8.3.1 General Introduction

The mycorrhizal fungi have gained considerably more prominence in the 1980s and 1990s. Whereas the understanding of their function in plant nutrition was tentative, this has now been largely confirmed by many hundreds of publications (Wilcox 1996). This is particularly true of the arbuscular fungi, so we deal with these first here, though other systems also occur (Pate 1994).

A mycorrhiza is a close symbiotic association of a fungus and a root of a higher plant. Four main types have been defined: ectomycorrhizal, arbuscular, orchidaceous and ericoid (Harley & Smith 1983; Smith & Read 1997), of which we only deal with the first two, because they are the most important for plant nutrition. The great majority of the world's plant species are mycorrhizal, which is the normal condition for a plant growing in the field, or in any non-sterilized soil. Most of the hosts obtain a benefit from the symbiosis, probably nutritional, and for some plants it is essential. In effect, mycorrhizal colonization may completely alter the nutrient-uptake properties of the root system. The mycorrhizal condition is consequently a core part of our understanding of plant and crop nutrition. The discussion we give here concentrates upon the movement of nutrients and the effect on plant nutrition, and extensive parts of the total subject of mycorrhizas are therefore omitted (see Harley & Smith 1983; Reid 1990; Smith & Read 1997).

8.3.2 Arbuscular Mycorrhizas (Previously Vesicular-Arbuscular Mycorrhizas)

The arbuscular mycorrhizas (AM) (or vesicular-arbuscular mycorrhizas, VAM) occur on grasses, legumes, herbs, most tropical trees and most crop plants. The AM fungi are classified in the Endogonaceae, but their taxonomy is still confused and complex. They are not very specific and most form mycorrhizas with a wide range of plants (Brundrett 1991; Wilcox 1996; Smith & Read 1997).

The AM fungi are obligate symbionts, so they need a higher plant 'host' if they are to be grown for extended periods. Spores (figure 8.7b) will germinate on agar media, but the fungus depends upon the spore for its carbon. Some authors have

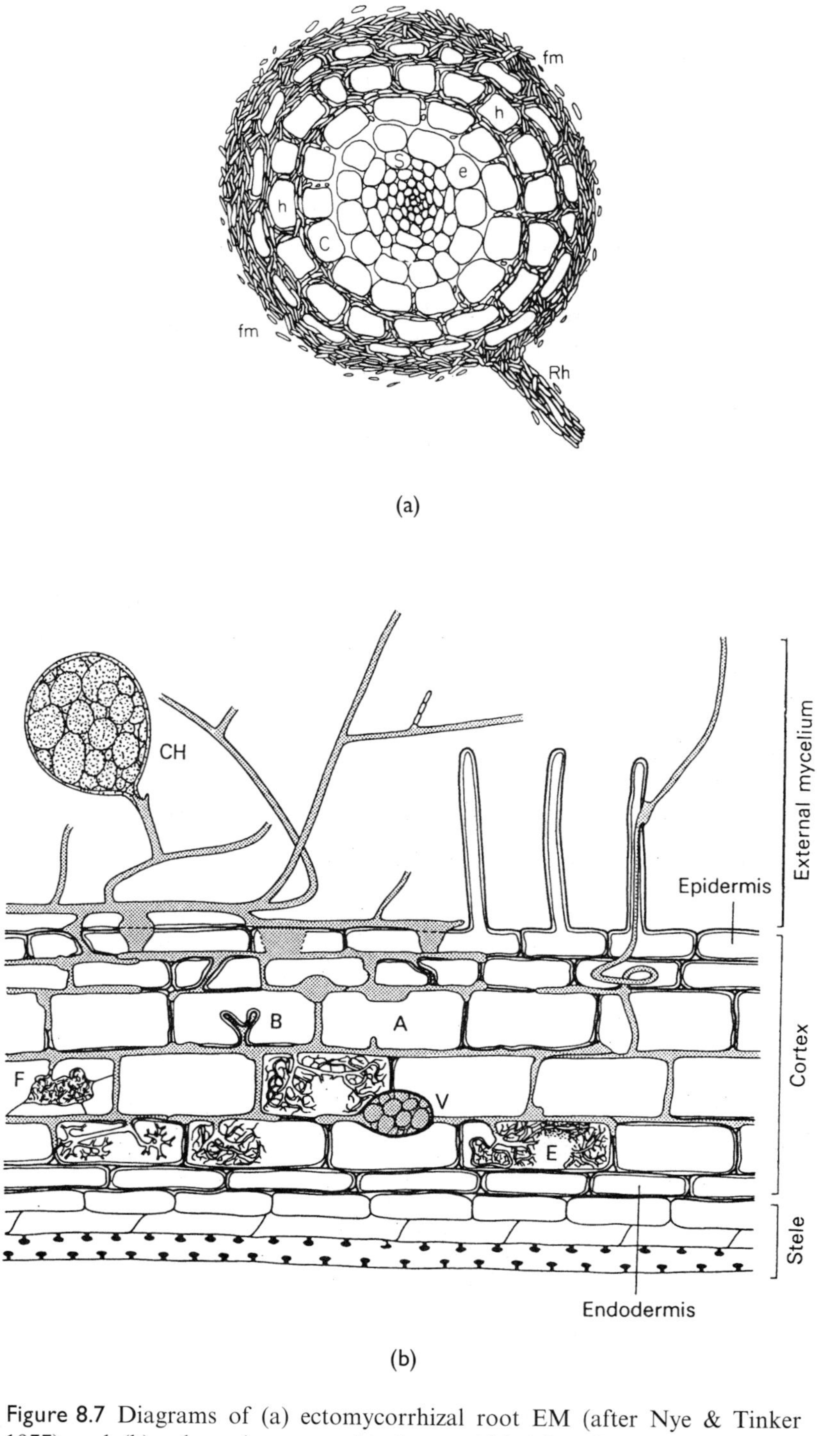

Figure 8.7 Diagrams of (a) ectomycorrhizal root EM (after Nye & Tinker 1977) and (b) arbuscular mycorrhizal root AM (after Scott Russell 1977). Main features include (a): h, Hartig net; fm, fungal mantle; Rh, external rhizomorphs; (b): V, vesicles; A, B, E, F, arbuscules; CH, external spores.

considered this to be a major disadvantage for AM research. However, a number of ectomycorrhizal (EM) fungi can be cultured, but this has not caused understanding of the function of EM to be more clearly understood than that of AM.

An external hypha, from a germinating spore or an existing infection, can invade a root, developing an internal mycelium in infection units (Cox & Sanders 1974) (figure 8.7) and producing specialized vesicles and arbuscules. The total infection thus consists of a considerable length of external mycelium that carries spores (Sylvia 1990), and internal mycelium with vesicles and arbuscules contained within the roots. The arbuscules penetrate into cortical cells, but, in fact, they never break the host-cell plasmalemma. The two phases are connected through varying numbers of thicker entry-point hyphae.

The fraction of the root length colonized is a useful measure of infection that varies with the species of the host and fungus, the inoculum density, the duration of the infection and the environmental conditions (Mosse *et al.* 1981). Sometimes, root lengths containing hyphae, arbuscules or vesicles are reported separately. In general, shading and waterlogging reduce colonization. Low pH may do so, but fungi vary in their pH optima in the soil (Wang *et al.* 1993). An existing infection will decrease the colonization by a different fungal species (Pearson *et al.* 1993). The extent and intensity of infection is usually depressed by high phosphorus supply under controlled conditions (figure 8.8), but this is less consistent in the field (Black & Tinker 1979; Jensen & Jakobsen 1980). At the present time, the

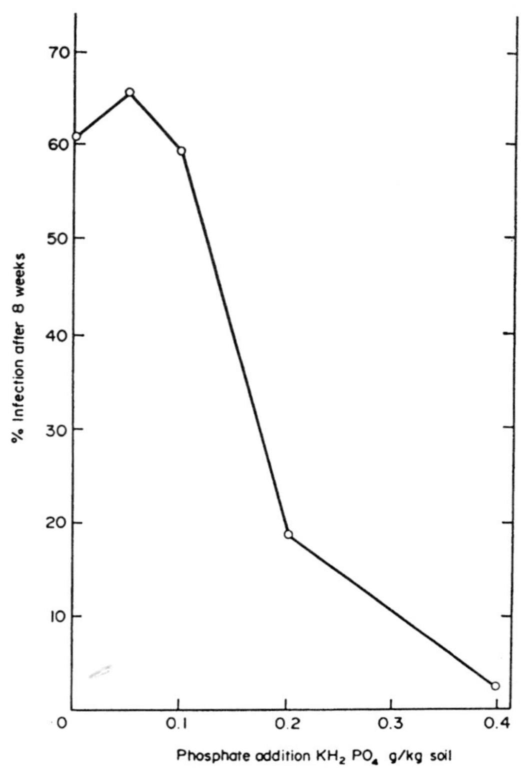

Figure 8.8 Effect of soil phosphate level on percentage of leek root length infected with AM (after Sanders & Tinker 1973).

mechanisms that control the degree of infection are still uncertain (Amijee *et al.* 1989, 1990; Koide & Li 1990; Smith & Read 1997; Gianinazzi-Pearson *et al.* 1996).

In the last 20 years, interest in the external mycelium has greatly increased, relative to the properties of the fungus within the root. The range of length is considerable (Smith & Gianinazzi-Pearson 1988; Sylvia 1990), with a typical value of around 100 m of hyphae per metre of colonized root. However, dead and live hyphae have not been separated in most work. Where these have been differentiated (Hamel *et al.* 1990; Sylvia 1990), only 40% or less of the total length was found alive; Jones *et al.* (1990) found the same for EM hyphae. The real inflows of nutrients into hyphae may therefore be two or more times larger than the mean values calculated from total hyphal length.

There is no consistent correlation between percentage infection and size of growth response of the host (Fitter & Merryweather 1992), though better correlations may be found with the length of root that contains arbuscules. The poor agreement is not surprising considering the enormous differences in the experimental conditions used, and the possibility of varying extent and viability of the hyphae.

8.3.3 Ectomycorrhizas

These occur mainly on forest trees of the temperate and boreal zones, and have major morphological differences from AM. The symbiotic fungi have a much wider taxonomic range than the AM fungi, many may be saprophytes under appropriate conditions, and many have been grown in culture (Harley & Smith 1983). Typically, the EM fungi form a dense sheath of hyphae around the colonized roots (figure 8.7a), which may be up to 40% of the total weight of the colonized root, with a web of hyphae (the Hartig net) between the cortical cells of the host root. They form no specialized bodies such as vesicles or arbuscules, but the branching of the fine root system and its internal morphology are often greatly altered.

The very large number of EM fungi produce a large variation in morphology and behaviour, and over time there may be a succession of different fungi on young mycorrhizal trees growing on sites without mature trees (Last *et al.* 1987). These differences from the AM have caused research on them to follow different lines. Work on AM has concerned itself with process and function more than has work on EM, which has tended to address taxonomy, development, and biochemistry (see Smith & Read 1997).

The frequent increase in plant growth following colonization with EM has been known for a long time. Exotic trees often fail when they are introduced on phosphate-poor soils without a suitable mycorrhizal symbiont, and the beneficial effects of introducing mycorrhizal fungi may be startling (Marx *et al.* 1989, 1991). The increase in internal phosphorus concentration following colonization (Harley 1969) suggested that plant nutrition, probably involving phosphorus, was the basis for these responses. It is now known that nitrogen nutrition may also be greatly improved by colonization with EM fungi in some conditions (Abuzinadah & Read 1989; Turnbull *et al.* 1995) (section 8.3.6).

Recently, experimental methods developed for AM have been applied to EM symbioses with *Salix*. The similarity of the results and those with many AM colonized species led Tinker *et al.* (1992) to suggest that nutrient uptake processes of EM fungi with fast-growing host species were basically similar to those of AM fungi. To allow this point to be considered, in the rest of this section we discuss EM and AM together.

8.3.4 Modelling of Infection Development and Function in AM

The percentage root length infected with the fungus at a particular time depends upon the extension rates of both the whole root system and the colonized part of the root. The development of infection is therefore a highly dynamic process, and there have been several attempts to model it. Figure 8.9 shows the usual form of the increase in colonized length of a plant in soil that contains inoculum.

Tinker (1975a) assumed that the rate of production of new colonized root length depended on the length of infected root (inoculum) and the length of uninfected root (target), giving

$$\mathrm{d}L_i/\mathrm{d}t = SL_i(nL_t - L_i) \tag{8.2}$$

where L_i is infected root length, L_t is total root length at time t, S is the specific spread rate of infection in the root (days^{-1}), and n is the maximum fraction of the root system that can become mycorrhizal. The mechanisms controlling the last factor are not known (figure 8.9).

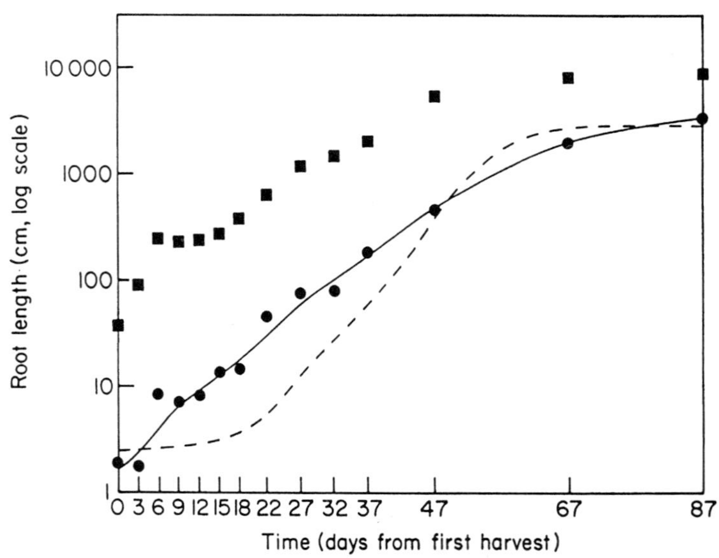

Figure 8.9 Progress of AM colonized length (circles) with time in a growing leek root system (squares). Simulation using equation (8.3) is indicated by full line, that with equation (8.2) by the dotted line (after Buwalda *et al.* 1982).

The approach of Smith & Walker (1981) was different and focused on internal rather than external hyphal extension, and their equation did not include the term n. Later, Buwalda *et al.* (1982) developed equation (8.3) that is closely similar to that of Smith & Walker (1981), by the argument that the total extension rate of the infection units of colonized root was proportional to the length of infected root, but that the extension does not occur or is not observed if the adjacent root is already colonized (equation (8.3)). The chance of this happening is L_i/L_t. The rate of extension of L_i is given by

$$dL_i/dt = SL_i(1 - L_i/nL_t) \tag{8.3}$$

Direct tests with different plant densities showed that equation (8.3) is superior (figure 8.9). Equation (8.3) is useful for comparing the development of a colonized root system, in terms of standard parameters, with different species or under differing environmental conditions (Sanders 1993), and it usually gives a good fit to data from young plants in pots. The fact that a very similar equation was produced by two different lines of reasoning suggests that underlying mechanisms need more investigation.

Sanders & Sheikh (1983) produced a more complex infection model which simulated the processes inside the roots in greater detail. Sanders (1993) also developed a full plant growth/colonization model, based upon the concepts that the phosphate inflow is proportional to the fractional infection, there is a carbon cost for the infection, and the R_w of the host is proportional to the percentage phosphorus (section 10.2.5). This plant growth model was then combined with the simple infection model in equation (8.3). This is, so far, the only fully integrated plant–fungus model, and it is able to reproduce some of the features found in host–fungus relations (figure 8.10). The use of models is the best way in which the dynamic nature of the mycorrhizal colonization of a growing plant can be properly captured, and the *ad hoc* one-off measurement avoided.

Little modelling has so far been done with the EM, where morphological changes in the root system often make modelling more difficult than for AM (Sanders *et al.* 1983). However, there seems to be no reason why modelling of this type should not be applied to rapidly growing angiosperms where colonization occurs without root modification, as compared to the short morphologically modified roots found with colonized conifers and other species (Jones *et al.* 1991).

8.3.5 Enhancement of Host Growth by Phosphorus Uptake

8.3.5.1 Phosphorus Uptake by Mycorrhizal Plants

The main theory of the essential nutritional function of the AM mycorrhizal colonization is simple. Nutrient ions of low mobility in the soil (e.g. phosphate) may diffuse to roots with moderate or high demand so slowly that the concentration at the root surface becomes effectively zero (section 6.2.3). If the inflow then is not sufficient to allow maximum growth rate (section 10.1.7), the plant is deficient. Fungal hyphae are able to take up ions externally, translocate these to the internal mycelium, and then transfer them to the host, thus bypassing the diffusional impedance. The mechanism is similar to that of root hairs (section

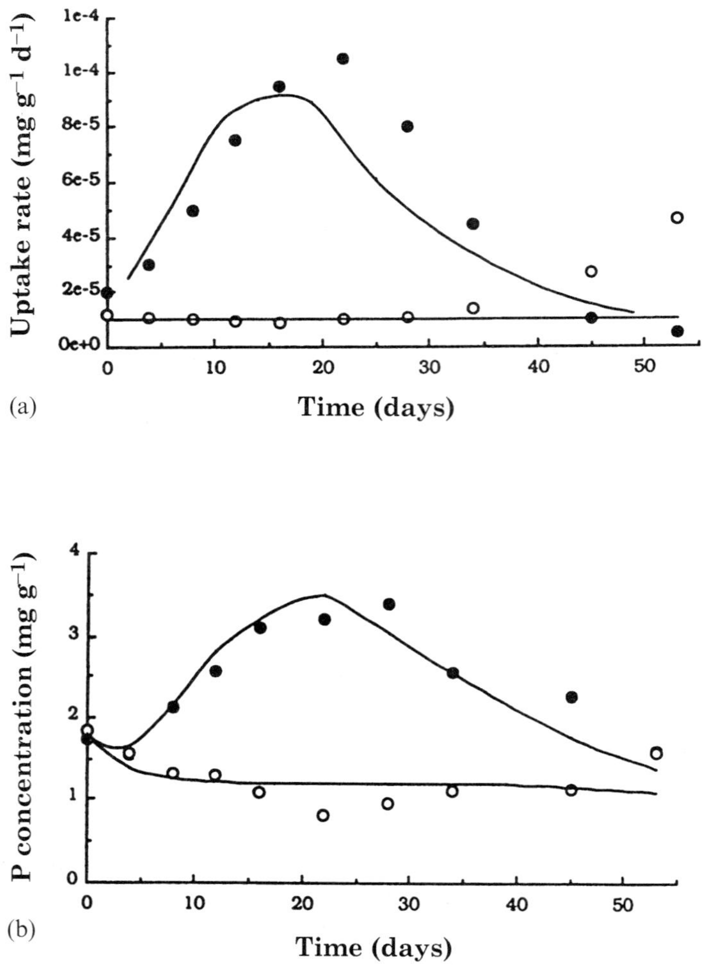

Figure 8.10 Lines are results of simulations with the model of Sanders (1993). This shows transient peaks in (a) uptake rate and (b) P concentration shortly after infection has developed, as found in a number of experiments. Observations for ●, AM; and ○, for non-AM plants (after Sanders 1993).

6.3.1), but the hyphae spread further and have a greater length per unit length of root (Tinker 1975b). A response to mycorrhizal infection can be expected to occur when the required mean inflow of a nutrient to the uninfected root system cannot be supplied by that soil. Plants with sparse, thick roots ('magnolioid') would therefore be expected to depend upon mycorrhizas (Baylis 1970).

The idea that colonized plants could absorb phosphorus more rapidly than non-infected ones was stated quite early (see Harley 1969, p. 281; Mosse 1973). Subsequently, the mycorrhizal effect has been expressed more precisely by comparing the phosphorus response curves of colonized and non-colonized plants. A range of different soils with different histories of cropping and different phosphorus amendments, and sterilized to eliminate indigenous mycorrhizal fungi, were used to grow leeks. When shoot weight was plotted against Olsen extractable phosphate, the mycorrhizal and non-mycorrhizal plants fell on two well-defined separate lines (figure 8.11), so the mycorrhizal and non-mycorrhizal plants had different phosphorus uptake properties (Pairunan *et al.* 1980; Stribley *et al.* 1980b). The response to infection may be quite small at extremely low phosphorus levels, increase to a maximum at intermediate levels, and then disappear at high levels. In this work, infection changed the soil critical level for maximum yield from 150 to 50 mg kg^{-1} P in soil (figure 8.11).

Comparable experiments with EM plants (Bouger *et al.* 1990; Jones *et al.* 1990) showed that infection altered their uptake properties and changed the critical level for P deficiency in the same way. It would be simple to repeat the experiments with other elements with low mobility on suitably chosen soils, but we are not aware that this has been done.

A number of other mechanisms were proposed to explain the effect, before the importance of external hyphal length and distribution had been fully appreciated (Harley & Smith 1983). Sanders & Tinker (1971, 1973) showed that (a) the inflow to infected roots could be several times larger than to the roots of uncolonized plants in the same soil, so a change in root system morphology or duration could not explain the effect (tables 8.3 and 8.4); (b) these enhanced inflows were greater than the calculated inflow to a zero sink root in that soil, so changes in root uptake characteristics could not explain it; (c) after ^{32}P labelling of soil, the specific activity of P in both colonized and uninfected plants grown in it were equal to that of the soil solution, hence both were drawing phosphate from the same isotopic pool in equilibrium with the soil solution (Tinker 1975b). This showed there was no 'solubilization', or transfer from organically bound or inor-

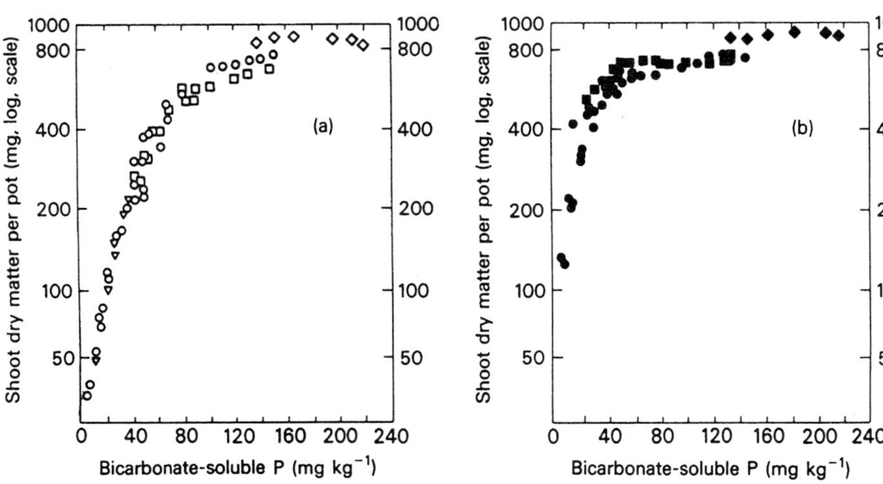

Figure 8.11 Effect of soil phosphate level on yield of leeks, when grown (a) without or (b) with AM colonization in a series of 10 soils all given five rates of P fertilizer, in relation to Olsen bicarbonate-soluble phosphate. Note the difference in the response curves (after Stribley *et al.* 1980b).

Table 8.3 Phosphorus inflows for roots with and without AM infection, and inflows to roots coming only via the hyphae, and inflows into the hyphae themselves, with calculated expected maximum inflow to hyphae.

Mean inflow into uninfected onions, measured	4×10^{-14} mol cm^{-1} s^{-1}
Mean inflow into infected onions, measured	17×10^{-14} mol cm^{-1} s^{-1}
Mean maximum inflow into uninfected onions, calculated	4×10^{-14} mol cm^{-1} s^{-1}
Mean inflow into hyphae, from measured whole plant uptake and hyphal length	2×10^{-15} mol cm^{-1} s^{-1}
Mean maximum inflow into hyphae, calculated approx.	8×10^{-15} mol cm^{-1} s^{-1}
Hyphal inflow to root (see text)	26×10^{-14} mol cm^{-1} s^{-1}

Source: after Tinker (1975b)

Table 8.4 Some published values of inflows of phosphorus with and without AM or EM mycorrhizal infection.

	Root inflows (mol m^{-1} s^{-1} $\times 10^{-12}$)		
	Mycorrhizal (M)	Non-mycorrhizal (NM)	Ratio M:NM
Vesicular-arbuscular mycorrhiza			
Sanders and Tinker (1973) (onion)	12.0	3.7	3.2
Sanders *et al.* (1977) (final harvest) (onion)	5.0	2.0	2.5
Smith (1982) (av. 30–38 days low P) (clover)	0.3	0.1	3.0
Smith *et al.* (1986) (onion)	7.0	2.0	1.0–4.0
Jakobsen (1986) (peas)	7.8	3.0	2.6
Jones *et al.* (1998) (eucalyptus)	0.8	0.3	2.5
Ectomycorrhiza			
Jones *et al.* (1991) (willow)	2.3	1.0	2.3
Jones *et al.* (1998) (eucalyptus)	1.2	0.3	4.0

Source: after Tinker *et al.* (1992).

See original publication for non-listed references.

ganically crystallized phosphate (section 3.5.2). However, exudation of organic acids could possibly alter the equilibrium between phosphate in the soil solution and exchangeable phosphate on colloid surfaces, to give an increased soil solution concentration and hence a larger zero sink inflow (Tinker 1975b). Organic acid efflux is known to happen with some roots (section 7.3.2), but the specific activity of the absorbed phosphate is usually changed when this occurs. There is no significant additional excretion of exudates by AM roots (Ratnayake *et al.* 1978), and the changes in the acid–base relationships of a root due to infection are small (Buwalda *et al.* 1983), so this mechanism seems unlikely. Li *et al.* (1991) have shown that the presence of hyphae extends the phosphate depletion zone well beyond the roots (figure 8.12) and so supports the theory of hyphal uptake.

At the present time this appears to be the best theory. Where there is a growth response to infection, inflows increase by a factor of 2–5 (Smith & Gianinazzi-Pearson 1988; Tinker *et al.* 1992; Smith & Read 1997) (table 8.4). The work with ^{32}P labelling has been repeated on many species, and the great majority gave a similar result to that described above (Smith & Gianinazzi-Pearson 1988; Tinker *et al.* 1991).

For the EM, a great deal of work on the uptake of ions by the fungal sheath showed conclusively that it could absorb a variety of ions (Smith & Read 1997). In particular, phosphate was absorbed at a greater rate than by true roots from solution culture, followed by transfer to the roots as P_i, so that most phosphate would be expected to pass through the sheath. With EM, a wide range of hypotheses were considered to explain the enhancement of the host growth. The sheath could be an absorbing or a storage organ, but the increase in the radius of the

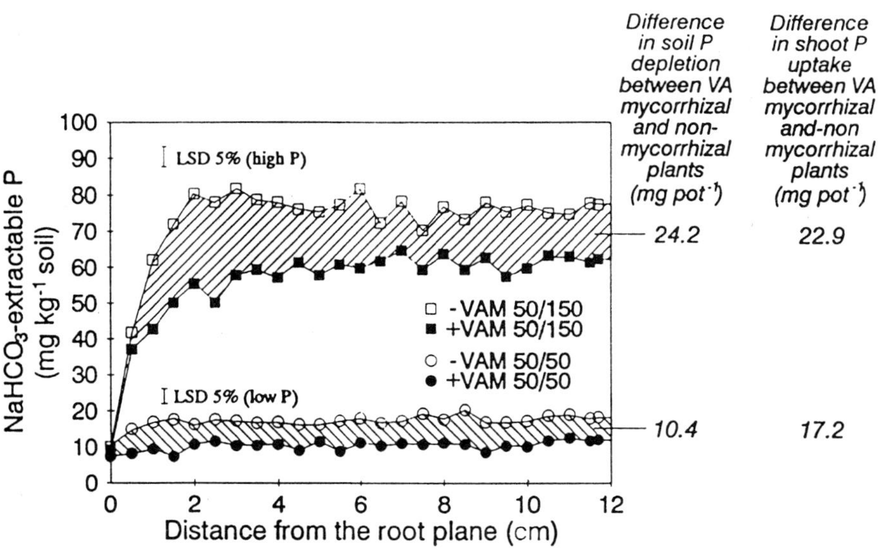

Figure 8.12 Depletion of extractable phosphorus in soil adjacent to a clover root plane, with two levels of soil phosphorus, and with and without AM colonization. Note the agreement in depletion and in uptake of P for mycorrhizal and non-mycorrhizal plants (after George *et al.* 1992).

mycorrhizal root would be of little advantage in diffusion-limited uptake (section 6.1.2), and the combined storage capacity of the sheaths for phosphate would probably be small compared with that of the whole plant.

The theory outlined above has been tested with quick-growing *Salix* species infected with two species of EM, in which changes could be followed much more easily than the usual slow-growing EM test plants such as *Pinus* (Jones *et al.* 1990). The inflow to roots was increased several-fold by infection (table 8.4), and there were amounts of external mycelium comparable to those in the AM experiments, distributed well away from the sheath (table 8.5). Tinker *et al.* (1992) therefore suggested that the essential phosphorus-supplying processes were basically similar in the EM and AM.

However, EM may also improve host nutrition in other ways, probably because the temperate forest trees that mainly carry EM often grow in soils with most of their surface roots in a litter layer, and with much of the soil phosphorus in organic forms. There is now good evidence that EM fungi can hydrolyse proteins and organic phosphates and absorb amino acids and phosphate (Abuzinadah & Read 1989; Turnbull *et al.* 1995). Indeed, the known properties of saprophytic fungi might suggest this in the absence of any direct evidence, and it is now known that AM can also increase uptake of phosphorus from organic compounds (Jayachandran *et al.* 1992). These processes still require quantitative work to estimate their relative importance in typical field situations. The fungal sheath may have other non-nutritional functions that are useful for the host plant, such as protection against pathogenic organisms or increasing root duration.

8.3.5.2 Uptake and Translocation

This theory requires that the fungal network can absorb, translocate and transfer ions at the necessary rate. The external hyphae can absorb ^{32}P several centimetres

Table 8.5 Percentage infection of mycorrhizal willow roots, and the length of hyphae per unit length of infected root, for six levels of soil phosphorus and two species of ectomycorrhizal fungus.

	Soil phosphorus concentration (mg kg^{-1})					
	4	6	10	21	60	90
Length (m) of mycorrhizal hyphae per m of mycorrhizal root						
Laccaria proxima	289	208	313	193	n.c.	n.a.
Thelephora terrestris	319	185	106	103	n.c.	n.a.
Percentage infection						
L. proxima	44.0	44.3	43.3	30.2	0.59	0
T. terrestris	50.6	41.7	49.8	43.0	0.06	0

Source: after Tinker *et al.* (1992).

n.c. = not calculated; n.a. = not applicable.

away from the root and move it into the host (Hattingh *et al.* 1973). More recently, George *et al.* (1992) and Pearson & Jakobsen (1993) used systems that allowed discrimination between ion uptake via external hyphae or through the true root, by separating the two by a fine mesh. All work of this type has confirmed considerable uptake by hyphae alone, but infection may also alter the uptake characteristics of the root (Pearson & Jakobsen 1993) (figure 8.12).

By assuming that the true root absorbs at a rate equal to that of a non-mycorrhizal control plant, Sanders & Tinker (1973) estimated the uptake rate per unit length of infected root via hyphae (the hyphal inflow) as

$$I_h F_i = (I_m - I_{nm}) \tag{8.4}$$

where I_h, I_m, and I_{nm} are, respectively, inflow per centimetre of colonized root through its external hyphae, mean inflow to this colonized plant, and inflow to roots belonging to a non-colonized plant; and F_i is the fraction of the root length that is mycorrhizal. They also calculated the minimum mean flux of phosphorus at the entry points as 3.8×10^{-8} mol cm^{-2} s^{-1}, and the mean hyphal inflow as 2.2×10^{-15} mol cm^{-1} s^{-1} (table 8.4). The latter was below the calculated inflow to a zero sink of hyphal radius in this soil (Tinker 1975b), so the model is credible in a physicochemical sense. Li *et al.* (1991) also found a similar value for hyphal inflows of 3.3–4.3×10^{-15} mol cm^{-1} s^{-1} in different conditions. More specialized systems (Pearson & Tinker 1975; Cooper & Tinker 1978, 1981) were used to measure directly the phosphate fluxes in AM hyphae (table 8.6). These were well below the measured flux in an entry hyphae (Sanders & Tinker 1973), but the conditions were very different.

Flows of nutrients in EM hyphae were observed at a very early stage by Melin and collaborators, who determined that P, Ca and N were translocated, though they did not measure fluxes (see Harley & Smith 1983). Finlay (1992) measured fluxes of phosphorus, rubidium and calcium in ectomycorrhizal mycelial strands (aggregated bundles of hyphae) (table 8.6) and found that fluxes of P were similar

Table 8.6 Measured fluxes of different nutrients in the mycorrhizal hyphae of various species of plants.

Plant/fungus	System	Translocation flux (mol cm^{-2} s^{-1})
Allium cepa/*Glomus* sp.	Soil/P entry points	3.8×10^{-8}
Trifolium repens/*G. mosseae*	Agar/^{32}P	3–10×10^{-10}
Trifolium repens/*G. mosseae*	Agar/^{32}P	2.0–20×10^{-10}
Apium graveolens/*Glomus* sp.	Soil/^{15}N	7.42×10^{-8}
Trifolium repens/*G. mosseae*	Agar/^{65}Zn	2.1×10^{-12}
Trifolium repens/*G. mosseae*	Agar/^{35}S	16.5×10^{-12}
Pinus sylvestris/*Suillus bovinus*	^{32}P	1.8–11.7×10^{-10}
Pinus contorta, Larix eurolepis/*Boletinus cavipes*	^{32}P	13–153×10^{-10}
Fagus sylvatica/*Paxillus involutus*	^{32}P	36–47×10^{-10}
Fagus sylvatica/*Paxillus involutus*	^{86}Rb	16–28×10^{-10}
Fagus sylvatica/*Paxillus involutus*	^{45}Ca	0.3–0.5×10^{-10}

Source: after Smith & Read (1997), table 5.3, p. 139; and Finlay (1992).

in magnitude to those found for AM. The information on nutrient translocation in mycorrhizal hyphae is thus quite extensive (table 8.4).

The translocation process in AM was inhibited by a cytoplasmic streaming inhibitor and increased by plant transpiration (Cooper & Tinker 1978, 1981). This suggested that translocation along the hyphae was partly due to mass flow induced by evapotranspiration by the plant, and partly by cytoplasmic streaming in the hyphae, which can be observed under appropriate illumination. Diffusion over such distances cannot account for the observed fluxes. Continuous bulk flow in response to host transpiration is impossible because carbon has to move in the opposite direction, so bi-directional streaming seems the best theory at present. This could provide net one-way translocation if nutrient is loaded in one location and unloaded in another.

It is possible that the transport of polyphosphate rather than P_i may make transport through the hyphae more efficient (Cox *et al.* 1975; Tinker 1975a; Ashford *et al.* 1994). The polyphosphate is concentrated in the small mobile vacuoles in the AM hyphal cytoplasm (Cox *et al.* 1975). Electron microscopy and staining reactions identified the material in small granules and the vacuolar sap in the vacuoles of internal hyphae and arbuscules, but less in the external hyphae. The electron-dense granules originally found are probably artefacts of preparation (Ashford *et al.* 1994; Smith & Read 1997). However, it is still possible that the vacuoles could function as transportable 'packages' of polyphosphate in the hyphae and arbuscules. Polyphosphate could be hydrolysed when required to produce high local concentrations of P_i in the arbuscules, but its function is not yet proven. Ashford *et al.* (1975) showed that polyphosphate was also formed in the EM fungal sheath when phosphorus was in good supply, and later identified (Ashford *et al.* 1994) a 15-monomer short-chain polyphosphate. However, the mechanism is not yet settled.

8.3.5.3 Transfer Processes to Host

The transfer of nutrients from a symbiotic organism to the host root is an analogous process to the uptake step in a non-infected root (section 5.2.3), and is a critical process in mycorrhizal nutrient supply. This is still not well understood.

The four transfer steps between host and fungus (figure 8.13) are (1) the uptake of phosphate by the host root cells, which is a normal process requiring no special explanation; (2) the uptake of carbon compounds (probably glucose) by the fungus, which again is normal; (3) the release of carbon compounds by root cells, which happens in the normal exudation process, but probably at a lower flux; and (4) the release of phosphorus from the fungus, which happens in normal efflux processes, but at a much lower rate. The steps (3) and (4) are consequently the central processes which allow the symbiotic process to function. It has been suggested that the phosphorus and carbon fluxes are linked, but there is no specific evidence of this (Schwab *et al.* 1991).

The arbuscules in the AM are thought to be transfer organs, mainly because their convoluted shape gives a large interface (Toth & Miller 1984). This is not a decisive argument, as early growth responses may occur before the arbuscules are fully formed, and EM manage their transfers at comparable rates without them

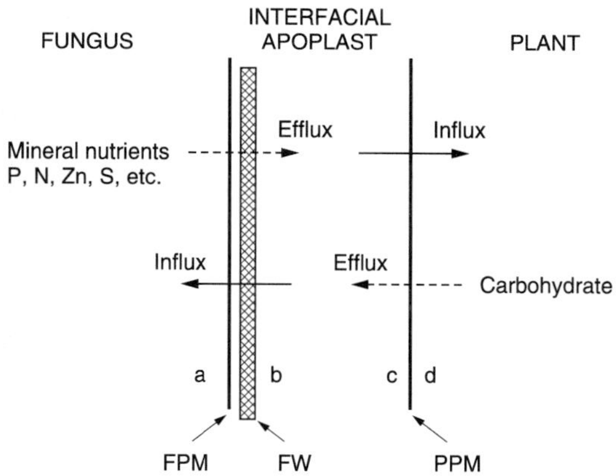

Figure 8.13 Schematic diagram of the fluxes of carbon and phosphorus in the arbuscular interface between fungus and host plasmalemma. FPM, fungal plasma membrane; PPM, plant plasma membrane; FW, fungal wall. Dashed lines are efflux; full lines are influx (after Smith & Read 1997).

(Schwab *et al.* 1991; Tinker *et al.* 1992; Smith *et al.* 1994b; Smith & Read 1997). It therefore seems there must be transfer across the walls of the intercellular hyphae, certainly in the EM. There are differences in the membrane phosphatase enzymes in the arbuscule and in uninvaded root cells (Gianinazzi-Pearson *et al.* 1991), which may suggest active transport in the arbuscules. The theory that the transfer of nutrients occurs by breakdown of arbuscules has been disproved (Cox & Tinker 1976).

On the assumption that all the phosphorus transfer that forms the hyphal inflow occurred across the surface membranes of living arbuscules, the mean flux has been estimated at about $12\text{–}13 \times 10^{-9}$ mol cm^{-2} s^{-1} (Cox & Tinker 1976; Smith *et al.* 1994a). This figure is of similar magnitude to uptake fluxes into cortical root cells growing normally, and suggests a normal uptake process by the root membranes.

These values are, however, very large compared with effluxes. Elliott *et al.* (1984) showed that at ambient concentrations above micromolar P, uninfected maize root influxes were some 10 times larger than effluxes. The very few results for fungi suggest that efflux rates are 10–20% of influx rates (Smith *et al.* 1994b), but a preliminary value for efflux from *Pisolithus tinctorius* was only $1.4\text{–}4.6 \times 10^{-12}$ mol cm^{-2} s^{-1} P.

Passive leakage depends upon the concentration gradient, dC/dx, between the cell contents and the space outside the membrane if it is diffusion-driven (figure 8.13), that is, it depends upon maintaining a very low concentration very close to the efflux surface. Even in well-stirred solution, the stationary layer close to a solid surface is about 50 μm thick, whereas the distance between the two plasmalemmas in an arbuscule is only around 30 nm (Cox & Sanders 1974). If the fungal surface has a high efflux rate and a low uptake rate, and the other surface has the reverse, the arbuscule is a uniquely efficient transfer organ due to the small intermembrane distance. If the polyphosphate in the fine arbuscular branches was steadily hydrolysed to P_i, it could maintain a very high internal concentration of P_i, which could enhance efflux and inhibit uptake (Cox *et al.* 1975; Tinker 1975a; Clarkson 1985; Ashford *et al.* 1994). It is doubtful if any work to date has really established the maximum theoretical rates of transfer in a functioning arbuscule.

Transfers of both phosphorus and carbon across the fungal–host interfaces are normally assumed to happen simultaneously over the same membranes (Smith & Read 1997), but it is not known whether this means that the fluxes interact (Schwab *et al.* 1991). The transferred materials are probably sugars (Shachar-Hill *et al.* 1995) and transfer may be aided by rapid conversion of these to other compounds — sugars that are specific to fungi in the EM fungi, or lipids in the AM fungi (Smith *et al.* 1994b; Smith & Read 1997). Both these processes and the depolymerization of polyphosphate suggested above could act as 'non-return valves' that would enhance transfer rates by stopping return flow.

It is important to consider these fluxes in the context of the bi-directional movement of organic exudates across root membranes discussed in section 5.1.4 (Darrah 1996). This suggests that the real gross effluxes of compounds may be much larger than the net outward fluxes on which attention is usually concentrated, so that the net efflux can be greatly increased if there is a strong sink immediately outside the membrane. The model of Darrah (equation (8.1)) might be developed in terms of diffusion and active uptake across two plasmalemmas, and diffusion across the interspace (figure 8.13):

$$\begin{aligned}\text{Flux of } P_i &= (C_a - C_b)U_f/x - F_{maxf}C_b/(C_b + K_{mf}) \\ &= (C_b - C_c)D_L/y \qquad\qquad (8.5) \\ &= F_{maxh}C_c/(C_c + K_{mh}) - (C_d - C_c)U_h/x\end{aligned}$$

In this C is the concentration of P_i immediately inside (a) and outside (b) the fungal plasmalemma, and immediately outside (c) and inside (d) the host plasmalemma (figure 8.13). U_f and U_h are the fungal and host plasmalemma permeabilities; D_L is the diffusion coefficient in water; x is the plasmalemma thickness; y is the distance between plasmalemmas; and F_{max} and K_m are the Michaelis–Menten constants for fungus (f) and host (h). Few of these parameters are known yet, but the approach may be useful.

The outlines of the three processes discussed in this section have reasonably good explanations now, except for the transfer step. This is particularly difficult to study in the AM because of the complex geometry of the arbuscules. The critical question is whether some new process is required to explain the fluxes, or whether the usual processes of membrane transfer can explain it, along lines such as those discussed here.

8.3.6 Nitrogen in the Symbiosis

Nitrogen has always received attention in relation to mycorrhizas (Weaver 1926, p. 255), but it has taken a long time to reach a consensus on its importance. This is partly because of the complexity of nitrogen chemistry in the soil, partly because nitrogen could occupy such different roles in the symbiosis. At one time, there was a belief that di-nitrogen gas could be fixed by EM fungi, but if this appears to occur, it is probably due to association of the root with N-fixing bacteria.

On average, plant tissue contains around 10 times as much nitrogen as phosphorus by weight, so a significant nitrogen contribution to plant growth by translocation along hyphae requires that the mean N/P flux mass ratio is near

10. Nitrogen is often growth-limiting in natural vegetation or in unfertilized crops, so a significantly enhanced N uptake due to mycorrhizal colonization may cause a growth response. It is important to distinguish the properties of ammonium and nitrate N in this context. Nitrate is, in general, not adsorbed on soil, and is highly mobile by diffusion or mass flow (section 6.1.2). It is also rapidly absorbed by plant roots at concentrations above micromolar, so that if there is a reasonable root density L_V (section 11.3.5), a non-mycorrhizal root system is able to extract virtually all nitrate from a soil. Little advantage would therefore be gained by uptake of nitrate into mycorrhizal hyphae. Ammonium N, being a cation, is adsorbed on soil (section 3.1.3), and its mobility is comparable to that of potassium, though at least an order of magnitude greater than that of phosphate in most mineral soils. The mobility of potassium can limit uptake into roots (section 10.2), and there will certainly be cases where this will hold for ammonium also, as the uptake rates are comparable when ammonium is the main N supply form. There is consequently a strong possibility that mycorrhizal colonization could aid N uptake in soils where ammonium is the main N form, as in acid forest soils. The alternative possibility is that the mycorrhizal fungi could hydrolyse proteins and other N-containing compounds in the soil, followed by uptake. It is important to distinguish these two quite separate mycorrhizal mechanisms for the enhancement of N nutrition.

Uptake and transfer of N from AM fungi to their hosts, to an extent that increases growth and nitrogen content appreciably, has only recently been fully proven (Ames *et al.* 1983; Bethlenfalvay *et al.* 1991; Barea *et al.* 1992; Frey & Schuepp 1992a, b; Hamel & Smith 1992) (figure 8.14). Some authors have found the amounts transferred to be small, and the effect is usually detectable only if the plant roots are excluded from parts of the soil which act as the source of labelled nitrogen for the hyphae. The practical importance in the field is still uncertain. Despite several suggestions, there is little hard information that indicates that AM fungi can break down organic N, though the evidence that organic phosphorus can be used by mycorrhizas (Jayachandran *et al.* 1992) suggests that the same may be true for nitrogen (section 8.3.5).

The EM fungi can, of course, absorb mineral nitrogen for their own metabolism and growth (Carrodus 1967). It was shown very early that nitrogen could be translocated into the EM root by its hyphae, and there be transferred to the host (Melin and collaborators; see Harley 1969). A transport model of growth enhancement is therefore possible. However, with EM fungi, interest has mainly concentrated upon the ability of mycorrhizas to use organic forms of nitrogen, since mineral forms of N, and particularly nitrate, are usually low in typical soils on which EM plants grow. This requires that the external fungal structures hydrolyse N-containing organic materials in the soil. Abuzinadah & Read (1986) showed clearly that proteins could be better sources of N for plants when they carried mycorrhizas (table 8.7). This is probably connected with the ability of some fungi to break down even recalcitrant organic polymers. In one case (Abuzinadah & Read 1989), a non-mycorrhiza-forming fungus also increased plant growth, presumably by breaking down the protein to ammonium, which could be absorbed by the plant. The four tree species tested were almost unable to use protein nitrogen when grown without fungi present.

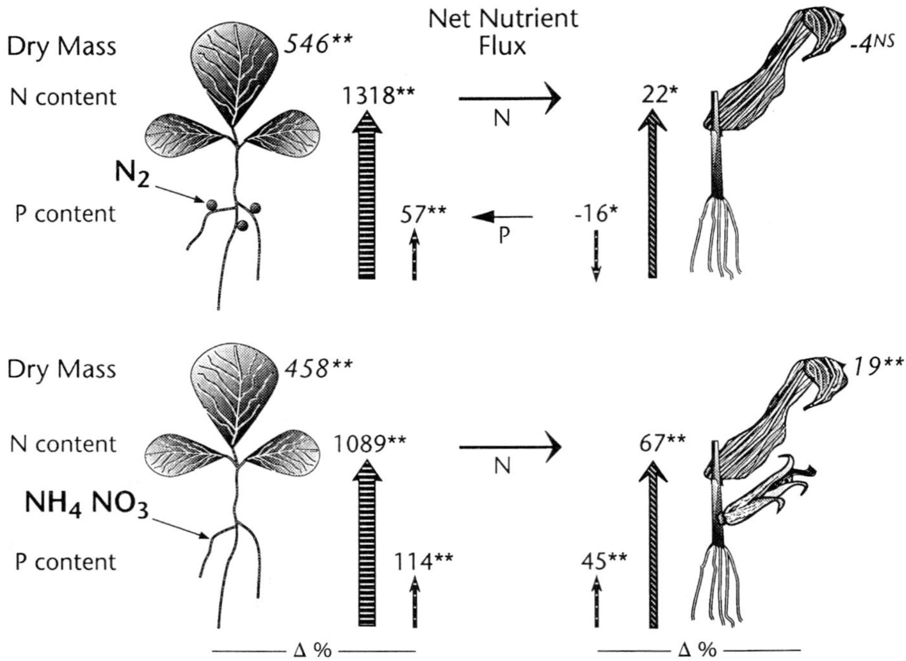

Figure 8.14 Nitrogen transfers in a system with soybean and maize plants growing with their roots connected by AM hyphae. Values are in all cases the percentage differences between N-sufficient (nodulated or fertilized) and N-deficient soybeans and their associated maize plants. Arrows show the transfers of N and of P (after Bethlenfalvay *et al.* 1991).

Bearing in mind that all mineral nitrogen in the soil is much more mobile than phosphorus, except possibly in highly organic soils, the conclusion remains that mycorrhizas are unlikely to enhance plant nitrogen uptake greatly, at least in mineral soils, where uptake into roots is as mineral nitrogen, and where plant roots are well distributed and have moderate L_V and αa values. Uptake of mineral N from the soil by hyphae can certainly happen, but the true roots themselves could possibly absorb sufficient nitrogen in most cases. The complicated experimental systems often required to prove that hyphal N uptake occurs suggest that it is normally a minor effect.

There could certainly be considerable benefit from the proven increased breakdown of insoluble or high-molecular-weight organic nitrogen compounds in organic soils or litter layers by EM fungi, though more quantification is needed to distinguish the effects of mycorrhizal and saprotrophic (organic matter decomposer) fungi in the field.

8.3.7 Carbon Demand by the Fungus — Costs, Benefits, and Mechanisms

A crucial part of the symbiosis is the flow of carbon compounds from host to fungus. This is essential in the AM, which are obligate symbionts, and highly

Table 8.7 Nitrogen uptakes of four tree species with or without mycorrhizal infection, and with protein as the sole source of nitrogen.

Species	Dry weight of seed (μg)	N content of seed (μg)	N content of plants at harvest				Proportion of N content derived from protein (%)		Proportion of applied N used by M plants (%)
			NM plants[a] (μg)	NM plants[a] (%)	M plants[a] (μg)	M plants[a] (%)	NM plants	M plants	
Betula pendula	129	3.8	3	0.21	670	5.15	0	99	53
Picea mariana	862	58	46	0.85	453	3.38	0	88	35
Picea sitchensis	2310	135	160	1.25	505	2.97	16	73	36
Pinus contorta	2100	200	132	1.4	676	3.1	0	76	47

Source: after Abuzinadah & Read (1986).

[a]NM, non-mycorrhizal; M, mycorrhizal.

probable in the EM, which grow only slowly on complex organic compounds (Harley & Smith 1983); it has been proved for both types by supplying radioactive CO_2 to the host, and observing activity in the fungus. Mechanisms for the supply of carbon are complex and it is still uncertain whether this loss of carbon can be detrimental to the growth of the host. The sugars transferred from host to EM fungus, and the 'fungal sugars' into which they are converted, are known (Smith & Read 1997), but this is still uncertain for the AM, within which the numerous lipid globules rapidly become labelled in ^{14}C experiments and are assumed to be the immediate sinks for carbon. However, the main question is of the total amount of the transfer and its consequences.

There have been a number of reported cases of apparent reductions in yield due to mycorrhizal infection for both AM and EM systems (e.g. Sparling & Tinker 1978) (table 8.8). This may be due to a diversion of carbon to the fungus, though there are other possibilities, such as a subclinical pathological effect. Stribley *et al.* (1980a) determined that mycorrhizal and non-mycorrhizal plants, given different levels of soil phosphate so that they were of the same size, nevertheless had appreciably different internal phosphorus concentrations (figure 8.15). This general effect has been confirmed in many other cases. Stribley *et al.* (1980a) suggested that this was due to the carbon being used by the fungus in the mycorrhizal plant, so that the plant needed a higher phosphorus status to reach the same size as an otherwise equivalent non-mycorrhizal plant. From this, they estimated that some 40% of the host carbon was used by the fungus, though the benefit due to the infection was always larger than this, so that there was a net gain to the AM plant. Basically the same approach has been used by Raju *et al.* (1990), though they expressed the idea in terms of 'P efficiency', or the inverse of P concentration in the tissues.

Direct measurements of carbon use by the various plant parts and fungal structures can be obtained by ^{14}C labelling (section 8.1.2). There is now a fairly extensive set of data (Durrall *et al.* 1995) that shows a diversion of carbon to the root in mycorrhizal plants (tables 8.9 and 8.10). Almost all studies have involved

Table 8.8 Reduced growth of Pennine grassland species infected with AM in sterilized soil

Grass species	Treatment	Root weight (mg)	Shoot weight (mg)	Shoot P (%)	Infection (%)
Anthoxanthum odoratum	Control	356	676	0.083	0
	Inoculated	168	524	0.122*	66
Cynosurus cristatus	Control	467	640	0.074	0
	Inoculated	115***	410***	0.106*	63
Festuca rubra	Control	581	694	0.103	0
	Inoculated	296*	417	0.165*	11

Source: after Sparling & Tinker (1978).

*, **, ***, Difference from the control significant at $P < 0.05$, 0.01, and 0.001, respectively. Tests were performed o[n] log-transformed data.

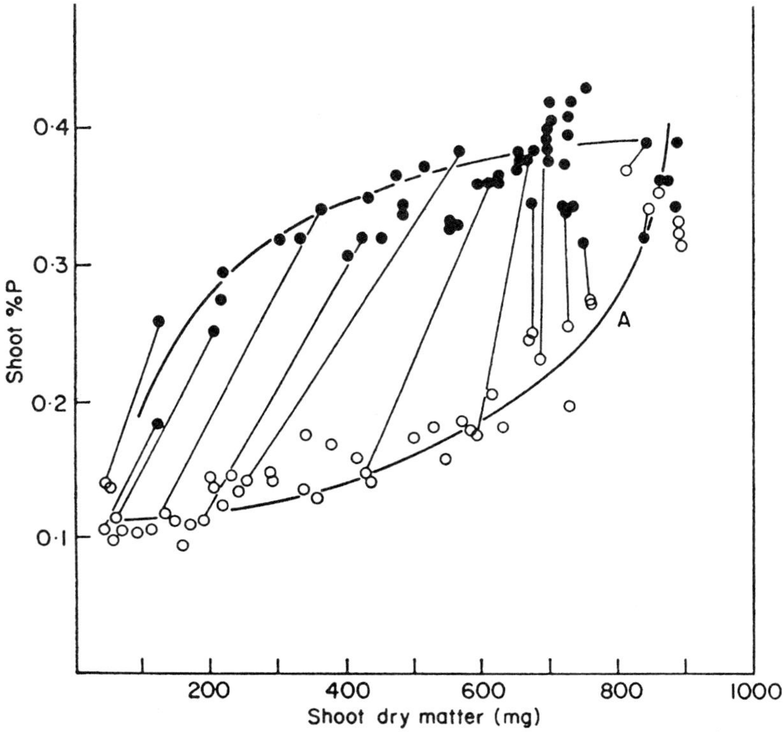

Figure 8.15 Different concentrations of P in leaves of leeks grown, ●, with and ○, without AM infection. Lines connect plants growing in the same soils at the same phosphorus levels. Percentage infection declined with P level, and was very small at the largest P percentages (after Stribley *et al.* 1980a).

supply of radioactive CO_2 to the plants in a pulse, followed by a chase period, with some use of split-root systems in single plants (section 9.1.2).

The increased demand for carbon by the root is usually taken as 6–10% of fixed carbon for the AM (table 8.10) though the full range may be 4–20% (Smith & Read 1997, p. 109). Data for EM are still too few to decide (Durrall *et al.* 1995).

Table 8.9 Percentage allocation of carbon below ground to roots of willow with and without colonization by EM, in relation to phosphorus supply (+ P or − P).

				Differences between treatments	
Harvest (days)	M[a] − P	NM[a] − P	NM + P	(M − P) minus (NM − P)	(M − P) minus (NM + P)
50	45.4	41.5	—	3.9	—
60	51.7	45.3	48.6	6.4	3.1
85	47.3	39.9	42.9	7.4	4.4
98	47.2	35.7	—	11.5	—

Source: after Durrall *et al.* (1995).

[a]M, mycorrhizal; NM, non-mycorrhizal.

Table 8.10 Changed allocation of carbon to root of leeks after infection with AM.

Treatment[a]	Shoot tissue (%)	Root tissue (%)	Below-ground respiration (%)	Shoot respiration (%)	Soil organic matter (%)	Root washings (%)	Total (%)	Shoot/root ratio
NM	57.2	16.2	18.4	5.7	2.3	0.2	100	3.54
	(22.49)[b]	(6.36)	(7.22)	(2.25)	(0.92)	(0.08)	(39.32)	
M	49.7	15.7	23.1	6.3	5.1	0.2	100	3.17
	(18.91)***	(5.96)	(8.80)**	(2.38)	(1.96)*	(0.07)	(38.08)	

Source: after Snellgrove *et al.* (1982).

[a]NM, non-mycorrhizal; M, mycorrhizal.
[b]Figures in brackets are mg ^{14}C.
Significant differences: *, $P = 0.10$; **, $P = 0.05$; ***, $P = 0.01$.

When comparing equally sized plants growing at the same rate, this may be expressed with a diagram such as figure 8.16, which indicates that a diversion of carbon must normally produce a slower growth, and allows a calculation of the growth that would occur otherwise (Tinker *et al.* 1994). However, this increased diversion to the root is usually found to be compensated by an adaptation, such as larger photosynthetic activity, lower weight per unit leaf area, or a larger percentage of water in leaf tissue (Snellgrove *et al.* 1982, 1986). These adaptations may be connected with the increased percentage of P in the mycorrhizal plant. The situation is particularly complicated in a triple symbiosis, such as soybean–*Rhizobium–Glomus*, where both symbionts require carbon from the common host (Harris *et al.* 1985). These authors found that the root nodules of *Rhizobium* used 9% of the photosynthate in non-mycorrhizal plants and 12% in mycorrhizal plants; the mycorrhizal fungus used 17% in young plants, 9% later. In comparisons with control plants given extra nutrients to compensate for the deficiency of nutrient, the plants with symbionts had up to 47% higher rate of carbon dioxide fixation. Thus there was again an adaptation to the extra carbon demand, and the mechanism of this adaptation is very interesting. It is probably not simply due to the extra nutrient present, because control plants also received extra nutrient.

Koide (1993) and Fitter (1991a, b) have introduced the concept of cost and benefit to the plant, with one measure of this being the 'incremental efficiency of P accumulation', or dP/dC_b, where P is the total plant phosphorus, and C_b the

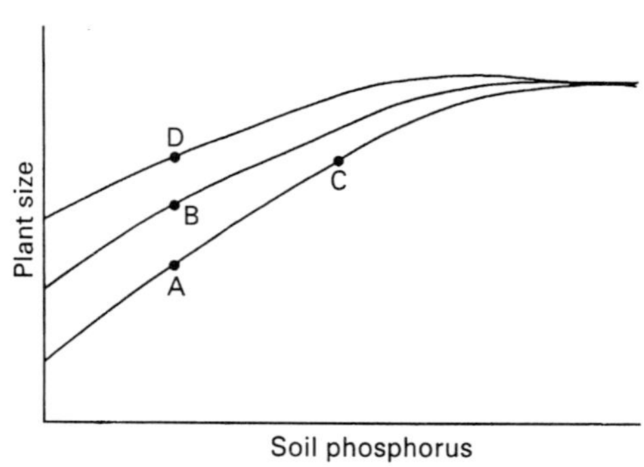

Figure 8.16 Diagram showing typical growth response curves of plants (B) with and (A, C) without mycorrhizal infection. The line with point D is a hypothetical curve that would have been found if the carbon allocation had remained as in the non-mycorrhizal plant at C (after Tinker *et al.* 1994).

carbon transferred below ground (table 8.11). The 'efficiency of below-ground carbon utilization' is also defined as dC_w/dC_b, where C_w is the whole-plant carbon (Jones *et al.* 1998). The very large below-ground allocation of C in ectomycorrhizal trees suggests that the mycorrhizas have a large demand (section 9.1.2).

At present, it is not possible to predict a growth-depression resulting from diversion of fixed carbon to the fungus because of the plant adaptations that may counterbalance the loss of carbon, and the possibility that the non-mycorrhizal plant has a surplus of carbon (sink-limited growth) which could meet any extra demand (section 9.1.2).

8.3.8 Interplant Hyphal Connections and Fluxes

Direct observation of mycelial networks on soil surfaces, such as in rhizotrons, shows that they are often connected to the roots of more than one host plant. Given the rate at which elements are moved through the hyphae (section 8.3.5.2), transfers between the two plants are obviously possible. This raises two related questions: first, can nutrients or carbon compounds be transferred from one host plant to another via a common hyphal network, and second, what determines how much different plants, of similar or different species, benefit from nutrients absorbed by the same hyphal network? These questions are obviously of vital importance in considering the ecological relations or the competition mechanisms between plants or species (Read *et al.* 1985; Newman 1988; Newman *et al.* 1992), and they introduce a whole new aspect to what are already very complex systems (section 11.4.2). If these processes are important, then the approach to resource capture and nutrient cycling in mixed-age and mixed-species plant stands has to be adjusted.

Most attention has been given to the first question. This presumes that the element can be transferred from host to fungus, translocated along the hyphae, and transferred from fungus to the second host (section 8.3.5). The four transfer processes occurring across the fungus–plant interface have already been discussed (see figure 8.13), and if both carbon and phosphorus, for example, are to move from plant to fungus to plant, two more efflux processes must be postulated, that of carbon from fungus to plant (Newman 1988) and that of phosphorus from plant to fungus. However, these are not necessarily at high rates, so that there is no theoretical difficulty, unless 'non-return valves' (section 8.3.5.3) operate by

Table 8.11 Phosphorus uptake efficiency values for two EM fungal species colonizing eucalyptus.

Treatment	ΔP^a (μmol P)	$\Delta C_b{}^a$ (mmol C)	$\Delta P/\Delta C_b{}^a$
+ *Thelephora terrestris*	67.4	6.0	11.2
+ *Glomus* 'E3'	34.8	3.5	10.0
NM − P	11.3	1.6	7.2
NM + P	69.4	11.4	6.1

Source: after Jones *et al.* (1998).

[a] ΔC is the marginal amount of C allocated below ground; and the ratio is the P gained per additional unit C invested.

the transferred element being converted to a different compound subsequent to transfer.

The most direct method of testing for any interplant transfer is to label one host plant with a radioactive or mass isotope, and to measure the label in the mycorrhizal network, and in host plants attached to it (Hirrell & Gerdemann 1979; Newman 1988). There is doubt about whether phosphorus can be transferred from plant to EM fungus (Finlay & Read 1986), and no clear and significant transfer of phosphorus from plant to plant has been shown in either EM or AM systems. Newman & Eason (1993) compared the rate of transfer of phosphorus from one ryegrass tiller to other tillers of the same plant, or to neighbouring plants. The rates of the latter were relatively very small and of little practical importance. In the special case where the donor plant was dying, phosphorus was transferred at significant rates to another plant in the same mycorrhizal network (Newman 1988), but this would be expected by uptake of free P_i released from the dying plant.

Read *et al.* (1985) and Francis & Read (1984) reported that labelled carbon moved freely between plants joined by mycorrhizal hyphae, for both EM and AM systems. Some acceptor plants were shaded, to ensure that they would be strong carbon sinks. However, the activity in the acceptor plants shoots was always much less than in its roots, and this difference was increased by shading (table 8.12). This may be explained if the transferred compound is not very mobile in the plant; for example, when soluble carbohydrates are supplied to plants via their roots, they move very slowly to the shoots (J. Farrar 1995, private communication). However, amino acids circulate freely in the phloem and xylem (section 5.3.3). The very great difference between the activity found in shoots and roots, and the similar activity in the shoots of mycorrhizal and non-mycorrhizal receiver plants, suggests that much of the activity remained in the fungus in the root. However, this experiment lasted for only 48 h. A longer experiment suggested that radioactive carbon was being transferred to the shoot, possibly following the turnover of the fungal structures, so work over longer periods is necessary (D.J. Read 1999, private communication). In another experiment, some ^{14}C was trans-

Table 8.12 Transfer of labelled carbon from *Plantago* to roots and shoots of *Festuca*, grown in the same container with or without mycorrhizal infection and under three light regimes.

	Mycorrhizal *Festuca*			Non-mycorrhizal *Festuca*		
	Activity in root[a]	Activity in shoot[a]	(%)[a]	Activity in root	Activity in shoot	(%)
Full light	8 900	363	0.015	650	150	0.002
Part shade	18 000	479	0.048	260	230	0.001
Dark	57 200	51	0.112	120	—	0.005

Source: calculated from Francis & Read (1984) and Read *et al.* (1985).

[a]Radioactivity in receiver plants expressed as dpm/mgm dry weight in roots and in shoots, and combined as percentage of the total activity in the donor *Plantago* plants at harvest.

ferred from hyphae into the host root cells inside the endodermis, but this was not quantified (Duddridge *et al.* 1988).

As shading had little effect on the label in the shoots of the acceptor plants, the transfer does not seem to depend upon sink strength there. The larger activity in the acceptor roots with shading is easily explained, because the fungal structures there were very largely dependent on carbon from the donor plant, and all this carbon would have a high specific activity.

More recent work has clarified the situation to some extent. Watkins *et al.* (1996) used the different ^{13}C abundances in C3 and C4 plants to determine the transfer between *Plantago* and *Cynodon* connected by an AM network. A deviation of the ^{13}C abundance in the root from that in the shoot was taken as a measure of that transfer. Most values for the C in the *Cynodon* roots indicated that $<10\%$ of the root carbon had been received from the other plant, with large variation among individual plants. There was no correlation of transfer with degree of mycorrhizal colonization. Graves *et al.* (1997) extended this by using ^{13}C-depletion labelling of grass turves, and following the level of depletion in the roots and shoots of neighbouring turves in ambient air. With stable isotopes, enrichment can be measured on the ^{13}C-depleted side, and depletion on the ambient-^{13}C side, so that flows in both directions can be quantified and the net transfer determined. With single radioisotope labelling, only the gross flow in one direction can be measured. In this work, 36% of fixed carbon was transferred to the roots of the mycorrhizal turves, and up to 41% of this C was translocated to the roots in the receiver turf — that is, 15% of fixed carbon. No change in ^{13}C was detected in the receiver shoots, and it seems possible that the transferred C did not leave the fungal structures of the receiver plants. Only 10% of the fixed carbon was transported to the roots in non-mycorrhizal turves, as compared to 36% with mycorrhizas, which suggests a very large carbon drain. The calculation does, however, assume that the root/shoot ratios were the same, which was not reported. The unexpected result in this work was that flows of 'label' in both directions were apparently not equal, indicating a strong net flux in one direction, despite the fact that the mycorrhizal turves were intended to be identical. It appears that further work on this is necessary. Fitter *et al.* (1998) have continued this line. They showed that considerable amounts of carbon could move between *Cynodon dactylon* and *Plantago lanceolata*; however, their evidence showed that transferred carbon remained in the roots, and probably in the fungal structures. These authors drew the interesting conclusion that this apparent between-plant carbon transfer is really part of the internal dynamics of the extended fungal network, and as such would have no major consequences for plant C budgets or fitness.

Simard *et al.* (1997) has reported the measurement of net and gross interplant carbon transfer through an EM network by reciprocal double labelling with ^{14}C and ^{13}C of species of *Betula* and *Pseudotsuga*. An AM tree species growing close by, that would have no hyphal linkage with EM species, received only 18% of the bi-directional transfer of labelled carbon between the EM trees, showing that some 80% of the transfer was via the hyphae. Between 3 and 10% of the total carbon fixed by both trees was transferred to the other, with the net transfer from

Betula to *Pseudotsuga* being 2–3 times larger to heavily shaded trees. This indicates that the process can be important. In this case, much of the labelled carbon was transferred to the shoot, the average transfer of the received isotope being 13% for *Pseudotsuga* and 45% for *Betula*. This level of transfer to shoots is an important finding, and so far has not been observed in AM systems. The reciprocal labelling used in this experiment raises the problem that the specific activity of the label transferred in the two directions may not be the same because of the difference in species; hence the comparison of the reciprocal fluxes is still uncertain.

There is conflicting evidence about whether significant amounts of nitrogen are translocated from one plant to another via the mycorrhizal hyphae. In particular, it appears that di-nitrogen fixed by *Rhizobium* in a legume is not easily transported in significant quantity into a neighbouring non-leguminous plant, even when it is known that they are connected by hyphal networks (Bethlenfalvay *et al.* 1991; Frey & Schuepp 1992b). Thus, in the work of Bethlenfalvay *et al.* (1991) (figure 8.14), it is clear that nodulated soybean contains more N than soybean with fertilizer nitrogen, yet the amount translocated into the maize is much greater from the latter than from the nodulated soybean, presumably because hyphae absorb ^{15}N from the soil in the latter case, so the initial transfer from plant to fungus is not necessary. In addition, the nodulated soybean actually decreased the amount of phosphate in the maize, whereas the soybean with fertilizer increased it, suggesting that the nodulation made the soybean into a much stronger sink for phosphorus. This interesting result again raises the issue of the allocation of nutrients to different plants within a single network. However, Arnebrant *et al.* (1993) showed that in laboratory experiments around 10% of the N fixed by an *Alnus* inoculated with *Frankia* was transferred to a *Pinus* with a joint ectomycorrhizal network. This direct transfer of symbiotically fixed nitrogen to another non-fixing species via another symbiotic linkage is important, but quantification in other systems is needed.

It appears, at present, that the common use of a single mycelial network by different plants is proven and a normal process. More needs to be known about how nutrients from this are allocated, and whether some plants are 'freeloaders' on the system by contributing a disproportionally low amount of carbon for the nutrients that they receive. Cost/benefit and dC/dP ratios for individual plants can now be measured, and within common networks this ratio may well differ greatly between individual plants, depending upon their sink strength for phosphorus and their source and sink strengths for carbon.

It seems that phosphorus is rarely transferred from a plant into a fungal network and then on to another plant. However, C and N transfers between plants can certainly happen, though there still are a number of questions about net quantities and the sink–source relationships that determine transfer. It is still not clear what controls the deposition of transferred carbon in the fungal structures in the root, the root itself, or the whole plant. If the average transfers in larger plants in the field are at the upper end of the available measurements, then this will give a new perspective on plant competition studies, with large implications for nutrient capture and distribution.

8.3.9 Other Mechanisms Benefiting the Host

8.3.9.1 Uptake of Other Elements

Elements other than P and N can also be taken up and moved into the host by mycorrhizal infections (Tinker & Gildon 1983; Smith & Read 1997). However, it is important to distinguish between the ability of the hyphal network to absorb, translocate and transfer an element, and the probability that this will result in a growth response by the host. On the general model given above, a response is only to be expected for elements that have low mobility in the soil — otherwise they will reach the true root sufficiently rapidly by diffusion or mass flow. The list of elements for which growth responses have been shown to occur is now P, N, Cu (Gildon & Tinker 1983a, b) (table 8.13) and Zn (Lu & Miller 1989). Clear evidence of transport into the host has also been obtained for Sr, Rb, Ca, Na and S. The uptake of zinc by three species of AM fungi that infect subterranean clover was detected up to 40 mm from roots by *Acaulospora levis*, up to 20 mm for *Glomus* species and up to 10 mm for a *Scutellospora* species, and the hyphal lengths were distributed in a similar way (Burkert & Robson 1994).

As a counterpoint to this, there is some evidence that mycorrhizal infection can help to prevent toxic uptake of heavy metals into plants (Smith & Read 1997). However, it is known that increased concentrations of phosphorus can alter trace metal uptake, and this effect may sometimes simply be a consequence of better phosphorus supply. Several authors have discovered mycorrhizal fungal strains that are well adapted to heavy-metal contaminated soils (Smith & Read 1997). There is evidence that mycorrhizas can produce siderophores (Haselwandter 1995).

8.3.9.2 Water Relations

There have been many reports that mycorrhizal infection can improve the water status of plants, but the interpretation has been unsatisfactory. There are several possible theories.

First, the apparent benefit of infection to plants in droughted situations may not be due to water supply, but to the greater impedance to phosphate diffusion in

Table 8.13 Influence of mycorrhizal infection with *Glomus mosseae* on the uptake of copper and the growth response to supply of copper to clover. (Note no effect on P concentration.)

	Copper treatment				
	Cu_0		Cu_1		LSD
	M[a]+	M−	M+	M−	
Dry weight (g)	1.33	0.79	1.31	1.41	0.16
Cu concentration ($\mu g\ g^{-1}$)	5.1	3.6	5.7	5.5	1.5
P concentration (%)	0.13	0.12	0.13	0.13	0.03

Source: after Tinker & Gildon (1983b).

[a]M, mycorrhiza.

dry soils. The normal mycorrhizal benefit would therefore be proportionately larger in dry soils, giving the impression of drought mitigation (Safir *et al.* 1972). Fitter (1988) also concluded that the effects on stomatal conductance were a consequence of changes in phosphate status in the plant.

Second, there have been repeated indications that hormonal levels are affected in infected plants. Ectomycorrhizal fungi can produce plant hormones in pure culture, and Allen *et al.* (1981) detected changes in cytokinins in certain grasses when infected with AM, which could have been the cause of the observed increase in stomatal conductance.

Third, there is the obvious possibility that fungal hyphae can absorb water from the soil, and act as an extension of the root system in drying soils, as reported by Hardie & Leyton (1981) with clover. However, Sanders & Tinker (1973) calculated that the flow rate of water to produce a significantly enhanced transpiration stream in onions would have to be unrealistically high in entry hyphae of around 10 μm diameter. Kothari *et al.* (1990) carried out similar calculations for maize, and found the mean flow velocity would have to be 0.9 m h^{-1} to explain the increased transpiration, which they considered to be unacceptably high. Soil that was only permeated by AM hyphae, but not by roots, did not appear to lose water even when the soil around the roots became dry (George *et al.* 1992).

There is better evidence that ectomycorrhizas may aid water relationships in host plants, because water can be transported through the rhizomorphs and hyphal strands formed from hyphae, which are characteristic of some ectomycorrhizas. These are comparatively large, and the central parts are empty, so that a hollow tube of up to 20 μm diameter is formed in the soil, which is comparable with smaller vessels in trees, in which sap velocities of up to several metres per hour occur (Fitter & Hay 1981, p. 166). The argument about the improbability of high flow velocities in AM hyphae cannot be applied to the EM fungi with such structures. Duddridge *et al.* (1980) showed that water is transported through these structures, in the direction of the host root, at a velocity of around 27 cm h^{-1}. However, it is not immediately obvious what process keeps these vessels water-filled under water tension in the soil, as 20-μm diameter tubes would normally empty at the very small negative matric potential of −0.015 MPa (section 2.1.2). The mechanism therefore still seems to need some investigation.

The situation with regard to water is therefore confusing. It is clear that infection changes various water-related parameters, such as stomatal conductance, transpirational flux and sometimes root conductivity, but the mechanism is unclear, and possibly hormonal. It may be associated with the changes in phosphate concentration and hydration of the leaves after infection, as found by Snellgrove *et al.* (1982, 1986) and others. At present, it is not clear whether plants in the field benefit from improved water relationships by mycorrhizal infection. Koide (1985) could find no differences in root hydraulic conductivity with AM infections, though stomatal conductance was greatly increased, and Koide (1993) considered that significant hyphal transport of water was unlikely.

The effects of salinity and drought are related. Despite earlier claims, more recent work (Fitter 1988; Graham & Syvertsen 1989) indicates that any effect of mycorrhizal infection on these factors arises only through nutritional interactions.

Much of the difficulty in this area of research is that phosphorus is such a central and pervasive element in plant functions, and most experiments involving infection with AM or EM will alter the phosphorus concentration or distribution.

8.3.10 Practical Applications of Mycorrhizas

It may appear obvious that mycorrhizas should have extensive and important applications in agriculture and forestry because many economically important plants are heavily dependent upon infection (Smith & Read 1997), and this dependence can be shown in sterilized soil in pots or the field (Yost & Fox 1979) (table 8.14). It is possible to produce pairs of plants with and without AM mycorrhizal infection that may differ in dry weight by several hundred fold. There are many authenticated cases of exotic tree species that are totally dependent upon appropriate EM fungi, and Le Tacon *et al.* (1992) and Marx *et al.* (1989) have found important field growth responses to EM fungi.

For AM, it is very difficult to find similarly clearcut field results. This is almost certainly due to the longevity of the large AM spores, the ubiquity of the fungi, and their low specificity as mycorrhiza formers. To put it simply, almost all soils are already inoculated. The only large responses have been found where there is little inoculum, due to soil disturbance or removal (Miller & McGonigle 1992), prior absence of mycorrhizal species (Thompson 1994), or artificial sterilization (Menge 1983). Smith & Read (1997) give a list of agricultural conditions that reduce the inoculum density of AM fungi.

Table 8.14 Mycorrhizal dependency of some agricultural and horticultural crops.

Plant species	Dry mass/plant: Fumigated soil	Dry mass/plant: Non-fumigated soil	REC[a] index (%)	RFMD[a] (%)
Currant	1.1	4.1**	39	74.6
Marigold	3.3	12.5**	29	73.6
Purple leaf sand cherry	0.3	1.2**	26	72.2
Spirea	5.2	17.2**	25	69.6
Carrot	0.07	9.2**	66	99.2
Garden pea	1.3	40.3**	89	96.7
Leek	0.5	11.9**	58	95.7
Kidney bean	0.7	13.3**	88	94.7
Faba bean	1.4	21.8**	62	93.5
Sweet corn	45.5	166.5**	69	72.7
Pepper	4.1	12.1**	42	66.1
Tomato	71.2	174.6*	50	59.2
Potato	107.5	185.3*	44	41.9
Oat	208.9	170.9	79	0
Wheat	155.5	155.6	55	0

Source: after Plenchette *et al.* (1983).

[a]REC, relative endomycorrhizal colonization index; RFMD, relative field mycorrhizal dependency. See original publication for precise definitions.

The economically important small grains and grasses vary in mycorrhizal dependency, perhaps because of their large specific root length (section 9.4.1). Intensive studies, including field fumigation of very phosphate-deficient soils, showed quite small responses to inoculation of cereals (Buwalda *et al.* 1983; Plenchette *et al.* 1983). They were extensively infected with AM fungi, but this did not appear to have any beneficial or detrimental effect (Buwalda *et al.* 1983). However, Baon *et al.* (1993) tested a range of 10 spring barley cultivars and found large differences, with some responding to AM infection.

So far, all attempts at field inoculation have used existing mycorrhizal strains that have been isolated from the field. It is likely that sexual recombination or genetic engineering will produce AM fungal types that have novel characteristics and may give larger field responses. There is little advantage in applying fungal inoculum that is already present in the soil.

8.3.11 Mycorrhizal Effects in Natural Ecosystems

Phosphorus and nitrogen deficiency is widespread in natural soils, and mycorrhizal infection is important or essential for many species there (Brundrett 1991). There is no doubt that the mycorrhizal status of natural vegetation may be critical (Hetrick *et al.* 1994), though Fitter & Merryweather (1992) noted the difficulty of proving this in the field. Several workers have shown that the absence or presence of mycorrhizal infection alters the competitive ranking of different species (Fitter 1991a). The interplant transfers discussed in section 8.3.8 have many ecological implications. Grime *et al.* (1987) concluded that carbon fixed by plants in full light can be supplied via fungal hyphae to understory plants in shade, and so maintain a much more diverse flora.

8.4 Effects of Other Organisms on Nutrient Uptake and Growth

8.4.1 Symbiotic Nitrogen Fixation

The microorganisms in question are the *Rhizobium* nitrogen fixers associated with the legumes, and a few other species such as *Frankia* with *Alnus*. This nitrogen fixation is given no further attention here, because the uptake of gaseous dinitrogen means that there is no significant impediment to transport through the soil. The physiological behaviour of the plant may be very different when using dinitrogen compared with nitrate or ammonium nitrogen (Marschner 1995), but this is not discussed here.

8.4.2 Effects of Larger Organisms

The rhizosphere is the habitat for a number of larger organisms which prey on the bacterial population, and which may graze on the roots (Brown & Gange 1990, 1991). The main bacterial grazers are the protozoa, which may comprise 5% of the total biomass below ground. Nematodes also consume bacteria, and the attempt to relate bacterial populations solely to the substrate supply, without reference to predation, is not really adequate. The rhizosphere organisms also graze on the roots themselves, especially root hairs and sloughed-off material. It has been suggested that grazing and predation is a major reason why laboratory studies on plants sometimes cannot be repeated in the field, but it is not possible to deal with this subject here. Larger organisms can also alter plant dynamics and nutrition (Thompson *et al.* 1993).

8.5 Conclusion

It is not surprising that the large literature on the rhizosphere in the past has produced rather little in the way of firm generalizations and mechanisms on the nutritional effects, bearing in mind its complexity, constant variation and difficulty of access. It is interesting to note the list of processes in the rhizosphere that have been proposed in the past, but that have only slowly and often partially been proven to occur. The critical question is whether they are important for plant growth, and it is only recently that we have gained some insight into this.

9

Root System Architecture, Density, and Measurement

The behaviour and properties of roots are central subjects in this book. A number of biochemical and physiological properties have already been described, for individual roots, in chapters 2, 5, 7, and 8. However, the macroscopic properties of root systems are of very great importance, to an extent that may not be immediately apparent from the point of view of the laboratory. These properties include the root/shoot ratio, the root system dimensions, its topological properties, and its distribution in the soil profile. The property of greatest practical importance is the way in which root length density (length per unit volume of soil) is distributed in the soil, because this defines the spatial limits to the efficiency of a root system in absorbing water and nutrients. For these reasons, we have collected material relating to root system properties here in a separate chapter. This may be particularly helpful to readers because there are very few single-part recent publications that deal with this subject. It appears logical to start with a discussion of how much root a plant possesses, its dependence upon the allocation of fixed carbon, and the efficiency with which this is used to form root tissue.

9.1 Root–Shoot Relations and the Allocation of Carbon into the Root System

9.1.1 Carbon Allocation

Carbon is the basic currency of plants, and the way in which they distribute and use it is part of their growth strategy. The allocation of carbon in plants has been extensively researched within the above-ground part, but not the below-ground part, because of the difficult access to the root system, and the difficulty of separating the root, root surface and soil processes. It is important to understand the way in which carbon is allocated to both the root system as a whole, and then to the different parts of the root system, its symbiotic partners, exudates and other root products.

Some broader issues are also relevant. Some of the carbon allocated to the root could be wasted, from the point of view of the plant or the farmer (Gregory 1994a). Could the plant grow even better if it allocated more of its carbon to more rapid shoot growth, if it had a more efficient root system? Part of this question relates to the use of carbon by symbiotic and non-symbiotic fixation of nitrogen (Zuberer 1990) and by mycorrhizas (Amijee *et al.* 1993) (section 8.3.7).

The carbon flowing to roots provides much of the food supply to the soil biota, and the activity in the rhizosphere depends upon it. It therefore affects soil-borne pests and diseases, and the formation, dynamics and properties of soil organic matter. It is difficult to understand the carbon balance of vegetation correctly, on either a field or a global scale, while the dynamics of carbon below ground are so poorly understood (IPCC 1996, p. 451 *et seq.*).

9.1.2 The Use of Photosynthate and the Efficiency of the Root System

Because most of the plant is composed of photosynthate, it is common to regard photosynthesis as the primary plant process, and to assume that the rate-limiting step for growth must be the supply of photosynthate. This may be wrong: first, because water or nutrients may be limiting, but second, because growth can be either sink-limited (limitation by rate of forming new plant structures) or source-limited (limitation by rate of production of photosynthate) (Gifford & Evans 1981). In the former case, the plant sinks are not large enough and growing rapidly enough to accept all the photosynthate that can be provided, and the surplus photosynthate inhibits the rate of further photosynthesis. Simple examples of this occur when storage organs such as tubers are detached from a plant. In source limitation, there is a shortage of photosynthate, and this state can be induced by removal of leaves or by shading.

The mechanisms that control the allocation of photosynthate are not fully understood (Farrar 1992). The Munch hypothesis states that photosynthate is transported in the phloem vessels by the osmotic potential differential caused by the loading of the photosynthate sugars. This draws water into the phloem, and drives flow along the phloem away from the source of sugar. This suggests that the physical arrangement of the phloem affects the destination of this carbohydrate. In addition, there appear to be short-term and long-term control systems dependent upon the state of both sources and sinks in the plant (Minchin *et al.* 1994). Gifford & Evans (1981) concluded that local sink demand was dominant in deciding the detailed allocations, even if the plant as a whole was source-limited. Unfortunately, there is no simple test that can be applied to determine whether a growing plant is in a source-limited or sink-limited state at a particular time.

Source versus sink limitation is an important issue because of the implications of diverting photosynthate from the growth of the shoot to that of the root. If there is surplus carbohydrate in the plant because of sink limitation, then the allocation of photosynthate to roots implies no loss to the plant, and there is no advantage in increasing efficiency. However, since plants, particularly crop plants with their high growth rates, are certainly often source-limited, we will assume here that diversion of carbohydrate from shoot expansion to the root normally decreases plant growth. A line of thinking has therefore developed in which the 'efficiency' of a root system is assessed in terms of the ratio of the benefit provided to the plant (the acquired water and nutrients) to the cost (the total carbon or photosynthate allocated below ground) (Bloom *et al.* 1985; Koide & Elliott 1989). It is therefore not surprising that in intensively grown and highly productive cereals, the root/plant ratio is frequently less than 0.10, whereas in wild grasses it may be above 0.5. This is regarded as necessary for the high net primary productivity and economic yield of the former, and the stress-tolerance and robustness of the latter (Grime *et al.* 1988).

We suggest three useful, but by no means infallible, generalizations about the root/shoot ratio. First, a plant in steady growth that has its root/shoot ratio changed by the removal of part of the root or shoot will tend to revert to the original ratio by adjusting the growth rates of these organs (figure 9.1). Second, the root/shoot ratio tends to decline if root-absorbed materials are in relatively better supply than shoot-absorbed materials (photosynthate), and vice versa (figure 9.2). Finally, most young plants have a high root/shoot ratio, which then declines with age (Barraclough 1984), and the root/shoot ratio may decline even more after flowering (Gregory *et al.* 1996). None of these rules is precise or totally dependable, but all are useful guides in the quantitative modelling of root development (section 9.5.3).

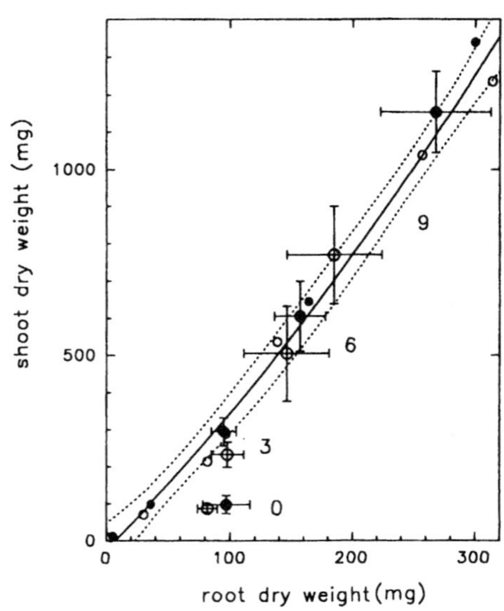

Figure 9.1 The shoot/root (S/R) relations of *Dactylis glomerata* after partial defoliation. The smaller symbols, without error bars, are means of control plants with fitted second-order regression and 95% confidence limits. Larger symbols, with standard errors, are plants partially defoliated when 29 days old; the number of days after defoliation is indicated by the number next to each pair of symbols. Plants were grown hydroponically in controlled environment cabinets at either 350 (open symbols) or 700 (filled symbols) ppm CO_2. Atmospheric CO_2 concentration does not alter root/shoot relations, but both re-establishment of control S/R and growth rate after partial defoliation are faster at 700 ppm CO_2. (Unpublished data of S. Gunn and J. F. Farrar, after Farrar 1996.)

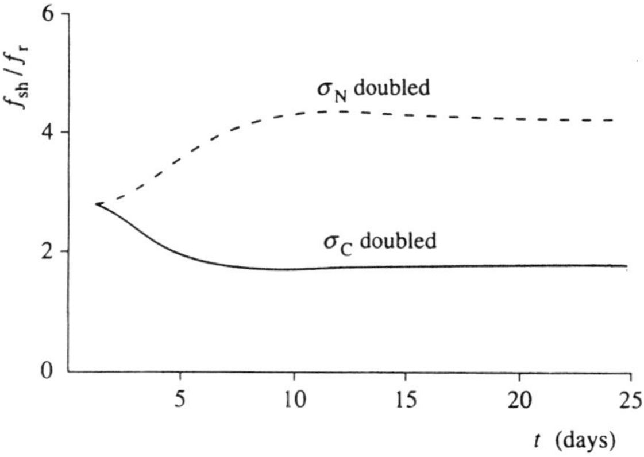

Figure 9.2 Control of root/shoot ratio (f_r/f_{sh}) in plants by the specific activities of carbon absorption (σ_C) and nitrogen absorption (σ_N) systems, as predicted by the teleonomic model (after Thornley & Johnson 1990, p. 383).

In general, labelling studies with ^{14}C indicate that with fairly young plants, between 30 and 60% of net carbon fixed goes below ground (table 9.1). Carbon dioxide loss from below ground is between 20 and 60% of the carbon that goes below ground, so that on average below-ground respiration is some 15–20% of total fixed C in young plants. Lambers (1987) has summarized the distribution of this carbon amongst formation of dry matter, growth and maintenance respiration, exudation and other losses (section 8.1.3).

More recently, pulse-labelling has been applied over longer periods, and in the field (Keith *et al.* 1986; Gregory & Atwell 1991; Swinnen *et al.* 1994a, b). Keith *et al.* (1986) and Swinnen *et al.* (1994b) found by the use of ^{14}C that some 20–30% of fixed carbon was translocated below ground with wheat, but Keith *et al.* (1986) found that at harvest only some 25% of carbon allocated below ground was still present in the living root system. Some of the dry matter formed in fine roots is lost into the soil (section 9.3.5).

The data of Swinnen *et al.* (1994a) for wheat show a particularly large allocation of carbon to the shoots, increasing with age, but the first measurement was made rather late, nearly 3 months after the sowing date (table 9.2). Most work has shown that the rate of carbon allocation to roots decreases with age and stage of development of the plant, whereas total below-ground respiration increases (Dormaar & Sauerbeck 1983).

The below-ground allocation declines to very small percentages near harvest, when nearly all the fixed carbon is going to the ear. By contrast, plant-derived soil organic carbon, and root decay increase gradually with time, as a percentage. Figure 8.2 shows the absolute values, as carbon fluxes. A full model of the carbon fluxes (Swinnen *et al.* 1994b) indicated that overall 5730 kg C ha^{-1} formed shoot growth, 920 was lost in root respiration, 500 was deposited in the soil, and root growth used 940, of which 370 was lost by root decay. In total, about 29% or some 2360 kg of fixed carbon was allocated below ground, of which perhaps about 1000 kg was left in the soil at the end of the season. As a contribution to the soil organic matter, this may be compared with the straw weight, which would be about half the shoot, or about 2900 kg C ha^{-1}. These data agree reasonably well with the compilation of Lambers (1987).

Table 9.1 Total allocation of carbon and its components below ground for four agricultural crops after long-term exposure to $^{4}CO_2$ in growth cabinets.

Conditions								C (%) transferred to root lost as:				
Plant	Age (days)	Temp. (°C)	Day-length (h)	Soil	CO_2 conc. (%)	Sterility	Net fixed C (%) transferred to root	Respiration (ignoring rhizodeposition in soil)	Respiration (including rhizodeposition in soil)	Rhizodeposition in soil	Respiration and rhizodeposition in soil	Reference
Wheat	28	21–17	16	1	0.03	NS	51	41	37	9	46	Merckx *et al.* (1985)
	35	21–17	16	1	0.03	NS	49	39	38	3	41	Merckx *et al.* (1985)
	42	21–17	16	1	0.03	NS	50	40	39	2	41	Merckx *et al.* (1985)
	28	21–17	16	4	0.03	NS	44	44	39	11	50	Merckx *et al.* (1985)
	35	21–17	16	4	0.03	NS	48	36	33	10	43	Merckx *et al.* (1985)
	42	21–17	16	4	0.03	NS	40	32	29	9	38	Merckx *et al.* (1985)
Maize	14	18–14	16	2	0.03–0.05	NS	28	32	16	52	67	Whipps (1985)
	28	18–14	16	2	0.03–0.05	NS	29	46	33	28	62	Whipps (1985)
	14	18–14	16	2	0.04–0.07	NS	26	50	33	34	67	Whipps (1985)
	28	18–14	16	2	0.04–0.07	NS	36	33	26	20	47	Whipps (1985)
	14	18–14	16	2	0.06–0.10	NS	36	38	31	19	50	Whipps (1985)
	28	18–14	16	2	0.06–0.10	NS	35	49	40	18	58	Whipps (1985)
Tomato	14	18–14	16	2	0.03–0.05	NS	43	67	20	70	90	Whipps (1987)
	14	18–14	16	2	0.03–0.05	NS	40	62	47	25	72	Whipps (1987)
Pea	14	18–14	16	2	0.03–0.05	NS	65	55	31	43	75	Whipps (1987)
	28	18–14	16	2	0.03–0.05	NS	44	74	53	29	82	Whipps (1987)

Source: after Whipps (1990).

Table 9.2 Carbon distribution in wheat plants at various growth stages after labelling with ^{14}C.

	Development stage when labelling was performed				
	Elongation	Ear emergence	Anthesis	Milk ripening	Dough ripening
	^{14}C Distribution, percentage of net ^{14}C assimilation				
Shoots	61.3	76.8	77.3	85.4	83.1
Roots	15.2	7.7	4.8	2.4	1.9
Soil–root respiration	14.1	9.4	9.4	6.7	8.6
Soil organic C	9.4	6.2	8.5	5.4	6.4
Total below-ground	38.7	23.2	22.7	14.6	16.9
	^{14}C Distribution, percentage of below-ground ^{14}C				
Roots	39.6	34.0	21.6	17.2	10.9
Soil–root respiration	35.2	38.9	41.4	46.4	47.8
Soil organic C	25.3	27.1	37.0	36.5	41.3
	^{14}C Distribution, percentage of shoot ^{14}C content				
Ears	7.8	25.2	54.4	85.3	83.9

Source: after Swinnen *et al*. (1994a).

Data for trees in the field are mainly obtained by manual methods, or 'crop study' (Whipps 1990), that give values for root weight. The totals for carbon delivered below ground, not including exudates or respiration, are remarkably large at up to 60–70% of fixed carbon (table 9.3) (section 8.3.7). This suggests that perennials invest more of their carbon below ground than do annuals, and that trees lose much carbon in high turnover of fine roots and to mycorrhizas. However, these high values may also be associated with the poor nutrient conditions in many forest soils. Thus, Linder & Axelsson (1982) used an alternative method to calculate a carbon budget from the measured net photosynthesis in tree canopies, and the biomass including that of coarse roots, with biomass of fine roots being found by difference. This showed that the control tree used 60% of its fixed carbon below ground, whereas a much larger fertilized and irrigated tree used only 40% — a value that is similar to those from many non-tree species. However, the amounts entering the rhizosphere cannot be found by such methods.

9.2 The Morphology and Measurement of Root Systems

9.2.1 Root System Development and Morphology

The anatomy and behaviour of the individual root has been described in chapter 5. Here, we discuss the form, description, and function of root systems belonging to isolated plants, crops, and associations of wild plants, mainly when growing in soil. Truly independent single plants are very rare in the field. They do occur in agriculture during the early stages of growth, when plants are too small to compete with each other in any way, and sometimes in natural conditions when scarcity of water or nutrients causes plants to be very widely spaced, but both can be considered as special cases, and the root systems of groups of plants is the important topic. This is well reviewed by Gregory (1988).

It is possible to describe a root system in two general ways: plant based or soil based. The plant-based method describes the way in which different parts of the root system develop and are interconnected in a branching system (Rose 1983; Klepper 1992). The frequency and length of the different orders of laterals have been studied, the order being defined as the number of branchings after the root left the plant stem (e.g. Hackett & Rose 1972b). Figure 9.3 shows typical data from the various orders of roots in cereals. Such data can be obtained from solution-grown or soil-grown plants whose root systems are carefully excavated, but the method would have to be applied to single plants, as it normally would be impossible to separate intertwined fibrous root systems. It is unlikely that the same results will be obtained with solution-grown and soil-grown root systems. Root structures determined in this way have some characteristics that are reproducible, but in detail the 'architecture' is not constant between individual plants, or with time or growing conditions. There are many diagrams of such root systems grown in soil or solution culture (Weaver 1926; Kutschera 1960; Rogers & Head 1969; Boehm 1979). However, illustrations such as figure 9.4 should only be taken as general guides to the structure and distribution of real root systems.

Table 9.3 Carbon movement to below ground (total allocation), 'rhizodeposition' (defined as in table) and loss of C as CO_2 (grasses) or as decomposition of root (trees) for perennial species growing in the field: grasses and trees.

Species	Net fixed C (%) transferred below ground	C (%) transferred to roots that is lost (CO_2)	Rhizodeposition (root C + soil C + CO_2C) (t C ha^{-1} $year^{-1}$)	Reference(s)
Grasses				
Agropyron/Koeleria-dominated mixture	35–50	20–28	1.3	Warembourg & Paul (1977)
Tall prairie grass (large mixture of species)	50	—	2.1	Kuccra *et al.* (1967)
Short prairie grass (*Bouteloua gracilis/Buchloe dactyloides*-dominated mixture)	80	—	—	Sims & Singh (1971)
	Net fixed C (%) transferred below ground (to roots + mycorrhizas)	**C (%) transferred to roots that is lost (white root decomposition)**	**Rhizodeposition (root C) (t C ha^{-1} $year^{-1}$)**	
Trees				
Deciduous forest				
Oak	—	52	—	Ovington *et al.* (1963)
Yellow poplar	40	42	—	Edwards & Harris (1977) Harris *et al.* (1980)
Coniferous forest				
Douglas fir	73	40–47	5.8–7.5	Sanantonio (1979), Fogel & Hunt (1983)
Scots pine	60	66	—	Persson (1978), Agren *et al.* (1980)
Pacific silver fir	60–71	—	—	Grier *et al.* (1981)
Pinus cembra	70	—	—	Tranquillini (1964)

Source: after Whipps (1990).

For some listed references, see original publication.

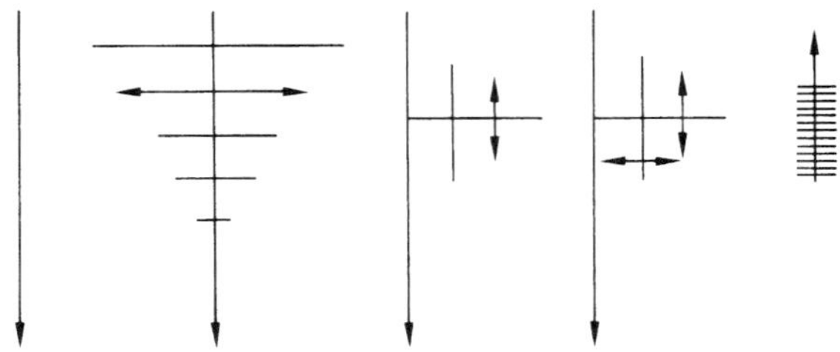

Order of root	Main	1st order lateral	2nd order lateral	3rd order lateral	Root hair
Diameter (cm)	5×10^{-2}	2×10^{-2}	1×10^{-2}	5×10^{-3}	1×10^{-3}
No. per cm of root of next higher order	–	2	1	5×10^{-1}	1×10^{3}
Length (cm) per c.c. of soil	1	5	2	5×10^{-1}	1×10^{3}

Figure 9.3 Representative dimensions and numbers of a cereal root system (after Barley 1970).

The most important morphological distinction is between monocotyledonous and dicotyledonous root systems (Klepper 1991, 1992). Dicotyledons and gymnosperms have a central tap-root from which other roots emerge. Monocotyledons usually have a fibrous root system based upon the system of seminal roots that arise from the apical meristem of the embryo, and the nodal (adventitious) roots that arise from the nodes of the monocotyledonous plant stem. In both monocotyledons and dicotyledons, the long laterals or adventitious roots extend and define the outer perimeter of the root zone, usually further out than the above-ground parts (section 11.4.3). The next orders of laterals then fill in this volume, producing the final pattern of root length density (Pages & Serra 1994). With adventitious root systems, main roots arise from progressively younger and higher stem nodes. These grow outward and downward, finally being almost vertical (figure 9.4C), and gradually increase the root volume for the individual plant.

A further important anatomical distinction is between the seminal and nodal (adventitious) roots in monocotyledonous plants. The seminal roots emerge directly from the apical meristem of the embryo in the germinating seed, whereas nodal or adventitious roots subsequently emerge from successive nodes on the stems (figure 5.1). Whereas these root systems are morphologically distinct, plants can subsist on roots of either type, and there is no consistent and general evidence that the individual roots differ significantly in their functional properties. However, different parts of the root system differ in nitrogen and potassium uptake (Kuhlmann & Barraclough 1987), and in the pH changes at the root surface (Nye 1986).

A great deal of descriptive work of this nature has been done (Rose 1983), but very little is known about the differences in the functional properties of the different orders or types of roots (Eshel *et al.* 1996). Such information will be needed for the next generation of detailed root modelling (Comerford *et al.* 1994a) (section 10.6). A closer connection between this external descriptive approach and the internal anatomy and physiology of the root system would also be useful. For

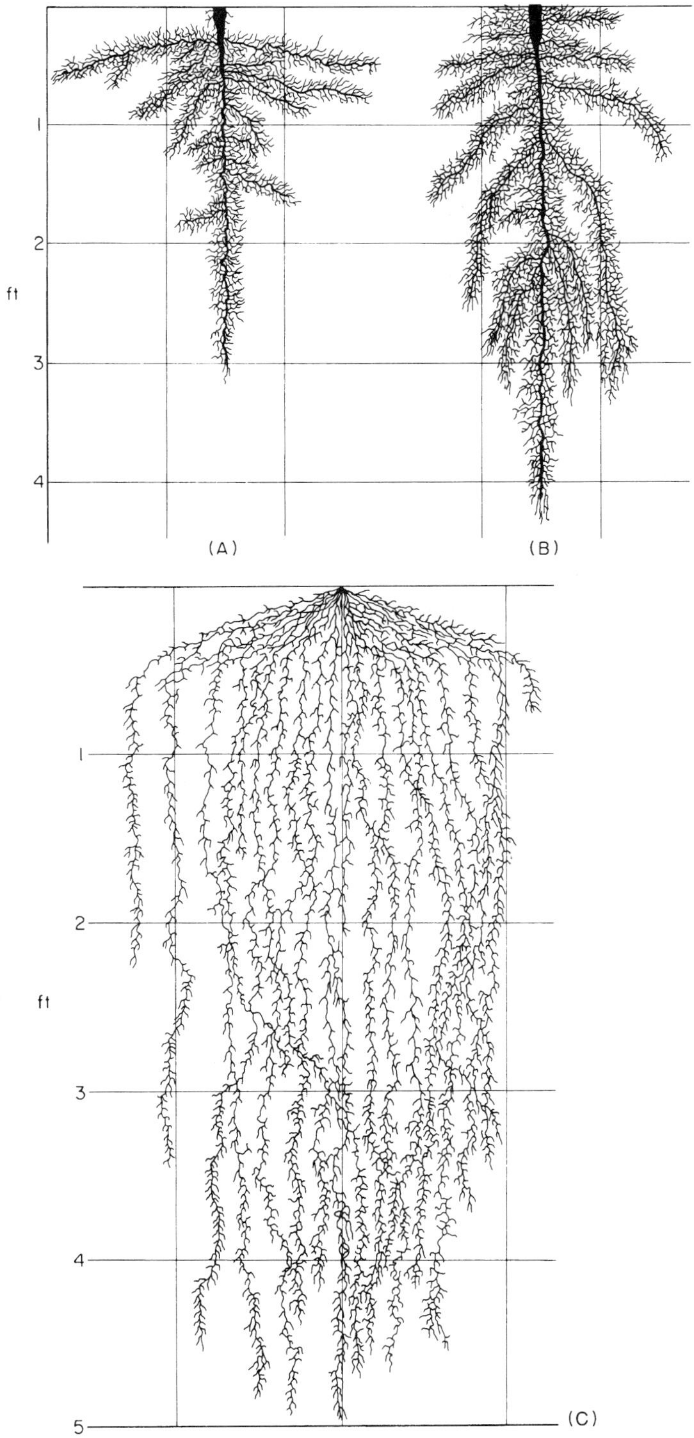

Figure 9.4 Root systems of sugar beet, a typical tap-rooted species at 3 months old: (A) without, and (B) with irrigation; and a typical monocotyledonous root system of rye (C) (after Weaver 1926).

example, McCully & Canny (1989) found that soil particle sheaths formed around maize roots only in regions where the large late-xylem vessels had not matured, suggesting that this anatomical change had some effect on exudation.

There are a few distinctive and specialized root forms, of which the most striking are the 'proteoid roots' on some of the Proteaceae of Western Australia. In these species, long and poorly branched main axes carry short zones with intense proliferation of branch roots at approximately regular intervals. The anatomical details have been studied by Skene *et al.* (1996). Their possible function in nutrient uptake is discussed in section 7.3.2. It is interesting that genetic transformation of plants with the *Rol* gene from *Agrobacterium rhizogenes* produces somewhat similar hairy roots (Mugnier & Mosse 1987).

9.2.2 Root System Distribution in Soil

The second and soil-based approach is to define root systems in soil in terms of the distribution of root length density ($m\ m^{-3}$) or mass throughout the rooting zone. The rooting zone depth is ill-defined, because root density declines gradually to zero (section 9.5.1), and the lower limit is, in practice, often defined by the practical constraints of sampling.

In this approach, no attempt is made to wash out root systems in a complete state, but root length or mass densities are determined, normally by coring, subdividing the cores by depth and washing out the roots before measuring the root weight and/or length. No distinction is made between different orders of root, all being regarded as functionally equivalent. In effect, 'average root' is measured, for which average properties will be assumed in calculation.

Examples of this type of work can be found in good descriptions of the winter wheat system by Barraclough & Leigh (1984), Vincent & Gregory (1989a, b), and Masse *et al.* (1991), and this is at present the standard way of measuring density distributions of roots in field soils.

Measurement of root length density in this way is, of course, applicable to single plants growing alone, and it has been applied to spaced-out individual plants, such as in maize crops, tree crops, or forest plantations. However, it is most often applied to uniform swards or crops, so that variation of density is only in the vertical direction. This is particularly apparent for SVAT-type models (section 11.1.2). With mixed vegetation, where the rooting zones and density distributions may vary with the species, there are serious problems in separating the roots of different species (section 11.4.7), unless there are very well-defined and consistent differences in appearance or colour.

9.2.3 Root Architecture

There is now interest in dealing in a more precise way with the distribution of roots in soil. Thornley & Johnson (1990, p. 495 *et seq.*) give a detailed exposition of the mathematics of branching structures, in relation to both shoot and root systems. Branching structures defined simply in terms of topological linkages do not have a three-dimensional structure; this comes only with geometrical models in which the angles between branches are defined, and such a full three-dimensional description is called the architecture of a root system.

If one defines both the way in which roots interconnect, and how they distribute themselves in the soil, this will integrate the approaches using root system structure and root length density that were discussed above, because the distribution of roots in three dimensions also defines root length density distribution (Fitter & Stickland 1991, 1992; Berntson 1994; Fitter 1996). In this terminology, roots are composed of links, connected to either one or two other links (figure 9.5). The branching behaviour is defined in terms of nodes, at which branches arise. The probability of this occurring has to be defined, and also the probability distribution of the angle of emergence. All roots have a developmental order, from 1 for the tap root or main axes, to 2 or 3 for laterals. In real systems, it cannot be assumed that a root continues in a straight line (see figure 9.4C), and rules for changes in direction need to be introduced also. For example, Tardieu & Pellerin (1990) found that nodal roots of maize tended to stay in a single vertical plane, but gradually became more vertical as they penetrated deeper. It is too early to say whether this rather complex approach will have a fully practical application, because the data requirements are formidable, and the various parameters change with nutrient and other environmental conditions (Fitter & Stickland 1991), but examples of its application to root modelling are given later (section 10.4.4).

There is a possibility that root systems have fractal properties, that is their spatial characteristics remain the same at smaller and larger scales, so that the quantitative relationships of main axes to secondary roots are the same as that between secondary and tertiary roots (Fitter & Stickland 1992; Fitter 1996). Spek & Van Noordwijk (1994) have also produced fractal models of root systems, but the practical value of this approach still remains to be proved.

Fitter *et al.* (1991) consider that the two characteristic architecture types are the herringbone and the dichotomous branching types (see figure 10.10). The

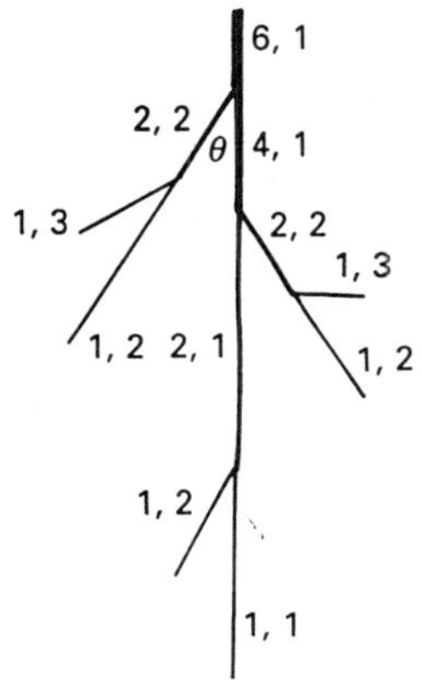

Figure 9.5 Components in the definition of the architecture of an idealized root system: each numbered length is a 'link'; and each is defined by the 'magnitude' (which determines the number of apices subtended) and the 'developmental order' (which defines the main root and branches). Branching angle is given by θ, and horizontal angles are 0 and 180°, as the diagram is in two dimensions (after Fitter *et al.* 1991).

efficiency of these and other rooting types at absorbing nutrients from the soil will be discussed later (section 10.3.4).

9.2.4 Measurement of Root Systems

We do not give a full review of root measurement methods here (see Nye & Tinker 1977). Several recent authors have reviewed these methods (Boehm 1979; Vogt & Persson 1989; Harper *et al.* 1991; Mackie-Dawson & Atkinson 1991; Anderson & Ingram 1993; Atkinson & Mackie-Dawson 1999). Abbott & Fraley (1991) have given a review of radiotracer methods. Very few truly innovative methods have been developed in the 1980s and 1990s, and the only one we know is the use of nuclear magnetic resonance (Bottomley *et al.* 1986). However, this appears to be insufficiently sensitive and is only used in specialized situations. Trenching and soil-face observation remains a dependable but labour-intensive method for obtaining quantitative data on root system development (see below). Soil coring and root washing continue to provide the most practicable way of obtaining quantitative data on root system length and its distribution in the field. It is the method closest to becoming accepted as a standard.

All methods in which roots are extracted depend upon some variant of Newman's (1966) method of measuring length by the use of intercepts. This is now normally done by commercial automated instruments, including a hand-held computer scanner (Kirchhoff 1992). However, the process of cleaning the root sample, and of ensuring that little of it is lost in the process, will always be laborious. Much of our later discussion assumes that dependable data obtained by coring are available, but work such as that of Escamilla *et al.* (1991b) or of Kucke *et al.* (1995), who compared several coring and trenching methods on different soils, show that there can be wide divergences between the results obtained with even slightly different root-extraction systems.

An attempt to reduce the labour of washing out and cleaning roots before measuring was made by the 'core-breaking' method, in which extracted cores were broken across, so that the number of roots crossing the exposed face could be counted. From this, root length density could be found by the formula of Melhuish & Lang (1968, 1971): $L_V = 2N$, where N is the number of cut root ends per unit surface area, assuming that roots are randomly oriented within the soil. However, tests of the method have rarely given dependable results (Bland 1991; Escamilla *et al.* 1991b), probably because roots are preferentially oriented in one direction.

Methods using radioisotope injection into soil or plants are still used, but with diminishing frequency, and usually for specialized purposes (Milchunas *et al.* 1992). It is clear that the interpretation of the information yielded by these methods cannot be relied on as a method of measuring root length. There are occasional reports of the use of the soil-face intercept method (Melhuish & Lang 1968), possibly modified to distinguish different root systems by the use of radioactive labelling (Baldwin & Tinker 1972), but again this has remained specialized, except insofar as the same principle is applied in the core-break method. There appear to be no easy methods of obtaining dependable quantitative data on root distribution in soil, and the practical problems in this area are inhibiting the development of the subject.

9.2.5 Measurement of Root Dynamics

Roots have a finite life, as is obvious from the death of the roots of annuals as the plants become senescent, and from the fact that woody perennials normally put out bursts of root growth each year, yet have much the same size of root system from year to year (Cannell & Dewar 1994). The amount of root measured under an annual crop will normally start to decline well before harvest time (figure 9.6). Root death or turnover has often been neglected in nutritional work, and this may be justified when working with young plants in a state of rapid growth and nutrient acquisition. If growth is exponential, the fraction of new young root will always be dominant, and dying root can be ignored (Brewster & Tinker 1970). When the full annual life cycle is being studied, this is no longer true, and serious errors may be caused by ignoring root turnover (Gregory 1994b). There is the associated interest in the amount of carbon returned to the soil and the consequences for soil organic matter (table 9.4) (Vogt *et al.* 1986, 1991), from the changes in which one could calculate the total annual input of organic residues (Jenkinson *et al.* 1992).

For perennials, root death is particularly important, especially in relation to the allocation of carbon in trees. The demography of individual roots is difficult

Table 9.4 Lignin and nitrogen percentages in tree roots of varying diameters, and the time taken for 99% decay to occur.

Species/location	Diameter (mm)	Lignin (%)	N (%)	99% Decay time (years)
Pinus sylvestris				
Sweden	< 1	51.8	—	—
	< 2	—	—	21.9–30.5
	1–2	22.3	0.57	15.7
	2–3	21.2	0.34	14.7
	3–5	21.6	0.29	18.2
	> 5	20.8–22.0	0.25–0.26	17.1–43.3
	20–30	21	0.31	12.5
White pine				
Wisconsin, USA	0.5–3	25.3	0.93	15.4
Hardwood				
Massachusetts, USA	0–0.5	21.9	1.32	36.7
	0.5–3	23.3	0.85	17.0
Sugar maple				
Massachusetts, USA	0.5–3	33.8	1.67	25.9

Source: after Vogt *et al.* (1991).

For sources, see the original publication.

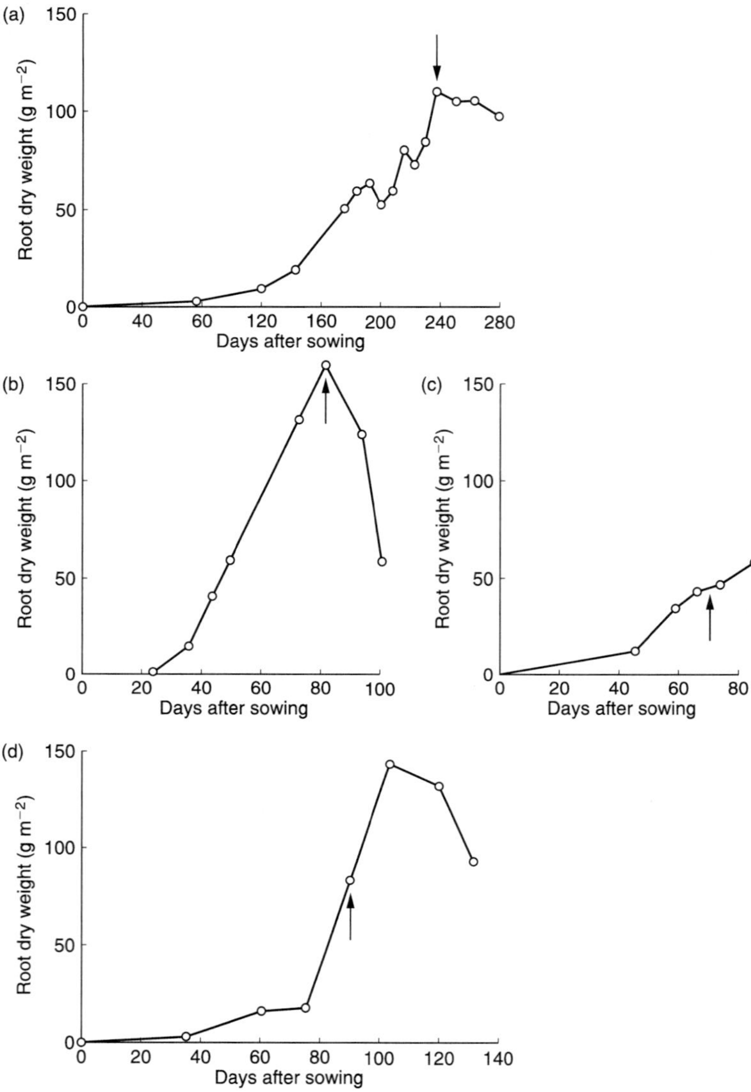

Figure 9.6 Examples of the progressive development of root weight for different crop species and locations in the field: (a) winter wheat, UK; (b) maize, Indiana, USA; (c) soybeans, Iowa, USA; (d) lupin, Western Australia. The arrows indicate approximate flowering times. (After Gregory 1994b; original sources given in this reference.)

to measure and often demands subjective judgments (Fogel 1991). However, it is certain that many fine tree roots live only a few weeks and death normally follows browning and loss of cortex, though at this stage some roots develop secondary thickening and become part of the plant's permanent root framework (Atkinson 1991; Bloomfield *et al.* 1991). (See also section 5.1.5.)

Vogt *et al.* (1986, 1991) questioned whether these short lifetimes are caused by poor growing conditions, or are genetically determined. Pregitzer *et al.* (1993) studied the effect of adding water and nitrogen to defined patches under trees. The additions both caused larger flushes of fine root, and caused these roots to live longer, but there was no information on whether pre-existing roots lived longer if supplied with water and nutrients. There may be analogues of these short-lived fine roots in crops also. Thus, in maize the very finest laterals are stated by Zobel (1992) to be of determinate habit and to have a lifespan of only 2–4 weeks.

The dynamics of roots are therefore important in plant and root system modelling, and in studies of the carbon demand of root systems. In annual species, the growth rate of the roots may decline sharply after flowering, and the measurement of the root dynamics during this period is difficult. The measurement of the dynamics of perennial root systems is particularly complicated (Fogel 1991). Much fine root is produced and dies off during the year, so that the standing crop varies with time, but both processes are proceeding simultaneously. There may be a flush of root growth in spring and autumn, with the largest death rate in winter, or in the dry season where wet and dry seasons alternate. The most frequently used method has been to measure root systems at different times, and to add all the increases in living and dead root length between each sampling and the subsequent one, to add an estimate of root that would have decayed during the period, and to regard this as the total root production. Similarly, the sum of all dead root increases between subsequent samplings, plus the estimated root decomposition, is the amount of root death (Persson 1983, 1990; Vogt & Persson 1989). Thus, root production F is given by

$$F = \text{Sum } B + \text{Sum } N + D \qquad (9.1)$$

where B are increments of live root during all periods between samplings, N are increments of dead roots during periods between samplings and D is the root decomposition occurring during these periods; the last is measured in a separate experiment. If different species in mixed vegetation have their root length maxima at different times of year, they may mask each other's changes (Persson 1990), unless they can be distinguished.

Estimates of fine root turnover losses range from 1.4 to 11.5 tons ha^{-1} $year^{-1}$ and account for 8–67% of primary production (Santantonio & Grace 1987), and the wide range suggests that there may be significant procedural errors. Santantonio & Grace (1987) therefore constructed a dynamic model of a pine plantation, based on the rate of loss by decomposition. Measured standing crops of fine root agreed well with predictions from this model, and this may be the best way forward.

Another method for observing root dynamics is the 'ingrowth technique', in which a net cylinder filled with local soil or with fine sand is fitted into a hole produced with a corer (Steen 1991). After a period of between a month and a year the cylinder is retrieved, and the roots in it are washed out and measured. Whereas this method may be useful for comparative measurements, the fact that all roots originally permeating this volume will have been cut off, so that the regrowth is largely from the branching of severed roots, must mean that the amount of growth is not typical.

There is a further practical problem in the considerable depth to which tree roots can penetrate (Vogt & Persson 1989; Hendrick & Pregitzer 1996), which makes sampling of the often low densities of root at these depths technically difficult. Hendrick & Pregitzer (1996) used mini-rhizotrons to determine flushes of deciduous hardwood forest trees, and found that they generally coincided with flushes of shoots, but a high water demand could cause a brief additional flush. Patterns of root mortality varied with depth.

9.2.6 Rhizotrons and Root Observation Systems

The easiest way to get detailed information about root system development is to use observational methods. Root observation chambers with flat glass walls have been used for many years (see Rogers & Head 1969; Harper *et al.* 1991). This general method was extended by the 'mini-rhizotron' (Sanders & Brown 1977), which is a transparent tube driven into a hole prepared in the ground, so that roots that grow against the tube wall can be observed as often as required by an optical system; the latter has now been superseded by a television camera. With this, extensive data on the number of roots that intercept the transparent face, the development of individual roots, and the total length of root that is exposed there, can be gathered easily. Useful observations on the microfauna of the rhizosphere can also be made (Sackville-Hamilton & Cherrett 1991), and rhizotrons and mini-rhizotrons have now become standard pieces of equipment in root studies, with a high level of technology (Box 1996).

The most recent installation is probably the 'Rhizolab' at Wageningen (Van de Geijn *et al.* 1994), which has excellent facilities for simultaneous above- and below-ground measurements on crops, so that integrated data-sets can be collected. The facility also has an automatic rainshelter, so that water supply is under full control.

The problem is how to convert observations at a surface into dependable data relating to the bulk soil (Mackie-Dawson & Atkinson 1991; Buckland *et al.* 1993). First, the presence of the transparent surface will itself alter the distribution of roots in its neighbourhood. Inevitably, there will be disturbance to the soil, and therefore to subsequent water flow and root growth, but this is probably least with the mini- or micro-rhizotrons, and may be lessened if the tubes are inserted at an angle to the vertical. However, the smaller the tube is made (so as to lessen disturbance), the smaller is the volume of soil effectively sampled and the more variable the results from each tube. In all cases, there will be a tendency for roots to grow alongside the glass or plastic surface.

The second and most fundamental difficulty is that there is no reliable theoretical relationship between the length or number of roots observed on the surface and the root length density in the surrounding bulk soil. Such a relationship holds if a flat cut is made in the soil and the cut root ends are counted (section 9.2.4). However, once roots grow along the surface and branch there, any relationship is lost, and such results do not agree well with those from coring (Meyer & Barrs 1991; Volkmar 1993). It is sometimes assumed that a volume of soil of definable depth behind the rhizotron surface is being observed. This is clearly doubtful, and the lack of a sound theoretical basis is a serious weakness in this approach. An empirical correlation factor, between observations of incidence or of root length at the rhizotron face against root length density behind it, can be found by simultaneous observation and coring, but this factor will not remain constant between sites or times. Andren *et al.* (1993) found that the apparent vertical distribution of barley roots differed with method, being apparently deepest with vertical mini-rhizotrons and shallowest with soil cores. This suggests that roots grew down alongside the tubes.

The most valuable use of these observational techniques is to assess single-root dynamics in detail, namely the sequences of root growth, branching, senescence and death. Observations of faunal behaviour can also be extremely useful. However, even with these, the quantitative statement of results is difficult, and most work is comparative in nature.

Measurement of root dynamics is particularly difficult where two or more species are intermixed (section 11.5.1). However, Campbell *et al.* (1994) obtained reasonably good results on cherry trees growing in a grass sward by using mini-rhizotron systems. There was considerable operator variance in measuring different root categories, but not between the two species. The authors suggest that tests of operator variation should always be made before using such methods, which depend upon individual judgments in identifying roots.

9.3 Factors Affecting Root Form and Distribution in Soil

9.3.1 Genetic Factors

Genetic factors control the gross differences between root system types. Kozlowski *et al.* (1991) considered the genetic effects to be predominant in trees, and O'Toole & Bland (1987) concluded that there was considerable genetic variability in some root characteristics, with a reasonably high heritability. Clark & Duncan (1991) reviewed the rather broader area of the improvement of mineral nutrition by breeding, and considered that there were good prospects for this, including the development of more efficient root systems. Gregory (1994a) has reviewed recent work on genetic diversity, and concluded that such studies are still at an early stage, and are rarely quantitative. Relationships between root characteristics and crop behaviour are usually assumed rather than proven. Gregory (1994a) noted that for a trait to be accepted as a breeding objective, three issues must be settled. First, the desired character must have high heritability; second, the character must be readily assessable during the breeding programme; third, the character must have been proven to improve crop yield or quality. Few root characters yet meet all these criteria.

It is difficult to know which root parameters to aim for in breeding for 'improved' root systems, and considerations of root system efficiency could aid the process. Larsson (1986) defined a 'drought index' to guide breeders in improving drought tolerance, in which

$$\text{Drought index} = 100(a/bc)/(a_s/b_s c_s) \qquad (9.2)$$

where a is the seminal root length at 10 days growth under standard conditions, b is the mean leaf length of the first and second leaves, and c is the leaf width; subscript s is for Seger, a standard barley variety. This formula has an obvious relationship to the amended equation of Davidson (section 10.2.2), in that both balance root quantity against leaf area, and this is assumed to indicate the ability of the adult plant to exploit soil water in relation to its water needs (but see section 11.1.3).

There is still a lack of consistent and extensive information about genetic control of root characteristics (Zobel 1992), partly because of the difficulty of measuring and observing roots, and partly because of the complexity of the results. Thus, there are genotypes and cultivars within the same species that react in different ways to soil conditions; for example, one tomato genotype had few fine roots and another had many fine roots irrespective of soil texture, but a third had few fine roots in sandy soil grading into many in a sandy clay (Zobel 1992). This behaviour is a good example of phenotypic plasticity (O'Toole & Bland 1987), that is the genetically determined ability of the plant to change its root architecture in response to different environmental factors (figure 9.7). Characteristics of this type may be more useful than fixed properties in developing genotypes with a wide range of uses, and more needs to be known about these.

There is thus scope for breeding for desired root characteristics, but past breeding programmes for growth characteristics appear to have had very varying effects on rooting systems. Thus, Welbank *et al.* (1974) found little difference between the root systems of conventional and semidwarf cereal varieties, whereas Hackett (1968) found differences of up to 50% in root length of two barley varieties grown in solution culture. Gregory *et al.* (1992) also found that the water use efficiency of barley varieties varied between 20 and 30 kg grain/ha mm water, and attributed this to differences in the root distribution. Sponchiado *et al.* (1989) showed a clear relationship between the depth of rooting of common bean varieties and their drought avoidance by the use of subsoil stored water. A major breeding programme on soybean (see O'Toole & Bland 1987)

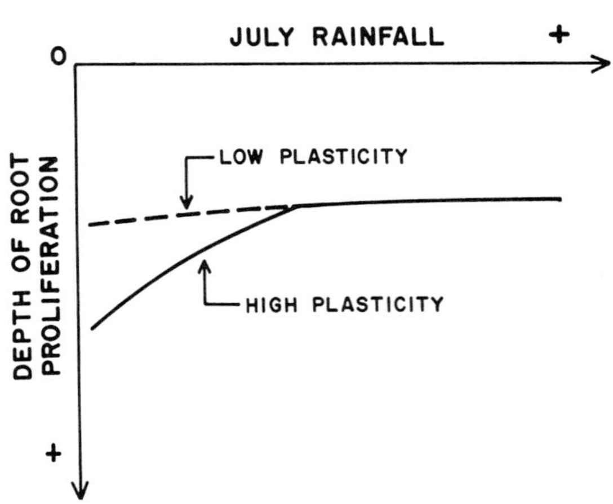

Figure 9.7 Genotypic plasticity to environmental change illustrated in terms of response of rooting depth to the amount of critical season rainfall (after O'Toole & Bland 1987).

attempted to produce deeper roots to prevent late-season drought, but had only partial success.

Rice has a very wide range of root characteristics and it is clear that upland rice has a deeper root system than lowland rice, as might be expected (table 9.5). Yan *et al.* (1995) measured nutrient uptake rates of three rice cultivars and compared them with root characteristics at several times up to heading. The results were difficult to use, because the ranking for N uptake differed from that for P and K. Such a result emphasizes the number of functions that roots carry out, and the difficulty of maximizing the efficiency for all of them simultaneously.

9.3.2 Time and Stage of Development

The root/shoot ratio of a plant changes considerably during its lifetime, with the root typically growing more rapidly at first (Barraclough 1984) (figure 9.8). In cereals, there is the particular shift from the seminal root system that initially supports the plant, to the larger and deeper nodal system. It is often difficult to separate the effect of age as such from that of phenology. There is often a close relationship between the number of main axes and the number of leaves produced by cereals (Porter *et al.* 1986), which is useful in maintaining the above-ground/below-ground balance. Klepper *et al.* (1984) found that this could be expressed by

$$Rn = 1.95Ln - 3.06 \tag{9.3}$$

where Rn and Ln are, respectively, root axes and leaves.

Roots often diverge from their original pattern of penetration with time. Mitchell & Russell (1971) suggested that soybean crop shoots and roots had three synchronized growth stages: (1) vegetative shoot growth, downward tap root growth, horizontal laterals; (2) flowering and pod set, with root development consisting of filling in with branching laterals; (3) maturing seed, and deep penetration of primary laterals at the end of their horizontal spread. The result was that root density spread and developed in a rather irregular way through the profile. It is desirable to have many such observations in different edaphic and

Table 9.5 Root length density distribution of seven rice cultivars from either lowland (flooded) or upland areas, all grown together on a non-flooded soil.

Designation	Ecological origin	Root length density ($cm\ cm^{-3}$) at varying depths (cm)							
		0–10	10–20	20–30	30–40	40–50	50–60	60–70	70–80
IR20	Lowland	14.4	2.8	0.9	0.4	0.1	—	—	—
IR2035-117-3	Lowland	22.7	5.8	0.8	0.1	0.1	—	—	—
IR442-2-58	Lowland	16.8	7.1	1.2	0.3	0.1	0.1	0.1	0.1
OS4	Upland	12.6	1.4	0.8	0.9	0.8	0.5	0.5	0.5
Moroberekan	Upland	11.8	2.3	0.9	0.8	0.6	0.8	0.4	0.2
Salumpikit	Upland	16.2	5.5	1.9	1.4	0.8	0.6	0.3	0.1
20 A	Upland	19.8	2.6	0.9	0.8	0.9	0.9	0.6	0.4

Source: after O'Toole & Bland (1987).

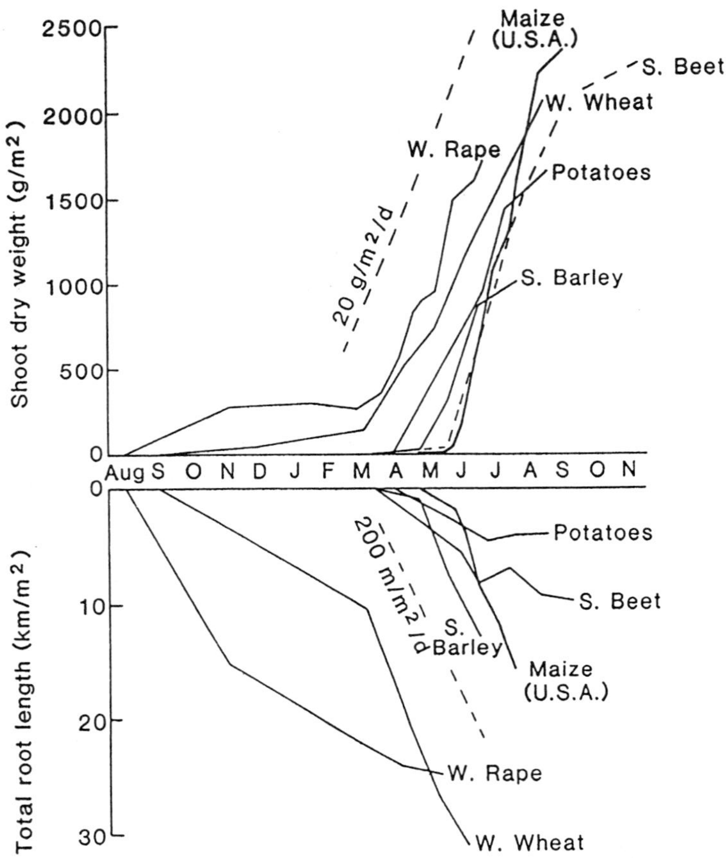

Figure 9.8 Development of root and shoot of winter wheat, winter rape, summer barley, maize, sugar beet, and potatoes. Root weight increases earlier than shoot growth (after Barraclough 1989a).

climatic regions, and with different varieties, to determine what is genetically programmed and what is a response to local soil conditions.

Flowering often has a major effect in slowing root growth rate, to the point where total root length may decrease. This is well substantiated for cereals, but Gregory (1994b) suggested that it may not occur with all legumes (figure 9.6).

9.3.3 Chemical Effects

9.3.3.1 General Nutrient Supply

Cakmak *et al.* (1994) tested the effects of nutrient deficiency on the root/shoot ratio in young bean plants in soil in pots. They also measured the sugar level in the root at the same time, on the theory that it may be related to root growth rate. Results were very variable for P, K, and Mg. When the last two elements were deficient, there was a low sugar concentration in the roots, and slow root growth.

When phosphorus was deficient, the root sugar level was very variable, but the root/shoot ratio was always very large (figure 9.9). Cannell & Dewar (1994) found that the effect of different nutrient deficiencies differed considerably in effects on root growth. They also found an accumulation of sugars in the root with deficiencies of some elements, but only with modest levels of nutrient stress, and it seems unlikely that there is a simple connection.

A detailed study of birch by Ericsson (1995), using solution culture in the laboratory, showed that the root/shoot ratio increased if N, P, or S were deficient, but the ratio decreased if K, Mg, or Mn were deficient, whereas deficiencies of Ca, Fe, and Zn had little effect. This confirms that the effects of nutrients differ widely, and generalizations are difficult.

Heavy metals at toxic levels may have considerable effects on root growth, though tolerance to the metal can increase even after short-term exposure (Davies 1991).

9.3.3.2 Local Nutrient Supply

A number of studies have shown that root growth is greatly altered by non-uniform distribution of nutrients (Drew & Saker 1975a, b; Robinson 1994). Not all nutrients have the same effect; in general, phosphorus concentrations cause rapid branching, which produces higher root length density in the enriched volume (figure 9.10). A similar but less pronounced effect occurs for nitrogen, whereas

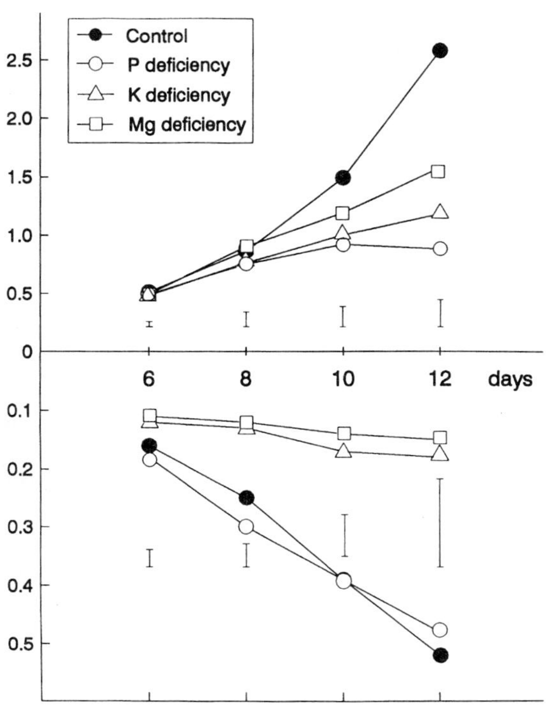

Figure 9.9 Shoot and root dry weights (g) of bean plants over a growth period of 12 days in nutrient culture, as affected by either deficient (0) or full (10 μM, 50 μM or 20 μM for P, K or Mg) nutrients. Bars indicate LSD at $P = 0.05$ (after Cakmak *et al.* 1994).

potassium and other elements have relatively little effect in this way (Drew 1975), though De Jager (1982) reported that all major nutrients had some effect.

The physiological mechanisms connecting local nutrient level, photosynthate supply and root growth are as yet not understood. It is interesting that phosphorus has the largest effect both in general and in localized application, and we may speculate that the benefit to the plant of such responses is greatest with a non-mobile nutrient like phosphorus.

The response by crops to local nutrient supply is clearly important for fertilizer placement. This includes 'sideband' placement, in which a fertilizer dressing is placed mechanically along and a little to the side of the plant row, the purpose being to supply phosphorus at high concentration to the plant early in its development, without the expense of fertilizing the whole topsoil to this concentration. The other possibility is to place fertilizer at around 50 cm depth, with the aim of putting nutrients into a soil layer that is expected to remain moist for much longer than the topsoil, and to encourage rooting at this depth.

Three different field soils were used by Zhang & Barber (1992) to test the effect of the initial anion-exchangeable phosphate on the maize root distribution obtained in fertilized and non-fertilized zones. They found a nearly straight relationship (figure 9.10) between the ratios of the phosphorus levels in the fertilized and the unfertilized soils, and the root density in the fertilized plot. The initial soil phosphorus concentrations in the three soils were intentionally quite widely separated, and the lowest initial concentrations gave the highest ratios, so the experiment could not discriminate between the effects of the concentrations and the ratios. Zhang & Barber (1992) have suggested that these results can be applied to fertilizer management in the field, so as to predict the soil phosphate level and the root length density needed in the enriched zone to prevent phosphate deficiency. This demands considerable precision in measurement and dependability in the model, and much field experience with the system is needed before it can be used with confidence.

Results appear to be different with the mobile nutrient nitrogen (Hodge *et al.* 1998). This work investigated uptake of labelled and unlabelled N, by five

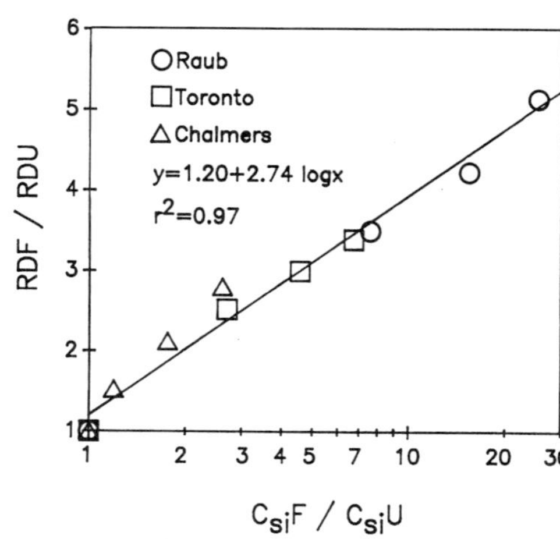

Figure 9.10 Relation between the ratio of root length density in P-fertilized soil compartment to that in unfertilized soil (RDF/RDU) and the ratio between initial exchangeable phosphate levels (C_{si}) in the same two compartments (after Zhang & Barber 1992).

different species, with N supplied in patches in organic or inorganic form. All species captured similar fractions of ^{15}N from a patch, as a fraction of that initially present. The organic material added to create a patch had a high C/N ratio, so this treatment might have reduced the mineral N supply. The overall result from this very complex experiment was that uptake of N was not closely related to root proliferation or root length duration. This was explained as being due to competition with microbial flora in the rhizosphere, but the relative amounts of plant and microbial masses would need to be known to substantiate this. The results appear to support the generalization that root length density is not of enormous importance for uptake of mobile nutrients, so long as the roots are well distributed.

9.3.3.3 Acidity

There are striking variations in acidity tolerance amongst crops and cultivars (Marschner 1995, p. 613). The effects of acidity are rarely due to the hydrogen ion, and in solution culture most plants will grow below pH 4. In soil, the toxicity is due to soluble (especially trivalent) aluminium (see section 5.3.4), and to a lesser extent to manganese, iron and other metals. However, higher levels of soil Ca or P can reduce the Al toxicity. At toxic levels, roots become thickened, 'coraloid' and poorly branched (Foy 1992; Marschner 1995). Balsberg-Pahlsson (1995) reported that acidity had an inhibiting effect on the root hairs of an acid-tolerant grass when grown in soil solution (section 5.1.3).

9.3.3.4 Organic Materials — Allelopathy and Agrochemicals

Allelopathy occurs when a chemical being lost from one plant has a specific effect on another, either damaging or advantageous (Rice 1984). It can therefore be important in competition (sections 11.4.2 and 5.5.3). Most of the allelochemicals studied so far are tannins and polyphenols, but a very wide range of organic chemicals have been implicated. Mostly, they are believed to be taken up by the roots from the surrounding growth medium into which they have been exuded from roots, but root grafts or common mycorrhizal networks (section 8.3.8) could also act as transfer processes. Sometimes, their action is so strong that they can be used as herbicides (Heisey 1990), and there is now considerable interest in the allelopathic properties of rice cultivars that give them a better competitive ability against weeds (Olafsdotter *et al.* 1995). Many possible mechanisms for their action have been suggested: inhibition of cell division, alteration of membrane function, inhibition of nutrient uptake (section 5.5.3), changes to growth rate and respiration, or blocking of xylem vessels. It is likely that there is an important effect on roots and nutrient uptake, but whereas there is a large literature on general effects by these plant compounds, there is little information on precise mechanisms. Nye & Tinker (1977, p. 276) discussed the mutual exclusion of roots by adjacent plants. This has usually been ascribed to allelopathy, but other mechanisms are possible. Ethylene also has various effects on roots (Jackson 1991).

9.3.3.5 Salinity

Salinity is a particularly damaging form of soil deterioration. Some plant species will tolerate high salinity, but the great majority are damaged. The subject has been very well reviewed by Marschner (1995), and the more practical aspects are detailed in Sharpley *et al.* (1992). There are many different physiological and biochemical effects throughout the plant, the main ones being water deficit, ion toxicity, and nutrient balance, and the roots are only one of several plant parts and processes that are damaged (Greenway & Munns 1980).

9.3.4 Water Content of Soil

Root/shoot ratios tend to be high in very wet and largely anaerobic soil, decrease rapidly as soil becomes fully aerobic, and then gradually increase as the soil becomes drier (figure 9.11). This agrees with the general principle that roots grow proportionately more when soil resources are limiting. It appears that roots extend more slowly as soil becomes drier, but this may be an effect of mechanical impedance rather than a direct effect of water potential (Drew & Goss 1973). Thus, root cells appear to adjust to small external water potential differences with little effect on their expansion rate, though applied external mechanical pressure will change extension rate sharply (section 9.3.5) (see figure 9.14). Cruz *et al.* (1992) reported structural changes in roots of sorghum following their exposure to drought, which reduced the root conductance; this seems likely to enhance the danger of any subsequent drought.

Roots will not penetrate completely dry soil. However, there are considerable differences in how the root systems of different species respond to dry soil. For example, in three species of oaks there was no change in architecture as the tap root began to enter dry soil, until the tap root tip eventually died. After this abundant laterals appeared in two species but not in the third (Callaway 1990). Conversely, if a main axis is penetrating from dry into moister soil, it will continue to extend. Thus, cotton root systems extended more and penetrated 20 cm deeper under drought conditions in the topsoil than when the topsoil was moist (figure 9.12).

An interesting experiment by Thorup (1969) allowed a separation of the effect of soil water content on root extension from that of nutrient uptake inhibition due to slow nutrient diffusion. Tomato plants were supplied with water by a split-root system, so that the water status of the plants was the same, and then allowed to root into soil at different water potentials that had been labelled with ^{32}P. Extension growth of roots in the dry soil was reduced by a factor of 7, but phosphate uptake by a factor of 30 (section 6.5.2).

9.3.5 Mechanical Resistance of Soil

Roots growing through solid media usually have to exert some physical force to compress the medium, or to move individual particles, because root tips cannot contract to pass through small holes (Wiersum 1957). The type of medium determines the magnitude of the force that is necessary, and in very compact and dense

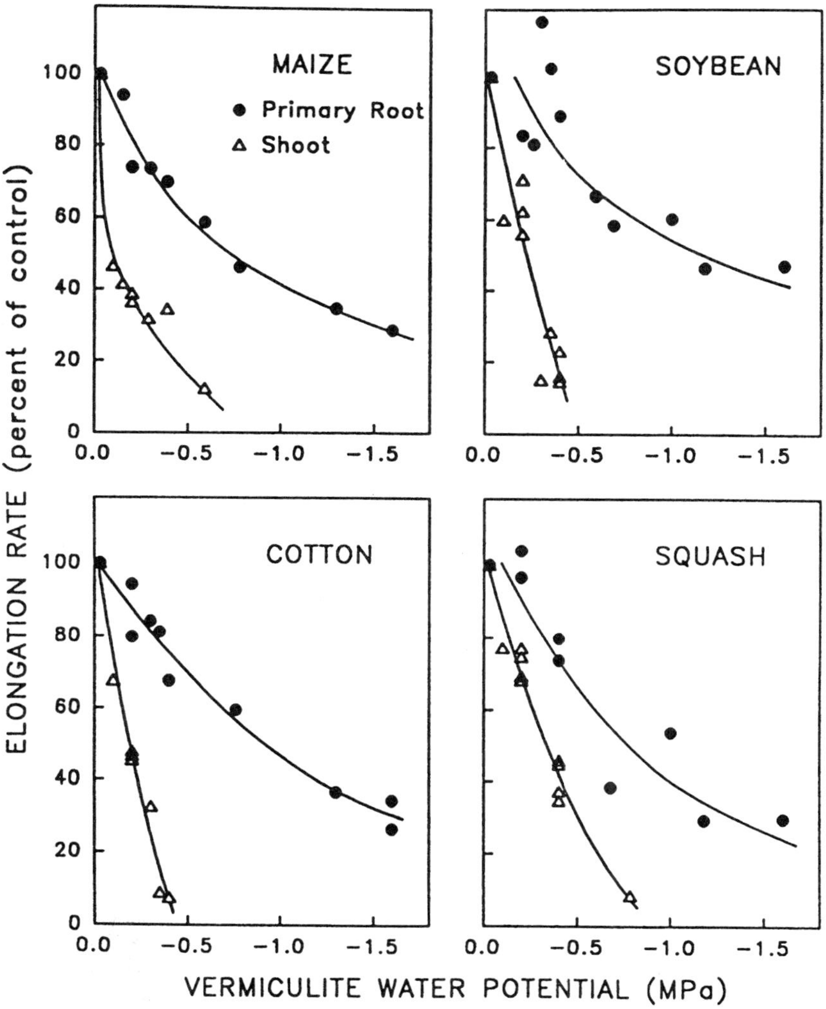

Figure 9.11 Elongation rate of primary root and of shoot growing in vermiculite at different water potentials, plotted as fractions of elongation rates at high water potential. Note the greater conservation of root rather than of shoot extension rates at low moisture potential (after Spollen *et al.* 1993).

media roots may not be able to penetrate at all. The mechanism whereby this force is produced, and the consequences for the growth of the root and the whole plant, are therefore very important. Bengough & Mullins (1990) have reviewed this subject well, including a careful comparison between the processes of penetration by a root and by a slender metal rod (penetrometer).

The origin of the force produced by the root is the turgor pressure of the cells, due to the osmotic pressure of the vacuole (section 5.2.2). This pressure causes the cell walls in the region of elongation to extend, and hence the root grows. In this state, the turgor pressure P is slightly greater than the limiting yield stress of the

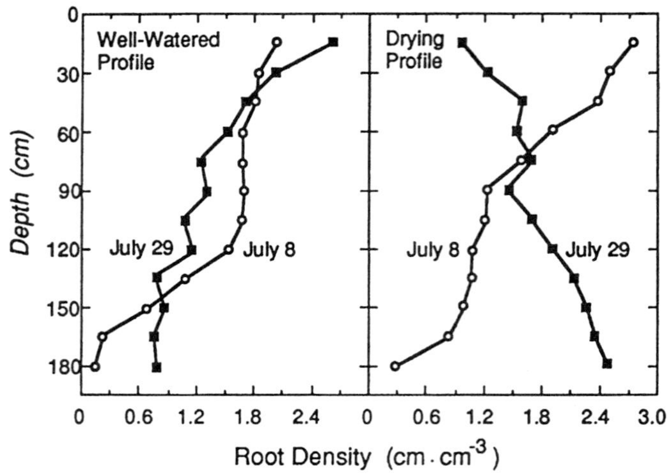

Figure 9.12 Root length density profiles for cotton growing in a well-watered and a drying soil. The change in slope in the drying profile occurred in only 3 weeks (after Kramer & Boyer 1995).

cell wall Y_l, which is the level of stress (Y) beyond which the cell wall deforms plastically to give a permanent increase in cell volume. If an external surface presses on the cell, part of the turgor pressure will then be carried by this surface, which can be another cortical cell, or a soil particle external to the root. In an equilibrium (non-growing) state,

$$P = Y + Z \tag{9.4}$$

so that the turgor pressure is divided between the elastically deformed cell wall, and the normal stress on the external surface Z. If $Y > Y_l$, the cell will grow, and if Z exceeds the yield stress of the external soil, then the soil will be deformed (Greacen & Oh 1972; Whalley & Dexter 1993; Gregory 1994a).

It has been suggested that there is a linear relationship between the rate of elongation of a root and the degree to which P exceeds Y_l and Z, so that

$$R = m(P - Y_l - Z) \tag{9.5}$$

where R is the elongation rate of the root and m is an extensibility constant. This equation is unlikely to be obeyed closely, because the terms in the equation may well vary during cell expansion, and the elongation rate of a root cell and the whole root may not be connected simply. Nevertheless, it may be useful in clarifying the concepts involved.

The value of the force that a root can exert on soil increases progressively when a root is completely impeded (Clark *et al.* 1996) (figure 9.13), both because of its deformed shape and an increase in P. This is a rather extreme case, but smaller degrees of impedance may also have analogous effects. The most important effect may be that Y_l decreases, because the cross-linking and orientation of the microfibrils in the wall (section 5.2.1) can change during extension, so that a greater fraction of P is carried by the soil.

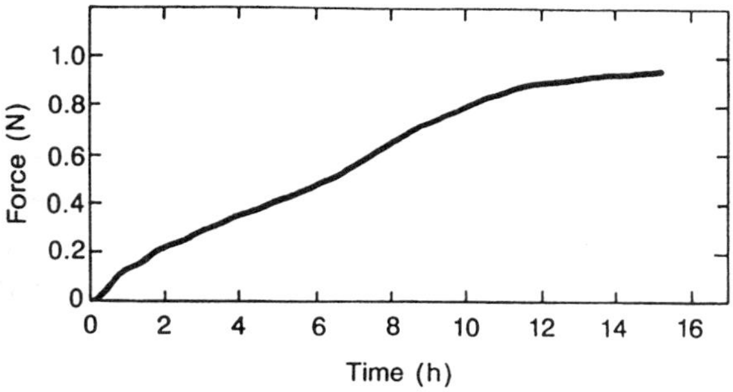

Figure 9.13 Increase in the axial growth force in a pea root with time after total impedance. The force might still increase after 16 h, but the growth pressure, σ, had always reached a maximum by then (after Clark *et al.* 1996).

Of course, Z will change with time as the adpressed soil is compressed and deformed. It is a rather more complex process than appears at first. A root penetrating a soil can be considered most easily if the latter is regarded as homogeneous, and the root as a smooth probe. As the root moves forward, the soil immediately next to it will be compressed with total mechanical failure, the next zones will be plastically deformed (figure 9.14), and the outer zone will be elastically deformed. The permanent changes in zones I and II will therefore produce a permanent root channel. If the root is small enough, or the soil is of sufficiently low density, zones I and II may not form.

In addition, a rigid body forced through soil has to overcome friction. The friction between root and soil will be less than with a penetrometer, because only the extension zone of the root moves, and it is lubricated with mucigel (chapters 6, 7 and 8). Additionally, the root tip follows a spiral path as the root extends (nutation), and this helps the tip to find the path of least resistance.

The lubrication by the mucilage is itself a complex phenomenon. Because the fully expanded mucigel loses water very rapidly as the tension increases (section 2.3.4), lubrication due to the presence of the compressed gel would cease as soon as the soil started to dry out. However, boundary-layer lubrication, between two layers of the mucilage solids, can remain effective even in quite dry soil (Mullins *et al.* 1997). It appears that the potential friction between root and soil is greatly reduced by these effects, and that this is the cause of the observed ratio of 3–8 between the resistance with a penetrometer and a root (Bengough *et al.* 1997).

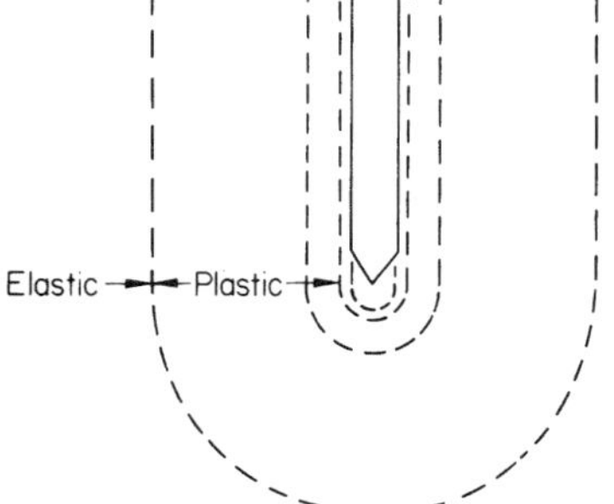

Figure 9.14 Compression zones of soil around an advancing root tip or penetrometer probe (after Barley & Greacen 1967).

Ballotini (tiny glass spheres) have often been used in measurements of the pressures that inhibit root extension (figure 9.15), but with this system also the friction between the ballotini causes the measured values to be much larger than those appropriate for roots (Bengough & Mullins 1990).

Several plant properties affect the ability to penetrate soil, and there is evidence for several quantitative trait loci (see discussion by Cook *et al.* 1997), so there are well-defined differences in the penetrating ability of roots of different plant varieties (figure 9.16).

The effects of mechanical impedance on the predictions of the ROOTMAP model (section 9.5.2) have been explored (Tezera Tsegaye & Mullins 1994; Tezera Tsegaye *et al.* 1995a, b) using separately measured parameters of the pea root system, and an actual growth experiment. When the effect of drying out of the clay soil by the roots on the soil impedance was taken into account, agreement with the model was satisfactory, whereas there was total disagreement if a constant impedance was assumed (figure 9.17). This is a useful example of how models can help to investigate practical problems, though much validation is needed to give sufficient confidence in the model.

Despite this well-developed body of theory (Greacen & Oh 1972), it is often difficult to explain quantitatively the response of plants to compaction in soils. There are interactions between soil compaction and water and oxygen contents (figure 9.18), and roots may find cracks and voids to exploit even when the bulk soil is heavily compacted. Further, if the root extracts water, that will tend to extend the incipient root channel as a crack if the soil has swelling/shrinking properties. This may aid root penetration, even if the drying of the soil simultaneously increases the mechanical strength of the soil (Tezera Tsegaye & Mullins 1994).

Mechanical impedance to roots is closely related to the soil bulk density. There is often little root penetration above a density of 1.7–1.8 g cm^{-3} (figure 9.18). The roots entering a dense soil will thicken and may be distorted. Often there is heavy branching behind the tip, or production of root hairs, but the hairs themselves may be unable to enter a very compacted soil. However, where root hairs do enter the soil, they effectively anchor the root a short distance above the extension zone, and so allow the exertion of the extension force without buckling the root. Some

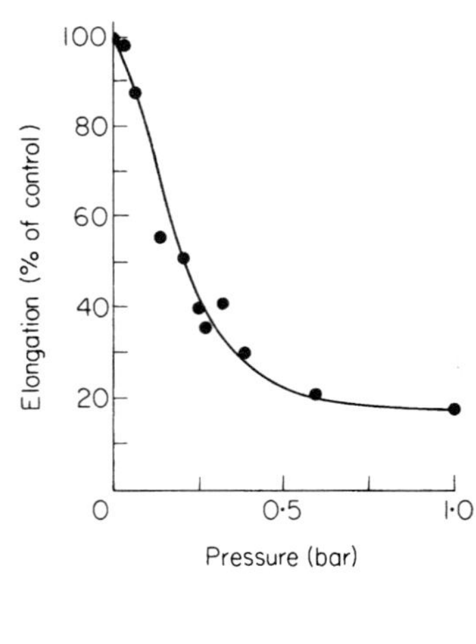

Figure 9.15 The relationship between the applied external pressure and the relative rate of elongation for barley roots growing in 1-mm-diameter glass spheres for 6 days (after Drew & Goss 1973).

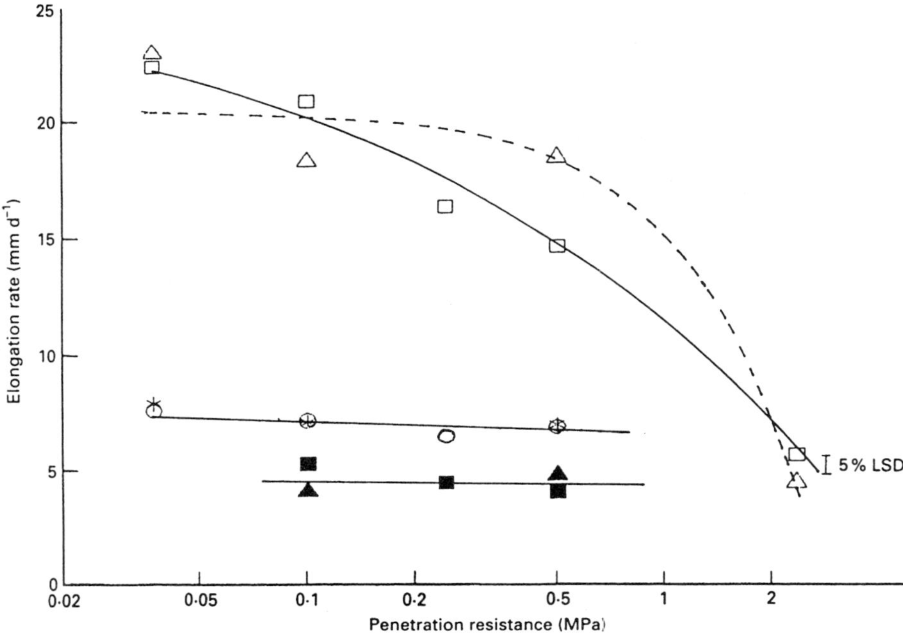

Figure 9.16 Mean elongation rate of roots of Progreta (square) and Solara pea varieties when grown in soil packed to give different resistances. Main axes (square, triangle), first laterals (circles), and second laterals (filled symbols) (after Tezera Tsegaye & Mullins 1994).

experiments have been reported with plant radicles, but Kierkegaard *et al.* (1992) showed that radicles and roots from similar plants did not respond similarly to mechanical impedance.

9.3.6 Oxygen Supply and Anaerobiosis

Oxygen deficiency in the soil atmosphere is usually accompanied by an increase in carbon dioxide, and is a consequence of excessive water content in the soil. The majority of species can tolerate only a limited amount of anoxia, but wetland plants have a series of adaptations that allow them to thrive in such conditions (Armstrong 1979; Drew 1988; Drew & Stolzy 1996; Armstrong *et al.* 1994). There is no exact boundary between wetland and other plants, because almost all plants can stand some anoxia, varying in time or degree. The acclimation process that roots undergo when they encounter oxygen deficiency includes the switching of biochemical energy production to an anaerobic process, and changes in root structure that finally lead to the formation of the spaces in the cortex called aerenchyma (Drew *et al.* 1994). De Willigen & Van Noordwijk (1989), on the basis of model calculations, showed that this air-filled porosity is important for transmitting oxygen into thick roots even in unsaturated soils, and noted that it is found widely in different species (table 9.6). Modelling and microelectrode studies

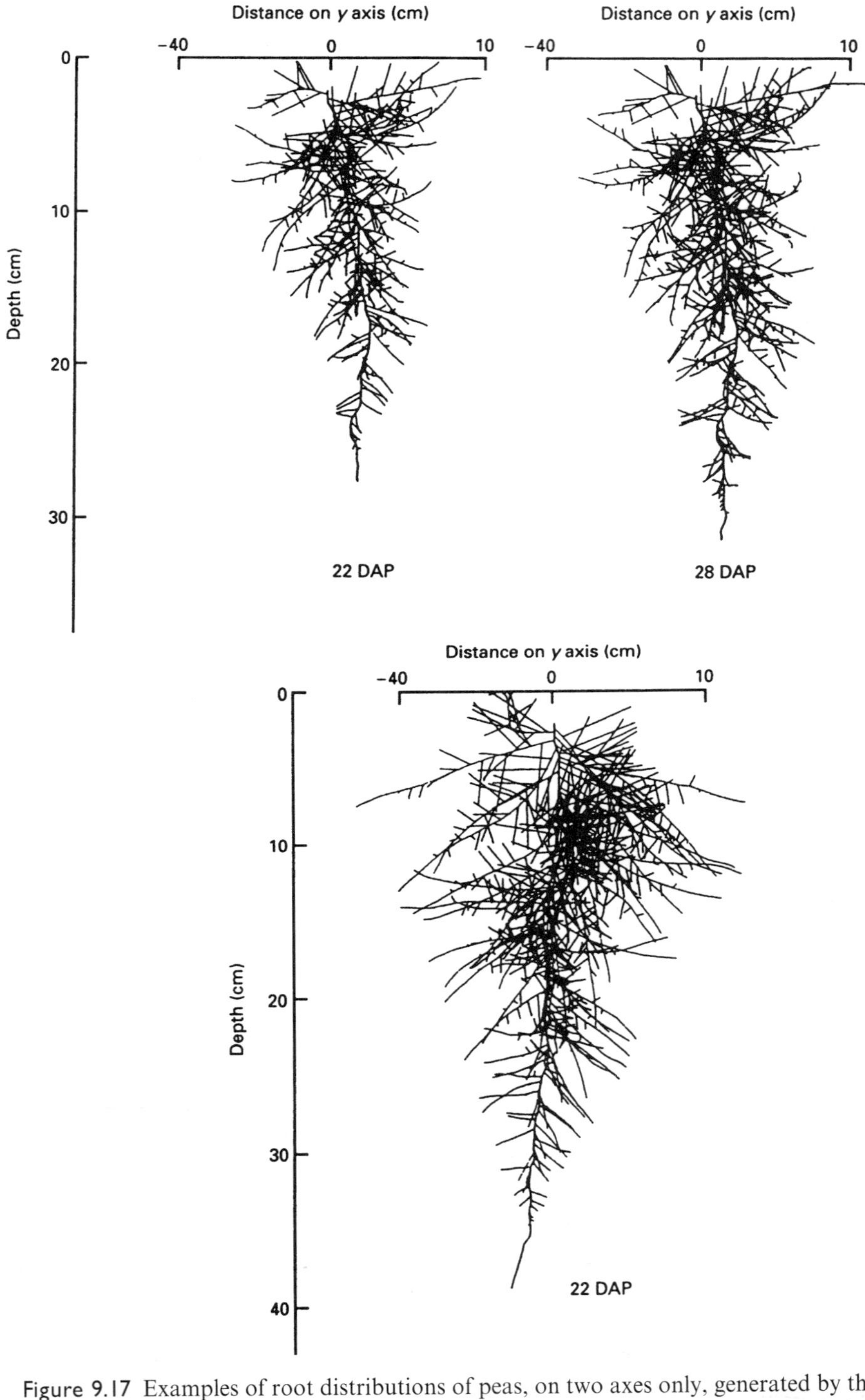

Figure 9.17 Examples of root distributions of peas, on two axes only, generated by the ROOTMAP model. The top two diagrams are for simulation allowing for the drying out of the soil; the lower diagram is for simulation without drying out. Days after planting are marked on the figure (after Tezera Tsegaye *et al.* 1995b).

Table 9.6 Wide span of measured root porosity in a range of species.

Crop	Root porosity (%)	Reference
Wetland grasses and rushes	8–45	Crawford (1982)
Rice	27–36	Jensen *et al.* (1969)
Maize	8–10	Jensen *et al.* (1969)
Maize adventitious roots	3–19	Van Noordwijk, unpublished results
Barley	4	Jensen *et al.* (1969)
	2–4	Yu *et al.* (1969)
	5–13	Van Noordwijk, unpublished results
Wheat cv. Inia, susceptible to waterlogging	3–8	Yu *et al.* (1969)
Wheat cv. Pato, tolerant	5–15	Jensen *et al.* (1969)
Wheat	3–5	Van Noordwijk, unpublished results
Onion	5	Jensen *et al.* (1969)
Wetland dicots	2–19	Crawford (1982)
Sugar beet	3–7	Van Noordwijk, unpublished results
Brussels sprouts	3–9	Van Noordwijk, unpublished results
Tomato	6	Jensen *et al.* (1969)
	4–9	Van Noordwijk, unpublished results
Lettuce	5–6	Van Noordwijk, unpublished results
Sunflower	5–11	Yu *et al.* (1969)
Bean, pea	4	Jensen *et al.* (1969)
Bean	1–4	Schumacher and Smucker (1981)
Gerbera	2–8	Van Noordwijk, unpublished results
Bouvardia	0–1	Van Noordwijk, unpublished results

Source: after De Willigen & Van Noordwijk (1989).

For listed references, see original publication.

by Armstrong *et al.* (1994) also suggest that the stele may often be nearly anaerobic. A causal relationship between internal root porosity and soil pore diameter has been suggested by Engelaar *et al.* (1993). This issue is of particular importance for wetland rice, which grows in wholly saturated soils. Each root is surrounded by a thin cylinder of oxidized soil, the oxygen being supplied by diffusion down the aerenchyma.

There is a complicated interaction between oxygen supply and mechanical impedance in effects on roots (Tackett & Pearson 1964) (figure 9.18). The effect of oxygen deficiency on extension rate is greatest with little compaction (i.e. mechanical impedance), whereas it has little effect where extension rates are already inhibited by mechanical resistance.

It appears that soil oxygen concentration usually has no effect on plants until it is reduced to less than 10% by volume, and there are a number of reports of metabolism and extension rates not being affected at even lower oxygen concentrations. However, it is difficult to know what the oxygen concentration is within the root in view of different rates of oxygen consumption, and different diffusion paths in both soil and root. The work of Eavis *et al.* (1971) shows how differing oxygen concentrations around shoot and root combine to produce a range of different results.

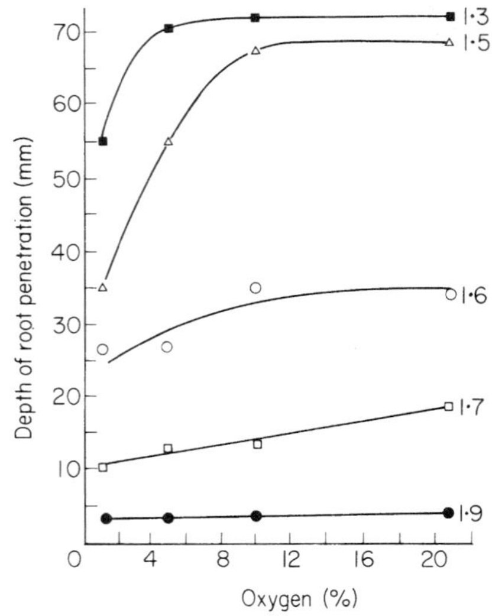

Figure 9.18 The combined effects of oxygen concentration in the soil air and the soil bulk density (numbers marked on curves) on the depth of cotton root penetration into a compacted subsoil) (after Tackett & Pearson 1964).

9.3.7 Temperature

Temperature affects plant development significantly and in many ways (Cooper 1973; McMichael & Burke 1996). For a given crop, the development of the root system is often very closely controlled by accumulated thermal time (figure 9.19). There are considerable differences in the effects on different species, even when they grow in roughly the same climatic zone. Root extension rate normally increases with temperature up to about 25–30°C, but maximum extension rates from 5° to 30° are possible (McMichael & Burke 1996). The interaction with the changes in the shoot growth rate produces a minimum in the root/shoot ratio at about 20–25°C. Desert plants can acclimate to high temperatures up to about 60° (Nobel 1989).

The effect of temperature on different root characteristics varies, though the temperature for peak values for tap root length, lateral root length and number of laterals are all fairly similar, and differ consistently between cotton and sunflower (McMichael & Quisenberry 1993). Branching is under temperature control, but the effects may vary widely between species, for example most branching with maize occurs at 20°C, but with pine at 34°C. The direction of the laterals is also temperature sensitive (Gregory 1994a), with the minimum angle between main root and lateral at 17°C in maize (Onderdonk & Ketcheson 1973). The architecture and size of a root system can therefore vary greatly with temperature, but there is as yet no fundamental understanding of how this happens.

The effects of temperature on roots can even vary widely among cultivars, and McMichael & Burke (1996) discuss the various ways in which breeding can alter these responses. Irrigation and water relationships can alter the soil temperature, so that water and temperature effects are probably often confounded in the field.

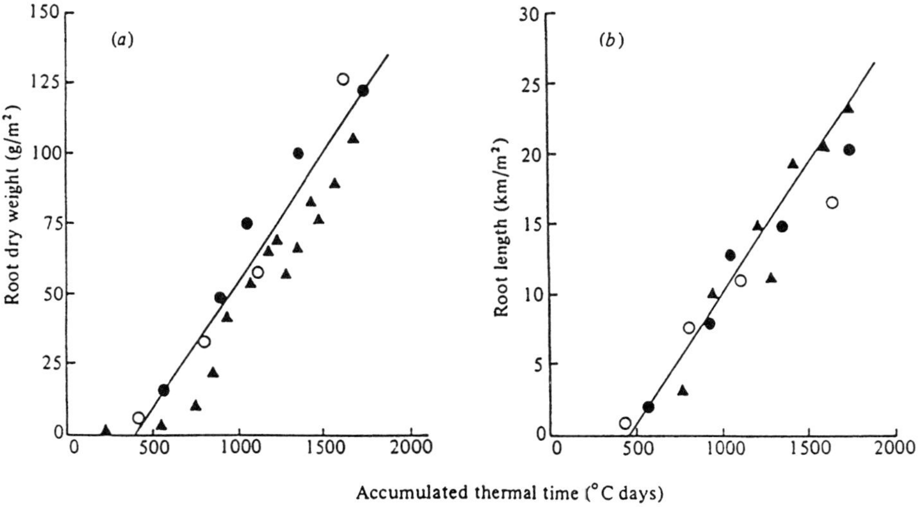

Figure 9.19 (a) Root dry weight and (b) length of winter wheat plotted against accumulated thermal time (day-degrees), for data from Welbank *et al.* (1974) and Gregory *et al.* (1978) (after Barraclough & Leigh 1984).

9.3.8 Pests, Diseases, and Other Biota

Some pests and diseases appear to exert their effects on plants by physically removing part of the root system, such as the free-living nematodes that infest sugar beet, barley and other plants. This is best regarded as simple grazing by the organisms (Brown & Gange 1991). In other cases, the pest alters the form and function of the root system by diverting photosynthate towards itself, such as the potato root knot nematode. Root fungi usually operate by invading the tissues and causing them to rot and die; thus, the 'take-all' fungus *Gaumannomyces graminis* causes lesions that isolate parts of the cereal root system. The damage due to fungi is partly due to loss of functional root, partly due to the production of toxins, and partly due to a spread from the roots into the shoot (Hillocks & Waller 1997). Paul & Ayres (1990) found that foliar infection of barley with rust fungi caused a decrease in root growth, but that root function was not affected.

This book deals in the main with healthy plants, but in the field almost all plants are to some extent attacked, even if only by grazing on root hairs or on mycorrhizal hyphae. A comparison under the microscope of solution-grown roots with the damaged and irregular roots extracted from field soils is instructive in showing the extent of this damage. The frequent observation of different experimental results in the laboratory and the field may be due to faunal grazing effects in the latter (Fitter 1991b), and it is important to bear these possibilities in mind in all experimental work. Mycorrhizal roots are often very different in their age and duration from the uninfected roots (section 8.3.3) (Hooke *et al.* 1995).

If the attack is simply by removal of parts of the root so that the root length density is decreased, better nutrition might overcome the effects of this damage, but it might be of little advantage where there is a production of toxins. In agreement with this, the damage caused by the free-living nematode *Longidorus* to sugar beet was found to be at least partly compensated by heavy applications of fertilizer (Whitehead *et al.* 1971).

Traditionally, the activities of earthworms have contributed to the fertility of soils, and also, it seems, to the ease with which roots develop in them. Much information can be found in Ketschmar (1992). Despite this impression, Hirth *et al.* (1997) found no evidence that ryegrass roots entered preferentially into holes in a subsoil filled with earthworm casts, ordinary soil, or left empty.

9.3.9 Carbon Dioxide and Ethylene Concentration

It is well known that the atmospheric carbon dioxide concentration is increasing steadily as a result of the combustion of fossil fuels (IPCC 1996, p. 53). This process has already raised the carbon dioxide level from about 285 ppm before the Industrial Revolution to about 350 ppm now, and it is forecast that it will reach twice the initial concentration, that is 570 ppm, sometime in the latter half of the twenty-first century. This process has important and extensive effects on roots, plants, and potentially on farming (Rogers *et al.* 1992; Tinker *et al.* 1996).

Most crop plants increase growth and yield when grown in elevated CO_2 (Rogers *et al.* 1992), though the effect is smaller for C4 than C3 species. The transpiration ratio normally is decreased, whether or not there is a yield increase, due to the greater closure of the stomata at elevated CO_2 levels. These processes have consequences that are important for our subject.

The below-ground effects vary widely among species and environmental conditions, but the following are possible: increased allocation of carbon to the root, so that the root/shoot ratio increases (Rogers *et al.* 1996); increased input of carbon to the soil, so that the soil organic matter level may rise; increased infection with symbiotic organisms; and increased levels of exudation (Tinker *et al.* 1996). There is still much uncertainty about the precise reactions to these changes, and some studies have suggested that the effects decline with time (acclimation), but it seems possible that this decline is an artefact of developing nutrient deficiency in some experiments.

The plant hormone ethylene has a pervasive effect on root development (Jackson 1993). Root tissues are sensitive to the concentration of ethylene, they can themselves produce ethylene, and the physical conditions around a root can greatly influence the local ethylene concentration, because the soil can entrap, oxidize or produce ethylene. Many results in this topic have been contradictory or variable because of this complex system. There is much evidence that ethylene can affect the growth, development and anatomy of roots, but there is as yet no clear and established theory of how it operates, or why its effects are so variable.

9.4 Root Distribution and Density in the Field

9.4.1 The Amount of Root per Unit Area

Whereas root data were scarce when Nye & Tinker (1977) was written, there are now several extensive data-sets of high quality, and the difficulty is to decide which data to select. It is reasonable to take L_a, the root length under unit area of ground, as the most useful characteristic (section 11.1.1). A crop or a uniform area of natural vegetation can be regarded as a horizontally extensive single plant. Root and nutrition studies should therefore use L_a as the simplest standard measure of the rooting behaviour of a crop or species. Table 9.7 gives a series of values for this. This table has been constructed with data from Nye & Tinker (1977) and relatively recent compilations by other authors (Van Noordwijk & Brouwer 1991; Barraclough 1989a; Robertson *et al.* 1993c; Gregory 1994a; Gregory *et al.* 1996) and some original papers. A few data are also given for the root length density, L_V, in the topsoil, where these have been reported in the same publication as an L_a value. It is recommended that readers consult these references for more detail.

Mostly, there appears to be no obvious cause for the different results, for the same crop, reported by different authors. However, it is obvious from table 9.7 that different species differ very widely in L_a, from about 30 km m^{-2} to less than 2. Hamblin & Hamblin (1985) made a direct comparison of several crop species, and found that the order of their L_a values was clover and medics > wheat > lupins and peas. More such general rules would be useful (Gregory 1994b).

These ranges for L_a are proportionately much greater than those normally found for the variations in leaf area indices of crops, which are the above-ground analogues of L_a values. These usually lie in a range from 2 to 8; presumably, the range is no greater because higher values of the leaf area index give little gain in terms of light interception. It may be that it is the greater variability of the below-ground environment, and the greater number of required inputs to roots than to leaves, that determine the greater range of L_a values.

The root length per unit dry shoot weight (table 9.7) is also an important parameter, based upon earlier discussion of the internal balance in the plant (equation (9.2)). This also varies widely, from 40 to 3 m g^{-1}.

9.4.2 Crop Root Systems and Effects of Agronomic Treatments

Root systems are affected by natural factors such as soil type, rainfall and temperature, but the agronomic treatment of crops can also be important. One of the most obvious differences in agronomic treatment is the sowing date, which affects temperature, irradiance, and possibly water stress. The development of root systems of various temperate crops are compared with the growth of the shoots in figure 9.8 (Barraclough 1989a). The advantage in root system development gained by autumn-sown crops is striking, with over double the final root length of spring-sown crops, but much less advantage in shoot growth. The maize crop was grown in the USA, with a considerably higher temperature and shoot growth rate than

Table 9.7 The origin, age, depth of measurement, length per unit surface area L_a, length per unit volume in the topsoil L_V (taken as 0–20 cm where possible), specific root length (s.r.l.) and root length per unit shoot weight L_a/Ws.

Reference	Crop	Country	Plant age (days)	Depth (cm)	L_a (km m^{-2})	L_V (km m^{-3})	s.r.l. (m g^{-1})	L_a/Ws (m g^{-1})	Source
Welbank *et al.* (1974)	Summer wheat	UK	94	100	20	70			1, 3
Barraclough (1989a)	Winter wheat	UK	215	100	31	70		29	3
Gregory *et al.* (1978)	Winter wheat	UK	230	150	25	60		22	2, 3, 4
Gregory *et al.* (1992)	Summer wheat	Australia	111	80	4	13	76	9	4
Bragg *et al.* (1984)	Winter barley	UK	247		27	68		25	3
Gregory *et al.* (1992)	Summer barley	Australia	83	80	5	25	60	7	4
Mengel & Barber (1974)	Maize	USA	79	75	15	35	100	12	1, 2, 3, 4
Robertson *et al.* (1980)	Maize	USA		150	8	44			2
Robertson *et al.* (1993a)	Sorghum (Watered)	US	104	190	5			3	5
Robertson *et al.* (1993c)	Sorghum (Kept dry)	US	102	180	4			6	5
Barraclough (1989b)	Winter rape	UK	280		25	97		26	3
Brown & Biscoe (1985)	Sugar beet	UK	179		10	25		5	2, 3, 4
Vos & Groenwold (1986)	Potatoes	NL	102	100	4	20		5	2, 3, 4
Barber (1978)	Soybeans	US	90		3	10			3
Taylor & Boehm (1976)	Soybeans	US	84	188	12	7			2
Brown *et al.* (1989)	Chickpea	Syria		100	6				4
Robertson *et al.* (1980)	Groundnut	US	95	150	8	16	46		2
Grimes *et al.* (1975)	Cotton	USA	133	183	18	23			2
Greenwood *et al.* (1982)	Onion	UK	126		2	10			3
Garwood & Sinclair (1979)	Grass	UK	730	60	125	370	356		2
Pearson & Jacobs (1985)	Clover	Australia	119	50	18		200		2
Barley (1970)	*Pinus radiata*	Australia	106		8	20			1

Sources: Compiled from lists in (1) Nye & Tinker (1997); (2) Van Noordwijk & Brouwer (1991); (3) Barraclough (1989a); (4) Gregory *et al.* (1996); Gregory (1994a); (5) Robertson *et al.* (1993c).

Numbers quoted may vary between these sources, as different data may be selected from the original papers.

the other crops, all of which were grown in the UK. Despite this, the root growth rate of maize was similar to that of spring barley in the UK, and the total root length was remarkably small for the total shoot mass.

Barraclough *et al.* (1991) have provided much information on how wheat root systems respond to different treatments in the field. The effect of omitting nitrogen was to increase the root/shoot ratio, but to decrease root length density in all soil layers. There was little interaction between nitrogen and water in terms of effect upon root distribution.

Boot & Mensink (1991) measured the effect of different nitrogen supply levels on the root length/leaf area ratios of five contrasting grasses (table 9.8), which responded in very different ways. It might therefore be expected that changes in nitrogen supply altered the relative drought resistance in these species.

Barraclough & Weir (1988) also investigated the effect of breaking up a compact layer (bulk density 1.8 g cm^{-3}) in a sandy loam on root development of wheat. Breaking up the pan greatly increased subsoil growth of roots, but the total root length was not changed because the roots simply proliferated above the pan if it was not broken. Of the eight treatments tested in this work (Barraclough *et al.* 1991), only three had important effects on distribution of roots in the profile (table 9.9). The deeper distribution with droughting is to be expected, but it is surprising that after deep digging the fraction of roots between 20 and 60 cm was considerably larger than in soils that had no compact layer in the first place. This suggests that roots are sensitive to even small levels of compaction (section 9.3.5), and that incipient compaction should receive more attention relative to fully developed pans.

Defoliation of shoots always reduces root growth. The effect on root uptake of nutrients is more variable, because there is a demand from regrowth of shoots, but also a decrease in root biomass, as the plant returns to the normal root/shoot ratio (Thornton & Millard 1996). The net result is usually increased inflow for N.

9.4.3 Root Interactions

Planting density and the possible interaction between neighbouring plants have received rather less attention. The possibility of root exclusion by neighbours was

Table 9.8 Differing leaf area/root length ratios of five grasses grown together under identical conditions, with low and high nitrogen supply.

	Leaf area/root length (cm^2 m^{-1})		
Grass species	Low N	High N	R(%)
H. lanatus	0.81	1.02	21
F. rubra	0.81	0.97	16
M. caerulea	1.53	1.97	22
F. ovina	0.57	0.78	27
D. flexuosa	0.92	1.40	34

Source: after Boot & Mensink (1991).

Results are means of 5 harvests and 6 replicates each. R is the relative difference between low and high N, as a percentage of the values in the high N treatment.

Table 9.9 Percentage distribution of wheat roots at anthesis in an experiment testing various parameters.[a]

Depth (cm)	SE	SL	+N+I	−N+I	+C	Mean	+N−I	−N − I	−C
0–20	65	66	70	67	67	67	51	55	48
20–40	18	14	12	11	15	14	15	17	31
40–60	8	8	8	9	10	9	14	10	14
60–80	6	7	6	7	6	6	10	9	5
80–100	3	5	4	6	3	4	10	8	2

Source: after Barraclough *et al.* (1991).

[a]SE, sown early; SL, sown late; +N, with nitrogen fertilizer; −N, without nitrogen fertilizer; +I, with irrigation; −I, without irrigation; +C, with subsoil compaction resulting from a subsoil pan; −C, without subsoil compaction. The three grouped treatments (right-hand side) all distributed roots further down the profile.

discussed by Nye & Tinker (1977, p. 276). The vertical orientation of roots after an initial downward sloping path, as in figure 9.4C, may indicate this effect. Mostly, it is assumed that a uniform root density in the horizontal direction is established rapidly. However, Pearson & Jacobs (1985) (figure 9.20) showed in a careful quantitative study that root density under subterranean clover varied markedly with distance between plants, even up to 153 days after emergence. Rooting depth was not affected by population density, though mean root length density increased sharply with population.

9.4.4 Root Systems under Forest and Natural Vegetation

Some natural vegetation is so uniform and homogeneous that it almost has the characteristics of a crop. Areas of Spartina grass, Norway spruce or mangrove forest may be close to monocultures, though there will be a considerable variation in the age of individuals, and the soil variation may also be much greater than in

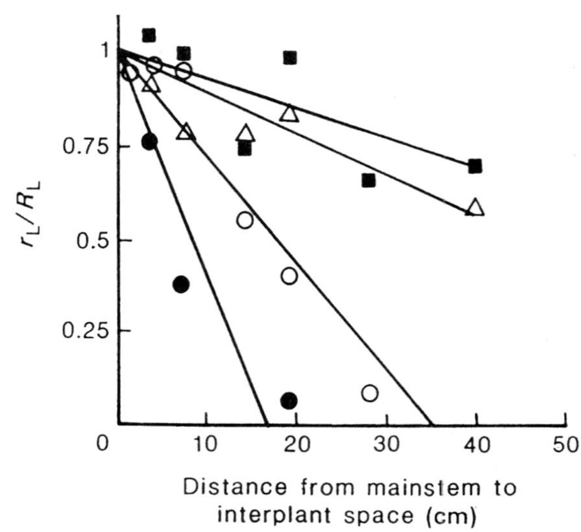

Figure 9.20 Relationship of root length, expressed as the ratio of root length density at a point to the root length density below the plant main stem, to distance from the stem, with a different slope for each date. ● = 64; ○ = 90; △ = 119; and ■ = 153 days after emergence (after Pearson & Jacobs 1985).

cultivated crops. On the other hand, some planted vegetation, such as intercropped peasant agriculture, will have large and random spatial differences, and the different species may be planted and harvested at different times. There is a very large range in heterogeneity in plants and soil across both planted and natural vegetation, but on average it will be largest in natural vegetation.

This heterogeneity is a major problem in applying the methods and concepts that have been discussed in this book. First, it is difficult to measure total root length, as it will vary both spatially and temporally in more complex ways than that under a uniform monoculture. Second, the total root length is of little interest under mixed vegetation, because questions in mixed vegetation concern competition between individuals and species, the use of resources, and the criteria for stable mixtures of vegetation. To answer such questions, the root length and variation of root density for each species must be known separately, all per unit surface area of land. These are very difficult technical problems, which have only been solved in a very few cases. Ong *et al.* (1991) used the tree species *Senna siamea* and *Senna spectabilis* together with cowpeas and maize in intercropping experiments. The roots of the tree species are black, and can therefore easily be separated from the roots of the crop species. There is not yet sufficient information to know how roots distribute themselves in such circumstances.

The problems inherent in measuring the root systems of forest trees are discussed by Vogt & Persson (1989) (section 9.2.5). Hendrick & Pregitzer (1996) found that with northern hardwood species, over 11% of all root length was below 75 cm, even though almost half of all new root was produced in the top 20 cm of the profile. Nepsted *et al.* (1994) showed that over large areas in the Amazon forest, groundwater was being used from up to 8 m below ground level.

9.5 The Modelling of Root System Growth and Morphology

9.5.1 Modelling of Root System Development

The modelling of nutrient uptake demands that the root system shall be specified from observations, or modelled within whole-crop nutrient models (section 11.3.1). A description of the general form of root systems in the field was given earlier, with a discussion of the various environmental and other factors that affect them. However, the mechanisms by which these factors alter root systems are poorly understood, and at present it is not possible to model a root system from a knowledge of the environmental conditions and the physiology of the plant. There are now very many vegetation models that deal with single plants, whole crops, and mixed vegetation. It is very obvious that vegetation models are always better developed and verified in relation to shoot than to root function. This is in part due to the greater simplicity of the above-ground environment; for example, carbon dioxide and other atmospheric gases are at almost constant concentrations, and CO_2, O_2, and water are the only major mass fluxes over leaf surfaces. The biological activity on the leaf surface is normally less than that on the root.

With so many models available, it is impossible to mention them all. Here, we select typical examples of particular conceptual approaches to root modelling (Klepper & Rickman 1990). These root models may be included in models of nutrient uptake or growth of single plants or whole crops as described in chapters 10 and 11.

Models of root systems can either deal with their morphological structure, the distribution of root density in soil, or a combination of these (section 9.2.2). Early root models were alternative methods of stating empirically gathered information about root morphology (Hackett & Rose 1972a, b; Lungley 1973). It is debatable how many of the current root models are simply more rigorous examples of this type, but none are wholly process-based. At present, the variables in some of these models are gathered empirically, but they may become predictable at a later stage. Wullschlager *et al.* (1994) have reviewed the whole topic, with emphasis on the effects of elevated carbon dioxide in the atmosphere.

Gerwitz & Page (1974) first proposed a simple empirical distribution of root quantity with depth that has surprisingly often been found to be approximately true for crops. The equation is

$$P = 100(1 - \exp(-kz)) \tag{9.6}$$

where P is the percentage of root mass or length between the surface and any depth z, and k is a constant. From this, $1/k$ equals the depth of soil containing 63% of the total root. Gerwitz & Page (1974) considered that this equation applied reasonably well to 70% of the sets of data they studied. This is, of course, a root length density model, and it does not address the root system structure at all. It does require that the plants are growing in a uniform soil without sharply defined horizontal layers or discontinuities because variation of texture, water or nutrients at depth will probably render it invalid. For a full discussion of this and other empirical equations, see Gregory (1994a).

Thornley & Johnson (1990, p. 472) have shown that the exponential distribution of Gerwitz and Page can be derived by assuming that roots develop according to the mathematics of a diffusion-like process, with a constant consumption of root mass for maintenance. This is interesting, but roots do not grow by diffusion, and the fact that we can approximate to the empirically determined distribution of root systems by such a means does not explain the real-world process.

Allan Jones *et al.* (1991a) have described a general dynamic method of simulating root growth, where root length density rather than the individual root is modelled; this is mainly intended for applied crop models. In this, the depth of the rooting front descends, at a rate that is controlled by thermal time (day-degrees), and that is related to a defined maximum rooting depth parameter for those conditions. Root length density increases behind the rooting front, also being dependent on thermal time and a maximum value of root length density for those conditions. A range of factors are defined as potentially limiting on the depth and the rate of extension, including acidity, calcium deficiency and coarse stones content in the soil. Soil strength and aeration are also detrimental, and are defined by equations combining bulk density, water content and texture. All stress factors are evaluated on a 0–1 scale, and the factor with the lowest value is regarded as limiting. The method is of clear practical value, but carries rather little mechanistic information. The output is similar in form to the Gerwitz and Page rule, but it gives a dynamic picture of the development of the system, and allows environmental conditions to be included.

9.5.2 'Growing Root' Models

The dynamic modelling of a developing three-dimensional root system will always be exceptionally complex. The uncertainty of the biological controls, the spatial heterogeneity of the environment, and the constant interaction with the shoot suggest that detailed mechanistic modelling of root systems will always be a scientific research task rather than a routine part of crop growth management (Lynch & Nielsen 1996).

Fitter & Stickland (1991), Fitter (1996) and others have used the 'root architecture' concepts (section 9.2.3) to develop simulation models based on a three-dimensional grid to 'grow' a root system in terms of defined dimensions of links, branching nodes and angles. This system can, in principle, be made very accurate, but so far the rules that govern the simulation are all empirical. The use of this type of model for investigating the efficiency of root systems is discussed in section 10.4.4.

A single-plant root model developed by Clausnitzer & Hopmans (1994) uses a three-dimensional grid, and feeds into it the coordinates of each root, artificially generated or from actual measurements. Soil strength and soil temperature can be used in the model to modify root growth rates and other parameters. The root model is linked to a water uptake model based on the Richards equation. The model of Porter *et al.* (1986) is more mechanistic because it 'grows' each root axis from a node, and then causes it to extend and branch according to a given set of rules. The speed of the process is determined by soil thermal time, namely the accumulated integral of soil temperature on time, in day-degrees. The growth from different age classes of main roots builds up root length density in 10-cm layers of the soil.

The 'ROOTMAP' model (Diggle 1988) also 'grows' a full root system, within a set of three-dimensional coordinates. The system is defined by growth period, number of axes, initiation time of axes, growth rate and branching characteristic of roots, and parameters defining the direction of root growth. The most important variables are the root elongation rate, the branching interval and branching lag time. It has similarities with the model of Porter *et al.* (1986), and also uses thermal time to drive the growth process. A random element is built into the orientation for each root, so no two root systems predicted in this way are ever identical (figure 9.17).

Another three-dimensional detailed model has been developed by Pages *et al.* (1989) for maize. The roots emerge from successive nodes, with seminal roots being from node zero. The number of primary roots is defined as 0.039 × day-degrees. Asseng *et al.* (1997) have recently produced a root model that has been incorporated into a full crop model. The root growth can therefore be driven by carbon supply; that is, it is also an allocation model (section 9.5.3). A number of soil and environmental variables can be taken into account; so far the best correlation between predicted and measured root lengths of wheat was $r^2 = 0.60$.

Most of these models are based directly on empirical measurements from the field, and need recalibrating for any new situation. It is to be expected that a more mechanistic approach will gradually be introduced. The application of some of these models is discussed in chapters 10 and 11.

9.5.3 Allocation Models

A rather more mechanistic approach is based on the availability of carbohydrate for root construction and maintenance. This approach predicts root/shoot ratio and the quantity of roots, but it does not give any information about distribution of the roots in soil.

The primary question is how much carbon is fixed by (net) photosynthesis, and what fraction of this is allocated to the roots. One simple approach is based on the idea that when a plant is in balanced exponential growth, any perturbation of the R/S ratio (e.g. defoliation, root pruning) causes the plant to adjust its growth pattern so as to restore the original ratio under those conditions. This was given simple mathematical form by Davidson (1969) for nitrogen, though similar ideas may apply to other nutrients also:

$$S_N W_R = P S_C W_S \tag{9.7}$$

where S_N and S_C are the specific nitrogen and carbon assimilation rates into root and shoot, respectively, W_R and W_S are the root and shoot weights, and P is an empirical parameter that is assumed constant over a range of conditions. In fact, this simply states that the ratio P between total net assimilation of N and of C is constant, which will be true over fairly short periods when plants are in steady growth. Over longer periods, the C/N ratio and the R/S ratio of the plant will alter so P is unlikely to remain constant.

This approach assumes that the root/shoot growth relationship depends upon the recent uptake of shoot (carbon) and all the root-absorbed nutrients. However, a detailed review (Ericsson 1995) concluded that the situation was rather more complex than this. The R/S ratio increased if N, P or S deficiencies constrained growth, but the reverse happened for K, Mg and Mn (Ericsson 1995) (section 9.3.3.1). Low irradiance did cause lower R/S ratios, but the effect was rather small.

The Davidson equation may therefore best be used with nitrogen. Indeed, almost all crop modelling that deals with nutrient uptake confines itself to this one element. There are good practical reasons for this in that nitrogen is the dominant nutrient in developing green leaf, and the practical problems of manuring the major agricultural crops tend to focus on nitrogen. However, this avoids some of the problems that would be encountered in dealing with other elements. In particular, nitrate is a very mobile element, and the local movement near the root is therefore usually ignored. Further, almost none of the complex additional processes that affect uptake of phosphate or trace elements are included (chapters 7 and 8). This strong focus on nitrogen may foster the impression that all plant nutrition problems are much simpler than is actually the case.

There are several ways of approaching carbon allocation between root and shoot (Cannell & Dewar 1994). Starting from the idea of the balance between

C and N in the plant, Thornley and associates (Thornley 1972, 1995; Johnson & Thornley 1987; Thornley & Johnson 1990) have produced two basic approaches for modelling. The first (Thornley 1972) depends upon transport of materials within the plant, and the internal resistances to these flows. The second is based on the idea that the plant is programmed to maximize its relative growth rate. Both models can describe a shifting balance between source-limited and sink-limited growth.

The first approach has been used by Dewar (1993) to develop an allocation model with explicit phloem and xylem transport processes, and including transpiration and plant water potential, for a plant in exponential growth. The interactions between the various uptake processes can thereby be explored more fully. The nitrogen uptake part of the model is very simple, and the effect of changes in nitrogen supply from the soil are not included. However, the model provides predictions of the effect of changes in the specific nitrogen uptake rate on the R/S ratio. This shows that models can be constructed, with a reasonably small number of parameters, that will behave in a way that appears intuitively reasonable.

The second idea has been elaborated in the 'teleonomic' or goal-seeking models (Thornley & Johnson 1990) (figure 9.21). In this, the plant is seen as two structural compartments, root and shoot, and two substrate compartments, for C and N, that are linked to both of them. The plant divides its photosynthate so that the single relative growth rate u for a plant in balanced exponential growth shall be maximized. The relationship between u and the supply of both photosynthate, C, and nitrogen, N, is taken to be

$$u = 1/2(k\ \mathrm{C}\ \mathrm{N}) \tag{9.8}$$

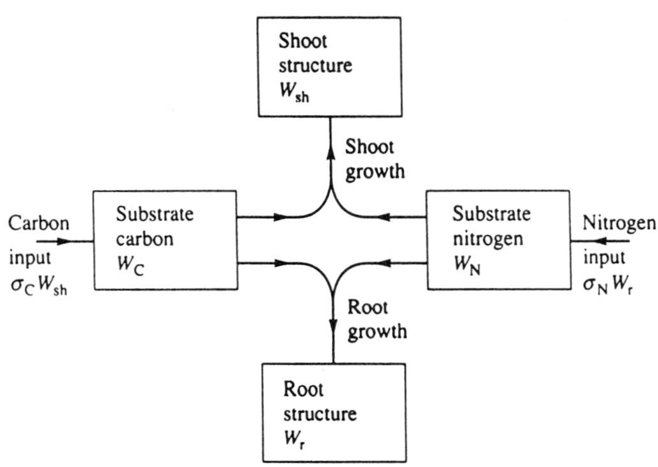

Figure 9.21 Basic structure of a teleonomic growth model, depending upon the interplay between shoot, root, and the two substrate pools to predict root/shoot ratio and other plant parameters (see figure 9.2). (After Thornley & Johnson 1990, p. 373.)

where k is a constant. This clearly must be a considerable simplification. The next step is to define the mass balances for C and for N. With this relatively small number of basic assumptions and concepts, a self-consistent model of root/shoot ratio and the internal nitrogen concentration of a plant can be made. Thornley & Johnson (1990) show the predicted response to an instant 95% defoliation of the plant, which appears intuitively reasonable. Makela (1986) has extended this concept to deal with plants with different turnover (i.e. loss) rates of material from root and shoot.

These two basic ideas have been compared by Makela & Sievanen (1987). They concluded that the earlier 'resistance' model (Thornley 1972) was the more general, and fitted normal plant behaviour better. Thornley (1995) agreed that the resistance model has a number of advantages.

Some of the basic ideas described above have been built into a plant model (Thornley & Johnson 1990) to describe the mass and distribution of a root system in much more detail than the earlier work, though the shoot modelling is kept simple. The root compartment is subdivided into a number of physical layers. In these, there is transport between layers (within the root only), a simple nitrogen uptake process and nitrogen recycling from senescent root. The forest model of Thornley & Cannell (1992) uses similar principles.

The conceptual gap between the 'specific uptake rate' of nitrogen in models of this type and the detailed root uptake models explained in chapters 5 and 6 is wide. As pointed out above, it is probably only for nitrate uptake that models of this type can function, because many soil processes are simply ignored. The most important parts of these models relate to the physiology of the plant, and particularly the nutrient concentration N/C, and hence go some way towards relating nutrient concentration to plant function (see chapter 10).

The assumption that the whole-plant relative growth rate is maximized is not necessarily true. It will probably apply to agricultural crops, because of the premium set on high yield during their selection and cultivation. However, in natural vegetation, growth strategies are not necessarily targeted towards this overall aim, as it could expose plants to fatal situations as resources change. In general, 'stress tolerators' (Grime *et al.* 1988, p. 2) grow slowly to avoid heavy demands upon limited or uncertain resources. A quite different set of assumptions will be needed to model these, but little has been done with such plants.

10

The Mineral Nutrition of Single Plants in Soil

Earlier chapters in this book have dealt with the various components of the soil–root system. In this chapter we aim to synthesize them into a unified treatment of a single whole plant growing in soil. Solute movement and root system uptake models are still the central subject, but we must also deal with the growth of the whole plant, which provides the growing sink for the absorbed solutes, and the expanding root system through which they enter. Here we deal only with homogeneous soils and constant growing conditions, usually in pot culture, and call this 'simplified conditions'. This is necessary in dealing with such complicated systems, so that essential principles shall not be obscured. In chapter 11 we apply these ideas, so far as it is possible, to crops and natural vegetation.

10.1 Types of Models

Models are often referred to in this book, because the ideas and concepts are most easily and precisely formulated in this way (Nye 1992a). Here, we outline the different types of models that will be dealt with, and their relationships with each other. Readers may consult Rengel (1993) and Silberbush (1996) for recent reviews of the modelling of nutrient uptake, and Penning de Vries & Rabbinge (1995) for general crop modelling concepts.

There are three basic situations:

(1) Models of single or few plants growing in pots under simplified conditions in greenhouse or growth chambers, in homogeneous soils, with ample supplies of water, constant temperature, etc.
(2) Models of monoculture crops. If a unit cell can be defined, only the vertical dimension need be considered, except possibly for light interception, and for radial transport around roots. These models are normally used for field situations.
(3) Vegetation models with mixed species. Separation of the uptakes by the different species can be extremely difficult. If the geometrical arrangement of the species is regular, it is possible to determine a recurring unit cell, which simplifies treatment.

Within each situation there is a hierarchy of complexity in the number of processes covered. All models may include water uptake as well as nutrient uptake. The scale of increasing complexity is:

(1) For single-root *uptake* models. Calculation of inflow and depletion for one root of constant length, from infinite soil, and with constant root uptake characteristics. (Dealt with in chapter 6, with special processes in chapters 7 and 8.)
(2) For root *development* models. Description or prediction of root system length and distribution in a pot or a soil profile in the field. (Dealt with in chapter 9.)
(3) For root system *uptake* models for simplified conditions, as defined above. Calculation of inflow and depletion with growing root length, and fixed or variable root uptake characteristics and soil properties, without root competition or special root surface processes.
(4) As for (3), but with inter-root competition.
(5) As for (4), but with root surface processes.
(6) For root system *uptake* models for the field. Calculation of inflow to roots, depletion around them and uptake properties, from nutrient uptake by whole plant, root length, and soil properties; with or without root competition.
(7) For single whole-plant *growth and uptake* models for simplified conditions. Calculation of growth, nutrient uptake and depletion from atmospheric and soil properties, and physiological relationships (section 10.2.1).
(8) For crop *growth and uptake* models for the field. Calculation of growth, nutrient uptake and depletion from atmospheric and soil properties, and physiological relations (section 10.2.1).
(9) For mixed vegetation uptake models. Calculation of uptake and depletion separately for two or more interpenetrating and growing root systems of different species.

The further classification of the models described here reflects the way in which the continuity equation for nutrient flow around the root is solved (section 6.1.2), for the various situations described above.

(1) Transient state models without competition, with analytical solutions.
(2) Transient state models without competition, with numerical solutions.
(3) Transient state models with competition, with analytical solutions. This is very rarely used.
(4) Transient state models with competition, with numerical solutions. This is a standard method.
(5) Steady-state models, with competition. This is a standard method.

Van Noordwijk & Van de Geijn (1996) have classified root models rather differently from our system, but they agree on the underlying distinction between root uptake models and root–shoot growth models. They describe one approach as 'models without roots', and there are, indeed, a number of models that make a link directly between the plant demand and the available nutrient in the soil, so omitting the activity of the root system, as will be seen in the discussion of crop models in chapter 11. These may be of practical use, but they clearly ignore an important mechanistic step in plant nutrition. All the models discussed here are mechanistic.

Silberbush (1996) classifies models as 'empirical' (not discussed further), and single-root or one-dimensional vertical (field crop) 'mechanistic' models. This classification seems to omit the numerous mechanistic models that have been applied to single plants, usually growing in pots.

Models without competition have rarely used transient state single-root models with analytical solutions (Brewster & Tinker 1970), but a number have been

Table 10.1 Selected list of plant growth models, mainly for simplified conditions, and the main aspects or developments in them.

Year	Authors	Plant growth model features
1969	Nye & Marriott	Mass flow and diffusion, transient-state, single root, M–M, infinite medium, numerical solution
1970	Barley	Mass flow and diffusion, transient-state, root competition, α constant, irregular parallel roots, numerical solution
1971	Sanders *et al.*	Electrical analogue
1973	Baldwin *et al.*	Mass flow and diffusion, steady-state, root competition
1975a	Nye *et al.*	Whole-plant model, diffusive supply according to Baldwin *et al.*
1976	Claassen & Barber	As Nye & Marriott + uptake by complete root system
1976	Bhat *et al.*	As Nye & Marriott + sink term for root hairs calculated according to Baldwin *et al.*
1978	Nielsen & Barber	As Claassen & Barber + C_{min}
1979	Van Noordwijk & De Willigen	Diffusion, constant P-uptake rate
1981	Barber & Cushman	As Claassen & Barber + root competition
1983b	Itoh & Barber	As Barber & Cushman + root hairs according to Bhat *et al.*
1986	Claassen *et al.*	As Nye & Marriott + root competition
1987	De Willigen & Van Noordwijk	Diffusion, incomplete root–soil contact
1989	Bouldin	Multi-ion uptake
1990	Van Noordwijk *et al.*	P fertilizer prediction
1991	Bar-Tal *et al.*	Zn transformations and uptake
1993b	Smethurst & Comerford	Competition between species
1993	Silberbush *et al.*	Multi-ions, salinity
1994	Yanai	Root competition
1994a, b	De Willigen & Van Noordwijk	Mass flow and steady-state

Source: after Amijee *et al.* (1991) and Silberbush (1996).

proposed that use numerical solutions (table 10.1). Root competition also is rarely modelled with transient state analytical solutions (Youngs & Gardner 1963; Van Noordwijk & De Willigen 1979), since these use the Laplace transform, resulting in solutions expressed in Bessel functions. The electrical analogue (section 10.4.3) is effectively a method of solving the diffusion equation for roots with competition. The numerical solutions of the transient state (Nye & Marriott 1969) can be applied readily to roots with competition by defining a zero-transfer boundary (section 10.5.1), and this has been widely used (table 10.1). A number of variants can be constructed, including mass flow, root hairs and pH change, because of the flexibility of the numerical techniques.

With root competition and a limited soil volume, it becomes possible to use the steady-state approximation (section 10.5.2). This method, together with the numerical solution of the transient state with competition, are the main methods used now including application to different soil horizons (Ibrikci *et al.* 1994).

Most of the single-root and plant uptake models published over the 1970s to the 1990s are given in table 10.1. There is only space to describe a small number of them in this chapter, following section 10.2 on whole-plant nutrient physiology. Multiple-plant models, dealt with in chapter 11, are usually one-dimensional vertical models (Silberbush 1996).

10.2 Relationships between Nutrient Uptake, Plant Composition and Growth, and Soil Supply

10.2.1 Pattern of Basic Relationships

The basic whole-plant physiological relationships that we need to deal with plant nutrition in the whole root–soil system are as follows (all symbols are defined in the Table of Symbols):

(a) The relation between root and shoot growth, or how photosynthate is allocated:

$$L \text{ and } W = f(R_W, X) \tag{10.1}$$

(b) The relation between the external concentration at the root surface and the inflow to the root, namely the uptake characteristic of the root:

$$I = f(C_{La}, C_{min}, I_{max}, K_m) \qquad \text{or} \qquad I = f(C_{La}, \alpha a) \tag{10.2}$$

(c) The relation between the nutrient composition of the plant and the mean uptake characteristics of the roots (see chapter 5):

$$\alpha = f(I_{max}, K_m) = f(X) \tag{10.3}$$

(d) The relation between nutrient concentration and the relative growth rate:

$$R_W = f(X) \tag{10.4}$$

(e) Finally, all these have to be combined to satisfy the requirements of equation (10.5) (section 10.2.6),

$$XR_W = \bar{I}L/W \tag{10.5}$$

where $\bar{I}$ is a mean value over a root or a root system, and possibly over a defined time period also.

Equation (10.5) expresses the conservation of the nutrient elements during the growth of a plant, and thereby defines the relationship between plant growth rate, nutrient composition, inflow and root/shoot ratio (section 10.2.6).

The five relationships discussed above are all physiological in nature, so that the variables could be measured in a solution culture experiment. However, in soil the concentration of a nutrient in the soil solution at the root surface cannot normally be measured, and has to be calculated from a knowledge of the soil's transport characteristics, the bulk soil concentration and the plant demand.

(f) The final relation is therefore that between the concentration of nutrient in the bulk soil and that at the root surface, and the consequences for the inflow in the context of relationship (b) above:

$$C_{La} = f(C_{Li}, D, I) \tag{10.6}$$

This relationship was discussed for a single root in chapter 6. In this chapter, we extend this to a root system, using the various types of model defined above. Ideally, the special processes at the root surface should also be included (chapters 7 and 8), but this is still difficult.

The importance of the inflow, I, is clear from this explanation (Robinson 1986). It defines the essential nutrient uptake step quantitatively, and it is linked to the uptake mechanisms of the root by relationship (b), to the general physiological state of the plant by relationship (e), and to the transport processes in the soil by relationships (f). It thus ties together all the essential parts of the system. Relationships (a) to (f) will now be discussed in turn.

10.2.2 Balance of Acquisition Above and Below Ground

Chapter 9 dealt with the way in which the plant grows by allocating assimilated carbon and nutrients, and thereby defines the root/shoot ratio. The general idea that a plant grows in a balanced way, and that its absorbing organs above and below ground must therefore function in balance (section 9.5.3), is most simply stated by Makela & Sievanen (1987):

Leaf area × leaf uptake activity = constant × root area × root uptake activity

This is not very helpful, because it does not define 'activity', and ignores the possibility that the two activities do not need to follow each other exactly, because of changes of composition within the plant. Equation (10.7) (Davidson 1969) is more specific, but is best stated in terms of the length of root rather than its weight (section 6.1.2). We then get

$$A_l \times sC = \text{constant} \times L\bar{I} \tag{10.7}$$

where A_l is leaf area, sC is net assimilation rate, L is root length and $\bar{I}$ is mean inflow. This defines conditions over a short period, for a plant in steady growth. However, there is as yet no simple way of predicting the root length/leaf area ratio of a plant in particular environmental circumstances (section 9.5.3). Table 9.8 shows how sharply this ratio can alter with nitrogen supply.

10.2.3 The Relation between the Concentration of Nutrient at the Root Surface and the Inflow to the Root

This has been discussed in chapters 5 and 6. The uptake characteristics can be defined as a Michaelis–Menten style relationship:

$$I = I_{max} C'_{La} / (C'_{La} + K_m) \tag{10.8}$$

where $C'_{La} = C_{La} - C_{Lmin}$. This relationship is wholly empirical, and will not allow sufficiently for the slow increase in I as C_{La} is increased progressively towards high values (the low-affinity pathway (section 5.2.3)). In some circumstances, a straight line relationship is a sufficiently good approximation (section 5.3.2) (see figure 5.8), so that $I = 2\pi\alpha a C_{La}$, up to the critical level X_{crit}, after which a constant inflow condition holds.

De Willigen & Van Noordwijk (1994a, b) have suggested that, in practice, it is sufficient to treat the root surface condition as either a constant flux (above X_{crit}) or a zero sink, but we do not wish to ignore the intermediate situation in which changes in C'_{La} alter I significantly.

10.2.4 The Relation between Plant Nutrient Composition and the Uptake Characteristics of the Roots

In section 5.3.3, we discussed the process of regulation of uptake characteristics. For a given external concentration of a nutrient, the inflow into a plant is greater when the concentration within the plant is low than when it is high. The mechanisms are not understood, and are specific for each ion. Nevertheless, we have to assume some quantitative relationship to express this process, which is particularly important when the external supply concentration changes abruptly, so that the internal concentration and the external nutrient concentrations are not in their normal balance.

To determine the effect of nutrient composition of the plant on uptake characteristics, it is essential to bring the experimental plants to a uniform and stable nutrient percentage first. Jungk *et al.* (1990) did this by growing soybean and maize in flowing solution culture at different phosphate levels for periods of 3 or 4 weeks. Inflows of phosphorus were measured by sequential harvests, and groups of plants were placed in 'depletion experiments', in which the progressive uptake of phosphate was measured. From this the I values were found, and using a Hanes plot (see figure 5.11), the Michaelis–Menten parameters were derived (table 10.2).

I_{max} was more strongly affected by the phosphate pre-treatment than K_m, with a change of about 4–5 fold against 1–2 fold. The effect on I_{max} was close to linear (figure 10.1), the equations being $I_{max} = (42.8 - 30.4\ (\%P)) \times 10^{-14}$ for maize and $I_{max} \times 10^{-14} = 23.3 - 25.8\ (\%P)$ for soybean. In the first instance, it therefore

Table 10.2 Influence of flowing solution culture concentrations of P on the shoot and root P concentration at equilibrium, and the Michaelis–Menten uptake characteristics, as determined in short-term depletion experiments for P, of soybeans and maize.

Solution P concentration (μmol L^{-1})	P (%) Shoot	P (%) Root	I_{max} (mol cm^{-1} s^{-1} 10^{-14})	I_{equil} (mol cm^{-1} s^{-1} 10^{-14})	K_m (μmol L^{-1})	C_{min}[a] (μmol L^{-1})	C_{min}[b] (μmol L^{-1})
Soybean							
0.03	0.22	0.23	17.6	1	1.6	0.06	0.01
0.3	0.34	0.30	16.9	3	1.7	0.10	0.03
3	0.59	0.56	6.5	8	1.2	0.16	0.08
30	0.66	0.90	3.7	8	1.0	0.16	0.06
Maize							
0.1	0.22	0.20	37.0	3	6.1	0.17	0.01
1	0.68	0.45	21.4	9	3.9	0.24	0.02
10	1.08	1.00	6.6	13	1.9	0.28	0.04
100	1.16	1.35	7.1	14	3.4	0.24	0.02

Source: after Jungk *et al.* (1990).

[a]Solutions unfiltered.

[b]Solutions filtered through 0.45 μm membrane filter. I_{equil} is I during the flowing solution treatment.

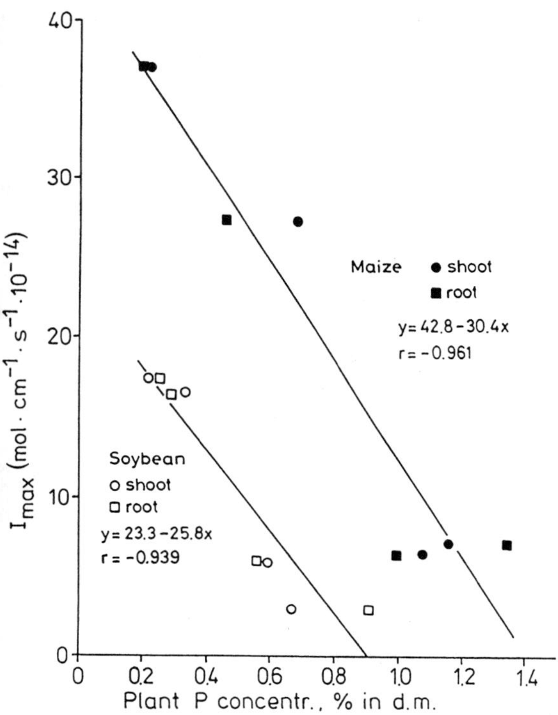

Figure 10.1 Relationship between internal P concentration in root and shoot of soybean and maize and the I_{max} uptake characteristic (after Jungk *et al.* 1990).

seems reasonable to regard regulation as occurring through I_{max}, and ignore changes in K_m.

Based on this, a suitable empirical relationship is

$$I_{max} = I^*_{max} - G(X - X_{min}) \tag{10.9}$$

Here, I^*_{max} is the upper limiting value of I at internal concentration X_{min}, the lowest concentration for growth, and G is a constant specific for each nutrient subject to regulation. In this work (Jungk *et al.* 1990), a growth response was obtained only at the lowest concentration in the flowing culture. The data thus span the X_{crit} concentration level and go well into the luxury concentration region; hence, this relationship seems to hold over a large part of the plant concentration range.

A similar type of result for potassium is given by Claassen & Barber (1977) (figure 10.2), according to which I_{max} changes by a factor of about 3 whilst the percentage potassium changes by a factor of 4 in a near-linear relationship.

Greenwood & Draycott (1989) also found that the uptake rate of nitrogen by a range of arable crops in the field was depressed by the N concentration in the crops by a factor of $1.42 - 0.525(X/X_{crit})$. Potassium concentration was found to depress the I_{max} for K according to the formula

$$I_{max}/I^*_{max} = \exp[-2.8(X/X_{crit})] \tag{10.10}$$

where I^*_{max} is the I_{max} when $X/X_{crit} < 0.3$ (Greenwood & Karpinets 1997). This equation is very similar to those of Siddiqui & Glass (1982) (section 5.3.3).

All these methods of describing regulation of I_{max} are similar in defining the maximum value of I_{max} (I^*_{max}) at very low concentrations of the nutrient, and then modifying this by the value of the internal concentration. All these methods are strictly empirical, and care is necessary with slightly different definitions.

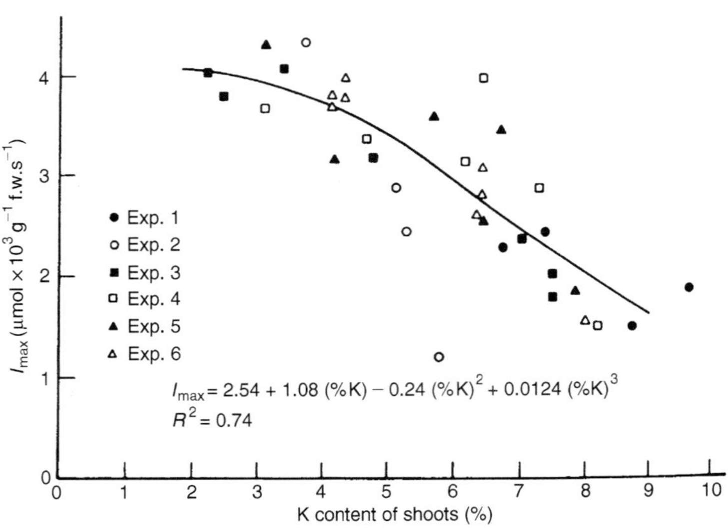

Figure 10.2 Relation between the K concentration in the shoots of maize and the I_{max} for K uptake (after Claassen & Barber 1977). Note that I_{max} is defined in terms of root weight here.

10.2.5 The Relation between Plant Nutrient Composition and Growth

10.2.5.1 The Nutrient Concentration in Plants

The 'mean nutrient concentration in the plant' is mentioned in many places in this book. It is both a consequence of the processes of growth and nutrient uptake, and a controlling variable for growth and further nutrient uptake. It is important to understand that changes in this average value include at least four separate processes as follows.

(1) There are some basic differences in the nutrient concentration in the cytoplasm of different species. Thus, the concentration of nitrogen in the leaves was shown by Field & Mooney (1986) to be related to whether the plant is a fast- or slow-growing species. The main compound of nitrogen in leaves is the enzyme 'Rubisco' that catalyses the carbon-fixation step, and the amount of this controls the potential rate of photosynthesis. The nutrient concentration of the cytoplasm in a given species is highly conserved and buffered against changes due to external supply, but it differs between species.

(2) Changes in the nutrient concentration of the cell vacuoles occur much more readily (Leigh & Wyn Jones 1986), and they may affect the physical equilibria in the plant, such as turgor (section 2.1.2). The mean nutrient concentration of the whole cell thus depends upon the fraction of its mass composed of cytoplasm, vacuole and cell wall, and the nutrient concentration in each.

(3) The nutrient concentration of a plant organ, to either wet or dry tissue, depends upon the type and relative number of cells, and their nutrient concentrations. Structural tissues, such as dead xylem, contain much lower concentrations of nutrients than living cellular tissue, such as root cortex, and changes in the ratio between different types of tissue will give differences in the mean nutrient concentration of the organ. The change in nutrient concentration with age is largely caused by the increase in the structural materials cellulose and lignin. A similar effect is found if the tissue deposits a specific material that increases the dry weight and so dilutes the nutrient concentration, as in the seasonal deposition of starch in the cells of the needles of spruce (Linder 1995) or of calcium oxalate in the leaves of sugar beet (Last & Tinker 1968).

(4) Finally, whole-plant concentration will depend upon the proportion of the various organs. The mean plant concentration is therefore a complicated variable to deal with, and all its applications must be handled with care. It is scarcely surprising that simple plant analysis techniques of fertilizer prediction are frequently unreliable.

10.2.5.2 Nutrient Composition and Growth

The relationship between growth rate and composition, which is needed to define the critical level, X_{crit}, has only been touched upon so far. It is not surprising that different species have very different relationships between nutrient concentration and growth rate (Marschner 1995, p. 463 *et seq.*), so that the value of X_{crit} is very different. This is well seen in the nitrogen concentration in C3 and C4 grasses (table 10.3).

The concept of using the nutrient concentration as a measure of deficiency—that is, of the degree to which growth and yield are increased by adding the nutrient—has a very long history in agriculture and horticulture, where tissue analysis is used as a practical guide to fertilizer use in many cropping systems. However, for reasons given above, it has repeatedly been found that the percentage nutrient value alone is an unreliable predictor of the size of a plant response to fertilizers, and much work has gone into attempts to improve on this practice (Walworth & Sumner 1988; Marschner 1995). The concept of a nutrient requirement for growth based solely on a single nutrient concentration is oversimplified, though it may be a useful first step in practice, as in the 'limiting element' doctrine of Liebig (Russell 1973, p. 49).

The effect of nutrient deficiency on total net photosynthesis is a fundamental process in the plant (Marschner 1995, p. 187). Many authors have found a linear relation between the N concentration in a leaf and the photosynthetic capacity (Vos 1995). However, a deficiency can cause both a decrease in cell and leaf expansion and in net photosynthetic rate per unit area. Where the nutrient is directly involved in the processes in the chloroplast, a deficiency certainly decreases net photosynthetic rate, as some 75% of the leaf nitrogen is in the chloroplasts. Despite this, deficient plants may contain excess levels of carbohydrate, so the deficiency effect is not a simple reduction of photosynthate at source, but also of low sink activity. Deficient leaves may senesce earlier than sufficient ones, thus acting as a further source limitation. These complex processes (Marschner 1995, p. 199) mean that we cannot expect a simple and predictable relationship between nutrient concentration in leaves and growth rate, and still less in the weight of whole plants grown to maturity.

The empirical relationship between plant growth and nutrient concentration within a single experiment is usually of the Michaelis–Menten type. The relationship should be measured with plants in flowing solution systems of precisely known concentration C_{La} (section 5.3.1). The critical level, X_{crit}, defined as the internal concentration that just allows maximum growth rate, can be found from these curves, though it is difficult to determine a precise value. Unfortunately,

Table 10.3 Differing relationship between dry matter production and percentage nitrogen in shoots of C3 and C4 grasses.

Nitrogen supply (equivalent to kg ha^{-1})	Dry matter (g per pot)		Nitrogen content (% dry wt)	
	C3	C4	C3	C4
0	11	22	1.82	0.91
67	20	35	2.63	1.18
134	27	35	2.77	1.61
269	35	48	2.78	2.00

Source: after Marschner (1995, p. 474).

C3 Grasses: *Lolium perenne* and *Phalaris tuberosa*; C4 grasses: *Digitaria macroglossa* and *Paspalum dilatatum*.

there is no fundamental way yet of predicting these relationships, since they will depend on many variables. The Michaelis–Menten type of relationship predicts a continuing rise of yield with increasing nutrient concentration, but in most cases yield finally reaches a true plateau, and is ultimately reduced by high concentrations of nutrients, especially the economic yield. It may therefore be useful to break this type of curve into three sections:

$$\begin{aligned} W &= W_{crit}(X - X_{min})/(X_{crit} - X_{min}) \qquad && X < X_{crit} \\ W &= W_{crit} && X > X_{crit} \\ W &= W_{tox} && X > X_{tox} \end{aligned} \tag{10.11}$$

where W_{crit} is the yield at the critical point on the curve, X_{crit} is the corresponding level of nutrient concentration, X is the actual internal concentration, and X_{min} is the nutrient concentration below which no growth occurs. An expanded statement could be made for X_{tox}, or the nutrient concentration above which toxic effects appear. Both X_{crit} and W_{crit} are functions of the age and size of the plant, and are important values in practical plant nutrition. Greenwood *et al.* (1986) concluded that growth rate of a field crop is proportional to the average nitrogen concentration when this is less than the critical concentration, in agreement with this.

The critical concentration will vary to some degree with environmental conditions, but, in particular, it will vary with the stage of development of the plant (Caloin & Yu 1984). This variation of the critical level with time makes it difficult to use it as part of a plant or crop nitrogen model (Vos 1995). There is no fundamental way of deriving the critical level, but Greenwood *et al.* (1991) and Justes *et al.* (1994) have developed various relationships for field crops to allow it to be calculated on the basis of plant size.

Greenwood *et al.* (1990) proposed that $X_{Ncrit} = 5.7(W_{crit})^{-0.5}$ and $X_{Ncrit} = 4.1(W_{crit})^{-0.5}$ would apply to C3 and C4 crops, respectively, because the latter tend to have lower concentrations of N in their tissues. Other approaches produced an exponential and a hyperbolic relationship (Greenwood *et al.* 1991). Such equations have been tested on various crops (Greenwood *et al.* 1990), but agreement is not always adequate (Justes *et al.* 1994; Vos 1995).

Justes *et al.* (1994) studied winter wheat crops grown in France. They stressed the large range of possible nitrogen concentrations in the 'non-limiting' or luxury uptake zone above X_{Ncrit}, that the critical N percentages determined by different authors vary widely, and that all the relationships over time depend upon nitrogen nutrition being smooth and regular. They proposed the relation

$$X_{Ncrit} = 5.53(W_{crit})\mathrm{e}^{-0.442} \tag{10.12}$$

unless the plant mass was less than 1.55 t ha^{-1}, when plant cover was incomplete.

The concept of 'steady rate growth' formulated by Ingestad and colleagues (for example, Ingestad & Agren 1988, 1992) is an alternative way of approaching this topic. In their experiments, nutrients were supplied at an exponentially increasing rate, and they found that the growth of the plant was also exponential, and the internal concentration of the nutrient remained constant. Under these rather special conditions, they found that the relative growth rate was directly proportional to the nutrient concentration in the plant. This remained true even if the light level changed, so long as the nutrient concentration was expressed as a 'normalized' value, referred to the value of the optimum nitrogen level (X_{Ncrit}).

The question is whether this offers a major new insight into the behaviour of deficient plants. The exponential growth with nutrient given at an exponential rate is to be expected: if the plant grows more slowly than the rate that maintains a constant concentration, its nutrient concentration will rise, and the growth rate will increase, and vice versa. Such steady exponential growth with constant concentration of nutrients cannot normally be continued for long, because of self-shading and changes in the ratio of structural to other tissues, as described above. In this special case where the total quantity of nutrients is limiting, the normal decrease in nutrient concentration with stage of development will allow the efficiency of the nutrient to increase, which will counteract the self-shading effect and may help exponential growth to continue longer than it would do if the plant were grown with a constant external concentration of nutrient. However, X could not remain constant indefinitely.

The linear relation between X and R_W, over a wide range of R_W, is important if it is substantiated in many species and conditions. The relationship must necessarily fail as X_{crit} is approached, and the complexity of both X and its linkage with total net photosynthesis, as described above, makes a simple general relation unexpected, though it is highly convenient for modelmakers. However, if X = constant $\times R_W$, then from equation (10.14), constant $\cdot X^2 = I \cdot (L/W)$, which is difficult to understand, Also, a linear relation for R_W would not agree with the proposed linear relation with W in field crops mentioned above (Greenwood *et al.* 1986), as there is a logarithmic relation between these.

10.2.6 The Nutrient Mass Balance and Plant Growth

A general and simple equation that defines the underlying relationships between growth, composition and uptake was given by Nye & Tinker (1969), based upon the conservation of nutrient mass. This relates relative growth rate R_W, nutrient concentration X, mean uptake per unit root length $\bar{I}$ and the root length/plant mass ratio L/W, as follows:

$$\mathrm{d}U/\mathrm{d}t = \mathrm{d}(WX)/\mathrm{d}t = W\,\mathrm{d}X/\mathrm{d}t + X\,\mathrm{d}W/\mathrm{d}t = \bar{I}L \qquad (10.13)$$

This equation is rigorous, being based upon a mass balance. Some related approaches (Garnier 1991; Ingestad & Agren 1992 and references therein) are usually dealing with special cases of this equation. In particular, this equation holds under conditions where the internal nutrient concentration X is changing.

Equation (10.6) can be simplified if the period of observation is reasonably long, and there are no sudden changes in the availability of nutrient or the relative growth rate. Over an appreciable period, $W\,\mathrm{d}X/\mathrm{d}t$ will normally vary little compared with $X\,\mathrm{d}W/\mathrm{d}t$, because X will remain within a range of no more than 2–3 fold, whereas W may vary by a factor of 1000 over the whole growth period. Accordingly, $\Delta W/W \gg \Delta X/X$, and $X\,\Delta W \gg W\,\Delta X$, for reasonably short Δt. Hence, the simple approximation from equation (10.13) is

$$X/W \times \mathrm{d}W/\mathrm{d}t = XR_W \approx \bar{I} \times L/W \qquad (10.14)$$

Equation (10.7) is probably the simplest statement of the approximate relation of growth and nutrient uptake over a defined time-span. The most important consequence is that the mean inflow is proportional to the *relative*, rather than to the *absolute*, growth rate, if other terms remain constant. A plant in steady exponential growth, with the root/shoot ratio remaining constant, then has a linear relationship between $\bar{I}$ and X. The definition of the conditions for nutrient deficiency is that the maximum possible value of $\bar{I}$ (the zero-sink value) is too low to support an X value above the critical level.

Equation (10.13) can easily be changed into a form comparable to those used in the general equations of Davidson (1969) and Makela & Sievanen (1987), by setting

$$\mathrm{d}W/\mathrm{d}t = A_l \times sC, \qquad \text{so that } \bar{I} \times L = \text{constant } XA_l \times sC \qquad (10.15)$$

where A_l is the leaf area and sC the mean net assimilation rate.

Over a short period, the value of X will normally show little change unless there is, for example, a heavy addition of fertilizer. However, over a longer period, this constancy of X will not hold, even with a constant environment (section 10.2.5). The mean concentration of a nutrient in a plant almost invariably declines with age, because of the increasing fraction of structural tissues (figure 10.3). As the plant ages, it will remain functionally the same if the concentration in its cytoplasm remains constant (section 5.4.3), even though the mean concentration declines. The relative growth rate will, of course, also tend to decline with age, both because of the smaller fraction of photosynthesizing tissue being formed and because of self-shading. Long-term exponential growth with constant composition is therefore unlikely, even with a single isolated plant (section 10.2.5.2).

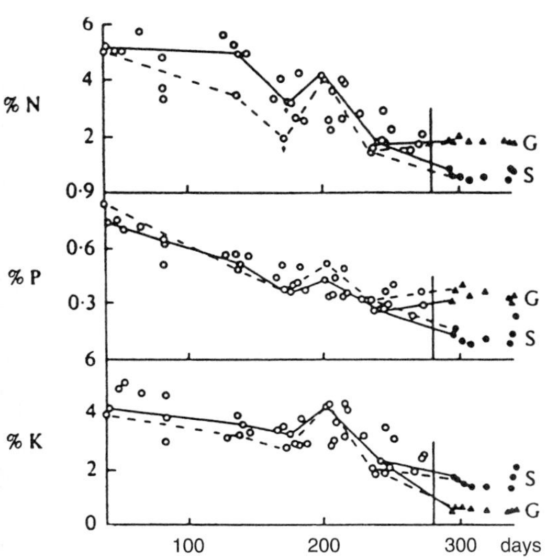

Figure 10.3 Nutrient composition of crops grown in multifactorial winter wheat experiments in 1980 and 1981. Note the increase in concentration of all nutrients after 180 days, when nitrogen dressings were applied, and otherwise a steady decline. Filled symbols represent (▲, G) grain and (●, S) straw at final harvest. Solid and dashed lines represent late sown crops in 1981 at Rothamsted and Woburn, respectively (after Barraclough 1986a, b).

10.2.7 The Quantification of Plant Demand for a Nutrient

The term 'plant demand' is used in a general sense in the plant nutrition literature, and it is important to define what this means in terms of the concepts outlined here.

In a whole-plant sense, we regard the basic *growth demand* as plant need, namely as the uptake rate $\bar{I}L$ that will just maintain growth at its optimum rate, and so maintain $X = X_{crit}$. This definition draws a line between plants that are in a deficiency ($X < X_{crit}$) or sufficiency ($X = X_{crit}$ or $X > X_{crit}$) condition. In the first state, a temporary extra part of the demand arises from the deficiency condition and the need to correct this by a larger $\bar{I}L$ value. We call this *deficiency demand*, but it cannot be defined precisely in terms of uptake rate, because there is no natural time limit within which X has to be returned to the critical level (see section 11.3.3). Similarly, there can be a temporary negative deficiency demand, if the plant is already in an 'excess' state following luxury uptake, with $X > X_{crit}$.

This is basically similar to the ideas outlined by Godwin & Allan Jones (1991), in which demand at any instant contains the same two components. However, equation (10.13) gives these ideas precise form. The term with $\mathrm{d}X/\mathrm{d}t$ in equation (10.13) may be regarded as an expression of 'deficiency' or 'excess' demand, and the term with $\mathrm{d}W/\mathrm{d}t$ as an expression of 'growth demand'. The former indicates the demand arising from the concentration change involved in moving to the critical level state, the latter from the increasing size of the plant.

'Plant demand' must, of course, be a single value for each nutrient at any one moment, and there is an interplay between the two components. Thus, if the nutrient supply were removed, growth would continue at a reducing rate, as the new growth diluted the existing nutrient stock within the plant. The 'growth demand' would therefore gradually become 'deficiency demand'. The plant uptake mechanisms certainly sense 'deficiency demand' in a very real way, because the root uptake characteristics change with plant composition (sections 5.3.3 and 10.2.4). On the other hand, there is no evidence that uptake systems are directly and immediately influenced in this way by the relative growth rate, except insofar as the growth causes a lower nutrient concentration over time, and so generates deficiency demand. There is some evidence that potassium uptake follows upon recent photosynthesis (Wild *et al.* 1974), but it is not known whether this acts as a signal to alter the root uptake properties over a longer time.

10.2.8 The Root Demand Coefficient

The above definition of plant nutrient growth demand, as $\bar{I}L$ when $X = X_{crit}$, is in terms of a whole plant. Alternatively, the intensity of demand on a unit-root basis can be given by the value of $\bar{I}$ alone, with the same condition that $X = X_{crit}$. A further step in the analysis of demand can be made by using the term αa (section 5.3.2) as in equation (10.17).

$$WR_W X = \bar{I}L = 2\pi\overline{\alpha a} \times C_{La}L \tag{10.16}$$

$$\overline{\alpha a} = \bar{I}/(2\pi C_{La}) \tag{10.17}$$

where $\bar{\alpha}\bar{a}$ is termed the root demand coefficient, because it expresses in a simple way the state of the plant uptake characteristics, which are linked to the demand of the plant. It is important to remember that this value is only indicative for the plant, and may not apply precisely to any part of the root. Here, $\bar{I}$ is a true average of the inflow over the whole root system. If C_{La} is constant, as in flowing solution culture, then $\overline{\alpha a}$ is also a true average over the root system. However, in uptake from soil, depletion zones will develop (chapter 6), so that C_{La} is reduced most where αa is largest. Calculating average values for C_{La} and for αa does not, therefore, give a true mean value. Nevertheless, it should be a reasonable indicator of the average root demand coefficient of the plant for a nutrient, if most of the root system is active in uptake. We have then expressed 'demand' in terms of the whole plant, of the unit length of root, and of the properties of the root surface. In chapter 6 and in the next section, we show that the value of αa has an important practical role in linking the concentration of a nutrient at the root surface to the concentration in the bulk soil.

The complexities of expressing 'demand' of the plant for nutrients have an interesting analogy in the problems of defining 'sink strength' for the movement of carbon around the plant. Though different sink organs in the plant can, in principle, be compared, so that the stronger sink can be identified, it is difficult to put a formal definition on the term (Farrar 1993).

The method of modelling the growth and nutrient uptake of a plant in solution culture is discussed in section 10.6.2. The parameters in the equations given there vary with species and environmental conditions, but they can be found by experiment. We can then, in principle, predict the nutrient uptake and growth of a plant for a given soil nutrient concentration at its roots with a suitable mathematical model. The rest of this chapter deals with this subject. Table 10.1 lists most of the models that have been published to date.

10.3 Root System Uptake Models for Simplified Conditions without Competition

10.3.1 Uptake Models Using Analytical Solutions without Competition

A root system uptake model can be constructed with any single-root uptake model of the type discussed in chapter 6 by summing the inflows from unit lengths of root over this increasing root length, possibly in a changing environment. The use of analytical solutions for this purpose is now rare. Brewster & Tinker (1970, 1972) used the simplified Passioura (1963) equation (Nye & Tinker 1977, p. 216) in this way to interpret plant growth data:

$$F = (C_{Li} - C_{La})bD\gamma/a + v_a C_{Li} \tag{10.18}$$

(where γ is a term tabulated by Jaeger & Clark (1942) depending on Dt/a^2). Brewster and Tinker grew leeks in containers outdoors, but shielded from rain. Plant dry weights, nutrient concentrations, root lengths, and transpiration were measured at intervals, together with soil solution and soil diffusion characteristics. A modified version of the Williams (1946) formula,

$$I = (U_2 - U_1)(\ln L_2 - \ln L_1)/(t_2 - t_1)(L_2 - L_1) \tag{10.19}$$

where U is nutrient content in a plant, L the root length and t time, for two harvests 1 and 2, then gave the mean inflow. During exponential growth inflows and nutrient concentrations were very constant, as predicted by equation (10.11). Using equation (10.8) and the simplified Passioura equation, the mass flow contribution, the mean depletion (C_{La}/C_{Li}) and the value for $\overline{\alpha a}$ were calculated (table 10.4).

As expected, Ca, Mg, Na and S were oversupplied by mass flow, whereas it was inadequate for potassium and phosphate, and $\overline{\alpha a}$ was much larger for potassium than for the other ions. Interestingly, the inflow for phosphate appeared to be much larger than the inflow to a zero sink (Brewster 1971), and later it was shown that this was due to the enhanced inflows of phosphate produced by infection with vesicular-arbuscular mycorrhizas (section 8.3.5).

This is the simplest possible way of modelling nutrient uptake by plants, but it does not verify the model, as $\overline{\alpha a}$ was not measured independently. However, the method tests whether the plant is likely to be deficient, because it shows whether it is close to the zero-sink limit for any nutrient. The use of this method on field crops, with a different single-root uptake model, is described in section 11.3.1.

If the root uptake characteristic is known, the procedure can be reversed. The mean inflow is then calculated from the single-root model, and the predicted uptake by the whole root system can be compared with that measured.

10.3.2 Numerical Uptake Models without Competition

Some of the original numerical models were for roots in an infinite soil. These were based on the Nye & Marriott (1969) model (section 6.4.1), using the Crank–Nicholson implicit numerical method and the Michaelis–Menten root uptake property, for application to a whole root system. Claassen & Barber (1976) constructed a similar numerical model that was improved by Nielsen & Barber (1978), who introduced the term C_{Lmin}, the concentration below which no net uptake occurred, into the root uptake property.

Table 10.4 Calculated inflows, mass flow, relative concentration at the root surface and αa values for leeks grown in large pots outside, for potassium.

Interval	Water inflow (cm^3 cm^{-1} s^{-1} × 10^{-6}	Inflow (mol cm^{-1} s^{-1} × 10^{-13})	(Apparent mass flow)/ Inflow	$\frac{C_{La}}{C_{Li}}$	$\overline{\alpha a}$ (cm^2 s^{-1} × 10^{-6})
June 16–29	0.24	14.8 ± 0.73	0.07	0.14	3.9
June 29–July 7	0.23	11.8 ± 0.73	0.08	0.33	1.3
July 7–18	0.24	10.0 ± 1.4	0.10	0.40	0.9

Source: after Brewster & Tinker (1970).

Claassen & Barber (1976) measured and simulated potassium uptake by maize from four different soils, one with levels of added potassium. Potassium uptake was well correlated with model output, but the latter was about 1.5 times as large as the measured uptake, and this difference was ascribed to inter-root competition. However, as with all uptake models, the most important variable is the root length, and this was not simulated, but supplied as a data-set, so it is difficult to know how rigorous this type of test is.

Schenk & Barber (1979b) obtained good agreement between model prediction and measurement for phosphorus uptake by maize on six soils given a high level of added phosphorus. This is a little surprising, because of the impact on phosphorus uptake by pH changes, root hairs and mycorrhizas (see chapters 7 and 8), all of which would be expected to occur on maize roots, but which were not included in the model. The soils were incubated at 70°C for 4 weeks, and this may have killed part of the mycorrhizal spore population. However, at the low level of phosphorus the observed uptake of phosphorus was about twice that predicted. This discrepancy was probably due to these root surface processes which are, of course, most important at low soil phosphate levels.

10.4 Uptake by Competing Roots within a Single Root System in Simplified Conditions

10.4.1 Uptake Models for Competing Roots — General Points

The models described in section 10.3 ignore competition between roots, and this could potentially cause serious errors. In theory, competition between different roots for nutrients must always occur if they are present in the same volume, because diffusive depletion zones spread to infinity as soon as they are formed. Hence, competition is present in all multiroot systems, but the importance varies enormously for different nutrients and times. Most roots on plants growing in the field compete significantly with roots from the same and from different plants (section 11.4.1). Most uptake models used now include competition between roots, and this facility has been added subsequently to several earlier models (table 10.1). Methods of dealing with such systems are explained here, and the following points are relevant.

(a) The depletion zones of individual roots overlap, so that a non-symmetrical concentration distribution forms around each competing root. This is difficult to deal with mathematically, other than by numerical simulation or the use of an electrical analogue (see section 10.4.3). Here we cannot regard the medium around each root as infinite, because at some point between the two roots there will be a zero-transfer boundary.

(b) Larger scale concentration gradients arise in the soil due to variation in root density or root demand coefficients in different parts of the soil volume, and hence in uptake. The term 'the concentration in the bulk soil' becomes a very uncertain quantity if nutrients and water move throughout the rooting zone.

(c) If the whole exploited soil is of limited extent, the mean concentration of nutrients in it will vary with time.

(d) The continuing growth and development of the root system alters its size and geometry, and the mean inter-root distance will decrease progressively.

(e) If the simultaneous uptake of water and nutrients is being studied, the competitive effects on both mass flow and diffusion must be considered.

(f) It is mathematically extremely difficult to calculate uptake by irregular roots oriented at random in soil. Most approaches have made the arbitrary assumption that roots are straight and oriented in parallel.

At this stage, all the complications of the field, such as nutrient concentrations varying with time and space due to fertilizer applications, or irregular transfers of water following rain, are ignored. Here, we deal with some of the basic approaches.

10.4.2 Competition within Groups of Parallel Roots

The simplest competing multiroot system consists of two parallel roots. By making them parallel, all changes can be considered to occur in only two dimensions, which simplifies treatment greatly. Marriott (1972) has shown that the mean distance between any point chosen at random in the soil volume and the nearest root is the same, irrespective of whether the roots are parallel or not, so this assumption probably does not introduce important errors when applied to real non-parallel root systems. However, even this very simple system cannot be solved analytically, though it is intuitively obvious that the concentration contours will look something like figure 10.4, which have been obtained from the resistance–capacity network (section 10.4.3). The overlap of depletion zones around different roots is illustrated as a concentration–distance diagram in figure 10.5, and the effect of distance between two roots on their combined uptake is shown in figure 10.6.

We therefore consider the uptake of nutrients by a population of similar parallel roots in a uniform soil. We need to define the following variables:

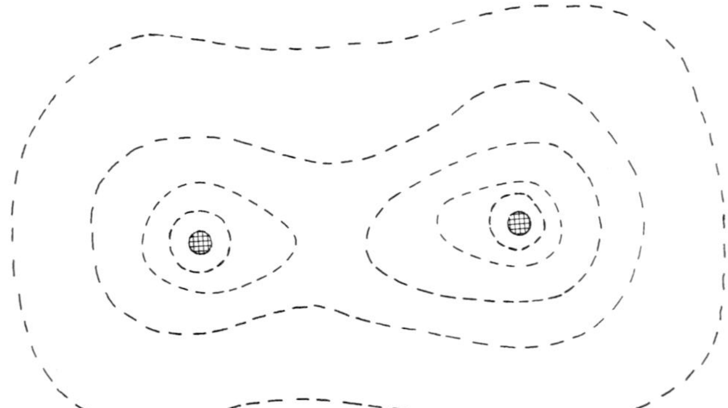

Figure 10.4 Schematic drawing of concentration contours around two absorbing and competing roots in soil (after Nye & Tinker 1977).

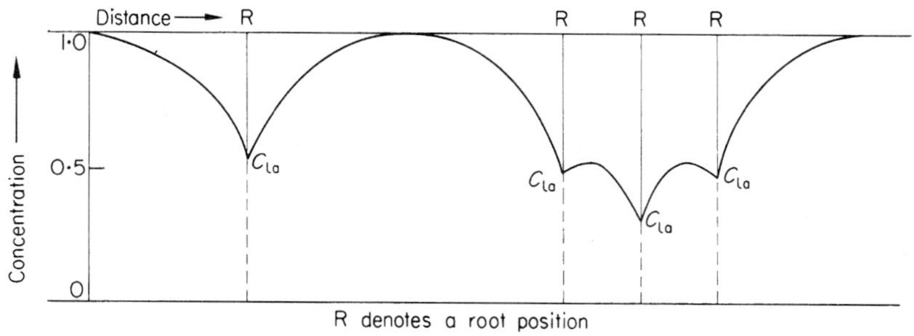

Figure 10.5 Profiles of concentration distribution around a single absorbing root in soil and three adjacent roots. Overlap of depletion zones intensifies the extent of depletion (after Baldwin 1972).

(i) The root uptake characteristics, αa or I_{max} and K_m, which are assumed to be similar in all roots.
(ii) The root radius a.
(iii) The mean length of root per unit volume, L_V, that is equal to the number of roots crossing the unit area of a plane normal to the root direction, and is the root length density.
(iv) The parallel roots will have an orientation in space, but it is often assumed that they are vertical. If not, the effect of percolating water on root depletion zones must be considered.
(v) Roots may be arranged in non-regular patterns on any plane cut through the soil and the pattern may affect uptake.

A non-regular pattern of the cut ends of the parallel roots may not be random, because various processes may encourage other patterns; for example, roots may ramify relatively less in nutrient-depleted zones, which would tend towards regularity. There is a large body of theory on pattern, and it has been applied frequently in plant ecology to define plant distribution (Pielou 1979; Kershaw 1985). A pattern of points may be regular (in which points are evenly spaced in a uniform pattern); overdispersed (in which the average nearest-neighbour distances are greater than those in a random distribution); randomly distributed; or

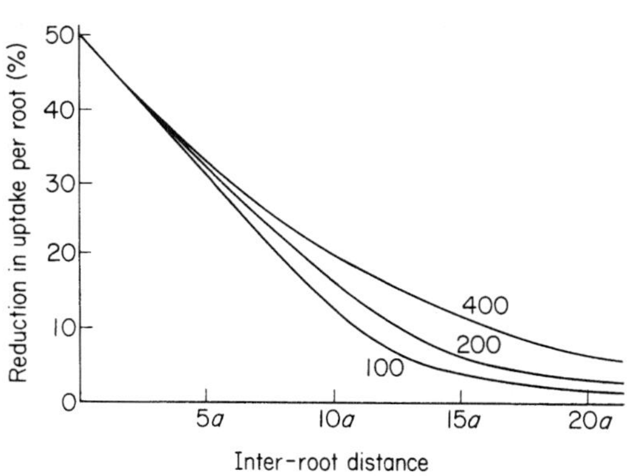

Figure 10.6 Reduction of total uptake by a pair of similar roots, as a percentage of uptake by widely separated roots, plotted against inter-root distance in units of root radius a. Numbers on curves are Dt/a^2, or dimensionless time. $\alpha a/Db$ was set at 15.9 for these runs on the resistance analogue, which simulates a root of high sink strength (after Baldwin *et al.* 1972).

contagious or clumped (in which the average nearest-neighbour distances are smaller than those in a random distribution) (figure 10.7).

In principle, pattern should be defined by a frequency distribution of distances between neighbours, but the best single-value parameter appears to be the *V/M* ratio. For the points on a surface that represent the cut root ends of a population of parallel roots, this is (variance of number of points in a specified area)/(mean number of points in the same area). The principle could be applied to non-parallel roots in soil, in which case the *V/M* ratio would be (variance of lengths of root in unit volume)/(mean length of root in unit volume). The *V/M* ratio for randomly arranged roots is 1, for more regular distributions it is less than 1, for contagious distributions it is more than 1.

Each root is considered to be surrounded by a cylinder of soil, such that the cylinder volumes add up to the total soil volume. For a regular arrangement, such as square or triangular, the radius x of the single equivalent cylinder will be $x = 1/(\pi L_V)^{1/2}$. The uptake process can be regarded as uptake by the total length of root from a single cylinder of soil of this radius. When the distribution is irregular, one may calculate the required result on the assumption of a regular distribution, and then apply a correction factor (figure 10.8) (section 10.4.3). Alternatively, one may consider a population of roots in cylinders of soil with a distribution of radii appropriate to the measured *V/M* ratio.

Barley (1970) used the latter method, and applied the idea of Voronoi polygons (similar in concept to the Theissen areas or Dirichlet tesselations mentioned later) to dealing with the patterns that represent cut root ends (figure 10.7). Each polygon contains all the area that is nearer to the point contained within it than to any other point. The soil cylinders have the same area of cross-section as the polygons. Barley used an analytical solution (Youngs & Gardner 1963) for cylinders of finite radius to determine total uptakes for different patterns of root ends. Later work suggests that this work underestimated the effects of non-regular distribution (figure 10.8) (section 10.4.3).

A similar approach, using Theissen areas and Dirichlet tesselations, has been used by Van Noordwijk & De Willigen (1987). Much attention was given to the non-circularity of the polygons and the distance of the root from the centre. These authors calculated the relationship of the period during which uptake is not affected by competition, to the eccentricity of the position of the root in the equivalent circle.

More recently, Comerford *et al.* (1994a) and Van Rees *et al.* (1994) have modified numerical methods to allow the automated drawing of Theissen maps. They have also investigated the shape of these areas, using the Theissen area equivalent radius/nearest neighbour distance ratios to indicate departure from circularity. Increase of this ratio, and of the *V/M* ratio, both indicate increasing departure from circularity, or that the root is positioned eccentrically within the polygon. Using these ideas, Comerford *et al.* (1994a) calculated uptakes of P and K for the equivalent circles of Theissen polygons, and confirmed the results found by Baldwin *et al.* (1972) with the electrical analogue (section 10.4.3).

Van Rees *et al.* (1994) investigated the pattern of distribution of roots in the field, most being close to random. Most results show that uptakes are little affected by differences between random and regular; thus, Escamilla *et al.*

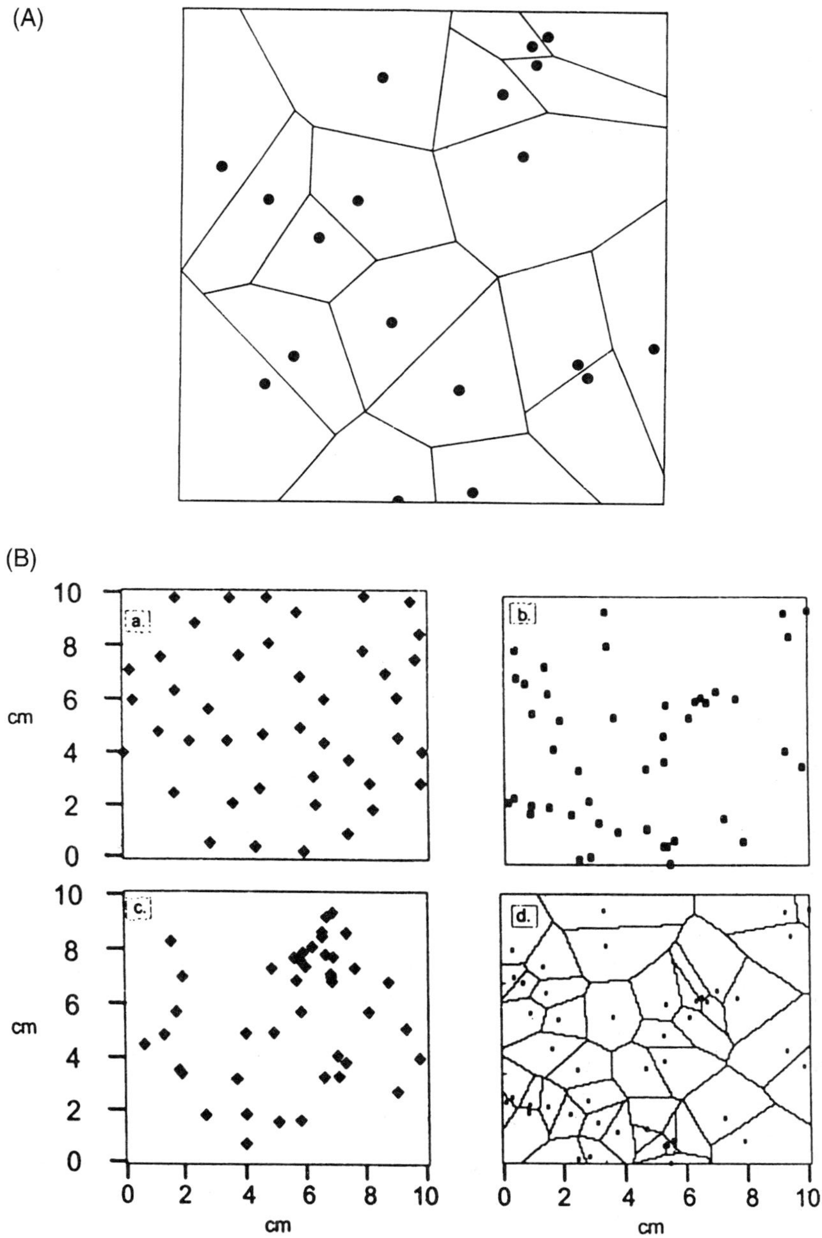

Figure 10.7 (A) Voronoi polygons (after Nye & Tinker 1977). (B) Examples of different degrees of dispersion, as for the ends of roots cut by a plane through the soil. (a) $V/M = 0.62$, tendency to regular; (b) actual field data, slightly clumped $V/M = 1.90$; (c) simulated clumped pattern, $V/M = 3.88$; (d) computer-drawn Thiessen map (same as Voronoi polygons) for (b) (after Comerford *et al.* 1994a).

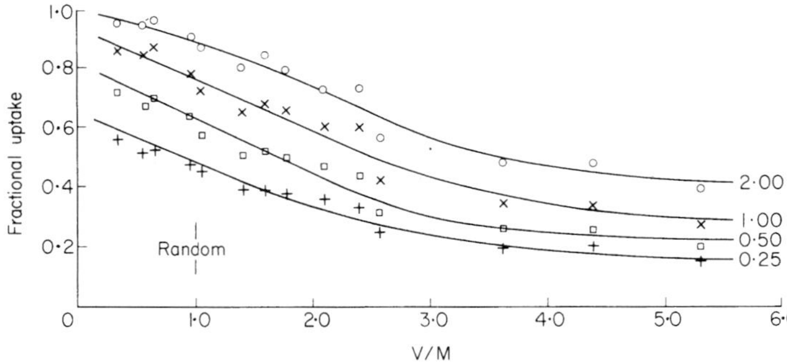

Figure 10.8 Fractional uptake of nutrient in a volume of soil by roots of high $\alpha a/bD$, plotted against a measure of dispersion V/M of a set of parallel roots. Numbers on curves are values of $Dt\rho$, where ρ is root density in cm^{-2} (after Baldwin *et al.* 1972).

(1991a) found that slash pine roots were randomly distributed, but considered that uptake was the same as that with a regular arrangement. However, differences in uptake between random and highly clumped arrangements can be important, especially in heavy soils.

10.4.3 Electrical Analogue of Root System Uptake

Sanders *et al.* (1971) simulated uptake by roots arranged in various patterns with an electrical analogue consisting of a two-dimensional resistance–capacity network. The flow of electrical charge in a circuit with resistance and capacity is described by equations that are analogous to the diffusion equations (equation (1.8)). The analogue had a square network of 525 units, each representing a slab of soil within which two-dimensional diffusion can occur. The capacitors are charged (representing nutrient) to a uniform voltage (representing solution concentration). A root is simulated by an earthing resistor. As Ohm's law is first-order in form, the resistor represents a root with $I = 2\pi\alpha a C_{La}$ characteristics.

This instrument was used to generate correction factors for non-regular compared to regular patterns of sink resistors. The simplest correction of this type is to determine the effect of the distance between two similar, parallel roots of high sink strength on their combined uptake (figure 10.8). Uptake per root can be significantly reduced if the roots are close together, but Baldwin *et al.* (1972) suggested that competition in uptake can, in practice, be ignored if $(Dt)^{1/2}$ is less than the distance between roots. If the sink strength of the roots is smaller, competition is naturally less, and if the sink strength is very low, significant competition only occurs via general depletion of nutrient in the soil.

Baldwin *et al.* (1972) also found that general correction factors for the effect of pattern (figure 10.8) could be significant with roots of high sink strength. The factors diminished with time, because the simulated uptake was from a limited volume, and gradually the uptakes converged towards total uptake of nutrient. The loss of efficiency of a random compared to a regular system was slight.

Heavily clumped systems were very inefficient, and where they occur, as in the proteoid roots (section 9.2.1), it seems likely that mechanisms other than simple diffusion operate in nutrient uptake.

10.4.4 Root System Architecture in Relation to Root Competition

The concept of root architecture has been explained in chapter 9 (section 9.2.3) as the distribution in space of root segments and the way in which these are connected. These are not models in the sense used in this book, but our emphasis on inter-root competition suggests that they are best discussed here. All the work discussed here has related to the length of root per unit soil volume L_V or per unit land surface L_A. Root architecture works on a different level of detail. Several authors have attempted to determine whether the different architectures have different efficiencies in uptake, in terms of nutrient absorbed per unit amount of photosynthate used to maintain the root system. Fitter *et al.* (1991) recognized two types of architecture: the herringbone and the dichotomously branched. The herringbone-type (section 9.2.3) has the highest efficiency, especially where nutrients have high diffusivity, because the well-defined main laterals penetrate into undepleted soil rapidly (Fitter & Stickland 1991). At the same time, Fitter *et al.* (1991b) calculated that it demands the largest photosynthate supply for a given total root length, because of the large root radii they had measured in typical root systems. The cost/benefit efficiency may not therefore be superior to that of the dichotomous branching type of architecture.

The detailed estimation of efficiency depends upon calculating the amount of overlap of the depletion zones around each root, as a measure of competition. Where two zones overlap, this is regarded as a loss to the plant of the nutrient present in one of the zones. The depletion zones surrounding each root are estimated very simply, as completely depleted cylinders of radius $(Dt)^{1/2}$ (Baldwin *et al.* 1972). The positions of the roots and their depletion zones are defined on a three-dimensional grid, and the total volume of overlap is calculated (figure 10.9). The total overlap is an estimate of the inefficiency of the root system. The most efficient root system is the one with the smallest investment of carbon substrate by the root per unit depleted volume.

Fitter *et al.* (1991) concluded that the most important factor in generating roots of high efficiency was large link length (section 9.2.3). Fitter & Stickland (1992) have tested the prediction that plants growing typically in low-nutrient conditions should have root systems with a herringbone type of topology and long root links. This was found to be true within a group of 13 dicot species, but within a group of eight grass species only the root link prediction was supported.

Berntson (1994) and Nielsen *et al.* (1994) have explored these ideas further. Nielsen *et al.* (1994) measured directly the biomass deposition, the respiration and the exudation from bean plant roots, so that they could calculate the full carbon usage of these roots. Both of these papers reported experiments at elevated CO_2, and found that the architecture could be changed by this, in addition to the expected increase in root mass.

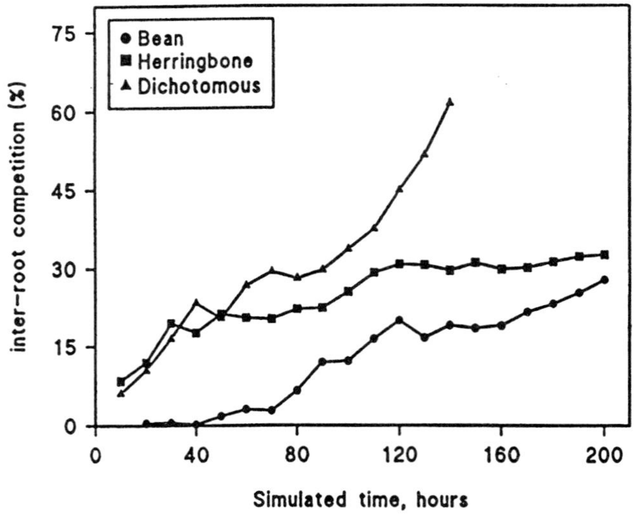

Figure 10.9 Estimation of degree of inter-root competition by root systems of different architecture, for the same carbon input, from the overlap of the computed depletion zones (after Nielsen *et al.* 1994).

Both papers defined three main architecture types: herringbone, dichotomous branching, and 'nodal' (figure 10.10), the last being a crown of nodal roots that first grow outwards at a sloping liminal angle, but gradually turn down. Berntson (1994) and Nielsen *et al.* (1994) estimated that the herringbone and nodal systems have similar efficiency, whereas the dichotomous branching model gave lower uptake and similar carbon use, hence lower efficiency (figure 10.9). Using the SimRoot model, Nielsen *et al.* (1994) produced a carbon/phosphorus budget for the three different architectures. The phosphorus acquisition was calculated by assuming that the radius R of a fully depleted zone was $R = a + 2(Dt)^{1/2}$. The use of these very simple formulae for uptake from soil is a major weakness, because depletion zones are much more complex than represented here (section 6.2.3), and the calculation of 'overlap' in this way may introduce significant errors. This work is very promising, but more sophisticated models of nutrient uptake and depletion zone structure seem to be needed.

10.5 Root System Uptake Models with Competition in Simplified Conditions

10.5.1 Uptake Models with Competition, Using Numerical Solutions

Barber and his collaborators (see Barber 1995) have used numerical solutions similar to that of Nye & Marriott (1969) in a series of papers (Claassen & Barber 1976; Barber & Cushman 1981; Silberbush & Barber 1984; Claassen *et*

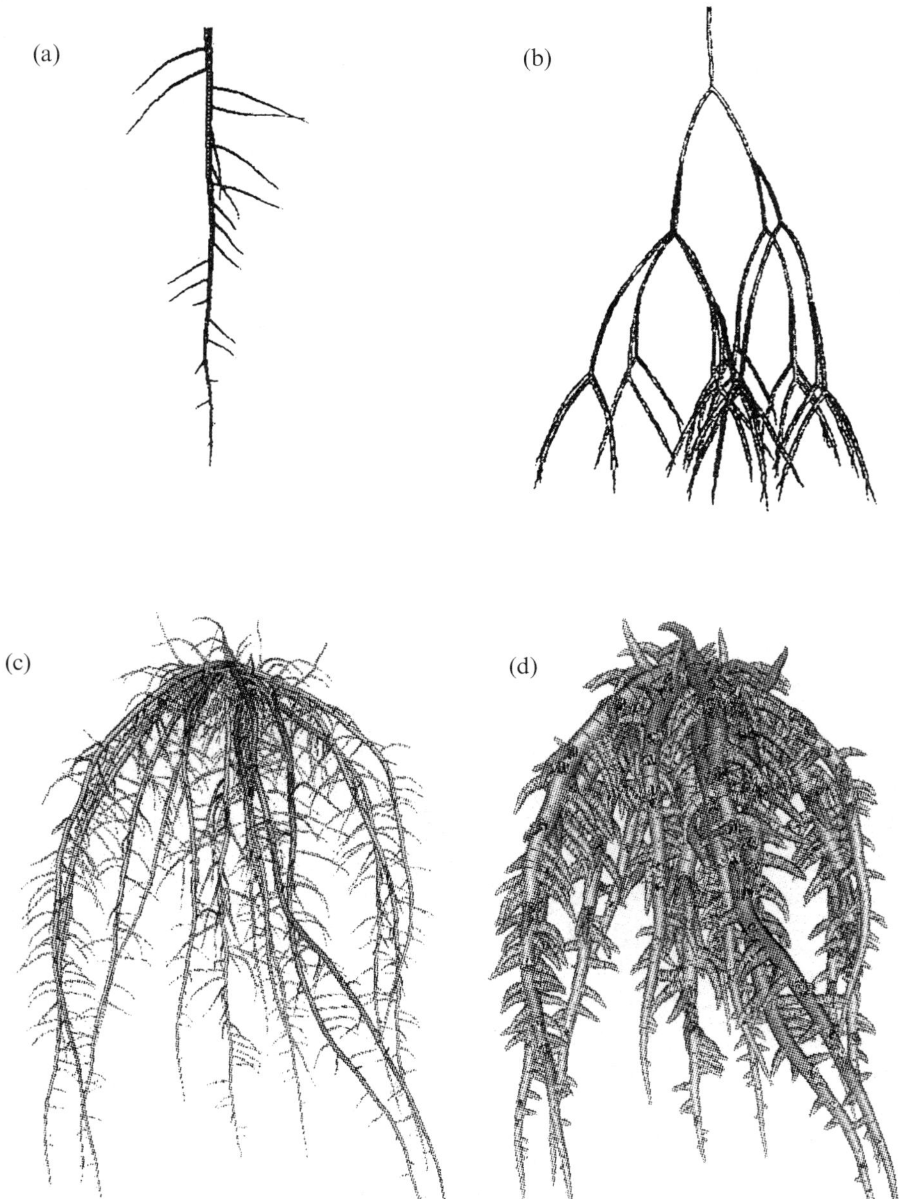

Figure 10.10 Simulation of a root system and its uptake by the SimRoot model. (a) Herringbone architecture, (b) dichotomous architecture, (c) nodal architecture and (d) simulated depletion zones for phosphate around the root system in (c), from the overlap of which the losses due to root competition can be estimated (after Lynch & Nielsen 1996, illustration supplied by Professor J. P. Lynch).

al. 1986). These all used a Michaelis–Menten uptake characteristic. The term C_{Lmin} was included in the model used by Nielsen & Barber (1978). The model by Claassen & Barber (1976) was for a growing root system in a soil of infinite extent (section 10.4.2). Barber & Cushman (1981) introduced competition between roots into this model by defining a zero-transfer boundary at a distance halfway from the next root surface. Figure 10.11 shows a sensitivity diagram obtained for potassium uptake with this model (Silberbush & Barber 1983), and the importance of the root length at all levels of change is striking. In contrast, the outer radius of the equivalent cylinder was of little importance for large radii, where there was no competition, but of strong importance for smaller radii, as competition appeared. In this model, the no-competition option was unusual in

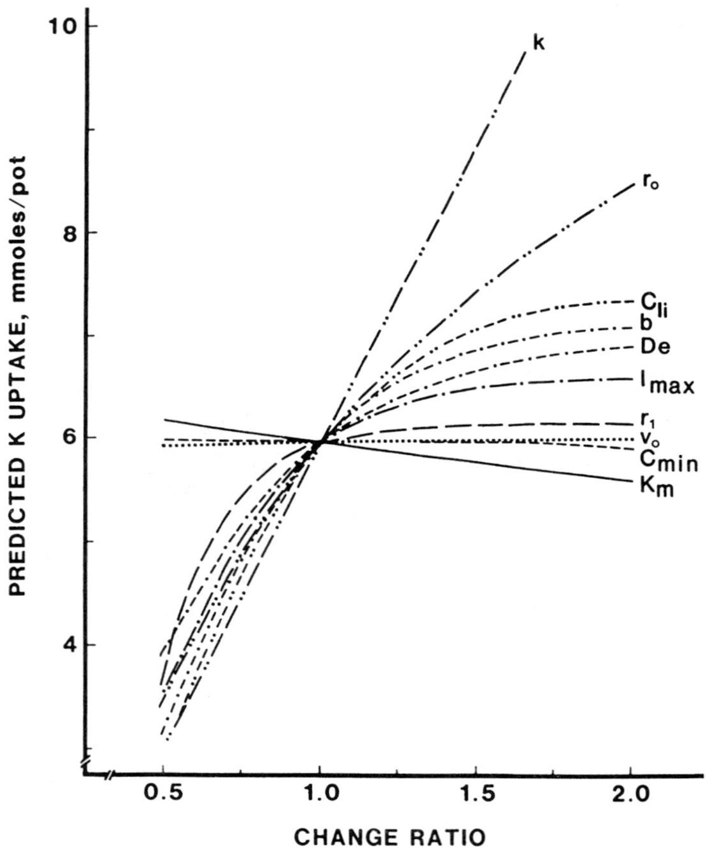

Figure 10.11 Sensitivity analysis of the results of simulating potassium uptake of soybeans on a silt loam soil with the Cushman–Barber model, showing the effect of variation of each parameter alone, whilst holding all others constant. Sensitivity was greatest to k (root growth rate), r_0 (root radius), with smaller effects of r_1 (half distance between roots) and the parameters C_{Li}, b and I_{max} (used as in this book), whereas uptake characteristics other than I_{max} had almost no effect unless very small (after Silberbush & Barber 1983).

defining a constant outer radius to an equivalent cylinder, since the latter is usually at infinity. Itoh & Barber (1983a, b) added a root hair submodel to the Barber & Cushman (1981) model, both of which have been widely used (figure 10.5).

Barber and co-workers have run a number of pot trials to verify various forms of the basic model, and Barber (1995, p.119) has listed 10 pot trial series in which the predictions of various forms of the models show generally excellent agreement with the observed uptakes. This was expressed as the correlation coefficient between the measured and simulated uptake data, and the regression coefficient to show the similarity in absolute terms. Four of these results are for potassium, where agreement is most likely for reasons given elsewhere (section 10.5.3). The agreement for phosphorus is less expected.

The model of Itoh & Barber (1983b), with a submodel for root hair uptake, has been used frequently. The uptakes of phosphorus by a series of different plant species with varying amounts of root hairs are given in figure 10.12, and show good agreement between predicted and observed uptake. However, it would be expected that the lack of a mycorrhizal submodel would cause uptake to be underestimated, as maize is easily colonized by mycorrhizas. It suggests that most of the tests were carried out with plants in which phosphorus was at above deficiency levels.

Brouder & Cassman (1994) used the Barber & Cushman (1981) model to investigate potassium uptake by cotton from a vermiculitic soil with high K-fixing capacity, on which cotton was usually potassium deficient. A pot experiment included factorial ammonium and potassium treatments, which gave a wide range of soil solution potassium. Uptake parameters were determined by the Claassen & Barber (1974) depletion technique. Model predictions did not agree with the observed results, but subsequent changing of the root uptake parameters improved the agreement. Further corrections to the buffer power gave a good correlation (0.87) and a regression coefficient of 1.16, but the complexity of the experiment and of the interpretation of the data makes the model verification uncertain (figure 10.13). This study is important because it is one of the few where the plants were clearly deficient; another was that of Van Rees *et al.* (1990), where verification was again difficult at low potassium levels. This is similar to the modelling of phosphate uptake, in that poor results tend to be found on low-nutrient soils.

This type of model has been recommended for use in practical agronomy (Barber 1995). This may be useful in closely defined conditions in which the soil and plant conditions are reasonably constant and well understood, but it seems hazardous in more varied conditions, especially in nutrient-deficiency conditions. The great sensitivity of the results to some soil and root parameters in such work is shown in figure 10.11.

Van Noordwijk & De Willigen (1979) also produced a model for multiroot situations, using a constant flux root uptake characteristic for $C_{Li} > C_{Lcrit}$, and a first-order uptake when $C_{La} < C_{Lcrit}$, but no increase in root length during the simulation. This model may be looking more towards the field than pot work, and we are not yet aware of any verification with plant studies. De Willigen & Van

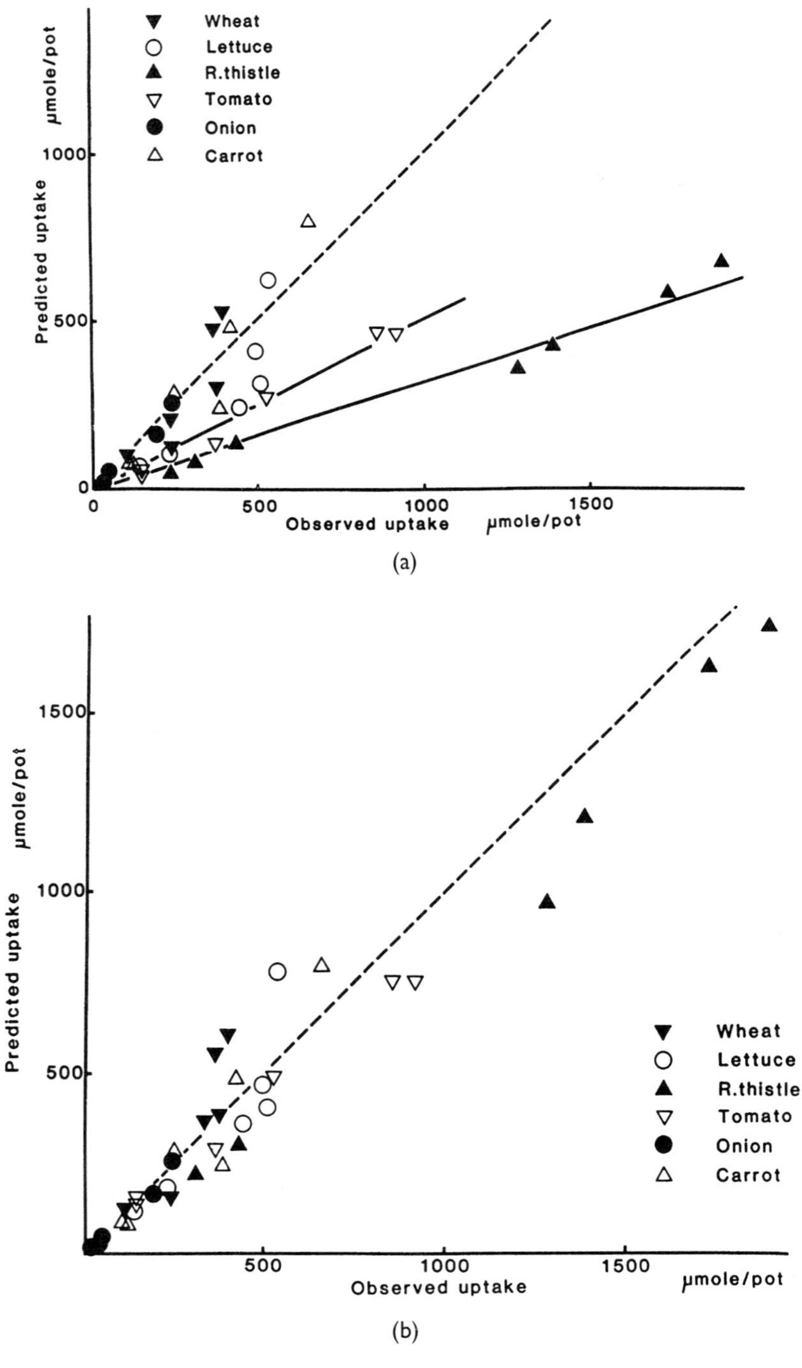

Figure 10.12 Comparison of predicted P uptake with observed P uptake for six species with varying amounts of root hairs. Calculated values obtained using (a) the Barber–Cushman model, and (b) the model of Itoh & Barber (1983b), assuming uptake characteristics the same for root hairs and for roots (after Itoh & Barber 1983a).

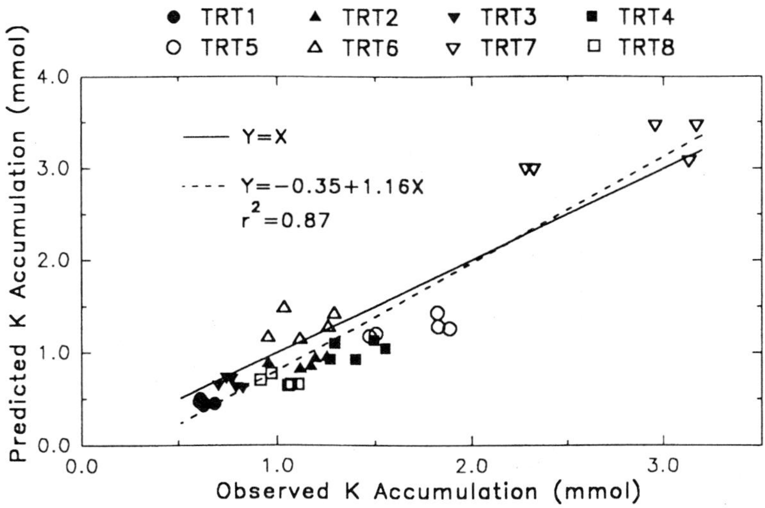

Figure 10.13 Relation between the observed potassium uptake of cotton plants in a K-fixing soil receiving various nitrogen and potassium treatments and the uptake predicted from the Cushman–Barber model, after modifications to uptake criteria and buffer power (see text) (after Brouder & Cassman 1994). (For details, see original publication.)

Noordwijk (1987) further developed this to include the effects of incomplete soil–root contact.

10.5.2 Uptake Models Using the Steady-State Approximation, in Simplified Conditions

10.5.2.1 Analytical Solutions

All real-world situations are transient; that is, fluxes and concentrations at a point vary with time. Youngs & Gardner (1963) and Van Noordwijk & De Willigen (1979) have published transient-state analytical solutions to the problem of a single root in a limited cylindrical volume, but neither is simple to apply since they use the Laplace transform.

Accordingly, simpler though less accurate methods have been sought. An early attempt to describe a transient situation by an approximate steady-state model was made by Gardner (1960) for transport of water to roots (section 2.3.4). Baldwin *et al.* (1973) followed this approach in developing a series of root nutrient uptake models. They started from the assumption that the concentration profile around each root has the same form as that of the steady state obtained when solute is taken in at the outer boundary of the root's equivalent soil cylinder at the same rate as it is taken up by the root. Thus,

$$I = 2\pi\alpha C_{La} = 2\pi r D b\, dC_L/dr \tag{10.20}$$

where r is the radius of a circular surface in the soil with the root at its centre.

Integration over the profile ($r = a$ to r) gives

$$C_{Lr} = C_{La}(1 + \alpha a/Db \ \ln \ r/a) \tag{10.21}$$

With a regular parallel array of roots, Baldwin *et al.* (1973) assigned to each root a cylinder of soil around it of radius

$$x = 1/(\pi L_V)^{1/2} \tag{10.22}$$

If roots deplete only their own cylinder, the perimeter at x is a zero-transfer boundary.

The average concentration of solute in solution within such a cylinder is

$$\overline{C}_L = 1/[\pi(x^2 - a^2)] \int_a^x 2\pi r C_{Lr} dr \tag{10.23}$$

Substituting the expression for C_{Lr} from equation (10.21) and integrating gives

$$\overline{C}_L = C_{La}\{1 - \alpha a/2Db + [x^2(\alpha a/Db)/(x^2 - a^2)] \ln x/a\} \tag{10.24}$$

If $x \gg a$, which will nearly always be true, this reduces to

$$\overline{C}_L = C_{La}[1 + (\alpha a/Db) \ln(x/1.65a)] \tag{10.25}$$

A weakness of the steady-state approximation lies in the assumption that a depletion zone, extending to the outer limit of the equivalent cylinder, is established instantly. Baldwin & Nye (1974) found that the accuracy of the method, at short times, improved if the boundary of the depletion zone around each root is considered to spread outwards with time, until it coincides with the boundary of the equivalent cylinder at $x = 1/(\pi L_V)^{1/2}$. This is achieved by setting $x = 2(Dt)^{1/2} + a$ until $2(Dt)^{1/2} + a > 1/(\pi L_V)^{1/2}$. Since x in equations (10.24) and (10.25) is increased at each time step, this automatically allows for the undepleted fresh zones being exploited.

Equation (10.25) gives a simple relation between the concentration at the root surface and the average solution concentration in the depletion profile, in terms of given parameters. It may now be used to calculate the uptake of solute by considering the change in the average concentration with time, assuming that there is no external supply of solute, but that the steady-state shape of the concentration profile is maintained. Equating the loss from unit volume of the soil with root uptake from it gives

$$-b \, \mathrm{d}\overline{C}_L/\mathrm{d}t = 2\pi\alpha a C_{La} L_V \tag{10.26}$$

Eliminating C_{La} between equations (10.25) and (10.26) and integrating from $t = 0$ when $\overline{C}_L = C_{Li}$ gives the approximate relation

$$M_t/M_\infty \approx 1 - \exp\{(-2\pi\alpha a L_V t)/\{b(1 + \alpha a/Db)/\ln \ (x/1.65a)\}\} \tag{10.27}$$

where M_t and M_∞ are the amounts of solute taken up at times t and ∞.

The integration assumes that L_V, and hence x, is constant, thus introducing an inaccuracy when t is small. However, when compared with a modified version of the accurate numerical method of Nye & Marriott (1969), it gave good agreement, especially at values of $\alpha a/Db < 5$. It provides a simple algebraic expression from which the effect of all the parameters can readily be seen. For example, αa and L_V

are seen to be partly interchangeable in effect, as would be expected. If $\alpha a/Db$ is very large, the expression gives the largest possible rate of uptake for that soil and that root density:

$$\text{Lim } \alpha a/Db \rightarrow \infty \qquad M_t/M_\infty = 1 - \exp[2\pi L_V tD/\ln(1.65a/x)] \tag{10.28}$$

10.5.2.2 Addition of Mass Flow

The method used in the last section may be developed to include mass flow and transpiration, though models of the type developed here are most interesting when depletion zones are well marked, and mass flow is then unlikely to be important. It was shown in chapter 6 that when mass flow is important, only slight depletion zones develop. Accumulations, with diffusion away from the root, are also small in most real cases (Wray & Tinker 1969). High accumulations require a high rate of mass flow compared with solute uptake—a situation most likely to occur in saline soils in arid regions.

Nye & Spiers (1964) showed from the continuity equation (1.6) that the steady-state concentration profile was given by

$$C_{Lr}/C_{La} = \alpha/v_a + [1 - \alpha/v_a][r/a]^{-(av_a/bD)} \tag{10.29}$$

If av_a/bD is small, as it usually will be, we can approximate

$$(r/a)^{-(av_a/bD)} = \exp(-av_a/bD \ \ln \ r/a) \approx 1 - (av_a/bD) \ \ln \ r/a \tag{10.30}$$

Hence,

$$C_{Lr}/C_{La} \approx 1 + \alpha a/bD \ \ln \ r/a - av_a/bD \ \ln \ r/a \tag{10.31}$$

Calculation of $\overline{C}_L$ as before (equation (10.23)) by integration over the profile gives

$$\overline{C}_L/C_{La} = \alpha/v_a + [(1 - \alpha/v_a)\{2/(2 - av_a/bD)\}\{(x/a)^{(2-av_a/bD)} - 1\}]/[(x/a)^2 - 1] \tag{10.32}$$

and, if av_a/bD is small, equation (10.32), by the approximation $x^2 \gg a^2$, reduces to

$$C_{La}/\overline{C}_L \approx [1 + av_a/bD \ \ln \ x/1.65a]/[1 + \alpha a/bD \ \ln \ x/1.65a] \tag{10.33}$$

that is, uptake ($\propto C_{La}$) increases linearly with water inflow ($\propto v_a$).

Equation (10.33) can be used to calculate uptake from a given soil volume by integration in the same way that equation (10.25) was used for diffusion without mass flow, if the unlikely assumption is made that the water content of the soil remains constant. Otherwise, numerical methods must be used. Initially, $\overline{C}_L = C_{Li}$. Thus, C_{La} can be found from equation (10.33), and uptake over the period Δt_1 is given by

$$2\pi\alpha a L_V C_{La}\Delta t_1$$

The new $\overline{C}_L$, $\overline{C}_{L2}$, is calculated from the uptake, and C_{La2} is calculated from $\overline{C}_{L2}$ as before. By this procedure, any time period can be covered. At each interval, the water uptake is calculated and v_a and D are adjusted accordingly.

One of the great advantages of equation (10.33) is that it relates uptake to the average concentration within any chosen volume or soil compartment. Thus, it may readily be incorporated in large-scale models that divide the soil into a series of compartments.

10.5.2.3 Parameters that Change over Time

The last model has the further advantage that the time-step procedure allows parameters to change during the simulated time, as we have seen for those depending on the water content. Another example of this advantage is where b alters greatly with C_L, as for phosphate, giving a concentration-dependent diffusion coefficient. Further examples are where L_V changes as a root system develops or α decreases as a root ages.

Baldwin *et al.* (1973) compared the uptake using this method with that of an accurate numerical method, and found the results in reasonably good agreement, with the largest errors (of around 12%) being for high αa, short times and high v, as would be expected. The agreement would have been much better had the outer boundary condition $x = 2(Dt)^{1/2} + a$ until $2(Dt)^{1/2} + a > 1/(\pi L_V)^{1/2}$ been employed in this study (section 10.5.2.1).

For rapid calculation, it may be sufficient to find the time-averaged root density $\overline{L}_V$ and use this value in equation (10.27). Time-averaged values can also be used for θ and other parameters. Any method of 'time-averaging' L_V causes an underestimate of uptake, because it assumes that one 'root day' is equally useful at any time and position, which is not true. A test by Baldwin *et al.* (1973) found that this method gave a 20% underestimate of uptake when the root system was increasing exponentially and rapidly.

Root systems may expand into fresh soil volumes. If all the soil were homogeneous, it could be treated as a problem with a clumped root system, which became more regular as it expanded. However, it is uncertain whether the correction factors explained above (section 10.4.3) would be adequate for radical changes in root density in different parts of a soil volume. In the field, the topsoil and the subsoil will usually be different, and it is probably better to subdivide the total soil volume into compartments, as is normally done for root modelling of field crops (section 11.1.2). There may be significant transfer of nutrients and water between compartments in such a system, but this can be simulated as a simple application of mass flow and diffusion.

Because they contain both assumptions and simplifications, approximate models need extensive testing with plant and soil data-sets of all important parameters. It is not sufficient to test only final uptake or yield because the increase of L can produce an increasing yield; as many of the time-varying variables as can reasonably be measured should be compared with experimental values throughout the model run.

Baldwin (1975) used pots for growing rape plants that were internally sectioned by silicone membranes. These prevented water transport between layers, and made it possible to determine how much nutrient had entered the root system in each compartment. Nutrient uptake measurements were compared with model calculations using two boundary conditions. The constant flux condition, when

$X > X_{crit}$, was not satisfactory for either nitrate or potassium. The first-order condition with $\overline{\alpha a}$ set at about 1×10^{-7} cm^2 s^{-1} fitted nitrate uptake quite well, but the uptake rate declined towards the end of the growth period, so that an αa that declined from about 2 to 0.5×10^{-7} gave the best fit. For potassium, the sink strength was very high, so the exact value of αa was not of critical importance. A value of 2×10^{-6} cm^2 s^{-1} gave a good fit. The uptake characteristics of the plant were not measured separately, so that there was no real comparison of 'predicted' and 'observed' data, but the general fit was considered satisfactory. The decrease of sink strength with development of the plant is now recognized as normal (section 5.3.3) (Jungk & Barber 1975). The agreement might have been better if C_{Lmin} had been included.

10.5.3 Further Development of Steady-State Uptake Models

Further work has concentrated on the following.

10.5.3.1 Comparisons with Numerical Models and Experimental Data-Sets

As with all uptake models, the root length has to be provided for the simulation. Nevertheless, if all the other parameters are measured independently, quantitative agreement over time gives some confidence in the model. In some cases, all parameters have not been measured; in particular, the root uptake characteristics are often estimated, derived from the experiment, or measured under conditions that are very different from the growth conditions of the main experiment. Such experiments cannot be trusted as real validations.

Van Rees *et al.* (1990) found that the model of Barber & Cushman (1981) and the model of Baldwin *et al.* (1973) gave closely similar results when applied to experimental data of potassium uptake by slash pine seedlings. However, the agreement between the simulations and the plant measurements was variable, which was attributed to very clumped root distribution or to mycorrhizal uptake. The soil contained little exchangeable potassium, so the assumption of a high αa was acceptable. The only plant factors to which uptake was highly sensitive were the root length and radius.

Smethurst & Comerford (1993a) developed a new model (COMP8) similar to that of Baldwin *et al.* (1973), but with provision for increasing the radius of the depletion zone as it developed and for varying α. This model gave similar results to that of Barber & Cushman (1981), with a maximum difference, for potassium, of 23%; the Barber–Cushman model was assumed to give correct simulations, and taken as the standard. By running the model with or without the additions to their model, Smethurst and Comerford were able to determine which approximations in the original Baldwin *et al.* (1973) model caused significant deviations, and the conditions under which this occurred.

10.5.3.2 The Depletion Zone Spread Correction

The correction to account for the instant appearance of the depletion zone when new root appears in the original model has been improved by Smethurst &

Comerford (1993a). This was basically similar to that suggested by Baldwin & Nye (1974) (Nye & Tinker 1977, p. 231). The errors were most serious for conditions in which the depletion zone only slowly extended towards the outer limit of the equivalent cylinder; that is, for low D values and low L_V values. Using this correction (section 10.5.2), Smethurst and Comerford found much better agreement with the numerical model of Barber & Cushman (1981). Yanai (1994) also dealt extensively with this aspect of the steady-state approximation. She pointed out that new roots enter the model having already attained a steady-state concentration profile, and that the solute taken up in this way is ignored. In her model, this solute in the 'instant depletion zone' is calculated and added in to the total absorbed by the new roots that enter the system.

Smethurst & Comerford (1993a) also investigated the importance of having low values of the dimensionless term $\alpha a/Db$ (Nye & Tinker 1977, p. 229). They found that the agreement of their steady-state model with the numerical Barber–Cushman model was affected differently by changing the individual parameters in the term, and that no generalization was possible for the dimensionless term itself, so that tests were recommended for each individual parameter. Smethurst & Comerford (1993a) showed that changes in b caused particular difficulty.

10.5.3.3 Sensitivity Analyses of Model Outputs

This has allowed the important variables to be identified (Silberbush & Barber 1983, 1984). As expected, root length and radius values usually dominate the output, but other important parameters are C_{Li}, b (the buffer power), x (the inter-root distance), and some soil parameters, depending upon the precise conditions for the simulation (figure 10.11).

All the uptake models discussed here are of the 'nutrient uptake' type (table 10.1), and they are verified by the agreement between the nutrient absorbed by the root system and that calculated using the model. A weakness of this approach is that the root lengths are measured, rather than developed as part of the simulation (section 11.1.5). This external data can be of overriding importance for the output (Van Rees *et al.* 1990; Rengel 1993), so that the increasing nutrient uptake follows automatically from the increasing root length. It is easier to model with potassium than other major nutrients, because it is not as affected by other root surface processes as is phosphate, or by microbiological transformations as is nitrogen. Continued verification using potassium is therefore necessary.

It seems that a steady-state model, especially if modified as by Smethurst & Comerford (1993a) and Yanai (1994), will give results that are acceptably near those of transient state models. They are much simpler and more rapid than the numerical models, and at the same time allow changes in various parameters such as water content, chemical composition and plant characteristics during the course of the simulation. Yanai (1994) noted that a number of recent models of forest growth or biogeochemical cycling do not include any mechanistic representation of the soil–root relationship; the same is largely true for most models of the growth of agricultural crops. The below-ground part of such models is now their weakest aspect, and simple but reliable soil–root submodels, such as the steady-state models, could correct this. In support of this, a number of studies that have been made on field crops have used the steady-state models (section 11.3.1).

10.6 Whole-Plant Growth and Uptake Models

10.6.1 Whole-Plant Models, Including the Shoot

So far, models have been discussed that focus almost wholly on the uptake and soil transport steps, with the exception of the root/shoot allocation models in section 9.5.3. In whole-plant models, photosynthesis, transpiration, allocation of photosynthate between root and shoot, and root and canopy growth and structure are simulated. From the last two, the light interception and the nutrient uptake models are updated. The essential input of the root model is the growth of root length and its distribution.

The interaction between shoot and root also includes the distribution of the absorbed nutrient throughout the whole plant, and the supply of water to meet transpiration demand. As seen in chapter 11, there are now a large number of whole-crop models that do this for field conditions, but with varying levels of sophistication in the root submodel. Here, we discuss some of the small number of such models designed for simplified conditions.

The details of canopy structure, light interception and phenology can be very complex. However, these parts of plant models are now well developed, being based on a large amount of work over the last few decades (Goudriaan 1996; Thornley & Johnson 1990). The root functions and the interactions of the root and shoot are less well understood (section 9.1.2). Net amounts of photosynthate and nutrients move up and down in the plant; in some cases, such as calcium and iron, these movements are always upward, whereas for most major nutrients there is a well-developed circulatory system. The way in which these fluxes are controlled, and the way in which they affect other plant functions, are not understood in detail.

The general approach for whole-plant models that incorporate a root–soil component is to generate a value for the growth rate from the light input, the temperature, and plant data relating to canopy structure and development; these must include a feedback from the nutrient concentration X. A fraction of the fixed carbon is directed towards the root, using allocation rules or mechanisms of varying complexity, based on purely empirical data, allometric constants or the models described earlier. These determine the amount of photosynthate available for root construction and maintenance.

In principle, the five relationships discussed in section 10.2.1 determine how the root increases in size, and its nutrient-absorbing functions. With some basic additional data that are needed to model the growth of the whole plant, this can be done. This is sufficient for totally constant environmental conditions, but, in reality, changing temperature, light regime and soil properties will affect root development and all the five stated relationships. In figure 10.14 we give a model structure that is internally self-consistent, that includes the essential nutrient relationships, and that has a reasonable balance between shoot and root in its

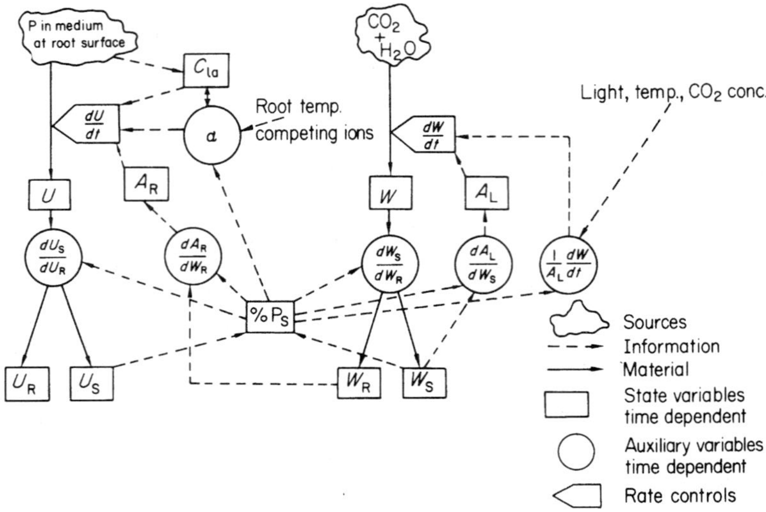

Figure 10.14 Flow diagram for phosphorus uptake from solution culture by a growing plant (after Nye *et al.* 1975b). A_L = leaf area; A_R = root surface area; C_{La}, α, X, U, and W have usual meaning, with R and S subscripts for root and shoot, respectively. Dashed line = information, solid line = material.

complexity and detail. It is a general rule of modelling that the degree of detail should be broadly similar throughout. We discuss some implications that follow from this later (section 11.6.1).

In such a model, the mechanism of nutrient deficiency or sufficiency is clear. If the root and soil system cannot produce an $\bar{I}L$ value that allows growth with $X > X_{crit}$, then growth at the maximum rate allowed by the radiation income is not possible (equation (10.13)). A new solution of equation (10.13) with a lower R_W and a lower X is then obtained. As X declines, the uptake characteristics of the root are enhanced, which will limit the decline.

10.6.2 A Single-Plant Phosphate Model for Growth and Uptake

The principles and relations discussed above can be seen in a single-plant growth and phosphate uptake model (Brewster *et al.* 1975a, b; Nye *et al.* 1975a). The conceptual model is similar to figure 10.14. The photosynthesis submodel generates plant weight and growth rate, which are subdivided between root and shoot. The central variable is the percentage composition for phosphorus of the shoot—the mean concentration throughout the plant may be a crude measure of all the effects of nutrient status (section 10.2.5.1), but, at present, more complex approaches are not possible. This value controls the root/shoot ratio, the distribution of absorbed phosphorus between root and shoot, αa, and the root specific surface area in the root submodel, by empirical relationships. In the shoot submodel, it controls the relative growth rate and the leaf area per unit shoot weight.

The phosphorus composition is determined by the relative accretion of dry matter and phosphorus in the shoot, and this then feeds back onto all the other variables.

The model was first applied to the growth of onions in solution cultures of various concentrations (Brewster 1975a), under similar conditions to those in which the relationships between the concentrations and the growth variables were measured. The variation of phosphorus concentration with time was sometimes larger than the effects of the differences in phosphorus supply, emphasizing the importance of growth stage; in particular, growth patterns appeared to shift at day 16 of the experiment. The model simulated growth and phosphorus uptake satisfactorily at four levels of solution phosphorus (figure 10.15).

The parameters in this model were measured over times and under conditions identical to those of the solution culture experiment, hence the satisfactory simulation of the plants grown in solution culture. The much more critical test was to grow plants in soil with different phosphorus levels (Brewster *et al.* 1975b), in a system that allowed water and electrolyte contents to be kept constant, and then to compare their growth and P uptake with model predictions using the parameters established in solution culture. However, when solution culture and soil solution concentrations were the same, the plants' root/shoot ratios and phosphorus percentages in solution culture and in soil differed, showing how difficult it is to produce identical plants in solution and in soil culture. Even the most careful measurements of plant parameters in solution culture may therefore not be adequate to ensure agreement of the simulation with the observed soil-grown plants.

Simulation of growth in soil was done in two ways. In the first, the measured root lengths were used as input to the model, so that it was used simply as a P-uptake model (sections 10.3.2 and 10.5.2). The second test was to run the full model with the parameters measured in solution culture, and allow it to generate root length, growth, and nutrient uptake. In this case, the measured uptake was lower than the simulated one at the high phosphate levels, similar at moderate levels, but greater at low levels (figure 10.16). Thus, the simulated uptake responded more to increased phosphate in the soil than the real plants did.

There were two possible reasons for the discrepancy. First, the method of measuring the buffer power for phosphate in the soil was, as usual, in soil suspension. However, measurements on the structured soil from the pots indicated that the real buffer power was much smaller than that measured in the traditional way in a soil suspension. Second, it is possible that the onions were infected with vesicular-arbuscular mycorrhizas, which can greatly increase plant phosphate uptake at low concentrations (section 8.3.5). Thus, at low P levels, mycorrhizal uptake could have prevented agreement with the simulation.

10.7 Conclusion

The major difficulties in modelling whole-plant growth and uptake still lie in the plant processes, and in the complexity of the relationships between different plant variables. These centre on our inadequate knowledge of how nutrient deficiency operates on growth, phenology, physiology and morphology of plants. The main aspects of the soil part of plant uptake appear to be well understood, and the

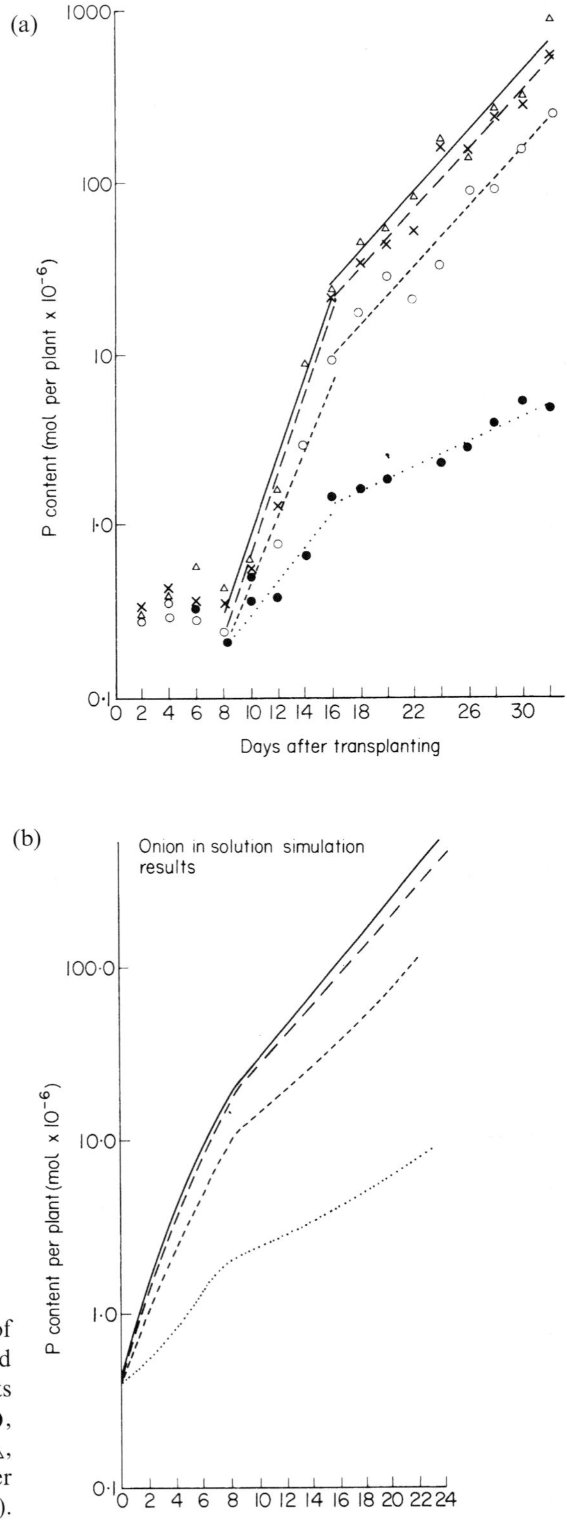

Figure 10.15 Comparison of (a) measured and (b) simulated content of P in onion plants grown in solution culture at ●, 10^{-6}; ○, 10^{-5}; ×, 10^{-4}; and △, 10^{-3} M concentration (after Brewster *et al.* 1975a).

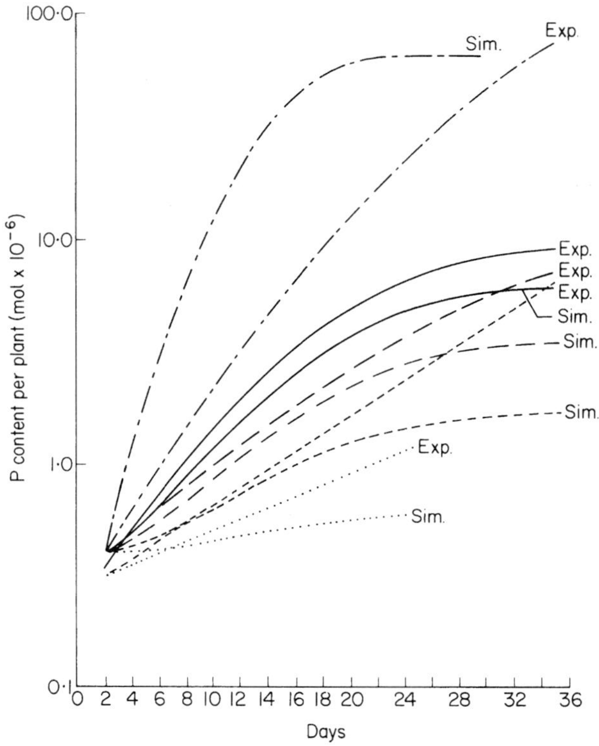

Figure 10.16 Comparison of experimental and simulated P uptake in onions grown in soil in pots, using the growth model in figure 10.14. Buffer power was taken as 2, from evidence that this was the effective value in undisturbed soil (after Brewster *et al.* 1975b).

models of this part of plant nutrition are probably dependable if soil parameters can be accurately measured. The root surface and rhizosphere processes discussed in chapters 7 and 8 are still very difficult to predict quantitatively. Fortunately, these are largely confined to some strongly adsorbed ions, such as phosphate and trace metals, that will therefore always be particularly difficult to deal with. Uptake of most other elements from soil, especially potassium, can now be modelled with some confidence, but the mechanistic modelling of whole-plant nutrition is still difficult. The final chapter, 11, deals with the development and behaviour of root systems in field soils, and a discussion of the nutrition of multi-plant vegetation, both managed and natural.

11

Solute Transport and Crop Growth Models in the Field

In this chapter we deal with vegetation growing in the field. This introduces new and challenging questions of scale and heterogeneity, in time and space, of the environment in which plants grow. It builds on the concepts and methods explained in earlier chapters, especially the movement of water and solutes (chapters 2, 3 and 4) and the distribution of roots (chapter 9) in field soils. In some cases, it requires changes and simplifications in the methods that we have used earlier.

The problems of dealing with water and nutrient movement and uptake at the field scale are discussed first. The modelling approach that we developed in the earlier chapters of this book, up to the end of chapter 10, logically resumes at section 11.3. This covers both uptake models and the more complex combined crop growth and uptake models that simulate the main interactions with the environment.

11.1 Uptake of Water and Nutrients by Field Crops in Relation to the Development of Crop Models

This chapter considers increasingly complex systems: first, uniform monocultures, including models of a 'green leaf crop', a root crop, a cereal, and a tree crop. At this level, the presence of weeds or groundcover is deliberately ignored. Interspecies competition is included later, with vegetation composed of more or less regularly spaced plants of more than one species. This occurs in many agricultural systems, such as mixtures of forage species and agroforestry systems.

The competition processes become even more complicated where there is no spatial symmetry, and models of crop/weed mixtures, grass/legume mixtures, and planted woodlands are used as examples. Progress with crops has been more rapid because of their more regular structure, so we deal mainly with these, but we believe that similar ideas will be applied to natural vegetation also, and this is discussed in section 11.5.

Most of these models have a water submodel, or, if not, one could be added. As the physical basis is normally rather similar for all water models, one model for water uptake is explained in some detail (section 11.1.2), but elsewhere water uptake is dealt with very briefly. For each model, the preferred order of discussion is water; growth, including economic yield; nitrogen; potassium; phosphorus; and other nutrients, unless the logic of the subject demands a different order. This sequence of elements is chosen because nitrogen is usually the main element most frequently limiting growth, and the most important fertilizer. Potassium is next because it is the element for which transport in the rooting zone is both important and relatively easily modelled. Phosphorus is dealt with last, because transport in the rooting zone is extremely important, but there are so many special mechanisms (see chapters 7 and 8) that detailed modelling may be very difficult.

11.1.1 The Structure of Field Vegetation

The study of field crops is made easier if we can define a small unit cell. Rabbinge *et al.* (1990) have discussed the question of scale within crops in terms of the minimum area that contains all the various above-ground subsystems of the crop, called the unit cell. It is assumed here that this unit cell would also define the minimum below-ground repeating unit, even though, on average, roots may penetrate further from the stem than shoot branches do.

In many crops, especially combine-harvested ones such as small grains or oilseed rape, the individual plants mix so closely that a square metre may be sufficiently typical of the whole if the crop is very uniform. In section 11.3, these monoculture unit cells are treated as identical plants, in which only the vertical dimension is considered (Silberbush 1996) (but see figure 9.20). With larger individuals planted in rows, such as maize, sugar beet or sorghum, the repeating unit is several square metres, and with tree crops or mixed trees and crops it can be hundreds of square metres. With peasant shifting cultivation or with natural woodland, it may not be possible to define a precisely repeating unit at all, and the area may have to be subdivided into irregular blocks within which vegetation is reasonably similar. Some forest models use the area covered by a single dominant tree as the spatial unit (e.g. Friend *et al.* 1997), but these do not deal in detail with processes in the soil (section 11.5.2).

It is customary to subdivide the canopy of a forest into different layers, each containing different ratios of the different species. Below ground, there must be analogous differences in the depth of root penetration, but so far little detailed work on this has been done, except for agroforestry (section 11.4.7).

11.1.2 The Modelling of Water Use

Crop water models aim to simulate the evapotranspirational demand, the rate at which water can be extracted from different soil layers by the root system, the consequential water stress within the plants and the effect of this on the crop growth and function. There are now a large number of models for water uptake by uniform crops. The above-ground parts of all of these are based on the well-established theory of energy, momentum, and water transfer across the canopy surfaces, usually expressed as a form of the Penman equation, and consequently they have a fair degree of reliability. The *potential* flux of water up through a crop is therefore usually well defined. The below-ground aspects and the responses of plants to water stress are less clear, and have to be dealt with by empirical methods. There has to be a model of the physical transport of water vertically within the soil profile as a result of water supply and evaporation at the surface, different rates of extraction by roots at different depths, and, possibly, movement of water up from a water table (see section 11.2). Few models deal explicitly with short-range transport to individual roots, as the majority take account of this local resistance to movement (see figure 2.5) in other ways. Most difficulties in modelling below-ground aspects arise from the poor definition of the rooting density and depth, the different properties of individual roots and crops, and the heterogeneity of the soil with respect to water-holding capacity and hydraulic conductivity.

Figure 11.1 shows the general design of a soil–vegetation–atmosphere–transport model (SVATS) and its linkage to a crop growth model (Feddes *et al.* 1978). The potential evapotranspiration rate, E_{pot}, per unit area is determined from a variant of the Penman–Monteith equation (section 2.3.1). The actual evapotranspiration is governed by the rate of supply of water from the soil, since the plant has negligible storage capacity. This restriction causes the water potential in the plant to fall, resulting in constriction of the stomata so that a balance of flow into and out of the plant is achieved (section 2.3.2). The water potential in the roots has to decrease sufficiently for the supply rate from the soil to equal the actual transpiration rate. When this is no longer possible, leaves — and ultimately the whole plant — dry out and die.

This balance is easy to state, but in practice the water potential differs both in different parts of the root system and in different parts of the soil profile (section 2.3.1). The soil is divided into layers for modelling purposes, and the uptake from the layers is summed. There are two general approaches (Feddes *et al.* 1978). One is to define the properties of a single root, and to assume that all roots are similar and are uniformly distributed within a set of soil layers. The summed water uptake over all roots in all layers is then set equal to the actual evapotranspiration at that time. This needs the definition of the root and soil properties (section 2.3), which is not easy. The second and simpler alternative is to consider water uptake as a sink term S per unit volume per day, and add this into the water continuity equation where F_w is the water flux and z the depth:

$$\partial\theta/\partial t = \partial F_w/\partial z - S \qquad (11.1)$$

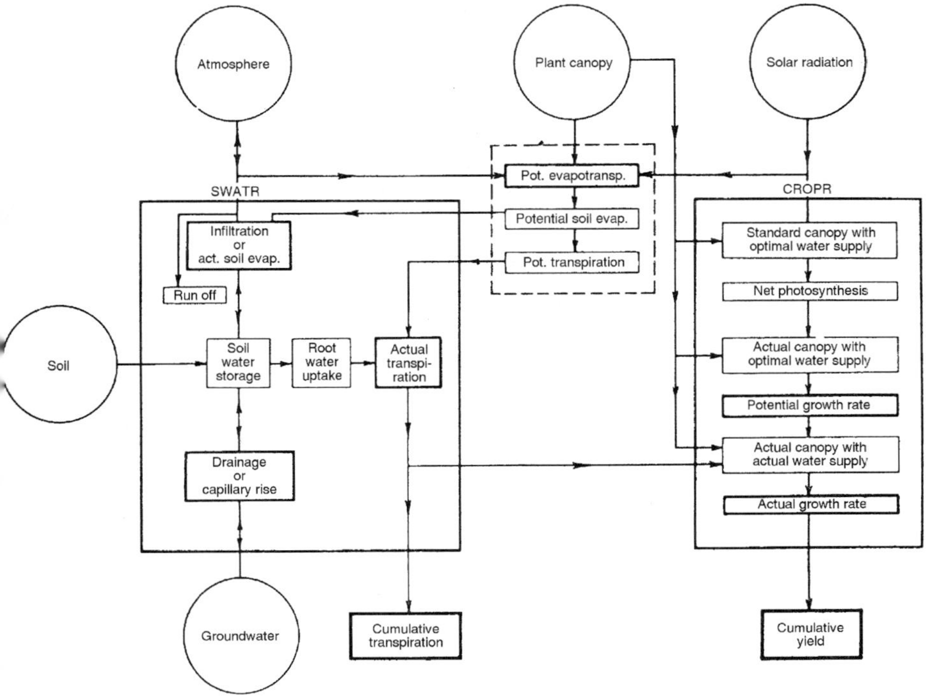

Figure 11.1 Outline scheme of a model for computing the influence of water use on crop yield (after Feddes *et al.* 1978).

The sum of this sink term has to equal the actual evapotranspiration $\Sigma S = E_{act}$, so S is determined by the potential evapotranspiration and the various impedances that constrain the water supply from the soil. Feddes *et al.* (1978) simply expressed S as a function of the soil-water pressure head (matric potential), ψ, (figure 11.2) in the bulk soil. It is assumed that S is zero in both waterlogged, anaerobic soil conditions and at the wilting point. It has a value S_{max} over a range of ψ, which is defined by the potential evapotranspiration rate, namely $S_{max} \times z_{eff} = E_{pot}$, where z_{eff} is the effective rooting depth. The value of S varies with ψ according to the results of measurements such as those in figure 11.3, and below ψ_2 the value of S, and hence of the transpiration rate, declines linearly with ψ. The value of ψ_2 is therefore determined by the root density, the rooting depth, the hydraulic conductivity of the root and the hydraulic conductivity of the soil, and thus includes most of the mechanistic detail of the uptake of water by roots. In the subsequent discussion of crop models, it will be seen that this macroscopic approach is used frequently, but that various additions are used to make the model more realistic (section 11.3.4).

The value of S is much larger in the topsoil than in the subsoil, at least initially, because of the much larger root density in the topsoil, and the smaller resistance within the root zone due to the shorter distance of flow and the larger root radii, together with loss of water from the soil surface. The effect is that an 'extraction

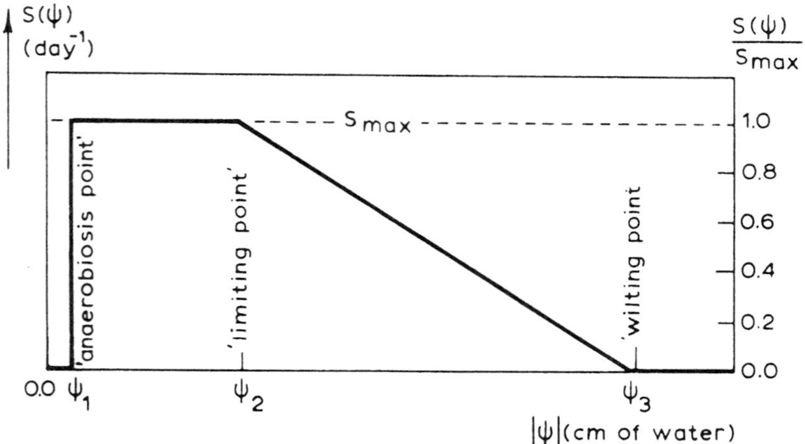

Figure 11.2 Relationship of the sink term S (equation (11.1)) related to the soil-moisture pressure head (matric potential) of the soil, with S expressed as a fraction of the potential transpiration rate (after Feddes *et al.* 1978).

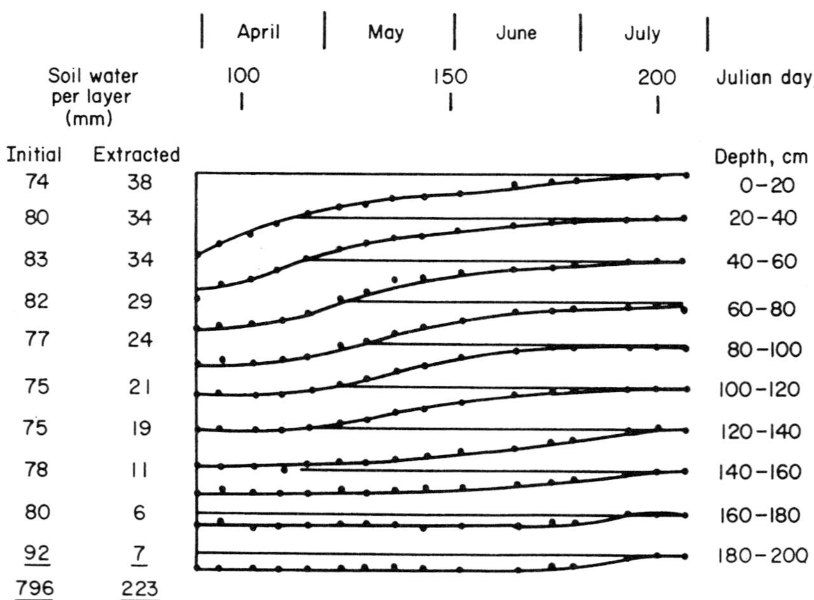

Figure 11.3 Amounts of water extracted from 20-cm layers of Hook series soil by strongly droughted winter wheat during spring and summer. Initial amount of total water, and amount extracted, given for each layer (after Weir & Barraclough 1986).

front' appears to move down through the soil (see figure 11.4). In agreement with this, Gardner (1991) has suggested that water uptake by a crop root system functions more like a distributed sink that moves down through the soil profile than a progressive drying out of the whole profile, and has produced a water uptake model on this basis (section 11.1.3).

11.1.3 Use of Water by Field Crops

It is important to be able to predict the rate at which water can be extracted from a rooting zone at a particular time, as in the general model briefly described above, but the total water relations of a crop during its growth period must

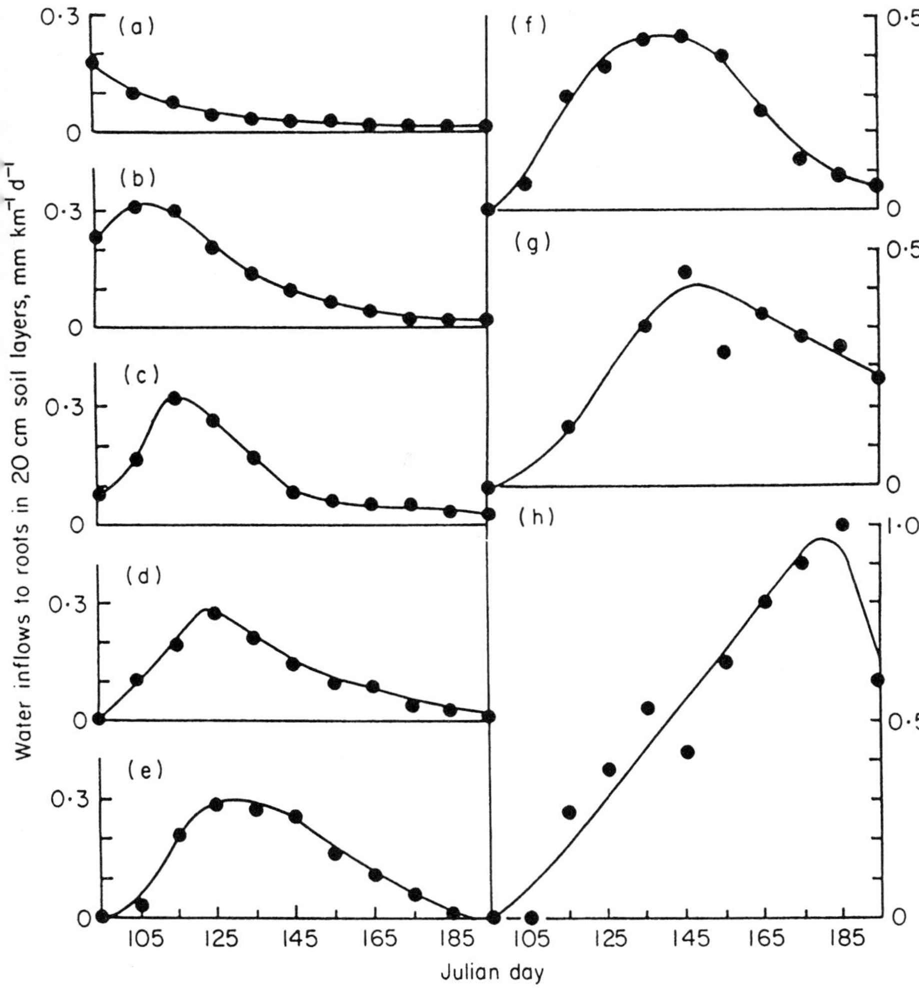

Figure 11.4 (a–h) Water inflows for each 20-cm soil layer between April and July for same experiment as in figure 11.3. Maximum values of water inflow are later and larger with increasing depth (after Weir & Barraclough 1986).

also be borne in mind. In particular if rainfall in the growing season is insufficient for the crop, it is dependent partly on stored water in the profile and partly on rainfall.

In chapter 9, it was implied that deep-rooting plants may extract more water than others, and therefore had an advantage over shallow-rooted crops. However, there are exceptions to this. If the rainfall is ample and regular, there may always be sufficient water in the upper layers, and the production and maintenance of deep roots may be wasteful. Conversely, if the rainfall over the whole year is too small, deep stored water that is used by one crop may not be replaced before the next cropping season, and a deep root system may supply little more water than a shallow one in most years (Ludlow & Muchow 1990; Gregory 1994a). The value of deep rooting for water extraction therefore is greatest in intermediate situations. Thus, in an agroforestry experiment in a semi-arid environment, McIntyre *et al.* (1996) found that recharge below 0.45 m occurred only rarely, so deep rooting was of only intermittent value. However, there is no doubt that many shrubs growing alone in arid situations have very deep rooting systems (Weaver 1926), presumably to utilize water at depth where rainfall and water infiltration patterns very occasionally allow deep recharge (see chapter 2).

The principal limitation on water supply rate to annual crops growing on stored water in the soil profile is the rate at which roots extend and proliferate in the moist horizons (Ong *et al.* 1996). Extension rates can be very high: for example, Squire *et al.* (1984) quote velocities of 'extraction fronts' of up to 70 mm day^{-1} in tropical climates, with the norm being from 10 to 40 mm day^{-1}. However, there are concerns (Passioura 1988) that if large amounts of rapidly growing root at depth supply the full evapotranspiration demand early in the season, from stored water, this may leave the crop without water during the subsequent ripening stage. It may therefore be better to have a smaller rooting density, or roots with a smaller hydraulic conductivity (Richards & Passioura 1989), so that the crop is maintained under moderate water stress for a longer time, but is able to grow through to harvest.

Designing crop root systems for the most advantageous rate of water extraction at all times is therefore difficult. Ludlow & Muchow (1990) considered that the most important issue in the efficient use of water by field crops was the matching of crop development to the timing of water supply, either as rainfall or stored water. This demands the right date of sowing and a cultivar with the right length of growing season. They ranked improved morphological rooting characteristics (Gregory 1994b) and better osmotic adjustment within the plant (Verma 1997) as equally important plant characteristics for the avoidance of water stress. The combined effects of soil structure on root development and on water relations is particularly complicated to predict (Hamblin 1985).

Weir & Barraclough (1986) studied the behaviour of a winter wheat crop growing on stored water in Britain. The crop was grown under a rainshelter; part was droughted from late March, the rest was irrigated. Root length developed as in table 11.1. Roots extended into the 160–180 cm layer, but the L_V there was never greater than 0.1 km m^{-3} under the droughted crop. The greatest L_a values were found at anthesis, up to 12.2 km m^{-2} in the top 20-cm layer, and 26.8 km m^{-2} in total to 180 cm depth. The latter value is comparable with the max-

Table 11.1 Root length (km m^{-2}) development of winter wheat grown under a rainshelter with and without irrigation.

Sampling date	3.3.82	4.6.82		9.8.82	2.8.82
Julian day	62	155		221	214
Treatment	—	Irrigated	Droughted	Irrigated	Droughted
0–20 cm depth	3.9	12.2	9.4	7.4	7.6
20–40	1.0	4.2	3.1	3.5	2.4
40–60	0.9	3.3	3.3	2.2	2.7
60–80	0.6	2.9	2.8	1.4	2.0
80–100	0.4	2.0	2.0	1.2	1.8
100–120	0.22	1.3	0.9	1.1	0.7
120–140	0.13	0.6	0.6	0.5	0.6
140–160	0.08	0.2	0.2	0.1	0.2
160–180		0.1	0.1	<0.1	<0.1
0–100	6.8 ± 0.68	24.6 ± 1.1	20.6 ± 0.62	15.7 ± 0.44	16.5 ± 0.91
0–180		26.8	22.4	17.4	18.0

Source: after Weir & Barraclough (1986).

imum value of 31 km m^{-2} found in the field by this group (see table 9.7), so this experimental crop was representative of large and productive crops.

The extraction front of water moved down the profile, so that the depth of maximum water inflow increased with time (figures 11.3 and 11.4). The droughted crop evapotranspired 223 mm in total, as against 340 mm for the irrigated one, so it suffered some moisture stress, and the full transpiration rate was only attained if the topsoil was moist. The mean water inflow at 140–160 cm reached a higher value than at any other depth, at 1 mm km^{-1} day^{-1} (or 0.01 cm^3 cm^{-1} day^{-1}). Mean water inflows in drying soil are usually rather less than 0.01 cm^3 cm^{-1} day^{-1} (Ong *et al.* 1996), so a larger root density at depth would probably have extracted more water than found by Weir & Barraclough (1986) (section 2.3.4).

Despite this suggestion that inadequate rooting in the subsoil caused or increased water stress in the droughted crop, which evapotranspired only two thirds of that of a fully watered crop, its dry matter and grain yield were, respectively, only 10% and 4% less than in the fully watered crop.

11.1.4 Global Change Effects

Several major changes are happening slowly, on a global scale, that will affect crops and natural vegetation in ways which are as yet ill-defined. The most important effect at present is the steady increase of carbon dioxide in the atmosphere. This generally leads to more rapid growth of plants, in the absence of other growth-limiting effects, but the extent of the effect varies greatly among species. It may cause changes in the root/shoot ratio, and the way in which carbon is allocated below ground (Tinker *et al.* 1996). In particular, there are complicated and variable interactions with both nitrogen supply and water (Walker & Steffen 1996). The production of a given amount of dry matter under elevated CO_2 normally requires a smaller amount of water transpiration than under lower levels. The important effects of increased carbon dioxide, and the interactions with water and nitrogen, have already been shown in field experiments with free-air circulation experiments (FACE) (Pinter *et al.* 1996).

The future increases in carbon dioxide levels can be predicted with fair accuracy. The greenhouse effect is expected to lead to changes in climate, including temperature, rainfall, and the frequency of extreme events. Together with the direct effects of elevated carbon dioxide levels on plant growth, these will considerably alter the way in which crops are grown, and crop models will be important in exploring the possible options (Tinker & Ingram 1996).

11.1.5 The Background to Crop Modelling

There has been an enormous increase in crop modelling recently. The general position of modelling in relation to other methods in plant science has been discussed by Nye (1992a, b). The scientific value of mechanistic as opposed to empirical (phenomenological) models is generally accepted, but whole-crop models necessarily still contain large sections that are empirical. Stochastic models that give a range of probabilities for outputs are still quite rare.

As in chapter 10, we distinguish here between crop nutrient uptake models, and crop growth and uptake models; the latter are referred to here as 'crop models'. However, the root and soil submodules of most crop growth models tend to be their weakest and least mechanistic parts. In modelling of complex crop systems, it is desirable that the degree of detail, process and organization in the model is reasonably uniform overall (Porter 1993). Much of the real-world variation in the growth or nutrient content of crops cannot yet be fully explained or predicted by these models, and this may in part be a result of oversimplification of the root and soil processes. Similarly, models of vegetation in ecosystems and biogeochemical cycling have been built that virtually ignore the root system, and these may need to be expanded in due course.

Several major schools of modelling have developed in the 1980s and 1990s, usually centred on a particular core model or module that expresses a particular view of how a crop functions (Goudriaan & Van Laar 1994). Bouman *et al.* (1996) have fully described the models that developed from the work of De Wit, such as the central SUCROS model, and the subsequent operational use of these. The CERES family of models is probably the other main school of this type, but many other models are now being developed by different groups.

11.2 Transfer of Solutes in a Profile

In any soil profile, there is a constant movement of solutes caused by addition of solutes and water at its surface; gains or losses by seepage at its side boundaries and losses by drainage at its base; the generation of solutes by decomposition of minerals or mineralization of organic matter; and uptake of solutes and water by plant roots. Here, we consider transport in a vertical direction only, though on any hill slope there will be horizontal movement, which must also be included if movement is on the scale of a field or water catchment area. These large-scale movements are outside the topics we are covering in this book. Kutilek & Nielsen

(1994, chapter 8) give a succinct account, with further references, of the problems of spatial heterogeneity encountered in studying water movement on the field scale.

The general principles of solute transport are discussed in chapter 4. They have been tested in the laboratory on columns of homogeneous soil. In the field, however, it is the exception rather than the rule that a given system can be accurately explained or predicted, because of unquantifiable variations with time and space in the controlling parameters and the complexity of the individual processes at work.

Nye & Tinker (1977) and many others showed that the movements of solutes in the profile could be described by the continuity equation (equation (1.6)) with chosen boundary conditions. Analytical solutions, showing the changes in solute concentration with depth and time, were available only for very simple situations in which the parameters in the equation, for example the water flux or the water content, were constant. With modern computers it is now easy to solve the continuity equation for a much wider range of conditions, using numerical methods. Most of the very numerous recent models describe a number of linked individual processes, such as the release of available nitrogen by mineralization of organic matter and its subsequent uptake by roots or its loss by leaching. Each process forms a submodel, and these are combined into the overall model. Since the main advances in this section since the mid-1970s lie in these developments, this book first examines the types of models that have been published, giving examples of their use and an account of their strengths and limitations in the solution of real problems.

Addiscott & Wagenet (1985) and Wagenet (1990) have classified the various leaching models as shown in table 11.2. We discuss examples of each in turn.

The primary division is between deterministic and stochastic ('determined by a random distribution of probabilities' — *Oxford English Dictionary*) models. Deterministic models presume that a given set of events result in a determined outcome. Stochastic models assume that the outcome is uncertain though the probabilities of different outcomes can be predicted. Stochasticity can arise either

Table 11.2 Classification of water and solute transport models for soil.

1. Deterministic models
   - A. Mechanistic (usually based on rate parameters)
     1. Analytical
     2. Numerical
   - B. Functional (usually based on capacity parameters)
     1. Partially analytical
     2. Layer and other simple approaches
2. Stochastic models
   - A. Mechanistic
   - B. Functional [Non-mechanistic] (transfer functions)

Source: after Addiscott & Wagenet (1985).

because the details of the processes or the values of the parameters are uncertain. Each class can be either mechanistic, or what is termed 'functional' ('serving a function: utilitarian' — *Oxford English Dictionary*). Mechanistic models incorporate the individual mechanisms of each process, as in the idealized and simplified model described by equation (11.2). They usually include rate equations for individual processes, such as equilibration of a solute between soil solid and the adjacent pore solution or uptake at a root surface. Functional models are based on simplified treatments of water and solute movements. They need fewer parameters and are much more useful for real problems. They usually work with the amounts of water and solute in a reasonable number of soil layers using rules (often empirical) for transfer from layer to layer.

11.2.1 Deterministic Mechanistic Models

The mass balance expression for a solute in unit volume of soil, which is the basis for all quantitative treatments, is, in one dimension (for derivation, see section 1.4, equation (1.6)):

$$\partial C/\partial t = \partial/\partial z(D^* \partial C/\partial z) - \partial/\partial z(vC_L) + S(z, t) \tag{11.2}$$

where, on the right-hand side, the first term covers diffusive movement and the second, convection. The third term, $S(z, t)$, which can vary with depth and time, is a source or sink term, which is either positive (for example for addition of a solute by mineralization), or negative (for example for loss from the soil by root uptake). The longitudinal dispersion coefficient, D^*, includes dispersion due to thermal diffusion and also hydrodynamic dispersion caused by non-uniform velocity of flow within and between pores and other effects (section 4.3). This equation is the basis for analytical and numerical solutions. The vertical flow velocity, v, must be found by solving a water flow model (section 2.2.1), if it is not deliberately kept constant.

Theories about solute leaching have mainly been tested on very simple systems; an example is the leaching of a thin band of solute, added to the surface of a homogeneous soil, by a constant flux of water. The governing equations for this idealized system can often be solved analytically. The analogy to movement in a chromatographic column at once suggests itself for this, and theory developed by Gluekauf (1955) and Helfferich (1962) has been freely used (Frissel *et al.* 1970b). Solutes that are not adsorbed by the soil are simpler to treat than adsorbed ones.

11.2.1.1 Non-adsorbed Solutes

We consider vertical movement of a thin band of an applied non-adsorbed solute as composed of two processes: a bulk transfer of solute in the moving water at a mean speed of v/θ, where v is water flux in ml cm^{-2} s^{-1}, and a spreading out of the band due to diffusion and hydrodynamic dispersion (figure 11.5). The situation has been dealt with simply by Gardner (1965), who expresses the change in concentration with depth and time by the equation

$$C_L = C_{L0} z_0 (4\pi D_L^* t)^{-1/2} \exp -[(z - vt/\theta)^2/(4D_L^* t)] \tag{11.3}$$

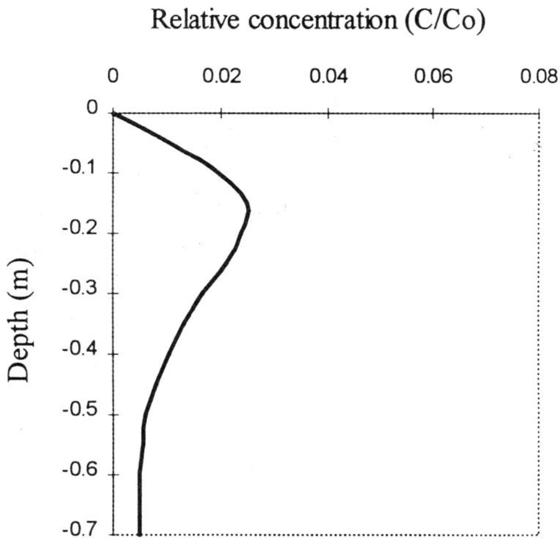

Figure 11.5 Solute concentration profile some time after the start of leaching in a macroporous soil (after Leeds-Harrison 1995).

where C_{L0} is the initial concentration of solute in the uniform band of depth z_0, and the other symbols have their usual meaning. Instead of applying a thin band of solute to the surface, a steady flux of water through a soil column may be abruptly replaced by a similar flux of solution and the solute concentration emerging from the base of the column can be measured. The resulting plot of concentration against volume of solution passing is known as a 'breakthrough curve' (figure 11.6).

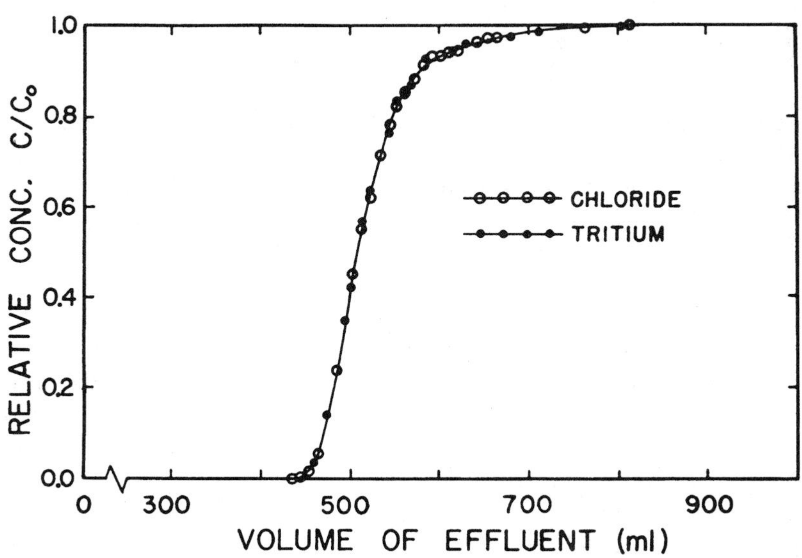

Figure 11.6 Breakthrough curve for tritium and chloride from saturated 200-μ glass beads. The average flow velocity and volumetric capacity of the sample were 2.11 cm h^{-1} and 512 ml, respectively (after Biggar & Nielsen 1962).

Such constant flux experiments have been made with saturated soils. In practice, drainage is commonly through unsaturated soils in which θ and v vary down the profile (section 2.2.2). If the variation of v and D^* with time and space is known, equation (11.2) may be solved by numerical methods, usually finite difference methods (Smith 1978). These will also allow for a known variation in S. Examples are given in Nimah & Hanks (1973a,b) and Nielsen *et al.* (1986).

In the field, it is most unusual for v or θ to remain constant for long periods, and D^* depends upon both v and θ. Truly mechanistic models include prediction of water movement and thus require a knowledge of the hydraulic conductivity, which is very sensitive to the water content (section 2.2.1). There is often a substantial term S in the equation due to removal of salts by roots, demineralization of nitrate or precipitation of sulphate, or production of nitrate or sulphate by mineralization. The removal of water locally by roots causes serious problems since it varies with time, and it is difficult to determine where it occurs. Under natural vegetation, the dominant anion is often bicarbonate, whose concentration alters considerably with changes in temperature and carbon dioxide concentration in the soil air. It is therefore not surprising that most treatments of these problems aim only at a rough agreement with field practice. It also accounts for the great use that has been made of numerical and computer simulation methods in this subject, since they allow great flexibility in solving equation (11.2) with specified boundary conditions.

Attention has been concentrated on models of leaching of non-adsorbed anions such as nitrate and chloride, since, with steady-state leaching, they present no special difficulties unless anion exclusion (negative adsorption) (section 3.5) is serious. The latter causes 'salt sieving', in which the soil — usually clay, with only very fine pores — acts as a leaky semipermeable membrane, allowing water to pass more easily than salts. Differences in salt concentration (osmotic potential) may then cause bulk movement of water (Letey *et al.* 1969; Bolt & Groenevelt 1972) (section 2.2.5). If the exclusion of anions is only partial, they will be concentrated in the large channels where water flow is rapid, and will move faster than the average water flow by a factor, which was determined experimentally to be 11% in an undisturbed sandy soil and 17% in a clay soil (Frissel *et al.* 1973).

11.2.1.2 Adsorbed Solutes

Adsorbed solutes introduce many further complexities. If the adsorbed solute has a linear isotherm, its mobility is reduced by a factor equal to the buffer power, so that the mean flux of the band is v/b. However, sulphate, phosphate, and organic compounds typically have curved isotherms. If D rises with C, the mobility of the solute is greatest at the centre of the band, where C is greatest, and a skew distribution must result, with a sharp leading boundary and a diffuse trailing one.

As discussed more fully in chapter 4, many solutes, particularly organics, equilibrate only slowly with the soil matrix. The reaction term S in equation (11.2) is therefore a function of t, z, and C and must include a rate equation. Hutson & Wagenet (1995) give several examples, including one by Van Genuchten (1981), who divided the sorption sites into two types: equilibrium, on which equilibration with the adjacent soil solution was rapid, and kinetic, which equilibrated with the solution slowly.

11.2.1.3 Exchangeable Cations

The movement of exchangeable cations is less simple, because the concentration of cations in the solution depends upon the concentration of free anions, as the cations do not move independently. Since electrical neutrality must be maintained on the exchanger, a cation moving down the profile may displace other species of cations. If the ion in question is a very minor fraction of all exchangeable ions, and if the concentration of competing ions in the percolating solution remains constant, the situation reverts to that of simple adsorption, since $C_L \ll C$. If the solution concentration varies, the situation is more complex. Thus, if for example potassium chloride is applied to the soil surface and worked into the top layer, the chloride will be washed down the profile as described for non-adsorbed species, but the nature of the cations associated with the chloride will change progressively towards that of the major soil cation present, and the added cation will move more slowly, and at a rate that will depend upon the subsequent anion concentration. If solid salt is supplied to the soil, the initial process of movement is very much more complex than simply leaching and diffusion as described earlier (section 2.2.5).

11.2.1.4 Transfer of Salt within the Profile

The leaching of sodium salt applications has received particular attention, because of the possible damage to structure caused by this ion. Bower *et al.* (1957) compared the rate of leaching in the laboratory of a band of surface-applied NaCl by a $CaCl_2$ solution, with a comparable band of $CaCl_2$ by NaCl solution. When the more strongly held Ca^{2+} leaches a band of Na^+, the Na^+ passes down the soil in a compact band with a sharp upper boundary. But when Na^+ leaches a band of Ca^{2+}, then the upper boundary of the Ca^{2+} band is diffuse and the band develops an elongated 'tail'. Theories of cation chromatography (Helfferich 1962, p. 421) usually deal with a constant concentration of total salts and consequently are not applicable directly to a salt applied to a soil surface. This situation is best dealt with by numerical solution of equation (11.2) (Frissel *et al.* 1970a, b; De Wit & Van Keulen 1972). For rough prediction, Tinker (1968) found it sufficient to assume the sodium to be moved down the profile at the same rate as chloride, divided by an estimated sodium buffer power for that concentration of chloride.

In the previous paragraph, the uniform movement of a single band or pulse of salt was considered, but such a simple situation is rare. In practice, salts are usually distributed non-uniformly within the whole profile, and water additions, drainage and evaporation can cause complex sequences of movement.

In reclamation of saline soils, the whole profile initially contains Na salts, and these have to be removed by leaching. Pure water will move the non-adsorbed ions down at a rate equal to v/θ, but with the upper boundary made unsharp by diffusion and dispersion. This situation has been discussed in some detail by Bresler (1972). Sodium held exchangeably on the soil must be displaced by another cation, and the course of its removal depends on what is supplied. If it is only pure water, structural breakdown may prevent further leaching, and for this reason gypsum is often added, so that leaching is with a saturated solution of calcium sulphate, which simultaneously prevents dispersion and provides a displacing cation.

In practice, nearly all theoretical treatments deal with the chloride ion rather than sodium chloride. The most thorough studies deal with upward and downward movement of salt by convection and diffusion (Bresler & Hanks 1969; Bresler 1972). The general equation is basically the same as equation (11.2) and numerical solutions for this, with the usual equations for water movement, gave results in fair agreement with laboratory experiment. Omitting the diffusion–dispersion term — that is, assuming that convective flow was dominant at the rather large velocity of $\sim 3 \times 10^{-3}$ cm s^{-1} — simplified the problem greatly, but gave very similar results.

A variant of the above problem of leaching of soil salts is that of a profile receiving saline water: a long-term steady-state distribution will eventually be set up, with the soluble salts eventually emerging at the bottom of the profile in a more concentrated solution, depending on losses of water by transpiration. Over a long period, this merges with the problem of the salt balance over a catchment (Kovda *et al.* 1973).

The position at which crop roots remove salts and water must be defined if a precise model for salt and water transfer is required. Accurate information on this is difficult to obtain in the field, since it will be a complex function of root distribution and water content. Bresler (1973) therefore put forward a simple mass-balance type of equation, in which diffusion and upward movement are neglected, that is similar in principle to the 'layer' models discussed in the next section. The equation used compares the increased salt in all layers from 0 to z with the difference between the input of salt in irrigation water and the salt flowing out at z, allowing for the loss of water by transpiration and the change in concentration with time at z. The time step is conveniently taken as that between successive irrigations.

11.2.1.5 Ammonium and Nitrate

Leached nutrients are always a waste, and may cause environmental problems — nitrate being a particular risk. Its movement will follow that of water, and may be considered to be composed of four processes: downwards movement, upwards movement during dry spells, uptake by roots, and gain or loss by nitrification or denitrification. However, water and ions are normally absorbed by roots at different rates at different depths in the profile, and nitrate is produced by nitrification of ammonium. It is difficult to describe this situation simply, unless it is near a steady state, or if one process dominates. During prolonged dry weather, upward movement may dominate in this way, and very high levels of salts have been found in the soil surface during the dry season in Africa (Wetselaar 1962). Gardner (1965) has given a simple steady-state treatment of this system, which predicted a logarithmic distribution of salt with depth that fitted the experimental data remarkably well. Complicating processes could be expected — for example, soil water movement may be altered by the intense osmotic gradient, and a very marked daily temperature cycle will occur — but these perturbations may be confined to a shallow layer near the surface, and therefore might not be detected in field studies.

The respective merits of ammonium and nitrate nitrogen in regard to leaching have been argued repeatedly (Bartholomew & Clark 1965). Because of its adsorption, it may be expected that ammonium ions will move much more slowly than nitrate, perhaps by a factor of 10, but in normal temperate agriculture it is unusual for nitrate to be applied so early, or rainfall to be so intense, that serious leaching occurs during the summer. Much of the leaching loss of applied nitrogen occurs in the following winter, by which time most ammonium will have been converted to nitrate. In tropical climates, the risk of nitrate leaching seems greater, though the literature is by no means consistent on this point. In practice, the microbiological oxidation of ammonium to nitrite, and then to nitrate, occurs simultaneously with leaching and transfer. McLaren (1970) has considered this situation in detail: starting from urea, his theoretical model predicts a steadily decreasing concentration of nitrate with depth, whereas the intermediates — ammonium and nitrite — have maxima at different points in the profile. Starr *et al.* (1974) included denitrification in their treatment of soil nitrogen transformations during leaching, and obtained fair agreement with laboratory experimental results.

11.2.2 Deterministic Functional Models

The simplest form of this type of model occurs in so-called piston flow, where one solution simply displaces another without mixing. In this case, $z = Q/\theta$, where z is the depth of penetration and Q is the amount of displacing solution applied. Alternatively, one can assume that the solution moves by piston flow, and at the same time disperses at the boundary between solutions as in a stationary medium. The most useful models of this form divide the profile into a number of horizontal layers, initially of known water and solute content. When rain or irrigation water is added to the surface layer it dilutes the solute concentration there. When the water content exceeds the field capacity of the top layer, the surplus, carrying its share of solute, is transferred to the next layer, for which the solute concentration is recalculated. After sufficient solution has transferred to fill its field capacity, solution is transferred to the third layer, and so on down the profile. Such a model can accommodate variable field capacities in the layers, variable rainfall, evaporation, transpiration, microbially mediated gains and losses (especially of inorganic nitrogen), and nutrient uptake by roots.

Such models work by calculating a mass balance for water and solute in each layer, together with a criterion for transferring water from one layer to the next. They are thus very flexible and call for only a limited number of readily measured parameters. In particular, they do not require the highly variable hydraulic conductivity to be estimated. Burns (1974) gave an example of this type of model, and Barnes *et al.* (1976) used one in their model for predicting fertilizer needs of vegetable crops (section 11.3.3). It allows for uptake from each layer by the crop according to its root growth and distribution (see section 11.3).

11.2.2.1 Bypass, Preferential, or Two-Phase Flow

In cracking and coarse-textured soils, it is often found that some solute is leached into the water table much earlier than would be expected if the percolating solution equilibrated in each layer in turn. The problem is caused by water flowing much more readily through wide spaces between structural peds, root channels, worm holes, and the like, than through the bulk of the soil (Cunningham & Cooke 1957; McMahon & Thomas 1974). This is variously known as bypass, preferential, and two-phase flow.

The retention of nitrate and other solutes within large peds or aggregates in coarsely structured soils during leaching of the profile can be regarded as an extreme case of the hydrodynamic dispersion that occurs in nearly all soils because of different flow speeds of water in their variously sized pores. However, breakthrough curves in such soils when they are leached with a steady flow of solution are very asymmetrical, with sharp leading edges and long trailing ones, whereas hydrodynamic dispersion alone should result in symmetrical breakthrough curves. In contrast, a typical bypass flow breakthrough curve is shown in figure 11.5. Leeds-Harrison (1995) (and other papers in the same publication), cited instances of the importance of bypass flow in accelerating the transport of surface-applied pesticides and other solutes to groundwater. Leeds-Harrison (1995) showed that the pattern of leaching is determined not only by the rainfall pattern and the initial distribution and release of the solute, but also by the size of the aggregates and the rate of diffusion of solute into and out of them.

Most of the numerous models that simulate bypass flow divide the water flow path into two phases: mobile and relatively immobile. This is clearly a convenient approximation, since, in reality, there are a multitude of flow velocities in the profile; and it is not surprising that none of the models satisfactorily reproduce field behaviour over a wide range of conditions. Addiscott (1977, 1992) and Addiscott & Whitmore (1991) adapted a simple layer model to deal with the two phases. Water held at a matric potential of less than −0.2 MPa is considered immobile, that exceeding −0.2 MPa is mobile. During each time interval, only the mobile water moves between layers, carrying solute with it. Then, in each layer, solute equilibrates between mobile and immobile phases, or, as a refinement, partially equilibrates slowly by diffusion or other slow reactions. Since there is no true sharp division between mobile and immobile water, the division between the two phases can, in practice, be adjusted empirically to improve the agreement between model output and field observations.

Biggar & Nielsen (1962) measured the concentration profiles in several small plots in one field when a pulse of chloride or nitrate was leached by water draining at a steady rate. Using a deterministic mechanistic model, they found that the values of D^* and v needed to fit the experimental data each differed between the plots by up to 2 orders of magnitude. Thus, even under the simplest conditions, an analytical solution using one set of parameters is unlikely to describe behaviour over a whole field, and conditions are rarely that simple.

11.2.3 Stochastic Mechanistic Models

Stochasticity can arise because the input parameters, for example hydraulic conductivity, vary unpredictably from point to point or because the details of some of the processes, such as the extent of equilibration between solution and solid over a time step, are uncertain. Stochastic mechanistic models attempt to deal with variability in the parameters. They describe each of the many input parameters needed to run a model by a mean and some statistical distribution. To find the statistical distribution, a number of runs are made with different combinations of the parameters to find the variation they cause in the output. Amoozegar-Fard *et al.* (1982), using Biggar & Nielsen's (1976) skewed distribution for v and D^* in equation (11.2), showed that more than half of a pulse of solute was predicted to have moved beyond 180 cm depth, whereas very little moved beyond this depth if v and D^* were assumed to be constant. Variation in convection, v, was much more influential than variation in dispersion, D^*.

11.2.4 Stochastic Non-mechanistic Models (Transfer Function Models)

Transfer function models are used by engineers and hydrologists to characterize systems governed by processes too complex to be treated mechanistically. In the field, soil drainage behaviour varies greatly from point to point over a field, so the time of arrival at the groundwater of a pollutant added to the surface may vary correspondingly. As a precaution, practice should be based on a worst-case scenario, so it is important to establish what this is.

These models abandon all attempt to treat the problem mechanistically. Instead they describe the solute leaching depth by a distribution function

$$P_z(I) = \int_0^I f_z(I)dI \qquad (11.4)$$

where $P_z(I)$ is the probability that a molecule in a thin band of solute, initially at the surface, will arrive at depth z after a net amount of water I has been applied to the surface. The term $f_z(I)$ is known as a probability density function. It is found by measuring the concentration at a particular depth, L, when a thin band of solute and a given amount of water have been applied to each of a wide selection of profiles in a field (Jury 1982). The probability density function is usually found to have a log-normal distribution. If the concentration of solute is known at a particular depth, L, after a given amount of water, I, has passed, then the expected concentration at any depth can be deduced from equation (11.4) (Jury 1982). This assumes that the soil characteristics that control water flow at depths greater than L do not alter. Jury (1983) has reviewed this approach. He has identified its main strength as a practical tool:

> Although in principle each depth, z, requires its own transfer function model and its own $f_z(I)$, in many cases of interest the distribution of factors giving rise to variable breakthrough volumes . . . may be reasonably unchanged over a large depth interval. In this case the frequency functions $f_z(I)$ for different z may be highly enough

> correlated with each other to be represented in terms of the reference distribution $f_L(I)$ . . .

and also its main weakness:

> The Transfer Function Model . . . has a domain of validity verified experimentally near the surface and projections outside this domain of validity to greater depth are only speculative.

11.2.5 Movement of Mineral Solids, and Solutes Sorbed on Solid Particles

Nearly insoluble solids, such as many phosphates or minerals that contain heavy metals, and organic pesticides or radioisotopes adsorbed on soil colloids, are potential pollutants because they can be carried into water courses in particulate form.

11.2.5.1 Mechanical Movement

The mechanical processes that may incorporate adsorbed solutes into the soil have been mentioned in chapter 4. They are mainly responsible for redistributing very strongly adsorbed solutes, such as trace quantities of cadmium and mercury. Unfortunately, they are difficult to quantify in any simple manner. Poelstra *et al.* (1974) have tested the use of a simple 'mixing function', which assumes that a given fraction of the mercury in the surface layer will be redistributed annually over all layers below. The actual distribution of mercury to a depth of 80 cm in 15 west European pastures was erratic, and could only roughly be reproduced by a simulation using the 'mixing function'.

Run-off and erosion are further very important methods of solute transfer, though they are difficult to predict. The movement of a strongly adsorbed solute like phosphate will, in practice, be a combination of all these processes, whose outcome has been reviewed by Wadleigh (1968).

Management of the soil also affects the destiny of pollutants and pesticides. Edwards (1974), for example, states that insecticides on the soil surface usually disappear 5–10 times faster than those cultivated into the soil. Exposure to wind is important, and several workers have shown that persistent volatile insecticides disappear more slowly when shaded by crops than in open soil. Many more examples of the importance of mechanical movement on the action of pesticides are given in the two encyclopaedic volumes on organic chemicals in the soil environment, edited by Goring & Hamaker (1972).

11.2.5.2 Colloid and Suspension Transport

In addition to their movement as solutes, nutrients and pesticides may be moved while adsorbed on the surfaces of very fine soil particles. These may be brought into suspension by the impact of rain drops or by cultivation, and subsequently washed through the soil in this form.

Pesticides like paraquat, and long-lived radioisotopes like ^{137}Cs, a product of the Chernobyl explosion, which are strongly adsorbed by the soil solid, may be little leached in solution form, but they may nevertheless be carried into the drainage system adsorbed on dispersed soil colloids. Also, since the organic matter content of suspended soil tends to be higher than that of the topsoil, and since many pesticides are preferentially adsorbed by the organic matter, colloidal transport of pesticides can be significant, as shown by Vinten *et al.* (1983) and Worral *et al.* (1995).

Vinten & Nye (1985) have discussed the mechanism of transport of colloidal suspensions through soils. Particles are removed from suspension by interception if they pass within one particle radius of a wall; by straining due to an accumulation of particles at narrow necks in the flow path; and by sedimentation if the particles are dense enough to be sufficiently affected by gravity (see figure 11.7). If a steady flow of suspended particles enters a soil, Vinten and Nye found, in agreement with Iwasaki's (1937) experiments on sand filtration, that the initial rate of deposition was proportional to the concentration of particles. Thus, $\mathrm{d}n/\mathrm{d}z = -\lambda n$, where n is the suspended concentration, z is the depth of a soil column and λ is a filter coefficient. As deposit accumulated, the flow rate decreased, but λ was little changed. This suggested that straining and interception were important, because (a) the consequent contraction of the pores would be expected to reduce flow rate, as observed, and (b) the fate of a particle should depend on the fluid pathway along which it is travelling rather than the fluid velocity.

In addition to solute leaching, we now mention some other well-tested practical models of solute movement that illustrate the general mechanistic principles in section 11.2.2.

11.2.6 Dissolution and Dispersion of Placed Fertilizers

In principle, it should be possible to predict how a placed band or granule of fertilizer will disperse itself in soil of defined properties, since all the contributing processes have been discussed earlier. We assume that the fertilizers have a fixed solubility, which will be correct for pure salts and urea, but not for materials such as superphosphate where there is non-stoichiometric dissolution, or where there are strong common ion effects. For such a constant concentration source, in a defined line, cylinder or sphere shape, there are well-known solutions of the diffusion equation (chapter 1) (Mokady & Zaslavsky 1967). The factors that complicate the situation are as follows:

(a) Very large changes in total salt concentration occur, leading to large differences in $\mathrm{d}C_L/\mathrm{d}C$, and hence D, for cations. The very first stage of diffusion from a fertilizer granule can probably be regarded as salt or co-diffusion without adsorption, but exchange of cations will rapidly become important.

(b) Very sharp differences in pH may arise, particularly with soluble phosphate fertilizers, or any that contain or produce ammonia. The pH of a saturated solution in contact with superphosphate is 1.0 (Huffman 1962).

(c) The osmotic potential of a saturated solution of a compound fertilizer is extremely large. As an illustration, liquid fertilizer mixtures may contain 30% by

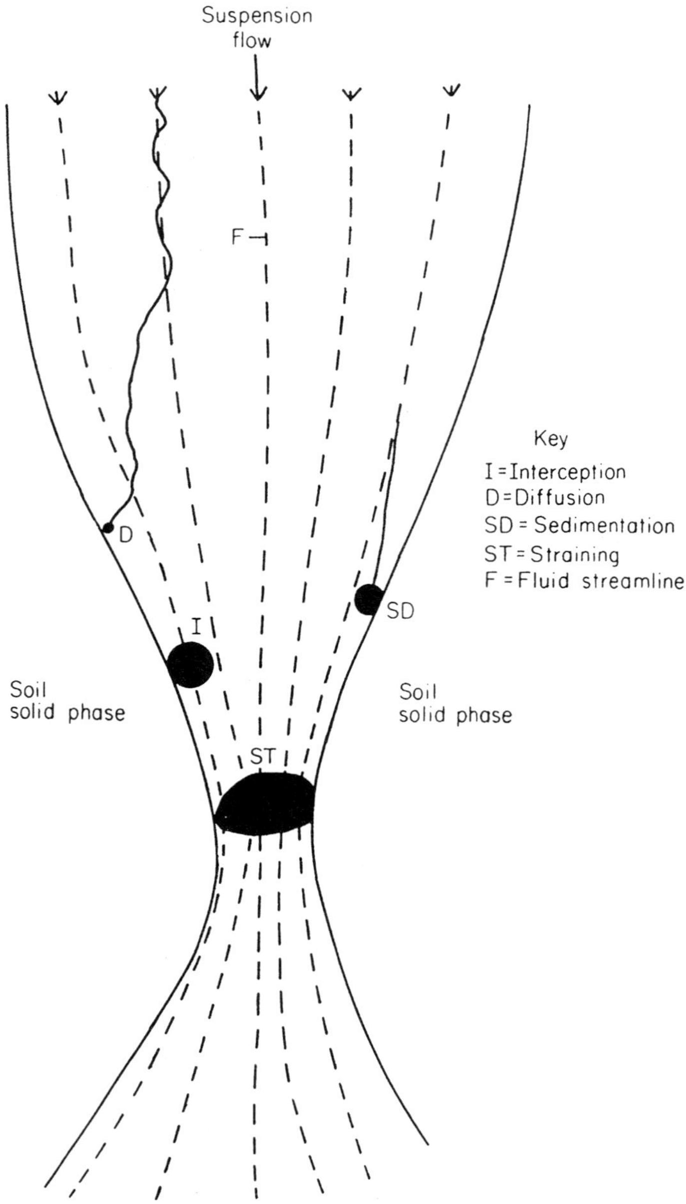

Figure 11.7 Mechanisms of particle capture from flowing suspension by a porous medium (after Vinten & Nye 1985).

weight of nitrogen, which corresponds to about 20 mol l^{-1}. This corresponds in theory to some −40 MPa water potential. The movement of water with osmotic potential as the driving force has been discussed in chapter 2 (section 2.2.5), and with such enormous gradients it could well be important for both vapour- and liquid-phase transport. The large changes in ionic activity coefficients at such high concentrations will also hinder exact calculations.

(d) Some components of a fertilizer mixture may react with and precipitate ions already in the soil — phosphate being the most important (Huffman 1962).

Theoretical predictions may be verified using simple examples. Kirk & Nye (1985, 1986) successfully modelled the rate of dissolution of sparingly soluble calcium phosphate particles in soil from first principles; and Nye & Ameloko (1987) have done the same for calcium carbonate. Rachhpal-Singh & Nye (1986) have also successfully predicted the rate of hydrolysis of a surface band of urea to the ammonium ion, the diffusion of this ammonium down the profile, and the proportion of it volatilized from the surface as ammonia gas. The effect on the losses of nitrogen by volatilization when urea was placed at depth (Rachhpal-Singh & Nye 1988), and the consequences of downward movement by leaching or upward movement by evaporation of water, have also been modelled (Kirk & Nye 1996).

Burns & Dean (1964) showed how nitrate moved out from a thin band of sodium nitrate in glass beads or soil (figure 11.8). The movement downwards was always greater than in other directions, and this was ascribed to the higher density of the solution produced from the fertilizer compound than the soil solution, after movement of water to the fertilizer under the osmotic gradient.

It appears that water may move to the salt in the vapour phase (section 2.2.4), and then out again as a liquid solution (Parlange 1973). A similar effect was found by Kemper *et al.* (1975), who investigated the placement of fertilizer in ridges, with irrigation water supplied in the intervening furrows. A distance of 5 cm in a sandy soil, and 10 cm in a clay soil, between the water level and the height of the fertilizer band prevented serious leaching, but less than this allowed water to move to the fertilizer to form a concentrated and dense solution, which then moved down the ridge and became dispersed in the irrigation water. Several groups (Lehr *et al.* 1959; Lindsay & Stephenson 1959; Gunary 1963) have investigated the movement of phosphate from superphosphate pellets; and the initial rate was always much more rapid than expected from the normal diffusion coefficient

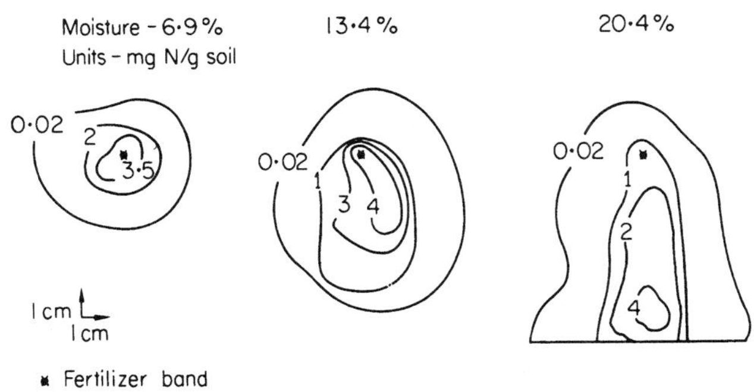

Figure 11.8 Movement of nitrate ions out from placed bands of fertilizer in a sand. Increased soil moisture content leads to more rapid downward movement of the dense solution formed by the fertilizer (after Burns & Dean 1964).

of phosphate in soil, presumably due to convective flow of water towards and solution away from the pellets, and also the fact that the phosphate adsorption sites, which normally lead to a low diffusion coefficient, were saturated.

11.2.7 Volatilization

Leistra *et al.* (1974) have given a model for the disappearance of the volatile herbicide propyzamide that allows for diffusion out of the soil in the gas phase, for varying rates of adsorption and desorption from soil solid, and for a first-order decomposition rate. The last was determined from the experiment, not independently. The form of the computed concentration profile agreed well with experiment. It is a virtue of such models that the importance of individual processes is easily tested. Figure 11.9 compares experimental data with that expected if assuming different loss processes.

Further interesting examples of modelling volatiles arise when dealing with liquid mixtures. Volatile mixtures of petroleum products are often washed or leaked to groundwater, where they float. They then gradually evaporate through the soil according to both their vapour pressures and the extent to which they are adsorbed by the soil, which acts like a fractionating column. Nye *et al.* (1994a, b) have tested a mechanistic model, which illustrates the way the individual processes involved may be combined.

11.3 Modelling of Monoculture Crops

11.3.1 Monoculture Crop Uptake Models

The modelling of uptake by uniform monoculture crops can be done in ways very similar to those explained for single plants in chapter 10, by ignoring interplant competition, and by treating the unit cells as single plants, as described in section 11.1.1. However, even if competition is not explicitly included in the treatment,

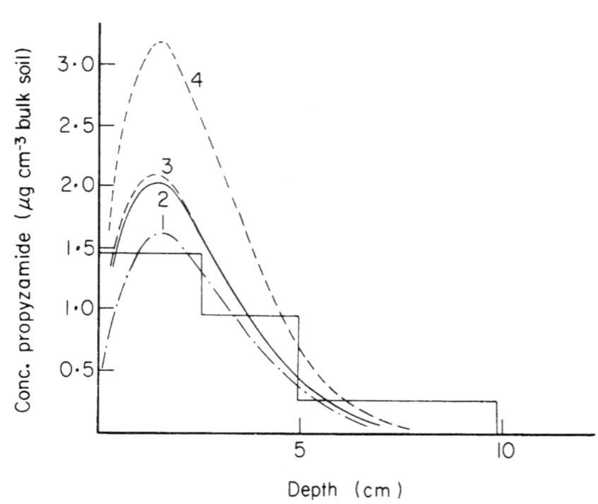

Figure 11.9 Comparison of measured and simulated distribution of propyzamide after 40 days using various assumptions about its behaviour. (1) With decomposition and free gaseous diffusion; (2) with decomposition, no gaseous diffusion from surface; (3) with decomposition, no gaseous diffusion at all; (4) without decomposition, full gaseous diffusion. Full line is the measured distribution (after Leistra *et al.* 1974).

intraspecies competition is, in fact, occurring. If the crop density is changed significantly, as by altering the sowing rate, the degree and type of interplant competition must change, and the crop morphology and yield will be altered (see sections 9.3.2 and 11.4.1). However, in field crops, the dry matter production and yields per unit area are not strongly dependent upon plant density near the optimum density for yield. In effect, the crop modeller accepts the type of plant produced by the established density, and does not attempt to model this aspect, as he would have to do if working upon a single plant basis.

It is usually assumed that the root systems of individual plants interpenetrate freely, and that roots leaving a unit cell equal those entering it, even though real crops are often so variable that this is not true. In some species there is root exclusion, so that the root systems of individual plants tend not to interpenetrate (section 9.4.3), but this does not affect the modelling of monocultures.

The simplest nutrient uptake model in the field is an extended 'nutrient flow analysis' (section 10.3.1) to test whether the root functions are likely to limit growth and yield. This approach has been used for wheat by Gregory *et al.* (1979) and Barraclough (1984, 1986a, b); for maize by Mengel & Barber (1974); for oilseed rape by Barraclough (1989b) and for barley by Soon (1988). The technique used by Van Noordwijk *et al.* (1990) is similar. The results of this analysis have normally been used to test whether the existing soil solution concentration was sufficient to provide the measured $\bar{I}$ by diffusion, but the calculations could be reversed, by starting from estimated values of the root uptake characteristics, and calculating a predicted plant uptake rate.

The mean I value ($\bar{I}$) is the most important measurement in this type of work. Figure 11.10 shows the general pattern of the $\bar{I}$ value over the life of an annual crop — very large in the early stage, then it settles to a steady value during rapid

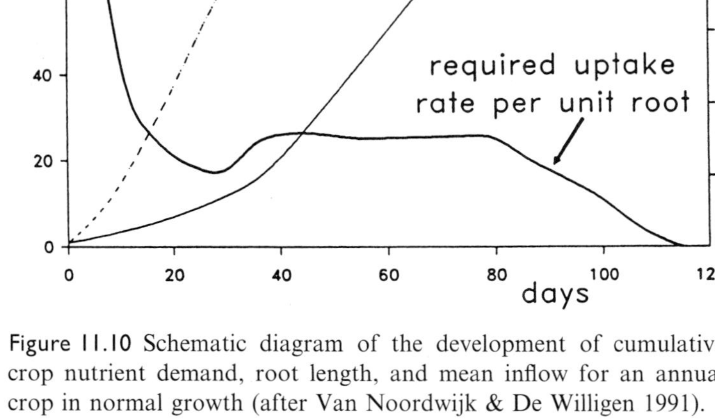

Figure 11.10 Schematic diagram of the development of cumulative crop nutrient demand, root length, and mean inflow for an annual crop in normal growth (after Van Noordwijk & De Willigen 1991).

growth and then declines towards senescence. Some typical $\bar{I}$ values are given in table 11.3.

Hunt (1973) proposed that these methods would be more accurate if small numbers of plants were harvested at frequent intervals, rather than a few large harvests, over the total harvesting period. A polynomial or other equation could then be fitted to these data to give nutrient content curves and root length curves, so as to obtain values of $\bar{I}$ at any time over the whole harvest period. The Williams (1946) formula (section 10.3.1) makes assumptions about the constancy of the nutrient concentration that are not necessarily correct (Soon 1988), and gives inflows that are averages over the period between two harvests.

These analyses of field crops have all used the model of Baldwin *et al.* (1973) (section 10.4.2) to calculate the depletion at the root surface, so that the effects of inter-root competition are taken into account. These steady-state models have also found wide application in other work, especially where factors and conditions change. This is valuable in field work, because the environmental conditions will always change during the study. Several groups (Burns 1980; Barraclough 1986b, 1989a, b; De Willigen & Van Noordwijk 1987; Robinson *et al.* 1991; Smethurst & Comerford 1993b; and others) have used the equations to predict the behaviour of solute in soils that contain absorbing roots, where their simplicity and ease of manipulation were attractive.

Barraclough (1989b) applied this analysis to winter oilseed rape (*Brassica canola*). Because it is a brassica, it cannot be infected with mycorrhizas (section 8.3.2), so phosphate diffusion calculations with this crop are probably more reliable than with most plants. However, it is known to have a specific hydrogen ion-releasing mechanism when severely phosphate-deficient (section 7.2.1). Root length of a commercially fertilized crop was measured on five occasions by soil coring; soil solution values were measured three times, and the nutrient content of shoots was measured 14 times. From this, inflows were calculated (figure 11.11) by both methods mentioned above, which gave somewhat different results.

The inflow values were compared with the diffusive inflows to a zero sink in the soil at that time, calculated by using the steady-state approximation (Baldwin *et al.* 1973) (section 10.5.2). This yields the minimum concentration in the soil solution that could have provided this measured inflow. As expected on this well-

Table 11.3 Examples of high values of mean I measured for different nutrients and species, from the field, with crops beyond seedling stage. See original papers for changes with time. The data here are intended solely to show high values that may be obtained in practice in the field, and a much wider range can be found in the literature. For comparable non-field data, see Robinson (1986). All data in the table are in mol cm^{-1} s^{-1} $\times$ 10^{-14}. (Note that some authors have used I as a symbol for flux across the root surface, in mol cm^{-2} s^{-1}.)

Nutrient	Barley[1]	Winter wheat[2, 3]	Maize[4]	Soybean[a]	Rape[5]
N	4.6	32	32	2.6	20
P	0.16	1.3	0.9	0.1	1.0
K	1.2	12	12	0.6	5

References: [1]Soon (1988); [2]Gregory *et al.* (1979); [3]Barraclough (1986b); [4]Mengel & Barber (1974); [5]Barraclough (1989b).

[a]Some N may be fixed.

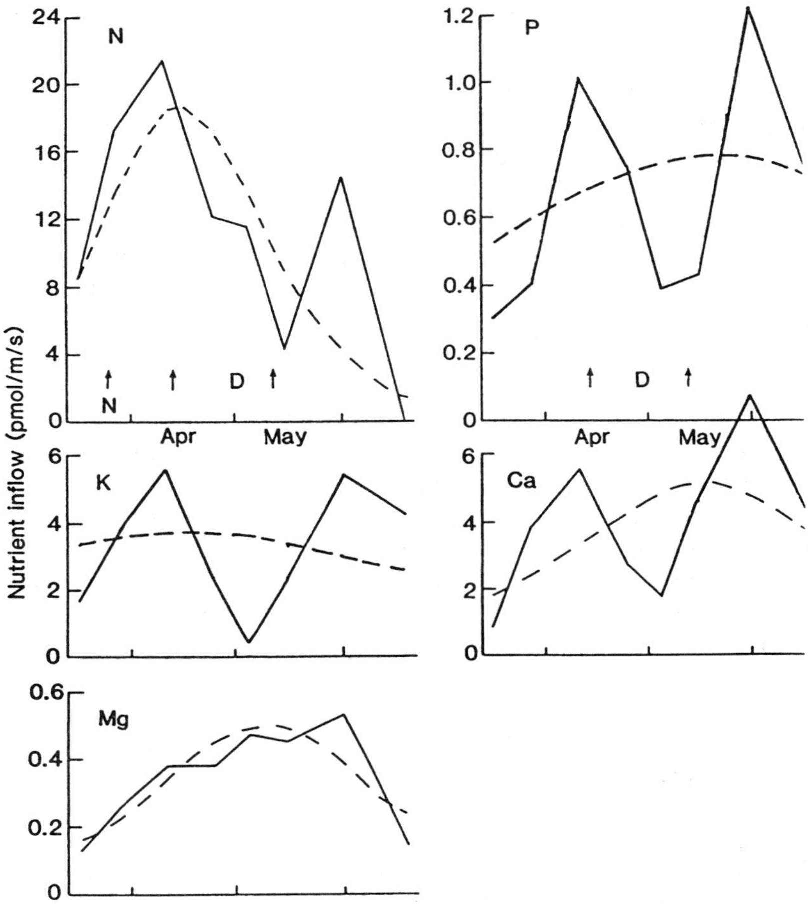

Figure 11.11 Nutrient inflows for rape grown in a flinty loam Rothamsted soil with adequate fertilizer. The period marked D, between arrows, was one of drought, causing an inflow decline. The point at which N fertilizer was applied is marked N. The dotted line is for same data, analysed by a fitted function, which did not show up the effect of drought (after Barraclough 1989b).

fertilized soil, the measured soil solution concentrations were appreciably larger than the minimum requirement. The phosphate concentration in the soil solution was only marginally sufficient when the diffusion coefficient was reduced by soil drying during a drought period. A contribution to the inflow by root hairs, and the presence of much of the root in the subsoil, were not allowed for in the calculations.

The calculated soil solution values to supply the measured inflows were compared with those expected from the soil chemical test values then recommended by the UK National Agricultural Advisory Service. They agreed reasonably well on the requirement for phosphate in moist soil, but when the soil was dry, the normal soil test value would not be sufficient to maintain the required phosphate inflows, according to the diffusion calculations.

This approach to uptake models can be applied with any single-root uptake model (section 10.4.1), applying it in the same way as the Baldwin *et al.* (1973) steady-state model. The most frequently used model of this type is that of Barber & Cushman (1981), and variants of this have been applied to several field studies. Generally, these studies have given good agreement between uptake as predicted by the model and that by measurement. Thus, phosphate uptake was well predicted for soybeans in the field (Silberbush & Barber 1984) (figure 11.12). However, Seward *et al.* (1990) found that this model overpredicted uptake of potassium by wheat in the field up to fourfold, being least accurate with nitrogen- and potassium-fertilized treatments. The total root length in the variously fertilized treatments was not greatly different at anthesis. The main origin of this discrepancy was that the root uptake parameters I_{max} and K_m, measured in solution culture with the depletion technique of Claassen & Barber (1974), were much too large in the heavily fertilized treatments. In the low-N low-K treatments, where control of uptake depended more upon the soil than the root parameters, agreement was reasonable (figure 11.13). The difficulties in measuring K_m and I_{max} in conditions quite different to those in the field have been discussed elsewhere (section 5.3.1). Van Rees *et al.* (1990) compared simulations of the uptake of potassium by slash pine seedlings, by the Barber–Cushman model and the steady-state model of Baldwin *et al.* (1973). The correlation between the methods

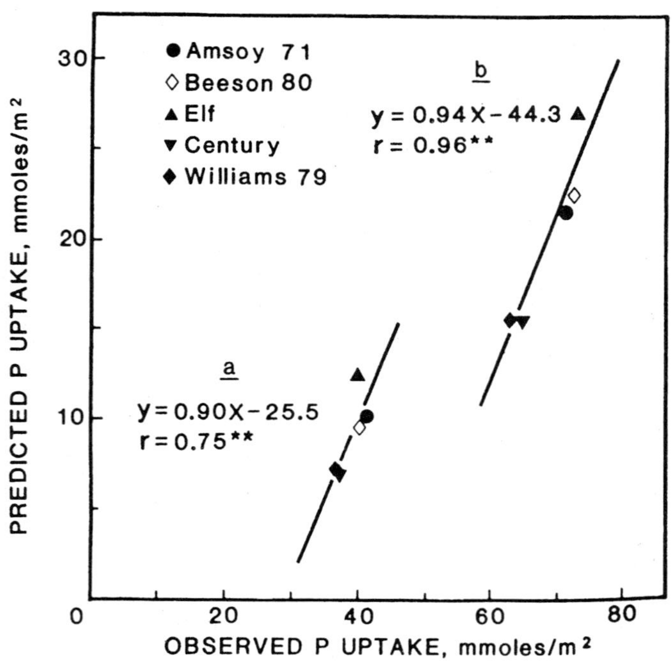

Figure 11.12 Comparison of observed P uptake by soybean cultivars grown in the field in a Raub soil and the simulated uptake by the Cushman–Barber model including uptake by root hairs (after Silberbush & Barber 1984).

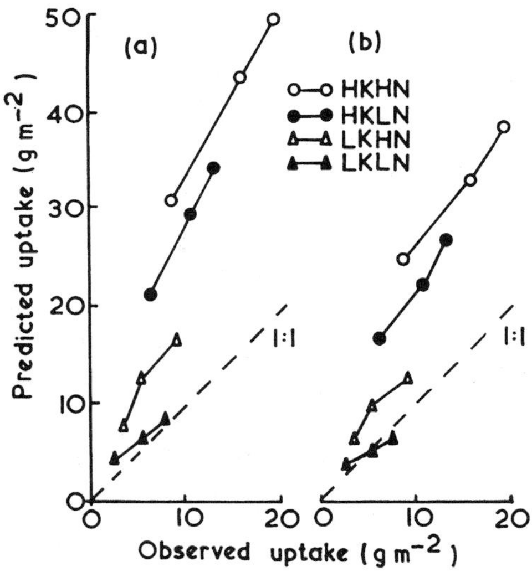

Figure 11.13 Comparison of observed uptake of K by wheat growing in the field on three occasions, with different fertilizer treatments (H = high, L = low, for N and K), with uptake predicted by the model of (a) Cushman–Barber and (b) Claassen–Barber (after Seward *et al.* 1990).

was very close ($r^2 = 0.99$), though the results with the steady-state method were always about 5% larger. Agreement between observed and predicted uptake was close for greenhouse and nursery experiments, but less good for seedlings in the field (figure 11.14).

The agreement between simulated and observed data has often, but not always, been found to be better for high nutrient levels rather than low ones, and for potassium rather than for phosphate (Schenk & Barber 1979a, b;

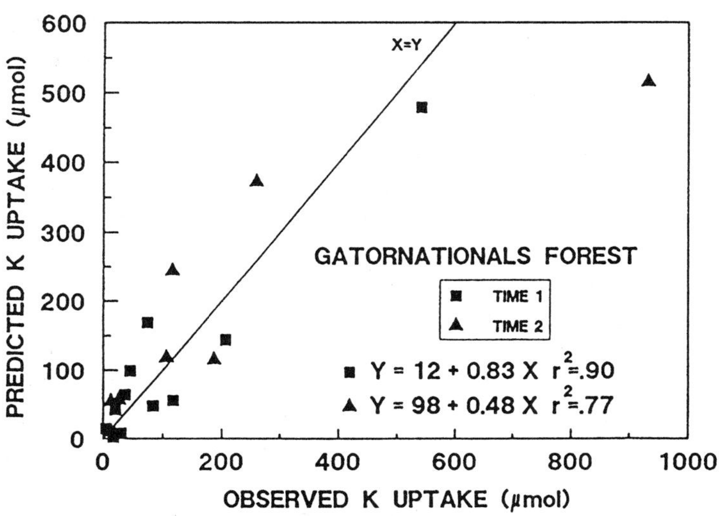

Figure 11.14 Comparison of K uptake by tree seedlings growing in the field, on two occasions (time 1 and time 2), as observed and as modelled with the Barber–Cushman model. Linear regression equations given (after Van Rees *et al.* 1990).

Silberbush & Barber 1984; Brewster *et al.* 1975, 1976; Smethurst & Comerford 1993b). This is not unexpected, because of the many root surface adaptations that can enhance uptake of nutrients (chapters 7 and 8). Most of these make their largest contribution at low concentrations. Until these have been reliably included in the models, the latter can be trusted most at moderate and high nutrient concentrations

These methods generally ignore any questions of root or uptake distribution within the soil, because inflows are obtained from the whole-plant uptake measurements. Discrimination between uptake from different soil volumes in field studies would be extremely complicated. Many of the studies have given reasonable agreement between modelling and measurement, but the necessity to supply the data for root length, and the problems of defining the root uptake characteristics in the field, leave significant remaining gaps. The following section deals with crop models that predict both growth and nutrient uptake.

11.3.2 A Basic Crop Growth and Nutrient Uptake Model

Penning de Vries & Rabbinge (1995) consider all the 'defining' factors for potential yield to be the above-ground factors, whereas the 'limiting' factors to attainable yield are water and nutrients. The above-ground factors such as light, temperature, and carbon dioxide are not controllable under normal field conditions, whereas water and nutrients can often be amended in developed agriculture. However, this would not be so in much tropical agriculture, and in forestry, where water and nutrients may, in practice, define the yield level, so we see no fundamental basis to this distinction, though it may be convenient.

It is not possible to describe and discuss all the various crop models that already exist and continue to be created. There are at least 14 wheat models (Goudriaan 1996), five rice models (Ingram 1997), and 14 models of nitrogen transfers in the soil–root system (De Willigen 1991) that may be used in the crop models. In this chapter, we will therefore describe a few of these models to illustrate some of the principles employed in the below-ground parts only for the various vegetation types. Published models of a vegetable, a cereal, a potato crop, and coppice woodland are discussed to cover the main types of single-species crop.

We outline first what we regard as the basic field crop growth and uptake model that contains the processes of greatest importance below ground, based on the analysis developed in section 10.1. The above-ground part is as simple as possible, because the interest of this book is below ground. We assume the model represents a green leaf crop with no sharp changes in growth rate or phenology, so that we can assume that the effect of nutrient concentration is solely and directly on growth.

Figure 11.15 defines the basic relationships that are used. They are listed and numbered here, and the relevant processes can be referred to when discussing other models. It is assumed that the models will simulate layered soils, though this complication is not shown here since it does not introduce any new principles. This contains the following processes:

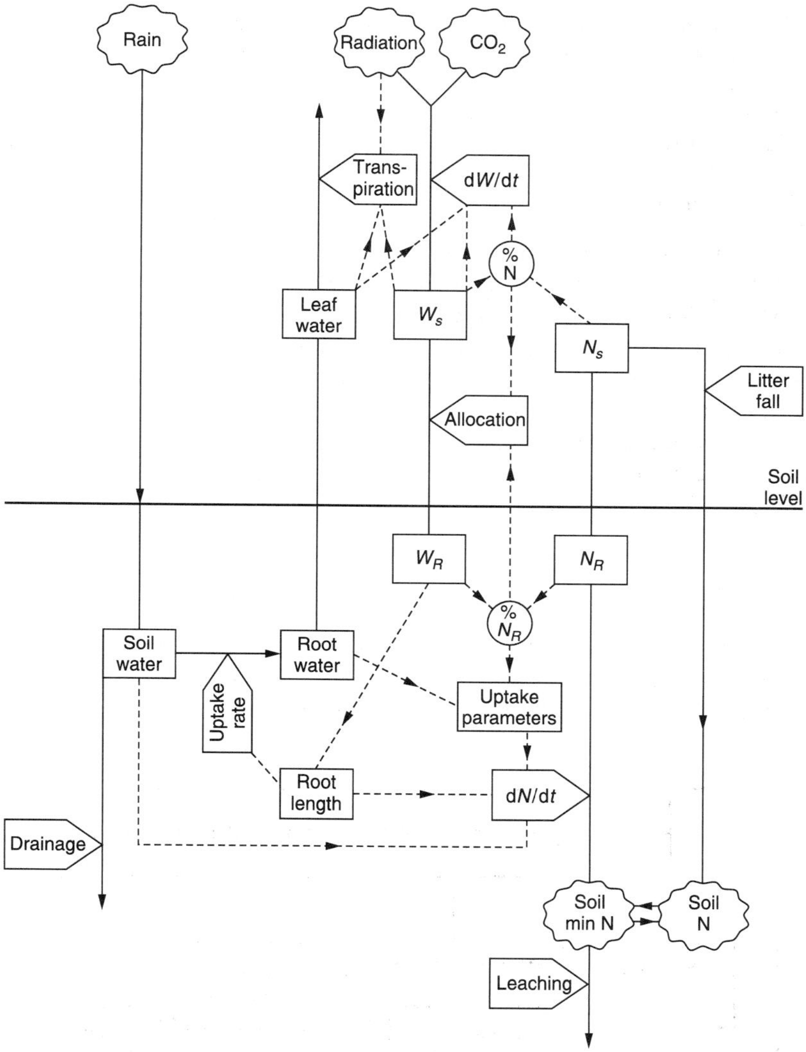

Figure 11.15 Schematic whole-crop model diagram for field crops with additional environmental processes to the model diagram in figure 10.14. Symbols are as in figure 10.14.

(a) Rate of photosynthesis; not discussed here. This depends upon radiation income and canopy interception. It needs coupling to the nutrient percentage and the water potential in the plant.
(b) Carbon allocation and root–shoot ratio. This can be set empirically, but will need to change with time, and with nutrient and water status. See chapter 9.
(c) Leaf growth and canopy structure; not discussed here.
(d) Root growth, radius and distribution. This needs coupling to photosynthetic rate and to allocation. See chapter 9.
(e) Water flow through soil, with solute transport (section 11.2).
(f) Soluble nutrient concentration in soil layers. This needs coupling to solute transport, possible mineralization processes, and nutrient uptake (section 11.2).

(g) Potential nutrient inflow, coupled to relative growth rate, root length/plant weight ratio, and nutrient percentage in plant. See chapters 5 and 10.
(h) Actual nutrient inflow, from root uptake properties, diffusion, and mass-flow processes in the soil. See chapter 6.
(i) Nutrient percentage in crop. This is coupled to growth and nutrient uptake rates. See chapter 10.
(j) Calculation of the new soil nutrient and water contents.

This basic nutrient uptake model is similar, in principle, to the model of Nye *et al.* (1975a,b) (figures 10.15 and 10.16). (See also Hanks & Ritchie 1986.)

Baldwin (1976) developed a rather similar model for uptake of nitrogen. This contains all the essential components given in figure 11.15, with some physiological elaborations, such as internal pools of carbohydrate and unbound nitrogen. The model also has a number of controls and feedbacks on root extension. The model has not been verified against plant data as far as we know. This model was used to investigate below-ground competition between plants of different species and is discussed in more detail in section 11.4.4.

11.3.3 A Model of Vegetable Growth and Uptake

As far as we are aware, the first uptake and growth crop model with a strong below-ground component that was applied to field crops was that of Barnes *et al.* (1976). Figure 11.16 shows the flow diagram of the model.

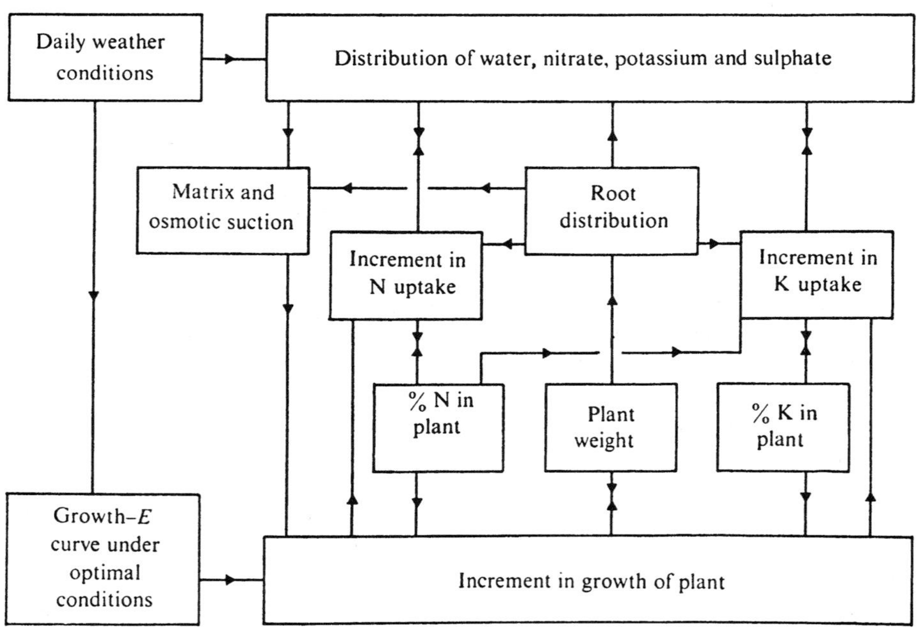

Figure 11.16 Flow diagram of model of growth and N and K nutrition of cabbage. Growth-*E* refers to the logistic growth curve modified by the evapotranspiration (after Barnes *et al.* 1976).

The test crop was cabbage, which was receiving the heavy rates of fertilizer normal for this crop. Therefore, a salt-transport model for the profile was included. In the early part of a simulation, each plant was assigned a progressively increasing area, above and below ground, until they filled the total area. The model included all the aspects of nutrient uptake that would be considered important now, including the five equations in section 10.1.1 and the factors listed in section 11.3.2, so it was a significant advance.

The use of cabbage as the crop had the advantage that the growth pattern was very simple with no reproductive phase, and using results of many field trials, growth was modelled empirically as a logistic equation with a maximum yield of about 10 t ha^{-1}. There was no photosynthesis submodel, so strictly it is not a growth model. Growth was advanced according to the cumulative evaporation from an open water surface, which was used as a proxy for time and radiation income. It was assumed that only one factor, water potential or a nutrient concentration, could limit growth at any one time. The root system weight was calculated from that of the whole plant by an allometric relation between shoot and root, and the root distribution down the profile was taken to be exponential (Gerwitz & Page 1974) (section 9.5.1).

Water was removed for evapotranspiration from each soil layer in turn, so that an extraction front moved down the profile (section 11.1.2). It was assumed that water stress had no effect on growth so long as the osmotic plus matric suction, averaged over the whole profile, was above −1.4 MPa (which seems a rather low figure). In the range −1.4 to −1.6 MPa, growth declined linearly to zero. In averaging over the profile, each layer was weighted by the amount of root there.

Two nutrients, N and K, were modelled, including soil organic matter mineralization and nitrification submodels. Another submodel calculated the interchange between exchangeable and non-exchangeable potassium and the potassium concentration in the soil solution. The nitrate concentration in a soil layer was taken to be uniform throughout the layer. For potassium, the steady-state approximation of Baldwin *et al.* (1973) (section 10.5.2) was used to calculate the transport to the root surface.

The effect of plant N concentration on N uptake, namely the 'regulation' of uptake (section 5.3.3), was expressed by equation (11.5):

$$\delta U_N/\delta t = \delta Y/\delta t\, N_m[1 + 12.0 \exp(-9N'/N'_m)] \tag{11.5}$$

where Y is yield, $\delta Y\, N_m$ is the incremental uptake with low internal and high external N concentration; and $N' = N - N_o$ and $N'_m = N_m - N_o$, in which N, N_o, and N_m are the actual concentrations of N in the plant, the minimum required for growth, and the maximum possible, respectively. This gives the time-step uptake of N not limited by the external N concentration.

This value was reduced by the N uptake characteristics in relation to the soil nitrogen concentration, which was expressed by a normal Michaelis–Menten equation

$$\delta U_N/\delta t = \delta U_{Nm}/\delta t\, C_N/(C_N + C_{1/2N}) \tag{11.6}$$

where C_N and $C_{1/2N}$ are the nitrate concentration in a layer, and that giving half the maximum uptake rate, respectively. The effect of potassium concentration on growth was described by a similar equation.

The third important relationship was between growth rate and N concentration in the plant, expressed as a relationship that multiplied the logistic equation that defined the maximum yield:

$$(\delta Y/\delta t) = (\delta Y_m/\delta t)(1 + N'_{1/2}/N')^{-1}(1 + N'_{1/2}/N'_m) \tag{11.7}$$

The agreement between modelling and the observed results of several cabbage experiments was good in general (figure 11.17), but there was much difference in detail, especially in the potassium percentage.

The model has been tested successfully with 12 other vegetable crops over a number of seasons (Greenwood & Draycott 1988). This model contains a number of empirical relationships with fairly large assumptions, and is biased towards soil

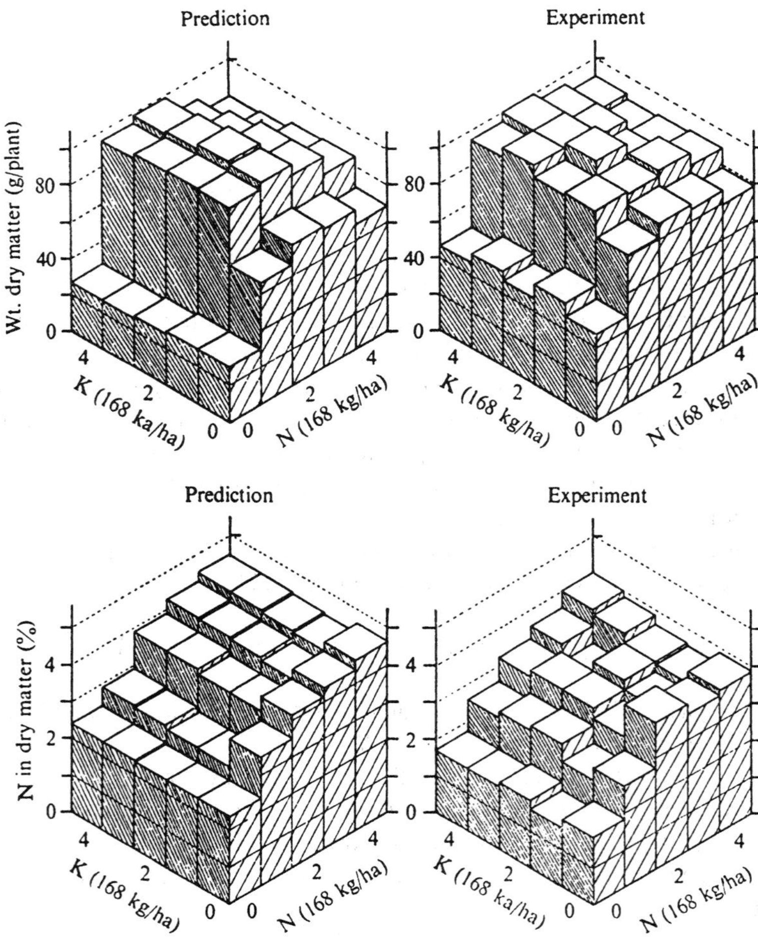

Figure 11.17 Observed and predicted values of dry weight per plant and percentage N in dry matter for cabbages in one experiment (after Barnes *et al.* 1976).

rather than plant processes. The number of parameters involved is very high, and their choice from limited basic information would usually be difficult, but subsequent work has led to the discovery of similar relationships, for widely different field crops, between plant mass per unit area, and their growth rates, root development, and critical nutrient concentrations (Greenwood & Stone 1998). These have been used for separate K and N models (Greenwood *et al.* 1996), and calibrated for potatoes, sugar beet, wheat and many vegetable crops (e.g. Greenwood & Draycott 1988). The main inputs are soil, cultural and weather variables. Other properties were found to be much less important, and the same values could be used in the model for all crops tested. Reasonably good agreement with field data has been found in several countries. A version for UK vegetable growers is available under the name WELL_N (HRI 1994).

11.3.4 Modelling Wheat — the AFRCWHEAT and CERES Models

There are now a number of wheat models, and they have been used in many parts of the world. The models of Van Keulen & Seligman (1987) have been influential in this development. The CERES model has also been used as the core of models of other annual crops. All the most used crop models at present are of the mechanistic/empirical type, and the many empirical factors make it absolutely essential to compare their results for widely different data-sets. The success of crop models under different environmental conditions, without *ad hoc* fine tuning of parameters, is a test of how well the underlying processes of root, shoot, soil and atmosphere are being simulated (Goudriaan 1996). It is now often quite difficult to determine exactly how different submodels work or fit into the whole, because models are constantly being changed and updated, and the documentation often lags far behind the actual state of the current model. This makes checking against field data and intercomparison of models even more important. If different models based on essentially the same thinking produce different results, it is important to determine why (Goudriaan 1996). A recent very useful paper (Jamieson *et al.* 1998) compared five different models in a joint study, and recent changes in these models, and comparisons between them, should be consulted there. In all of these models the main aim is to simulate and predict the effect of weather, that is, of temperature, radiation, and water.

In section 11.3.4.1, we discuss the AFRCWHEAT.2 model in detail, and then use this as a base for comparison with the widely used CERES model. The important issue for this book is how the nitrogen uptake, and to a lesser extent the water supply, are dealt with in the models, and other aspects of their design are dealt with very briefly. Modelling of phosphorus uptake is less frequently included in these models, but an approach can be found in Allan Jones *et al.* (1991b), though this does not include any detailed root mechanisms. Most models use fairly similar concepts for nutrient uptake, but that of O'Leary & Connor (1996a, b) is unusual in postulating 'active' and 'passive' processes.

11.3.4.1 AFRCWHEAT.2 (Porter 1993)

The origin of AFRCWHEAT goes back to the Yield Variation Programme of the Agricultural Research Council of the UK (Tinker 1985), and the main parts of the model are described in Porter (1984, 1993) and Weir *et al.* (1984a). The radiation interception process is mechanistic, because the principles are well understood, but the growth and development of the plant, and the connection of this with nutrient and water deficiencies, are very largely empirical. Photosynthesis is determined from the photosynthetically active radiation, the temperature and the canopy properties that determine how much radiation is intercepted. The rules governing allocation to shoot and root, and later to grain, are quite complex and depend upon the growth stage (Weir *et al.* 1984a). AFRCWHEAT has the ability to 'grow' seminal and lateral roots by allocating photosynthate in a way that produces a descending root front, and finally an exponential distribution of roots down the profile (Weir *et al.* 1984a).

The AFRCWHEAT.2 model includes a soil water/nitrogen transport model derived from the SLIM model of Addiscott (1977) and Addiscott & Whitmore (1987, 1991). This uses a chromatographic or layer approach (section 11.2.2), in which water in excess of the field capacity in one layer moves to the next lower layer, along with its solute load. Only half of the available water can move and transport solute down the profile in any one time step. Some mixing of solute and total water occurs in each layer as water passes down the profile.

Transpiration is determined from intercepted radiation and other weather variables by the Penman equation, but is reduced *pro rata* to the soil water content if the latter is less than 65% of field capacity in the rooting zone. Water can only be extracted from within the root zone, and the uptake is usually distributed between the soil layers in proportion to the amount of available water they contain (but not to root length). Available water (to -1.5 MPa) is absorbed by roots, up to the limit demanded by the calculated evapotranspiration. However, water inflow to roots cannot be larger than 0.3 mm m^2 km^{-1} root day^{-1} (equal to 0.3 cm^3 m^{-1} root day^{-1}), which limits the rate of uptake with low L_V. This rule can cause important water stress, because the limit is fairly small (see Jamieson *et al.* 1998). Root length density is an important parameter in AFRCWHEAT.2, whereas in some models, such as SUCROS2, it is ignored.

The ratio of the current soil water deficit below field capacity to available water remaining in the soil defines two empirical water-supply correction factors. As water supply decreases, these increase the rate of leaf ageing, leaf senescence, and specific leaf weight, but decrease the leaf expansion rate and the supply of photosynthate for grain-filling

The method of modelling nitrogen assumes that plants have a maximum and a minimum limit to their nitrogen concentrations (X_{Nmax} and X_{Nmin}) in both root and shoot. These values vary with stage of development (figure 11.18). The 'demand' for nitrogen is the difference between the actual concentration and the maximum concentration at that growth stage, multiplied by the dry weight of root or shoot, respectively, at that time, plus newly formed dry matter multiplied by X_{Nmax}. The relationship of this concept to growth demand and 'deficiency demand' (section 10.2.7) is clear. However, in chapter 10 we stated that uptake in

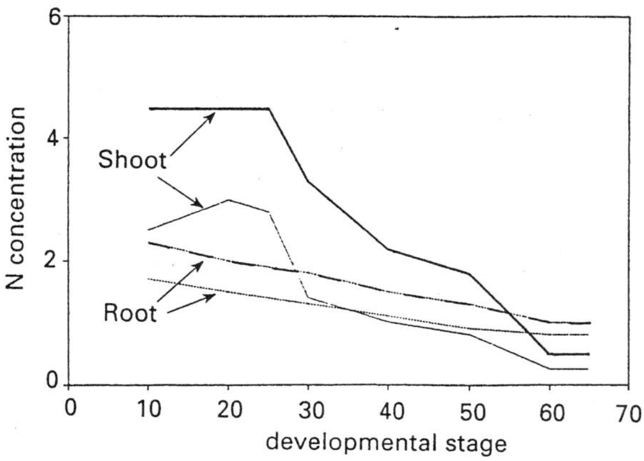

Figure 11.18 Upper and lower nitrogen concentrations limits for roots and shoots of winter wheat in the AFRCWHEAT.2 model, varying with development stage (10 is emergence, 60 is end of grain-filling) (after Porter 1993).

response to a 'deficiency demand' had no obvious time dimension — here, it is postulated that the demand should be satisfied in 1 day. If this demand cannot be met, $X_{\mathrm{N}act}$ remains below $X_{\mathrm{N}max}$.

The model sets a maximum value for nitrogen inflow of 1.7×10^{-13} mol N cm^{-1} roots s^{-1} for combined nitrate and ammonium, but does not use a formal uptake equation as discussed in chapters 5 and 10. The uptake rate per day is therefore the least of the calculated demand (as above), or the sum of the available nitrogen in all layers of the profile on that day, each of these being appropriately reduced so that the mean inflow in any layer is never greater than the stated limit to inflow. A separate demand rate is set for the wheat grains, at 1.7 mg N (grain)$^{-1}$ (°C)$^{-1}$ day^{-1}, that is supplied from the rest of the plant. The simulation of nitrogen movement and production in the soil proceeds at the same time.

The effect of nitrogen deficiency stress on growth is defined by FACN, where

$$\mathrm{FACN} = [(X_{\mathrm{N}act} - X_{\mathrm{N}min})/(X_{\mathrm{N}max} - X_{\mathrm{N}min})] \tag{11.8}$$

for roots and shoot separately (see chapter 10) and $X_{\mathrm{N}act}$ is the actual N concentration in each tissue. If $X_{\mathrm{N}act}$ exceeds $X_{\mathrm{N}max}$ there is luxury uptake, FACN exceeds 1, and this temporarily prevents further uptake. FACN determines four physiological correction factors (figure 11.19). These factors increase the time during which tillers are produced, increase tiller death rate, reduce tiller production rate, reduce leaf extension rate, hasten leaf ageing, and divert more assimilate to the roots. For both nitrogen and water, the effects of a deficiency are simulated in complex physiological detail, but the relationships are empirical.

Both water and nitrogen stress have their main effects on growth via leaf ageing and leaf expansion, thereby reducing the amount of green leaf per unit land area. Tests of the model in four field experiments showed fair agreement with crop data for yield and most other parameters. However, nitrogen percentage was frequently underestimated (Porter 1993), especially at the end of the season, when the simulated value was less than half the measured one. It seems that the uptake processes cause most trouble, as for the model of Barnes *et al.* (1976) (section 11.3.3).

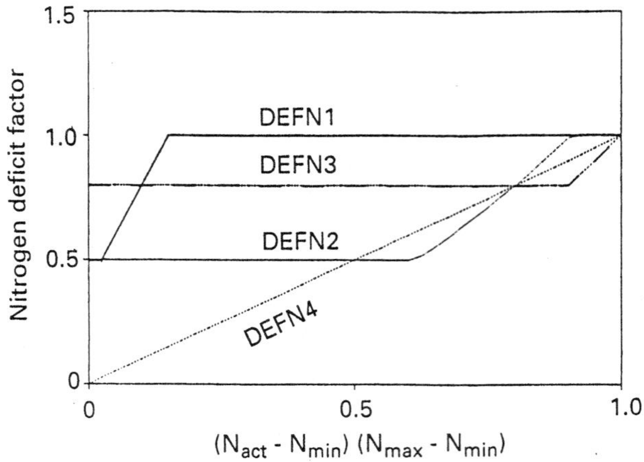

Figure 11.19 Correction factors DEFN for growth of tillers and leaves of winter wheat in the AFRCWHEAT.2 model, based on the ratio of actual minus minimum %N to maximum minus minimum %N (see figure 11.18) (after Porter 1993).

11.3.4.2 CERES Model (Ritchie & Otter 1985)

The CERES system of models covers a number of crops (Ritchie *et al.* 1988) and draws upon many other models for its components. Here, we discuss only the nitrogen component of the wheat model in detail (Godwin & Allan Jones 1991); other CERES references can be obtained from this paper.

The photosynthesis section of CERES is much simpler than that of AFRCWHEAT.2, and there are differences in the way in which phenology is simulated. These are not discussed here. The potential transpiration is calculated by the Penman equation. Water is extracted from each soil layer in proportion to the available water there, with a restriction on the rate of uptake when the root length density is less than the low value 0.25 cm cm^{-3}. The actual evapotranspiration per day is therefore defined by the available water content in the soil layers and to a lesser extent by root length density. The latter operates through a 'root restriction factor' $F_t = [1 - \exp(-8L_V)]$.

CERES has a layer submodel for water flow up or down the profile that also transports nitrate. A nitrogen transformation submodel predicts mineralization, nitrification, immobilization and denitrification in each layer. The products of mineralization, plus any added fertilizer, go into the nitrate, ammonium or urea pools in the soil.

CERES uses the concepts of both deficiency and growth demands (section 10.1.8). The 'deficiency demand' is defined as $(X_{\mathrm{N}crit} - X_{\mathrm{N}act})$ multiplied by the existing shoot biomass (using similar symbols as for AFRCWHEAT.2 for simplicity), and the same for the roots. The upper nitrogen limit in CERES is defined as the minimum at which growth is maximal $(X_{\mathrm{N}crit})$, whereas in AFRCWHEAT it is set at the upper limit of concentration $(X_{\mathrm{N}max})$. The term $X_{\mathrm{N}crit}$ has an empirical relationship with W (section 10.2.5).

It is assumed that the plant tends to maintain X_{Ncrit} in newly formed tissues, so that the 'growth demand' is given by X_{Ncrit} multiplied by the potential shoot growth over the next day (or other time period). If the potential growth rate is not attained for any reason, such as water deficiency, the calculated nitrogen uptake may cause X_{Nact} to exceed X_{Ncrit} temporarily, and so suppress further uptake. The internal recycling of nitrogen from vegetative tissue to the grain is separately simulated.

'Availability factors' F for nitrate and ammonium in the soil, based on empirical data-sets, are used to control how the uptake rate varies with the soil nitrogen levels:

$$F_{NO_3} = 1.0 - \exp(-0.0275C_{NO_3}) \tag{11.9}$$

where C_{NO_3} is the concentration of nitrate in the soil in that layer, and there is a corresponding equation for ammonium. Thus, F_{NO_3} can range from 0 to 1. This approximately exponential relationship is roughly equivalent to the relationship between uptake and solution concentration given by the Michaelis–Menten equation (section 5.3.2), and used in a parallel way by Barnes *et al.* (1976) (section 11.3.3) and other modellers. In effect, F_{NO_3} is a linearized statement of nitrogen availability in relation to uptake. There is a maximum for nitrogen uptake from within a layer, when F_{NO_3} is 1.0, of 6 g N ha^{-1} cm^{-1} depth day^{-1}, that is the uptake in a hectare × 1 cm depth if the root density is 1 cm cm^{-3}, which equates to an inflow value of 0.4×10^{-13} mol cm^{-1} s^{-1}. This is effectively I_{max}, though it is low compared with most measurements.

The potential uptake rate, $\mathrm{d}U_{NO_3}/\mathrm{d}t$, for nitrate from the whole profile is the sum over all layers of the equation for a single layer:

$$\mathrm{d}U_{NO_3}/\mathrm{d}t = \bar{L}_V wzF_{NO_3} \times 0.006 \text{ (kg N ha}^{-1}\text{ day}^{-1}) \tag{11.10}$$

where w is a water content correction factor and z is unit layer depth. In the terminology of our book, this is equivalent to stating

$$\text{Uptake rate} = I\bar{L}_V z = 2\pi\alpha a f(C_L) w\bar{L}_V z \tag{11.11}$$

The basic ideas about how nitrogen concentration affects physiology and growth are similar to those in AFRCWHEAT.2 but the four correction factors are applied rather differently. These four indices control photosynthetic rate, leaf area expansion, leaf senescence, tillering, and N in grain accumulation. The prediction of grain yield has compared fairly well with observed yields, but N uptake has sometimes been less well predicted (Godwin & Allan Jones 1991) (figure 11.20), as for other models discussed here.

It will be seen that most of the features described in this book as part of the uptake system for crops, and synthesized in sections 10.2.1 and 11.3.2, are present in these two quite similar models, but are expressed in very different ways. A number of empirical relationships are used, employing data from many sources. It is not surprising that different models, or even different versions of the same model, do not agree in their output. The purpose of these crop models is different to that of the 'scientific' models we discuss more extensively, and it may be impractical to include much greater mechanistic detail. However, there is a possible danger that they may be too complex for practical use, and also too empiri-

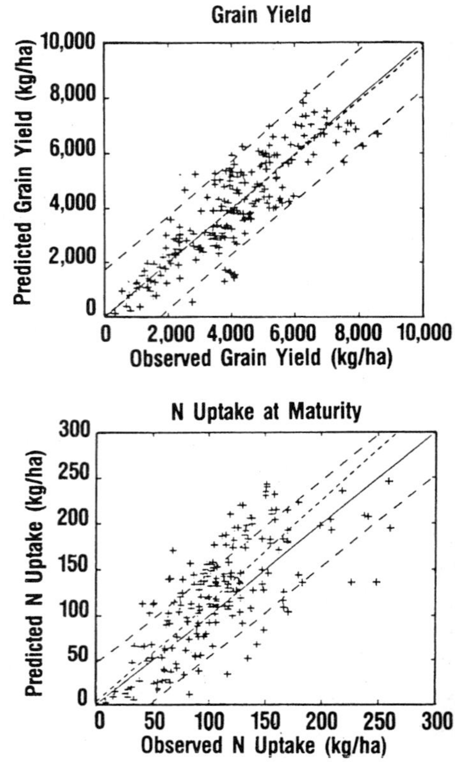

Figure 11.20 Comparisons of simulated yields and nitrogen uptakes of wheat with field experiment results, using the CERES–WHEAT model (after Godwin & Allan Jones 1991).

cal for scientific investigation. They should not be used outside the data field within which they have been validated, or beyond the assumptions they contain. They mostly assume that weather is the main determinant of yield, and they perform best where there is a strong weather signal (J.R. Porter 1997, private communication).

11.3.4.3 The Daisy Model (*Hansen* et al. *1991*)

This model includes a particularly large number of processes in the environment. It has a complex hydrological component, including a snowmelt model, and flow in the profile is modelled by the Richards equation to generate a water balance. Radial water flow to roots is modelled with an approximate steady-state equation (section 2.3.4). Soil temperature is modelled with a heat flow equation, including freezing and melting with associated water transfer.

The soil organic matter (OM) is considered to be in three pools: added OM, biomass, and original OM, and these can be further subdivided. These submodels deal with respiration and all the microbial reactions of the soil nitrogen cycle. There is, of course, a submodel for vertical movement of water and nitrogen.

Crop growth is modelled on the basis of a phenology driven by the sum of day-degrees, the canopy development being driven by this and by the rate of photosynthesis, respiration, and allocation.

The nitrogen uptake module is based on the usual concept of a demand generated by the growth of the plant, and a supply that depends upon the amount of mineral nitrogen in the soil layers. In this model there is a steady-state transport model (section 10.4.2) to control supply to the root surface, with a progressive increase in the radius of the depletion zone (Smethurst & Comerford 1993a). The nitrogen uptake rate is limited by the rate of diffusion to a zero sink when little nitrogen is present, or by the demand generated by plant growth when supply is ample. A single nitrogen concentration value is assumed to occur at all root surfaces at any one time, and inflow depends upon this soil nitrogen level, but soil layers that contain less than the common concentration of mineral nitrogen at the root surface do not contribute to the uptake. Nitrate and ammonium are simulated separately. This model therefore follows mechanistic principles reasonably closely in the uptake step.

11.3.4.4 A Wheat-Fallow Model

A further model with interesting features is that for a wheat-fallow system (O'Leary & Connor 1996a, b), and this is particularly suitable for simulating the interaction of water and nitrogen in semiarid areas. There are extensive soil and leaching submodels. The model is unusual in having both passive (transpiration stream) and active (to meet the current growth demand) uptakes; this terminology does not accord with that used here, and the distinction seems dubious. The amount of N translocated up the stem is the excess over that needed to give the optimum N concentration in the root, with total N in the latter acting as a buffer stock for the shoot. In other ways, it has a broad resemblance to the CERES model family.

11.3.4.5 A Comparison of Wheat Models (Jamieson et al. *1998)*

A comparison of results from five models on a single set of data gives some idea of the variety of wheat modelling (Jamieson *et al.* 1998). The data were obtained under an automatic rainshelter in New Zealand, so the main interest was on response to drought.

Despite the empirical nature of most parts of the models, they produced reasonably similar results (table 11.4). All but one predicted the final grain yield to within 10% on the watered plots. Surprisingly, all but one model predicted grain yield better than total biomass. The models predicted different rooting depths, and hence soil moisture reservoirs differed between 250 and 350 mm of water (Jamieson *et al.* 1998), which is probably the main cause of the differences between the model simulations. The AFRCWHEAT model underestimated the rate of actual evapotranspiration, because the root restriction on I_W (see above) was too severe, even though this feature was closer to a mechanistic simulation than in the other models. None of these models simulated nutrients other than nitrogen.

Table 11.4 Comparison of results of five crop models in predicting grain yields, and the observed results in a rainshelter experiment in New Zealand (tons/ha).

	Observed	AFRCWHEAT.2	CERES	Sirius	SUCROS2	SWHEAT
Grain						
1[a]	9.93	9.31	8.93	10.72	10.62	7.74
3	9.52	8.19	8.93	10.53	9.73	7.58
5	5.58	4.91	5.37	4.90	4.78	6.55
6	5.78	5.90	6.02	5.23	3.86	6.41
7	7.15	7.34	8.70	7.67	4.96	7.35
8	8.55	8.66	8.93	10.15	7.30	7.74
11	3.59	3.11	2.29	4.26	3.11	5.46
RMSD		0.64	0.90	0.90	1.28	1.42
r^2		0.94	0.86	0.94	0.89	0.91
a		0.06	−0.34	−0.90	−2.2	4.44
b		0.94	1.03	1.19	1.19	0.35

	Final root depth (m)	Size of the soil reservoir (mm)
AFRCWHEAT.2	1.95	351
CERES	1.60	288
Sirius	1.40	252
SUCROS2	1.67	301
SWHEAT	1.70	306

Source: after Jamieson *et al.* (1998).

[a]Code numbers refer to different irrigation treatments.
Correlation coefficients for the comparisons with observed results are given for each model, and the intercept (a) and the slope (b) of the regression of observed on simulated results. The final root depths and size of soil water reservoir simulated by each model are also given in the insert table.

11.3.5 Nitrogen Transport and Residual Nitrate after Cropping

Nitrogen has a complex soil chemistry and a moderate or high mobility, so that attention is focused on mineralization and related soil processes rather than on ion transport in many crop models (De Willigen 1991). However, the transport of nitrate needs to be included in a modelling approach if residual nitrate may be left in the profile after the crop has reached maturity. This is important for the nutrition of the crop, and also because of the pollution hazard for ground- or surface-water.

Robinson & Rorison (1983) used the model of Baldwin *et al.* (1973) to investigate the effect of L_V on nitrate uptake, and concluded that an L_V of 1 cm cm^{-3} was sufficient in moist soil to absorb almost all the soil nitrate from that volume during a cropping cycle, if there was demand for all the nitrogen by the crop (see section 11.3.2). This L_V value is exceeded by virtually all crops in the topsoil, but values may be much less in the subsoil (see section 9.2.2), so complete uptake cannot be considered as certain.

The possibility of residual nitrate being left in the subsoil depends upon the water relationships in the soil, because mass flow is important in nitrate movement and because of the effect of drying out of the soil on the diffusion coefficient. In a profile drying from the top, the demands for water from the less dense root system at the bottom of the rooting zone steadily increase (see section 11.1.3). High inflows of water must develop in the zone between the rooting front and the drying front above it. Mass flow will therefore favour nitrate uptake also. However, though roots just below the drying front might have an I_w value much greater than the $\bar{I}_w$, roots above might have much less, especially if the latter are not in complete contact with the soil (section 2.3.4) (Tinker 1976; Passioura 1988; Gregory 1994b). For this reason, regions with appreciable amounts of nitrate may be left in volumes with low L_V and low I_w after the drying front has passed (Robertson *et al.* 1993b; Robinson *et al.* 1991; Kage 1997). The intensity of soil drying in the front also decreases with depth because of the lower L_V (see figure 11.3).

Kage (1997) found that relatively large amounts of nitrate were held in the subsoil after the growth of faba beans, compared to that after wheat, and explained this as due to low root densities of the beans at depth. Using a steady-state model, he calculated that there was insufficient time for roots to extract most of the subsoil nitrate from the bean plots, and that 20–30 kg ha^{-1} more of nitrogen was therefore left after harvest of beans than after that of oats (figure 11.21). The difference was particularly marked in the 80–100 cm layer. With West European water relationships, this amount, if leached out during the

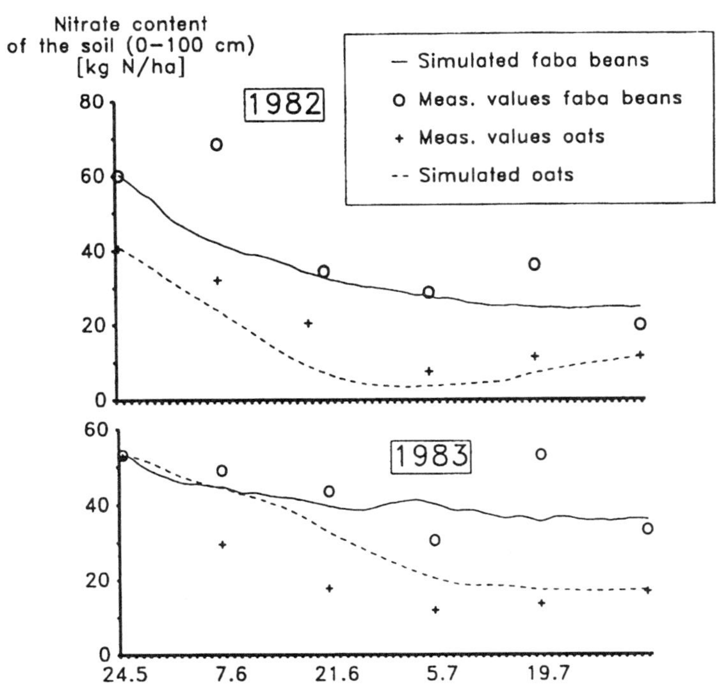

Figure 11.21 Measured and simulated nitrate content in the soil at 0–100 cm depth under faba beans and oats (after Kage 1997).

winter, may be enough to put the nitrate content of drainage water above the World Health Organization limit of 50 ppm NO_3.

11.3.6 Modelling Potatoes

Haverkort & MacKerron (1995) have reviewed the modelling of potatoes and their ecology, but mainly in relation to above-ground processes. The common practice of growing potatoes on soil ridges complicates the modelling of solute movement, and the presence of carbohydrate sinks in the root system complicates photosynthate allocation.

De Willigen *et al.* (1995) developed a model for this crop that assumed a root system of constant length, and used measured parameters from a field experiment. It is therefore a nutrient uptake model rather than a crop model, so it cannot be applied over long time periods, especially in the early phase of rapid vegetative growth. However, because of the soil ridges, the soil component is interesting. The water submodel has to be arranged for two dimensions — the vertical, and one horizontal dimension at right angles to the ridges. The two-dimensional field is subdivided into square cells (figure 11.22) and the movement of water in these is simulated with the Richards water transport equation, under the pressure head ψ (matric) and z (gravitational) driving forces (see chapter 2). The hydraulic conductivity $K(\psi)$ is dependent upon ψ, and the water transport equation is

$$\partial\theta/\partial t = \partial/\partial x(K(\psi)\,\partial\psi/\partial x) + \partial/\partial z(K(\psi)\,\partial\psi/\partial z) + S \qquad (11.12)$$

where S is a sink or source term for water. This equation is solved numerically.

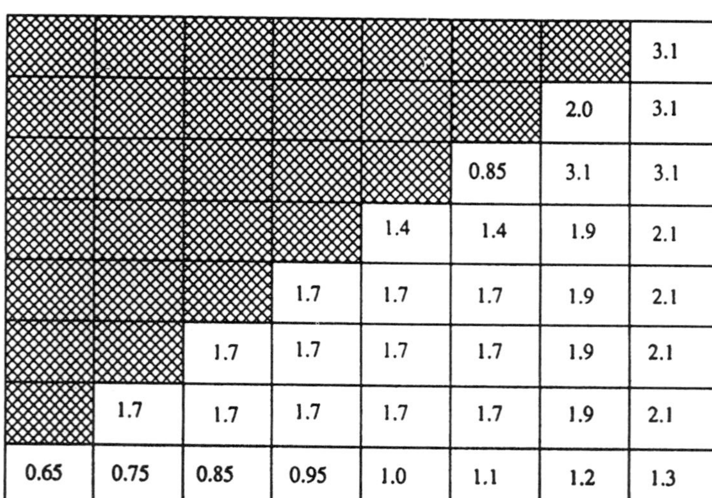

Figure 11.22 Model representation of a ridged potato field, as a two-dimensional plane across the ridge. The cross-hatched region is above ground, and is blocked off by setting the water conductivity to zero (after De Willigen *et al.* 1995).

The ridged soil surface was modelled by allocating zero $K(\psi)$ values to all the cells above the soil surface (figure 11.22). The water flux into the root in each cell is defined as the product of root conductivity (set to an average value of 5.10^{-6} cm day^{-1}) and the difference in pressure head between the root (set at a constant value) and the soil adjacent to the root. The latter is calculated by the usual steady-state approximation (section 2.3.4). The actual transpiration Ep, is defined as $Ep = ETp\,e^{-0.6\mathrm{LAI}}$, where the potential transpiration is ETp, and LAI is the leaf area index. There was very little root length (data of Vos & Groenwold 1986), and hence water uptake, below 70 cm depth.

The maximum nitrogen uptake rate was defined as 4.1 kg N ha^{-1} day^{-1}, which is taken up so long as it can be supplied from mineral nitrogen in the soil. The roots absorb at a rate appropriate to meet this demand until roots in a particular cell reach the 'zero-sink' condition (section 6.1.2), after which the uptake rate in this cell declines according to a diffusion equation (De Willigen & Van Noordwijk 1994a, b). In the simulations reported, uptake was maintained at the 'demand' value throughout the growth in a sandy loam, but a loamy sand dried out so much that diffusion limited uptake from this, and the simulation indicated that the initial distribution of nitrogen in the soil was then important.

11.3.7 Modelling Coppiced Willows

Eckersten (1994) modelled a short-rotation willow forest with emphasis on nitrogen turnover. Modelling of perennials is different from the annual crops dealt with so far, because the soil will almost certainly be more heterogeneous than with annual crops, and a layer of litter will develop that is an important part of the total nutrient cycle (figure 11.23). Growth rates were supplied from a photosynthesis submodel, with the allocation of photosynthate to roots depending upon the C/N ratio in the canopy, and roots dying off at a rate proportional to their growth rate.

Nitrogen was supplied from a pool of soil-available N, as the minimum of demand or amount available, with demand simply being the new tissue weight multiplied by the maximum nitrogen concentration ($X_{\mathrm{N}max}$). Rather unusually, roots got priority for nitrogen if this was in short supply, then stems, and finally leaves. The N-turnover model in the soil maintained pools of ammonium and nitrate. Demand for N from a particular soil layer was taken as the total demand by the tree, multiplied by the fraction of the root surface area present in that layer, which was considered to decline exponentially with depth in the soil. The total rooting depth was assumed to be proportional to the root biomass. Demand can be shifted from layers without available nitrogen to layers with remaining N.

The model simulation agreed well with variation in stem biomass between sites, but was less successful at simulating leaf weight growth rates. In this model, the use of layers in the soil had little effect on the plant simulation, but it did make a big difference to the prediction of leaching. The simulated values for soil mineral nitrogen were in poor agreement with the measured data. In comparing different years, a distinction could be made between years in which growth was N-limited and those in which it was CO_2-limited. A harvest routine is included in the model, in which stems are cut and harvested, but this part could not be tested. This model

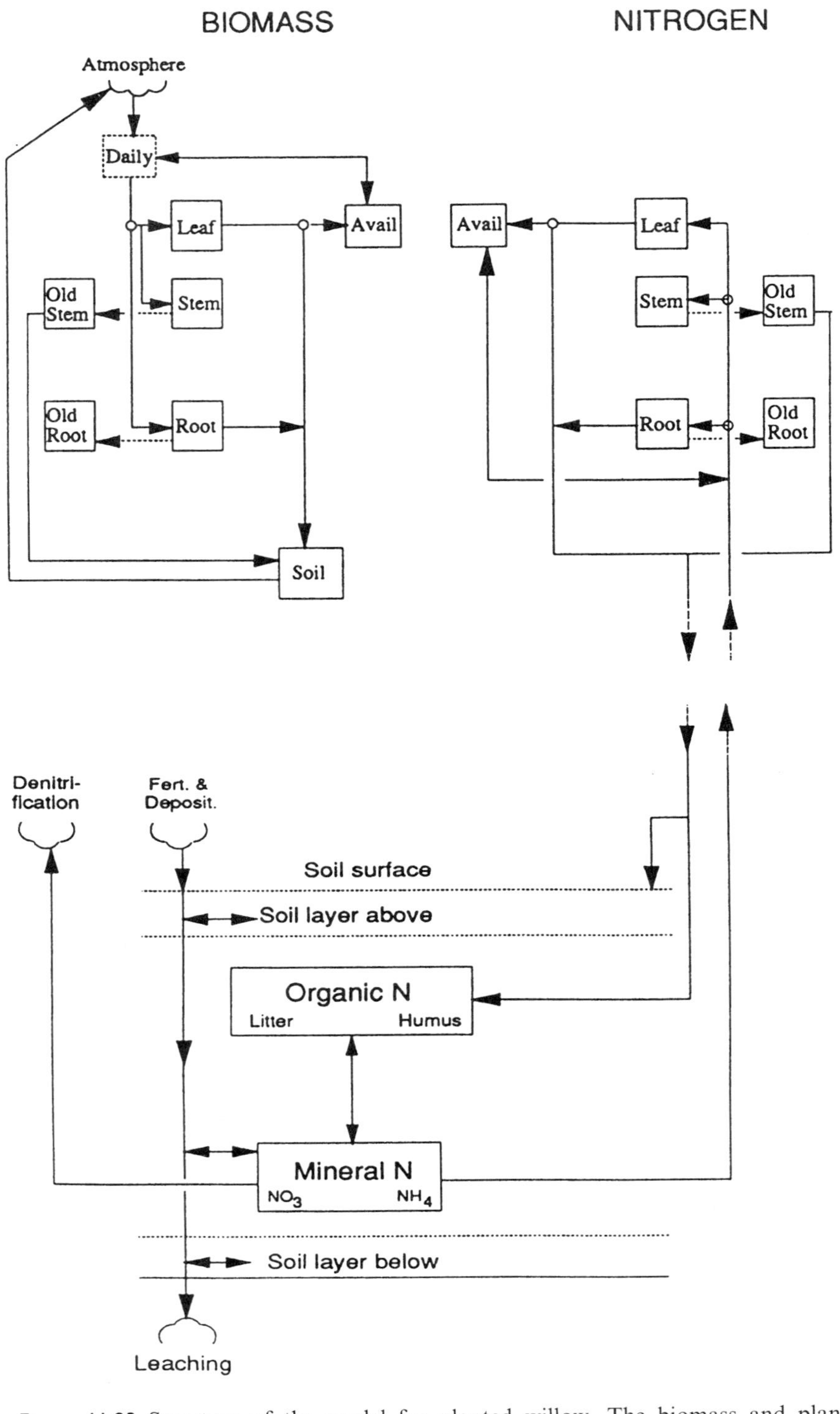

Figure 11.23 Structure of the model for planted willow. The biomass and plant nitrogen submodels are set up for each plant compartment (young and old tissues separately), and are connected daily by exchanging data. Solid arrows indicate daily mass flows, dotted arrows indicate annual mass flows (after Eckersten 1994).

should be relatively simple compared with most forest models, because it simulates a monoculture under uniform treatment.

11.4 Nutrient Uptake by Mixed Vegetation

11.4.1 Competition as a Concept

In this section, we discuss plant competition in general, relating to both managed and natural systems. After this, inter-root competition models for roots of different species are discussed, followed by examples of competing vegetation types.

The mechanisms underlying competition between individual roots below ground have been outlined in chapter 10, in relation to a single plant. In this chapter, we have discussed communities of plants, but so far they have been assumed to be dense and uniform single-species crops in which we ignore competition between plants. Here, we consider interplant competition. Competition for light may be important, and a full study of any competitive situation should use a model that incorporates both above- and below-ground processes. However, above-ground processes are secondary in this book, and we ignore them here.

Competition can be intra- or interspecific (Begon *et al.* 1986). Intraspecific competition between plants is important in agronomy, because it defines the best planting densities. As the number of plants increases, they start to interfere with each other and grow more slowly — in this state, the total biomass yield is almost independent of the density of the population. The economically important parts of crops usually reach a maximum yield at an intermediate density, and then decline. At densities larger than those used for commercial crops, the weaker individuals start to die off; this is particularly marked if the size distribution is asymmetric due to differences in germination date or non-uniform dispersion of propagules (Grace & Tilman 1990).

11.4.2 Interspecific Competition

Interspecific competition is more complex than intraspecific because root and shoot architectures may vary and processes such as allelopathy or physiological responses to competition may differ between species. Several different ways of expressing the data or conceptualizing the results are discussed by Begon *et al.* (1986) and Grace & Tilman (1990). The approaches of De Wit (1960) and Tilman (1987) deal specifically with nutrients and resources, but neither is sufficiently mechanistic to be useful in this context. The gap between the descriptive approach to plant mixtures and a detailed mechanistic analysis of their interactions still seems large.

Our earlier discussions concern soil-borne resource use, over times of less than one growing season. Plant communities develop over long periods and variable conditions. The component plants interact with many non-plant organisms, such as pests, diseases, or herbivores, and their long-term success or failure depends upon these interactions. Their reproductive performance is as important as their vegetative growth for long-term success, and this depends upon the plant strategy during its whole life cycle. Nevertheless, nutrients and water are essential for growth, and must be a basic part of any mechanistic explanation of plant competition and species fitness (Grime *et al.* 1988).

Species mixtures occur in both managed and natural vegetation. The managed situations include crop-plant/weed competition, where one competitor is to be advantaged and the other suppressed; tree crops with a groundcover, both of which have to be managed; agroforestry, where resource use by the two species needs to be carefully balanced; and peasant intercropping, where a varying suite of species is grown in almost random spatial mixture. The natural mixtures of species in woodland and grassland are even more complex, because of the irregular arrangement and age, and the difficulty of defining a 'unit cell'.

11.4.3 Nutrient and Water Uptake as Competition Mechanisms

Above- and below-ground processes are often difficult to separate (Donald 1963). The most obvious below-ground mechanism of competition is that the roots of one species deplete the soil so efficiently of nutrients or water that the other species absorbs less than it otherwise would have done (section 10.4.2). If this causes slower growth and smaller shoot and root systems, the disparity between the plants becomes greater, light competition develops, and eventually the less successful plant or species is eliminated. Second, the roots of one species may be able to exclude the roots of another, so that the rooting zones are sharply defined (Nye & Tinker 1977, p. 276). The size of the rooting zones for individual plants is thus limited, with obvious consequences for uptake of water and mobile nutrients, but there is not root-to-root competition in the sense we have described it in chapter 10. No full study of this effect appears to have been made, but it occurs within groups of plants of a single species, and a few experiments have reported intermediate degrees of exclusion between different species (Litav & Harper 1967) (figure 11.24). Exclusions could be caused by allelopathic effects, or by differential uptake of nutrients that change the root distribution in the zone of competition.

Models discussed here deal with partition of nutrients or water between two species by calculating the total uptake by methods similar to those outlined for single crops, and then use a rule for dividing the total uptake between the two species (e.g. Kropff 1993a; Kropff & Van Laar 1993). Conversely, a detailed model in our sense would calculate uptake by the two (or more) root systems separately, using the known distribution and properties of the root systems. The difficulty of this fundamental approach lies in modelling the uptake by interpenetrating root systems with differing uptake properties. An approach to such a model follows, based on the parallel-root steady-state approximation (section 10.5.2) (Nye & Tinker 1997).

We assume that roots belonging to one species have the same uptake properties, and that the root lengths are the same for both species. The important factors are the relative volumes of soil exploited solely by each species, and the relative speed of uptake in zones where the systems intermingle.

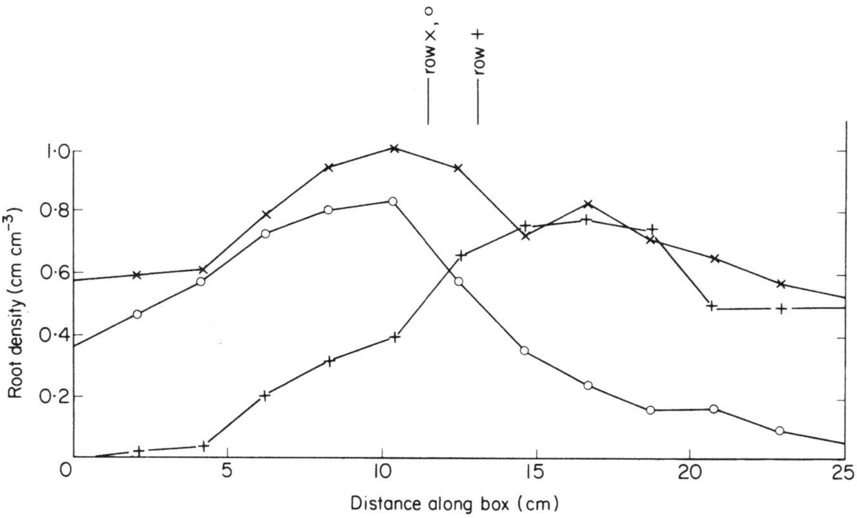

Figure 11.24 Lateral displacement of the roots of a row of onion plants by a similar adjacent row, planted across a 25-cm-long box: ×, row growing alone; ○ and +, rows growing together (after Baldwin & Tinker 1972).

The uptakes of mobile elements in unit volume of the mixed zone by roots from plant P and Q are (using L for L_V):

$$\frac{\mathrm{d}U_P}{\mathrm{d}t} = 2\pi(\alpha a)_P L_P C_{LP} \qquad \text{and} \qquad \mathrm{d}U_Q = 2\pi(\alpha a)_Q L_Q C_{LQ} \tag{11.13}$$

If the element is mobile, C_{LP} and C_{LQ} can probably be replaced by $\overline{C}_L$, the mean concentration in the soil solution throughout that zone. The relative uptake rates are then $(\alpha a)_P L_P/(\alpha a)_Q L_Q$. If the value of $\overline{C}_L$ is large (section 10.1.4), then $2\pi\alpha a$ may be replaced by I_{max}, and the relative uptakes will be $(L_V I_{max})_P/(L_V I_{max})_Q$.

For non-mobile elements, the situation is more difficult. The steady-state approximation (equation (10.27)) can be applied directly if the uptake characteristics of the two root systems are similar, and uptake is then proportional to root length within the equivalent cylinder, as is often assumed. However, if the αa values differ, the depletion zones of the two root types will be of different extent and intensity, and the zero-transfer boundary between the roots will not be equidistant from them. We assume a regular array as before, and that the two types of root are regularly intermixed with different equivalent cylinder radii:

$$1/\sqrt{[\pi(L_P + L_Q)]} = x \tag{11.14}$$

where x is the mean equivalent radius, using the same argument as in section 10.3.2. The concentrations must be the same at the zero-transfer boundary; if the latter is distance y from roots P, then from equation (10.21)) we have

$$\begin{aligned} C_{Li} &= C_{LP}(1 + (\alpha a)_P(\ln\ y/a)/Db) \\ &= C_{LQ}(1 + (\alpha a)_Q\{\ln[(x - y)/a]\}/Db) \end{aligned} \tag{11.15}$$

Hence,

$$C_{LP}/C_{LQ} = [Db + (\alpha a)_Q\{\ln(x-y)/a\}]/[Db + (\alpha a)_P \ln(y/a)] \qquad (11.16)$$

We have assumed so far that the position of the zero-transfer boundary remains constant, relative to the root P and the root Q, and allowing for the decrease in x with the development of the root systems. In the real transient state, it will move if the root uptake characteristics are different, and if uptake continues over a period. However, if this movement is ignored, so that x/y is taken to be constant, the mean decrease in concentration in each depletion zone will have to be the same over time, so that

$$I_Q y^2 = I_P(x-y)^2 \qquad (11.17)$$

Substituting $2\pi\alpha a C_L$ for I, we get

$$C_{LP}(\alpha a)_P/y^2 = C_{LQ}(\alpha a)_Q/(x-y)^2 \qquad (11.18)$$

If the two equations (11.16) and (11.18) are solved numerically, then C_{LP}/C_{LQ} can be found, though the final expression cannot be made explicit in y. The I values and the uptakes for each type of root can therefore be found.

The use of this approximation is only possible if the αa of the two root types are not grossly dissimilar, because if they are, the relative position of the zero-transfer boundary will change markedly during uptake. A more exact solution requires a numerical model using a very fine two-dimensional spatial grid, so that the true concentration distributions in the plane at right angles to the roots can be calculated. The electrical analogue system (section 10.4.3) can be used to solve particular problems of this type.

11.4.4 Interspecific Competition Experiments in Pots

Chapter 11 formally deals with crops and vegetation in the field. However, modelling of competing plant species is at so early a stage that it appears best to start with relevant pot experiments. Baldwin (1976) produced an early plant growth and nutrient uptake model with nitrogen as the only nutrient, probably the first mechanistic model to deal with two competing plants of different species. It focused on that type of competition in which all roots of both plants intermingled at random in the same space. A number of runs were made with 'species' with single characteristics that were different, and 'pots' with either all plants of one type or the other, or an equal mixture, all at three nutrient levels and with different pot sizes.

The results are too complex to summarize in full, but are shown in part in figure 11.25. Out of the 12 tests, three gave higher yields with the mixture (AB) than with plants of either species alone, which is a particularly interesting result of plant competition (section 11.3.2). Where one crop greatly exceeded the other in growth, this was due to more complete exploitation of nitrate at depth fairly early in the run; sometimes, this showed the advantage of having few seminal roots that extended rapidly. This model gives a valuable insight into the mechanisms of competition, but it needs to be compared with real plants to ensure that all the

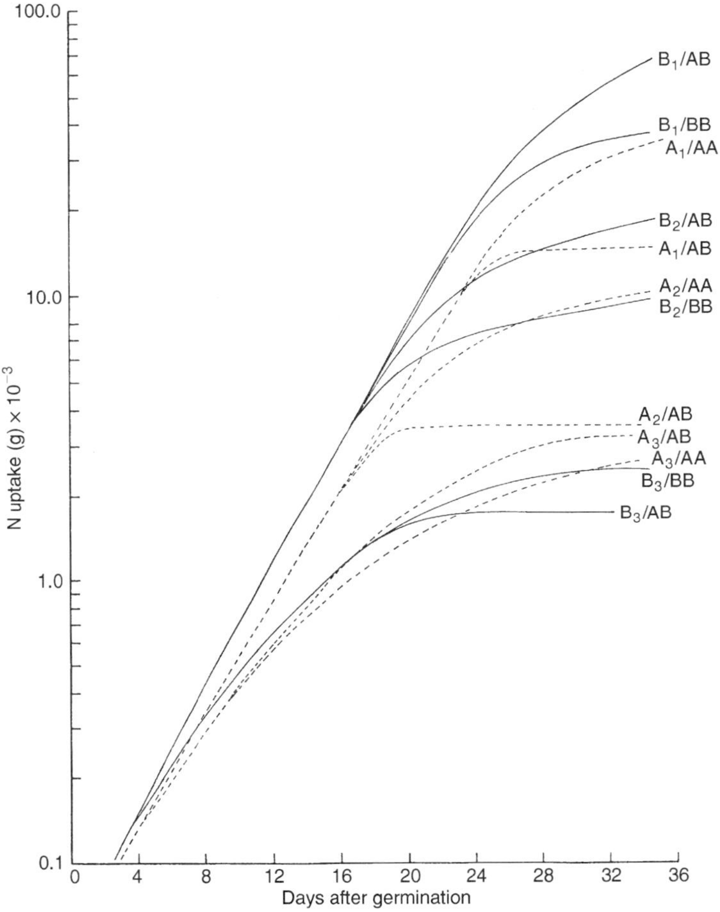

Figure 11.25 Simulated growth of two plant species (A, B) with properties defined separately (AA and BB) or together (AB). Coding A/AB is for the A plant in a mixed pot, etc., subscripts 1, 2, and 3 refer to high, medium, or low nutrient status, respectively. The 'competitor-type' plant B strongly outgrows A at high nutrient levels, but is inferior in growth at low levels (after Baldwin 1976).

important characteristics have been simulated. Ultimately, the root architectures of the two species will have to be included (section 10.4.4).

Smethurst & Comerford (1993a) used the steady-state approach to develop an uptake model (COMP8) (section 10.5.3) that can also be applied to competing root systems. Several variants of COMP8 were also tested, to determine which improved features were most important. Smethurst & Comerford (1993b) modelled the rather complex system of slash pine (*Pinus elliotii*) and a grass (*Panicum aciculare*), growing together in pots, in a careful experiment designed to gain an insight into the way understorey plants may compete with tree growth in forest plantations, at the early stages when L_V for the understorey may be much greater than L_V for the trees. The model COMP8 was applied to P and K uptake in various combinations of numbers of plants in pots in the glasshouse, with two levels of P and K (Smethurst & Comerford 1993b). The uptake parameters were taken from the literature. L_V for the pine was of the order of 0.1 cm cm^{-3}, whereas that for grass was 2–7 cm cm^{-3}. The potassium inflow and concentration for pine was always smaller when it was grown with grass than alone, and was usually smaller for phosphate (tables 11.5 and 11.6). Overall, grass absorbed much more nutrient than pine during the experiment. Pine roots were always less effective than grass roots in uptakes per unit surface. In the high-nutrient pots, simulated and observed uptakes were similar for uptake by pine and for potassium uptake by grass, but not for phosphate uptake by grass.

Where nutrient concentrations were low, the observed nutrient uptake was almost always considerably larger than the predicted one for both species (table 11.6). Only one simulation of potassium uptake by pine and one of phosphorus uptake by grass gave results similar to those observed. The uptake parameters for the plants were taken from the literature, and not measured as part of the experiments. Some of the simulation results could be made much closer to observation by changing these parameters, but other predictions for the low-nutrient treatments remained less than the actual uptake when using any reasonable value of

Table 11.5 The biomass, (L_V), P and K composition, and uptakes of P and K, in total and per 100-cm root, for pine and grass grown together in different ratios and nutrient levels.

	K Uptake		P Uptake					
Treatment[a]	Total (μmol)	Per 100-cm root (μmol)	Total (μmol)	Per 100-cm root (μmol)	L_V ($\times 10^4$ m m^{-3})	Biomass (g per pot)	Shoot K (g kg^{-1})	Shoot P (g kg^{-1})
Experiment 1: Pine								
L20	87	19	20	4	0.053	0.715	6.4	1.1
L22	15	4	13	4	0.038	0.470	2.2	1.4
Experiment 1: Grass								
L22	568		190		3.36	4.615	8.6	1.9
Experiment 2: Pine								
H20	167	15	134	12	0.129	1.34	4.2	3.5
H22	57	6	74	8	0.102	0.84	2.6	2.9
H28	36	4	29	3	0.091	0.72	1.2	1.7
H82	177	5	309	8	0.424	3.75	1.9	3.0
Experiment 2: Grass								
H22	1273 b		580		2.22	7.22	8.6	2.8
H28	2084 a		1132		7.25	12.39	7.0	2.6
H82	1210 b		379		2.07	5.19	5.8	1.2

Source: after Smethurst & Comerford (1993b).

[a]L and H indicate low and high nutrient levels, respectively. The first and second numbers in the code indicate the number of pine plants per pot and of grass plants per pot, respectively.

Table 11.6 Comparison of predicted and observed values for P and K uptake by pine only or pine plus grass in pots.

Treatment[a]	K Uptake		P Uptake	
	Observed (μmol)	Predicted (μmol)	Observed (μmol)	Predicted (μmol)
Experiment 1: Pine				
L20	87.3	86.5	20.6	6.2*
L22	15.5	24.6*	13.4	3.2*
Experiment 1: Grass				
L22	568.0	1043.0*	189.5	129.0*
Experiment 2: Pine				
H20	167.4	353.0*	134.5	131.0
H22	57.4	68.0	74.7	94.0
H28	36.5	31.6	29.6	47.9
H82	177.2	259.6	309.4	387.9
Experiment 2: Grass				
H22	1272.6	1291.3	580.3	1103.6*
H28	2089.7	1637.6	1133.2	1832.5*
H82	1280.3	1057.3	421.1	882.1*

Source: after Smethurst & Comerford (1993b).

Asterisk indicates a significant difference between observed and predicted uptakes. Some of the differences could be explained as due to inappropriate Michaelis–Menten uptake characteristics. Coding as for table 11.5.

the uptake characteristics. However, the *relative* uptake by pine with and without competition by grass was reasonably well predicted (figure 11.26).

Smethurst & Comerford (1993a) also carried out a sensitivity analysis using COMP8 to investigate the effect on species 1 of varying the properties of species 2. As expected, increasing the root length of species 2 had a strong negative effect on the uptake of species 1, but the important effect of water content was presumably a direct effect on uptake by both species, rather than a competition effect.

It seems likely that despite the occasionally poor agreement with observation, the steady-state model, as modified by Smethurst & Comerford (1993a), gives

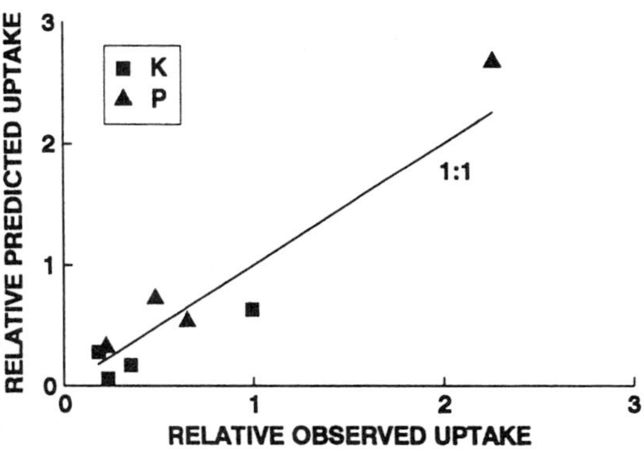

Figure 11.26 Comparison of the relative uptake of P and K by mixed pine and grass compared with uptake by two pines, as measured and as predicted by the model COMP8 (after Smethurst & Comerford 1993b).

results that are acceptably near to transient state models, that have great flexibility in terms of changing conditions during the simulated periods, and that may be able to deal with competition between species of plants, and other real complexities of vegetation in the field. The Smethurst & Comerford (1993a) model was added to a routine for calculating the effects of competition between dissimilar roots from different root systems, based on an equation from Nye & Tinker (1977) (section 11.4.3). This is the first model to attempt to deal with below-ground competition in a detailed mechanistic way, and as such is an important step.

11.4.5 Crops and Weed Competition

The balance between crops and their weeds can be of extreme economic importance. The competitive effect of weeds is very strongly dependent upon stage of development of the weeds relative to that of the crop (Radosevich & Rousch 1990). The results of experiments are usually stated in terms of yield loss related to weed population, compared with a weed-free crop. This yield loss is very highly dependent upon the period during which the weed infestation exists. For example, figure 11.27 (Azmi & Mashor 1990) shows how the yield of direct-seeded rice is affected by the weed grass *Echinochloa cruz-galli*. If the crop is kept weed-free for 30 days, infestation later has little effect. Similarly, infestation before 15 days causes relatively little loss. The critical period, which must be kept weed-free, is between 15 and 30 days. Weed competition is therefore a very dynamic and complicated process.

It is normal in field experiments for the crop plant to be at a constant density, and the weed plants to be additive to this, at a widely varying density that depends upon seed burden, germination conditions and intraweed competition (Radosevich & Rousch 1990). It is therefore possible to develop a matrix table of the weight of a single crop plant, with the rows and columns containing,

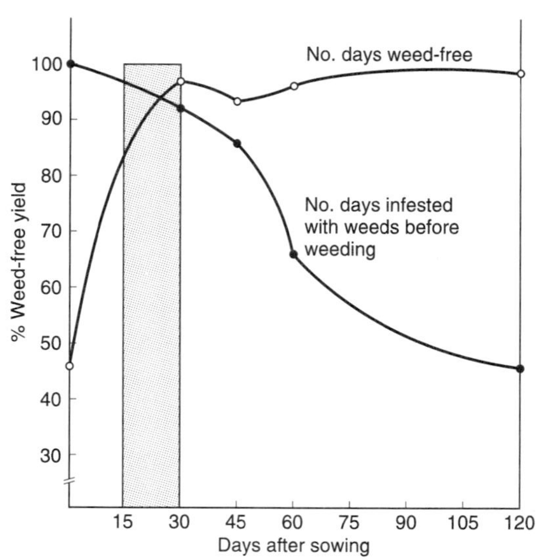

Figure 11.27 Effect of competition by barnyard grass (*Echinochloa cruz-galli*) on the yield of direct-seeded rice, relative to a weed-free crop. Plots were kept free of weeds up to various dates, or weeds were allowed to grow after various dates. Weeds growing before 15 days, or after 30 days, had little effect on yield. Cross-hatched area denotes the critical period of competition (after Azmi & Mashor 1990).

respectively, the ratio of the species and the total plant density. As the ratio of weeds to crop plants change at constant plant density, the weight of a single crop plant will alter, and the effect due to crop plants and to weed plants can be compared. In this way, Concannon (see Radosevich & Rousch 1990) found that one wheat plant was equivalent to seven grass plants in its competitive effect on another wheat plant. However, the data from this type of experiment give no indication of the mechanism of competition in terms of resource deprivation. A similar analysis is possible with data such as that in Grace & Tilman (1990, p. 345).

Whole-crop models will eventually have to include submodels for weed competition (as well as diseases and pests). A mechanistic/empirical model has been formulated for water and nitrogen uptake by a crop/weed mixture (Kropff & Van Laar 1993). The first step is the partitioning of incoming radiation between the two species in the mixed canopy, and the calculation from this of their potential growth rates. The application of these equations to a patchy and variable weed infestation requires that the model is run for each patch separately.

The general approach to competition for water is to calculate the potential evapotranspiration for a short grass cover by the Penman equation, deriving the radiation from the number of sunshine hours. The potential evapotranspiration of each species was calculated from the radiation intercepted by the two species in the mixed plant canopy, and that reaching the soil. The actual evapotranspiration was taken to decrease linearly with soil water content below a given critical value, which depended upon the plant species. The impact of water supply on the growth of the two species was taken to be their potential growth rate multiplied by the ratio of their actual to potential evapotranspiration. The division of water between the species thus depended upon their relative interception of radiation, though Kropff (1993a) suggested that root length density could be used as an additional control.

This model is greatly simplified by assuming that both the crop and weed species have the same rooting depth and distribution, as the model assumes a thin surface soil layer of 2 cm that contains no roots, and all the rest of the 'rooted depth' is a single layer. Soil evaporation from the surface layer, and capillary movement into it, is calculated separately. If differences in rooting depth were important, a multiple soil layer model would have to be used.

The method of dealing with nitrogen is rather similar to that in other crop models, in that it is based on deficiency demand (Kropff 1993b). The potential uptake rate of nitrogen by each species is related to their maximum N content by an equation of the form

$$\mathrm{d}U_{\mathrm{N}}/\mathrm{d}t = (WX_{\mathrm{N}max} - U_{\mathrm{N}act})/T \tag{11.19}$$

where $\mathrm{d}U_{\mathrm{N}}/\mathrm{d}t$ is the potential rate of uptake of N of that species, $U_{\mathrm{N}act}$ is the actual quantity of N in that species, W is the biomass weight, $X_{\mathrm{N}max}$ is the maximum concentration of N, and T is the time for this amount of nitrogen to be taken up (set arbitrarily at 2 days). As with water, spatial heterogeneity of the weed infestation could make the application of this equation difficult.

The actual N uptake is the lesser of these calculated potential rates of uptake for crop and for weeds, and the amount of mineral N in the soil layer that contains roots. If the mineral N in the soil is not sufficient to meet the joint demand, then the nitrogen available for uptake in that time step is divided between the species present, in proportion to their root lengths and to their calculated demands, so that

$$\delta U_{Ni} / \sum \delta U_{Ni} = L_i / \sum L_i \tag{11.20}$$

where δU_{Ni} is the uptake to species i (so long as it is less than its demand), L_i is the root length per unit area of species i, and both are divided by the respective sums over all species. The simple and convenient rule of allocating uptake by root length has not yet been sufficiently tested. It is important that it should be.

Such methods ignore differences in below-ground factors such as rhizosphere and root resistance to water flow, rooting depth, or differences in distribution of root length with depth. Results of applying the model to field crops are given in Kropff & Van Laar (1993), showing that reasonable success can be achieved, but the authors suggest that the formal models are better suited to determine simple practical rules to decide upon the damage caused to a crop by weeds, rather than being applied directly to field situations.

11.4.6 The Grass–Legume System

This is one of the best-known competitive systems in agriculture. The model of Thornley *et al.* (1995) deals with the nitrogen nutrition of this system. Basically, it addresses the well-known fact that there is a balance between the legume that can fix nitrogen from the air but is not an efficient uptake competitor for nitrogen, and the grass that is wholly dependent upon root uptake, at which it appears to be very efficient. Thornley *et al.* (1995) show that steady-state equilibria can be obtained, but also that oscillatory conditions can arise, especially concerned with the delay in transferring nitrogen from litter to soil organic matter and then to mineral nitrogen. The uptake are defined in terms of the mineral nitrogen level in the soil, and the specific rates of uptake and of fixation per unit amount of root (figure 11.28) (Thornley *et al.* 1995). This simple method of simulating the uptake system appears to be adequate to produce sensible results in accordance with general experience, but we are not aware that the model has been compared with plant data. The uptake step itself is not modelled, though root length is involved.

11.4.7 Mixed Cropping Systems

Mixed cropping systems include all those in which more than one species are deliberately grown together on the same land area. These systems include 'inter-cropping', which usually means a combination of two annual crops, and 'agro-forestry', in which a tree and an annual crop grow together. The latter is more difficult to deal with because the competing plants may have very different sizes and life cycles. Both systems have a long and somewhat contentious history, with strong arguments about whether they can yield more per unit area than mono-cultures (Sanchez 1995). It is important to be clear about the mechanisms that

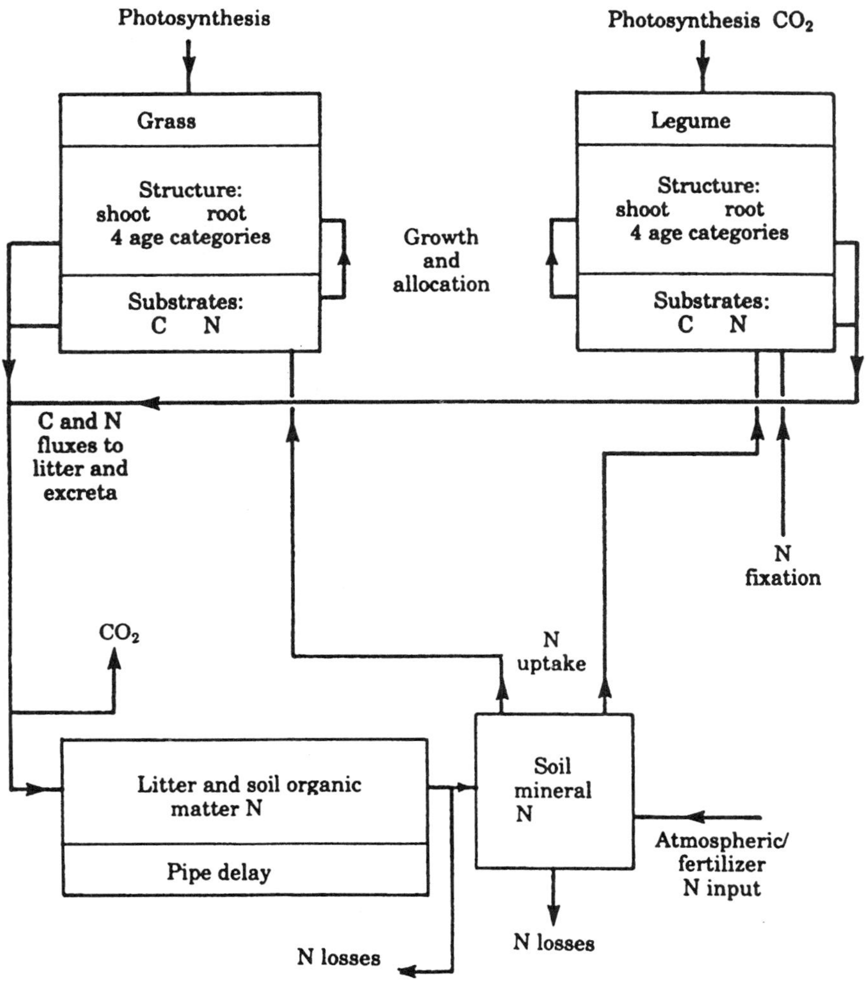

Figure 11.28 Grass–legume C–N competition model with N cycling. The N loss includes volatilization, denitrification, and leaching (after Thornley *et al.* 1995).

may control total yield (Ong 1995). First, there may be nutritional transfers between the crops. If one species is nitrogen-fixing, the nitrogen may be made available to the partner (Ofori & Stern 1987). However, the amount of such transfers may have been overestimated in the past (Ong 1995), and isotopic tracer experiments do not support the idea of large transfers in cereal–legume mixtures (Giller & Wilson 1991). Deep-rooted species may absorb various nutrients at depth, and deposit them on the surface in litter and prunings, to the advantage of a shallow-rooted partner.

Second, competition processes must reduce the growth of a non-dominant partner species. It is only in the last few years that the extensive root systems of some agroforestry trees, such as *Leucaena leucocephala*, have been recognized. There are cases where rooting of trees under sole-crop control plots in field experiments has depressed the yield there, and thus given an exaggerated idea of the benefits of intercropping (Ong *et al.* 1996). Experiments have to be designed with very large guard belts of several metres around plots to prevent this effect.

Third, there are a number of physical effects. If the system involves the spreading of prunings from the tree species around the crop, the physical condition of the soil will be improved and more soil moisture will be retained, in addition to any nutritional benefit. The microclimate may be improved by the larger species acting as a shelter. Finally, soil conservation will be aided by both the spreading of prunings and the physical prevention of erosion by trees planted as hedgerows. This last is a major and proven benefit in some areas (Huxley & Ong 1996).

With so many possible effects, it is difficult to predict what the final outcome will be. Too often, it has been assumed as a matter of course that the net effect will be beneficial in terms of total yield. It is now clear that this outcome can only be predicted on the basis of a clear physical and chemical understanding of the processes involved and the properties of the partner species (Sanchez 1995). Here, we discuss the below-ground competitive processes only.

A simple equation has been used to separate different mechanisms that influence yield in a set experimental design (Ong & Black 1994; Ong 1995):

$$I = F - C + M + P + L \tag{11.21}$$

where I is the total effect of intercropping in terms of yield per unit area relative to that of sole crops. This overall effect is then divided between the various contributing processes: F is general soil fertility, C is interplant competition, M is microclimate, P is soil physical properties, and L is effect of soil loss. The equation assumes that there are no interactions between the effects, which is probably not true. Values for F and C can be found from the yields on a set of plots with and without fertilizer and competition. Ong (1995) showed how competition against the maize crop increased over the years with one of the two tree species tested, whereas the fertility benefit for maize was about equal for both tree species. The size of the physical changes in M, P, and L can be measured directly on the agroforestry or intercropped plots, but the yield losses or gains that are likely to be caused by these changes have to be derived from further experimentation. The results are specific to species, soils, climate, and agronomic system.

The separation of competition into above- and below-ground effects must be done by separation of the root systems by a physical barrier. Thus, Singh *et al.* (1989) grew several crops between hedges of *Leucaena*, with a polythene barrier to root penetration down to 0.5 m on one side of the crops only. Above-ground competition was therefore the same, but with the root barrier, the yield of the crops was much higher than without (figure 11.29). A detailed study of competition for water resources (Ong *et al.* 1996) showed the importance of complementarity—unless the combined root systems can utilize resources that are not used by both single species alone, intercropping brings no advantage. In many cases, there is little difference in the total water use by an intercrop or its component sole crops, and hence there is no reason to expect better utilization of water or better total yield.

The matter is however complicated by the fact that the two species frequently have different durations, and some may be perennial, so that use of different resources occurs over time rather than over space (see section 11.1.3). Thus, an experiment with a millet–groundnut intercrop (Reddy & Willey 1981) showed that

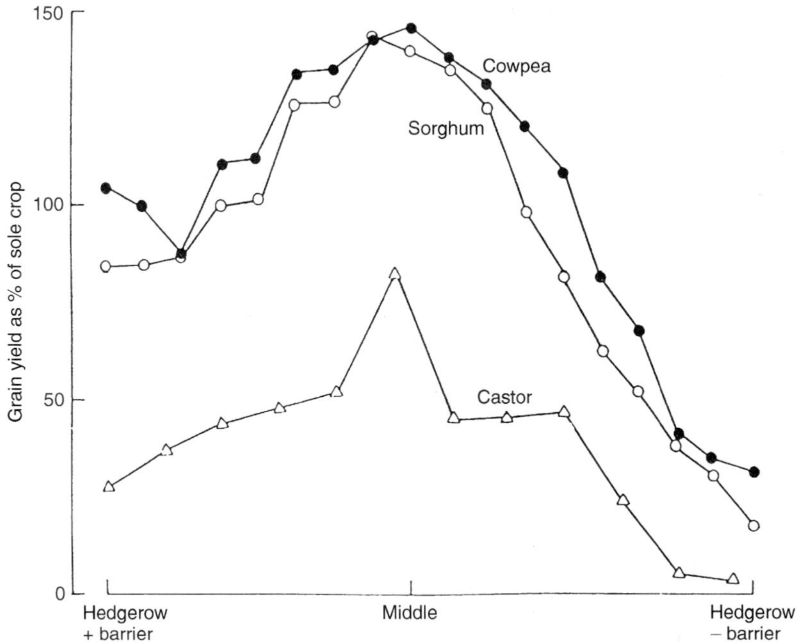

Figure 11.29 Grain yield of sorghum, cowpea, and castor growing between *Leucaena* hedgerows in Hyderabad, India. An 0.5-m-deep polythene barrier placed between the crop and trees on the left-hand side of the alley removed much of the competitive effect for the crops (after Singh *et al.* 1989).

this used 10% more water than the sole groundnut, and 34% more than the sole millet, on the same ground area. The reason, in this case, was that the mixed crop maintained Leaf Area Index above 2 for over twice as long as the sole millet, so that more of the soil water was tapped. Other examples with spatial complementarity are discussed by Ong *et al.* (1996).

The availability of the heat balance method (Ishida *et al.* 1991; Ong *et al.* 1996) of determining fluxes of sap in individual stems or roots should now allow measurements of the extraction of water from different layers and compartments in the soil, so that water uptake in intercropping systems can be precisely measured from different soil zones. The design of mixed cropping systems with higher yields than the component species should now be possible, though measuring the behaviour of mixed root systems in the field will always be difficult (section 9.4.4).

11.5 Natural Vegetation

11.5.1 Competition in Natural Vegetation

It is as yet impossible to apply the concepts in this book accurately to mixed natural vegetation that contains an irregular mixture of different species (Chapin 1980, 1988; Tinker 1990). The basic practical difficulty lies in separating the fluxes of nutrients to different plants when there are no replicate plants or areas, in deciding where these fluxes originate from in the soil, and in measuring the dynamics of different interpenetrating root systems. However, at the least the methods and concepts described here can help us to understand these systems in a semiquantitative way (Tinker 1990).

Most natural vegetation results from a competition between species. Caldwell (1987) has described the complex interactions in multispecies vegetation, including allelopathy and infections by the same fungal network (section 8.3.8). Below-ground competition is important in most conditions, and Caldwell (1987) quotes seven studies similar in type to the classical work of Donald (1958, 1963) with plants that grew together, with partitions between either roots or shoots, all of which found that inter-root competition was more important than intercanopy competition.

Common sense suggests that competitive advantage depends upon the root density, and the work of Ennik & Baan Hofman (1983) showed that the yield reduction caused to the grass *Elytrigia repens* by competition with various clones of ryegrass was in direct proportion to the root biomass produced by the clones when grown alone. However, root density alone cannot always explain the outcome. Caldwell *et al.* (1985) measured the uptake of ^{32}P and ^{33}P from labelled soil volumes equidistant from a central plant of *Artemisia tridentata* and plants of either *Agropyron spicatum* and *A. desertorum* (figure 11.30). Despite the fact that the *Artemisia* had equal densities of roots in both volumes (0.7 cm cm^{-3}), it took up over six times more labelled P from the soil in which it was competing against *A. spicatum*, implying that *A. desertorum* was a stronger competitor. The *A. desertorum* had a larger root density than *A. spicatum* (2.7 against 1.6 cm cm^{-3}), but this seems insufficient to cause such a strong effect. The total root density was so low that inter-root competition for phosphate should have been fairly small. The *A. desertorum* had a somewhat heavier mycorrhizal infection than *A. spicatum*, with much more frequent entry points. If the mycorrhizal network was common to all plants, the competitive effect could operate through the relative sink strengths for phosphate of the grasses (section 8.3.8).

Allelopathy (section 9.3.3) may also be a major mechanism of competition and interaction in natural vegetation (Rice 1984), though the exact processes are often uncertain. It is difficult to distinguish between nutrient/water competition and allelopathy in the field, and the processes need to be better understood in simplified systems.

Many experiments show that nutrient uptake competition occurs, and differs markedly between similar competitor plants, but it does not identify precisely how the competitive mechanism operates. It is thus often difficult to explain even quite straightforward experiments, though understanding the behaviour of natural systems under different environmental conditions must depend on understanding the competitive processes. Models of natural vegetation that use the detailed concepts described here might be suitable for this, but it will always be difficult to obtain the necessary plant information for a set of intermingled species (section 9.2.5). Much more research is necessary on simple competitive systems before moving into truly natural mixed vegetation.

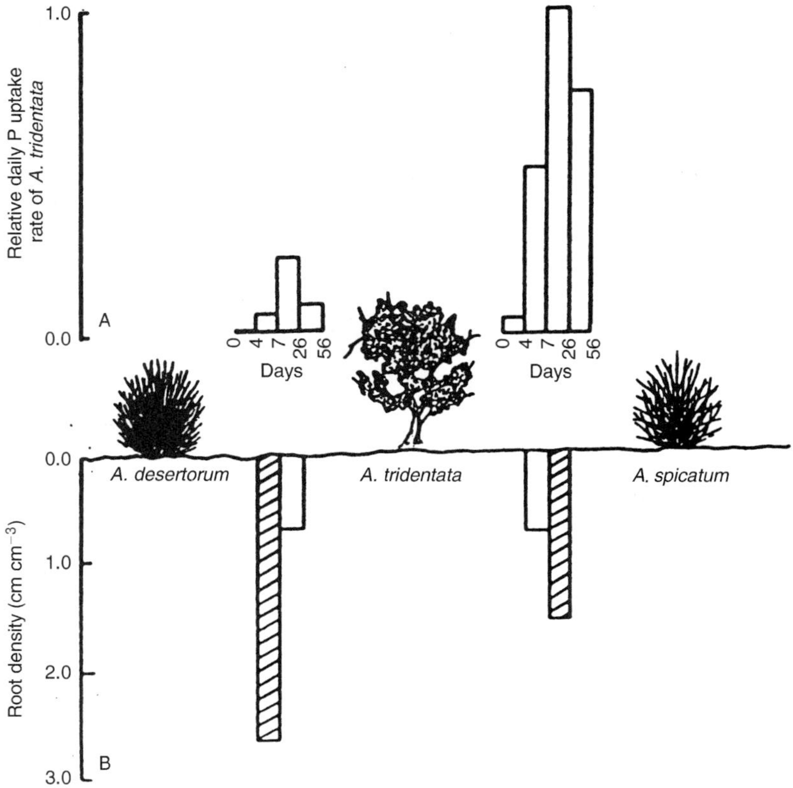

Figure 11.30 (A) Relative rate of uptake of ^{32}P as measured by activity in shoot tips of *Artemisia tridentata*, when the isotope was injected in the interspace between *A. tridentata* and either *Agropyron desertorum* or *Agropyron spicatum*. (B) Root length density of *Agropyron* species (cross-hatched) and *Artemisia* (open) in the interspaces where the isotope was placed. The *Agropyron* density was greater than the *Artemisia* density, but differences between the two interspaces were not significant (after Caldwell *et al.* 1985).

11.5.2 Models for Growth and Water and Nutrient Uptake by Natural Vegetation

It is not easy to summarize the many models that are applicable at very different scales and complexities. A few are mentioned here to indicate the variety. Where a nutrient is included in a model, it is almost always nitrogen, and many of the larger scale models work on the principle of a carbon–nitrogen balance or the carbon and nitrogen cycles (Tiktak & Van Grinsven 1995). Mohren *et al.* (1994) have discussed the contrast between mechanistic biological models and the more empirical management models for woodlands.

Thornley & Cannell (1992) constructed a carbon–nitrogen model of a forest plantation by combining a transport-resistance model of growth (section 9.5.3) with a soil organic matter model. The nitrogen uptake submodel is of interest, and the central uptake equation, in a slightly simplified form, is

$$dU_N/dt = [W_R f(T) C_{Neff}/(C_{Neff} + K_n)] \times 1/[1 + (K_c/C_f)(1 + N_f/J)] \quad (11.22)$$

where dU_N/dt is uptake rate per tree stem; C_{Neff} is the effective soil concentration of mineral N; W_R is the mass of fine root and mycorrhizas per tree stem; C_f and N_f are the carbon and nitrogen substrate concentrations in the fine root; $f(T)$ is a temperature function; and K_n, K_c and J are parameters, K_n being set at 0.005 kg N m^{-2}. The whole equation therefore contains a general Michaelis–Menten uptake relationship for mineral nitrogen in the soil (not expressed as solution concentration) (section 5.3.2), the amount of fine root and mycorrhizas, and the internal C/N balance within the fine root and mycorrhizas of a tree stem, which can act as a regulator (section 5.3.3). All the terms in the equation can therefore be recognized from discussion in this book, but in an empirical form.

The G'DAY model of Comins & McMurtrie (1993) for forests is a combination of several other model structures (McMurtrie *et al.* 1992), dealing with growth and soil processes that use and produce mineral nitrogen (figure 11.31). Growth is predicted, usually over a period of several years, from a basic growth rate that results from absorption of all incident photosynthetically active radiation under standard conditions. Functions for leaf N/C ratio, *P*hotosynthetically *A*ctive *R*adiation (PAR) fraction absorbed, temperature and atmospheric CO_2 level modify this basic growth rate.

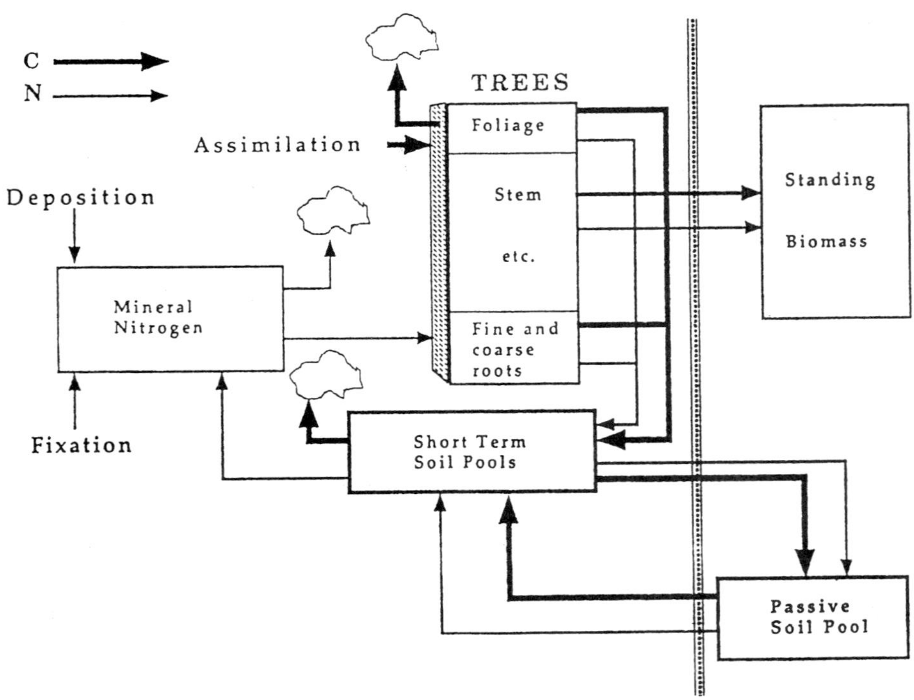

Figure 11.31 Model diagram of C and N dynamics in a forest. The model is updated daily, but the two pools outside the line are updated at much longer intervals (after McMurtrie *et al.* 1992).

Nitrogen in the soil is modelled according to the Century model (Parton *et al.* 1987), with seven different carbon and nitrogen pools in various forms of litter, microbial organic matter (biomass), 'slow' organic matter and 'passive' organic matter. Decay rates are first-order and depend upon temperature and moisture level. The N/C ratios of all pools are fixed values, except for two of the litter pools, where they depend upon the leaf N/C ratio. A set fraction of carbon is metabolized at each transfer between pools, and the mineralized nitrogen is computed from the surplus nitrogen left after all transfers have been evaluated. It is assumed that all mineral nitrogen is absorbed by the growing trees, except for 5% lost to the atmosphere. The effect of nitrogen is simulated by making the growth rate linearly dependent upon the N/C ratio, when the value of the latter is less than a set critical level (McMurtrie *et al.* 1992).

Nitrogen uptake and other flows within the plant–soil system are modelled as a mass balance. The 'input' is partitioned amongst stem growth via the N/C ratio, sequestration into a soil carbon pool, and atmospheric loss following on breakdown of leaf and root litter. The model has a long timescale, so it ignores the nitrogen uptake step into the root. It can therefore be described as a 'model without roots'.

Certain general ideas underlie natural vegetation models. The growth rate and water requirement are set by the incident radiation and the canopy structure. The demand or deficiency of a nutrient is related to its percentage composition or to the nutrient/carbon ratio. The supply of mineral nitrogen is generated in the soil, which is often represented by a single layer, and uptake is the least of demand or supply. There may be a term that relates to root efficiency. Nitrogen is normally the only nutrient considered. Compared with the approaches described earlier in this book, the representation is extremely simple. However, the scope of some of these models is enormous; some include socioeconomic modules with a biophysical core (Alcamo 1994). There is a huge range between fine-scale precision and wide-scale comprehensiveness, but it is to be expected that as computers become more powerful and the relative success of models better identified, greater complexity in submodels will be introduced.

Shifting cultivation is a particularly interesting system, because it rotates from woodland competition to intercrop competition, and perhaps something like agroforestry. The dominant process is the build-up in nutrients in the soil and the vegetation during the bush fallow stage. This process is the core of a model (ROTATE) of the N cycle in this form of cropping (Mobbs & Cannell 1995).

11.5.3 Modelling of Vegetation at Large Scale

First, we discuss a large biome model, the Hybrid.3 (Friend *et al.* 1997). This deals with the linked water, carbon, and nitrogen cycles between soil, plant and atmosphere, with many canopy layers and one soil layer. The growth of individual trees is simulated on an annual, and grass on a daily time step. All calculations are based on a 'plot', which is intended to be approximately the size of the canopy of a dominant tree. Vegetation can be classified as species or as particular 'generalized plant types', functional types as normally used in ecology.

Competition occurs for light, water, and nitrogen. The model is mainly intended to examine questions of global change, and can simulate transient changes as well as step changes. A sensitivity analysis indicated the importance of the ratio between C/N in the foliage and in the fine root. The size and complexity of the model necessitates that the nitrogen uptake step is simplified, but it contains elements similar to those we have seen:

$$\mathrm{d}U_N/\mathrm{d}t = k\, f(T) C_R C_{\mathrm{N}mineral} C_v/N_v \tag{11.23}$$

where $\mathrm{d}U_N/\mathrm{d}t$ is daily uptake of nitrogen, k is a constant, $f(T)$ is a temperature correction factor, C_R is the carbon mass in the fine root, $C_{\mathrm{N}mineral}$ is the soil mineral nitrogen content, and C_v and N_v, respectively, are the carbon and the nitrogen contents in the non-heartwood parts of the plant. Thus, $C_{\mathrm{N}mineral}$ represents soil supply, C_v/N_v is plant demand, and C_R is rooting efficiency. The approach is related to that of Thornley (1991).

The often quite primitive below-ground parts of large-scale models, some of them otherwise very sophisticated, needs some comment, as appropriate root–soil models have been available in the literature for many years (Yanai 1994). Measurements below ground are difficult to make, but there may be more fundamental reasons. Most major species include complex phases in their growth cycles, such as reproduction, tillering or storage organ formation, that demand priority attention. More complex root modules may follow when these developmental priorities have been fully met.

There are also good reasons why the mechanistic detail in a model should decrease with the scale of its application (Jarvis & McNaughton 1986; Luxmoore *et al.* 1991; Nye 1992a). As the scales shift upwards from large pot experiments to biome models, some of the reductionist detail becomes difficult to use or irrelevant. We then get 'models without roots', in which the link between the nutrients in the soil and the plants growing in it is expressed in a simple mathematical relationship (Kaufman & Landsberg 1991).

However, in the long run, more complex below-ground models will be required. The need for a scientific understanding of plant growth, as well as current demands for more sustainable agriculture must require more accurate control of plant nutrient use, and better control of the below-ground environment.

11.6 Conclusion

11.6.1 The Application of Uptake and Crop Models

The applications of crop models is not fully agreed. Most are verified to some extent and partially validated ('verify', establish the truth of; 'validate', make valid — *Oxford English Dictionary*) by comparison with plant data, but some argue that it is not essential to 'fit' complex models to real data-sets, because many models are so flexible that a fit can normally be obtained by changes in the parameters. However, unless this comparison is made, and unless the results give some encouragement that the model approximately simulates the real world, it is difficult to see how we make progress. It is also essential to compare the output of different models of the same system, using the same data-sets. A large-scale comparison of 10 different wheat crop models organized by the International Geosphere–Biosphere Programme gave widely different results for biomass and grain yields (Goudriaan 1996), though a comparison of five rice models (Ingram 1997) gave much closer agreement. The reasons for disagreement must be investigated.

11.6.2 Practical Uses of Nutrient Models in Farming

Much of the variation in yield from field to field of well-fertilized and managed crops is still not explained (Tinker 1985; Weir *et al.* 1984b), and this suggests that the effect of soil conditions on crop growth is still not properly understood or taken into account. This variation requires explanation, and proper below-ground models may assist in this.

Nitrogen is often the only nutrient that is regarded as sufficiently important to model for practical purposes. In contrast, potassium and phosphate are usually in ample supply in intensive agriculture, and their application rates may be guided by simple element balances or by leaf analysis systems. The modelling of their uptake may therefore not be seen as urgent.

Potassium is tractable for modelling in the soil, but potassium is a cheap fertilizer, and there is not a great premium for high precision in its use. It is likely to be more relevant to model potassium uptake for tree crops, where the potassium uptake is relatively high, and the root density is relatively small (Nye & Tinker 1977, p. 271).

For phosphate, the variety of plant mechanisms for enhancing the uptake cannot be dealt with easily. However, phosphate deficiency is frequently growth-limiting in tropical soils, and modelling may be an efficient way to make progress, if we can learn how to deal with these mechanisms.

Models for agricultural application have already been produced, aimed at determining more accurate rates, positions or times of application of fertilizer. Barber (1995) has strongly advocated the use in this way of the various models that have been produced by his group. Van Noordwijk *et al.* (1990) have produced a model expressly to help in deciding phosphorus fertilizer rates. At present, there is much interest in 'precision farming' and the use of models of this type could be valuable.

However, there are several questions regarding such applications. First, it is difficult to obtain all the data necessary to run some models, particularly on spatially variable soils and crops. Second, site-specific model predictions may turn out to be less reliable than the rough but well-tested methods in use now. Finally, the extra accuracy obtained with models may not warrant the expense of their use. Only repeated testing can decide these points.

References

ABBOTT M.L. & FRALEY L. (1991) A review: radiotracer methods to determine root distribution. *Environ. Exp. Bot.* **31**, 1–10.

ABUZINADAH R.A. & READ D.J. (1986) The role of proteins in the nitrogen nutrition of ectomycorrhizal plants. III. Protein utilization by *Betula, Picea* and *Pinus* in ectomycorrhizal association with *Hebeloma crustuliniforme. New Phytol.* **103**, 506–514.

ABUZINADAH R.A. & READ D.J. (1989) The role of proteins in the nitrogen nutrition of ectomycorrhizal plants. V. Nitrogen transfer in birch (*Betula pendula*) grown in association with mycorrhizal and non-mycorrhizal fungi. *New Phytol.* **112**, 61–68.

ADAMS F. (1974) The soil solution. In *The Plant Root and its Environment*, ed. Carson E.W., pp. 441–450. Charlottesville, VA: University Press of Virginia.

ADDISCOTT T.M. (1977) A simple computer model for leaching in structured soils. *J. Soil Sci.* **28**, 554–567.

ADDISCOTT T.M. (1992) Simulation modelling and soil behaviour. *Geoderma* **60**, 15–40.

ADDISCOTT T.M. & WAGENET R.J. (1985) Concepts of solute leaching in soils: a review of modelling approaches. *J. Soil Sci.* **36**, 411–424.

ADDISCOTT T.M. & WHITMORE A.P. (1987) Computer simulation of changes in soil mineral nitrogen and crop nitrogen during autumn, winter and spring. *J. Agric. Sci., Camb.* **109**, 141–157.

ADDISCOTT T.M. & WHITMORE A.P. (1991) Simulation of solute leaching in soils of differing permeabilities. *Soil Use Manage.* **7**, 94–102.

AGUILAR A.S. & VAN DIEST A. (1981) Rock-phosphate mobilization induced by the alkaline uptake pattern of legumes utilizing symbiotically fixed nitrogen. *Plant Soil* **61**, 27–42.

AHARONI C. & SPARKS D.L. (1991) Kinetics of soil chemical reaction — A theoretical treatment. In *Rates of Soil Chemical Processes*, eds. Sparks D.L. & Suarez D.L., Spec. Pub. No. 27, pp. 1–18. Madison, WI: Soil Science Society of America.

ALBRECHT W.A., GRAHAM E.R. & SHEPPARD H.R. (1942) Surface relationships of roots and colloidal clay in plant nutrition. *Am. J. Bot.* **29**(3), 210–233.

ALCAMO J. (ed.) (1994) *IMAGE 2.0: Integrated Modelling of Global Climate Change*. Dordrecht: Kluwer Academic.

ALEXANDER M. & SCOW K.M. (1989) Kinetics of biodegradation in soil. In *Reactions and Movement of Organic Chemicals in Soil*, eds. Sawhney B.L. & Brown K., Spec. Pub. No. 22, pp. 243–270. Madison, WI: Soil Science Society of America.

ALLAN JONES C., BLAND W.L., RITCHIE J.T. & WILLIAMS J.R. (1991a) Simulation of root growth. In *Modelling Plant and Soil Systems*, eds. Hanks J. & Ritchie J.T., Agronomy Series 31, pp. 91–123. Madison, WI: American Society of Agronomy.

ALLAN JONES C., SHARPLEY A.N. & WILLIAMS J.R. (1991b) Modelling phosphorus dynamics in the soil–plant system. In *Modelling Plant and Soil Systems*, eds. Hanks J. & Ritchie J.T., Agronomy Series 31, pp. 332–339. Madison, WI: American Society of Agronomy.

ALLEN M.F., SMITH W.K., MOORE T.S. & CHRISTENSEN M. (1981) Comparative water relations and photosynthesis of mycorrhizal and non-mycorrhizal *Bouteloua gracilis* HBK Lag ex Steud. *New Phytol.* **88**, 683–693.

ALLISON F.E. (1965) Evaluation of incoming and outgoing processes that affect soil nitrogen. In *Soil Nitrogen*, eds. Bartholomew V.W. & Clark F.E., Agronomy Monograph 10, pp. 573–606. Madison, WI: American Society of Agronomy.

AL-NAJAFI M.A. (1990) Root shrinkage in relation to water stress. D.Phil. Thesis. Bodlean Library, Oxford.

AMES R.N., REID C.P.P., PORTER L.K. & CAMBARDELLA C. (1983) Hyphal uptake and transport of nitrogen from two ^{15}N-labelled sources by *Glomus mosseae*, a vesicular-arbuscular fungus. *New Phytol.* **95**, 381–396.

AMIJEE F., STRIBLEY D.P. & TINKER P.B. (1989) Development of endomycorrhizal root systems. VII — A detailed study of effects of soil phosphorus on colonization. *New Phytol.* **111**, 435–446.

AMIJEE F., STRIBLEY D.P. & TINKER P.B. (1990) Soluble carbohydrates in roots of leek plants in relation to phosphorus supply and VA mycorrhizas. *Plant Soil* **124**, 195–198.

AMIJEE F., BARRACLOUGH P.B. & TINKER P.B. (1991) Modelling phosphorus uptake and utilization by plants. In *Phosphorus Nutrition of Grain Legumes in the Semi-arid Tropics*, eds. Johansen C., Lee K.K. & Sahrawat K.L., pp. 63–76. Hyderabad, India: ICRISAT.

AMIJEE F., STRIBLEY D.P. & TINKER P.B. (1993) The development of endomycorrhizal root systems. VIII The effects of soil phosphorus and fungal colonization on the concentration of soluble carbohydrate in roots. *New Phytol.* **123**, 297–306.

AMOOZEGAR-FARD A., NIELSEN D.R. & WARRICK A.W. (1982) Soil solute concentration distributions for spatially-varying pore water velocities and apparent diffusion coefficients. *Soil Sci. Soc. Am. J.* **46**, 3–9.

ANDERSON G. (1980) Assessing organic phosphorus in soils. In *The Role of Phosphorus in Agriculture*, eds. Khasawneh F.E., Sample E.C. & Kamprath, E.J., pp. 411–433. Madison, WI: American Society of Agronomy.

ANDERSON J.M. & INGRAM J.S.I. (1993) *Tropical Soil Biology and Fertility. A Handbook of Methods*. Wallingford: CAB International.

ANDERSSEN R.S., HALE R.P. & RADOCK R.R.M. (1969) The simulation with mathematical models of ion uptake by growing roots. *Plant Soil* **30**, 271–289.

ANDREN O., HANSSON A.C. & VEGH K. (1993) Barley nutrient uptake, root growth and depth distribution in two soil types in a rhizotron with vertical and horizontal minirhizotrons. *J. Agric. Res.* **23**, 115–126.

ARAMBARRI P. & TALIBUDEEN O. (1959) Factors influencing the isotopically exchangeable phosphate in soils. I. The effect of low concentrations of organic anions. *Plant Soil* **11**, 343–354.

ARMSTRONG W. (1979) Aeration in higher plants. *Adv. Botan. Res.* **7**, 226–332.

ARMSTRONG W., STRANGE M.E., CRINGLE S. & BECKETT P.M. (1994) Oxygen in roots. *Ann. Bot.* **74**, 229–236.

ARNEBRANT K., EK H., FINLAY R.D. & SODERSTROM B. (1993) Nitrogen translocation between *Alnus glutinosa* L. Garett. seedlings inoculated with *Frankia* sp. and *Pinus contorta* Doug. ex Lond. seedlings connected by a common ectomycorrhizal mycelium. *New Phytol.* **124**, 231–242.

ARNOLD P.W. (1970) The behaviour of potassium in soils. *Fertil. Soc. Proc.* **115**, 3–30.

ASHER C.J., OZANNE P.G. & LONERAGAN J.F. (1965) A method for controlling the ionic environment of plant roots. *Soil Sci.* **100**, 149–156.

ASHFORD A.E., LING-LEE M. & CHILVERS G.A. (1975) Polyphosphate in eucalypt mycorrhizas: a cytochemical demonstration. *New Phytol.* **74**, 447–453.

ASHFORD A.E., RYDE S. & BARROW K.D. (1994) Demonstration of a shortchain polyphosphate in *Pisolithus tinctorius* and the implications for phosphorus transport. *New Phytol.* **126**, 239–247.

ASLYNG H.C. (1954) The lime and phosphate potential of soils; the solubility and availability of phosphate. *Royal Veterinary Agricultural College of Copenhagen Yearbook*, pp. 1–50.

ASSENG S., RICHTER C. & WESSOLEK G. (1997) Modelling root growth of wheat as a linkage between crop and soil. *Plant Soil* **190**, 267–277.

ATKINS P.W. (1986) *Physical Chemistry*, 3rd Edition. Oxford: Oxford University Press.

ATKINSON D. (ed.) (1991) *Plant Root Growth: An Ecological Perspective*. Oxford: Blackwell.

ATKINSON D. & MACKIE-DAWSON L.A. (1999) Root growth: methods of measurement. In *Soil Analysis: Physical Methods*, eds. Smith K.A. & Mullins C.E. New York: Marcel Dekker.

ATKINSON R.J., POSNER A.M. & QUIRK J.P. (1971) Kinetics of heterogeneous isotopic exchange reactions: derivation of an Elovich equation. *Proc. R. Soc. London, Ser. A* **324**, 247–255.

AZAIZAH H., GUNSE B. & STEUDLE E. (1992) Effects of NaCl and $CaCl_2$ on water transport across root cells of maize (*Zea mays* L.) seedlings. *Plant Physiol.* **99**, 886–894.

AZMI M. & MASHOR M. (1990) Competition of barnyard grass (*Echinochloa cruz-galli* (L.) Beauv) in direct seeded rice. *Proceedings of the 3rd International Conference on Plant Protection in the Tropics*, pp. 424–449.

BACHE B.W. (1963) Aluminium and iron phosphate studies relating to soils. I. Solution and hydrolysis of variscite and strengite. *J. Soil Sci.* **14**, 113–123.

BACHE B.W. & WILLIAMS E.G. (1971) A phosphate sorption index for soils. *J. Soil Sci.* **22**, 289–301.

BAEHR A.L. & BRUELL C.J. (1990) Applications of the Stephan-Maxwell equations to determine limitations of Fick's law when modeling organic vapor transport in sand columns. *Water Resour. Res.* **26**, 1155–1163.

BAGSHAW R., VAIDYANATHAN L.V. & NYE P.H. (1972) The supply of nutrient ions by diffusion to plant roots in soil. VI Effects of onion plant roots on pH and phosphate desorption characteristics in a sandy soil. *Plant Soil* **37**, 627–639.

BAILEY G.W., WHITE J.L. & ROTHBERG T. (1968) Adsorption of organic herbicides by montmorillonite: role of pH and chemical character of absorbate. *Soil Sci. Soc. Am. Proc.* **32**, 222–234.

BAKER J.J., WRAITH J.M. & DALTON F.M. (1992) Root function in water transport. *Adv. Soil Sci.* **19**, 53–71.

BALABANE M. & BALESDENT J. (1992) Impact of fertilizer-derived labelled N to soil O.M. during a growing season of maize in the field. *Soil Biol. Biochem.* **24**, 89–96.

BALDWIN J.P. (1972) Nutrient uptake by competing roots in soil. D.Phil. Thesis, Oxford.

BALDWIN J.P. (1975) A quantitative analysis of the factors affecting plant nutrient uptake from some soils. *J. Soil Sci.* **26**, 195–206.

BALDWIN J.P. (1976) Competition for plant nutrients in soil; a theoretical approach. *J. Agric. Sci.* **87**, 341–356.

BALDWIN J.P. & NYE P.H. (1974) A model to calculate the uptake by a developing root system or root hair system of solutes with concentration variable diffusion coefficients. *Plant Soil* **40**, 703–706.

BALDWIN J.P. & TINKER P.B. (1972) A method of estimating the lengths and spatial patterns of two interpenetrating root systems. *Plant Soil* **37**, 209–213.

BALDWIN J.P., TINKER P.B. & NYE P.H. (1972) Uptake of solutes by multiple root systems from soil. II. The theoretical effects of rooting density and pattern on uptake of nutrients from soil. *Plant Soil* **36**, 693–708.

BALDWIN J.P., NYE P.H. & TINKER P.B. (1973) Uptake of solutes by multiple root systems from soil. III. A model for calculating the solute uptake by a randomly dispersed root system developing in a finite volume of soil. *Plant Soil* **38**, 621–635.

BALKE N.E. (1985) Effects of allelochemicals on mineral uptake and associated physiological processes. In *The Chemistry of Allelopathy*, ed. Thompson A.C., pp. 161–178. Washington, DC: American Chemical Society.

BALSBERG-PAHLSSON A.M. (1995) Growth, radicle and root-hair development of *Deschampsia flexuosa* L. Trin. seedlings in relation to soil acidity. *Plant Soil* **175**, 125–132.

BAON J.B., SMITH S.E. & ALSTON A.M. (1993) Mycorrhizal responses of barley cultivars differing in P efficiency. *Plant Soil* **157**, 97–105.

BARBER D.A. (1969) The influence of the microflora on the accumulation of ions by plants. In *Ecological Aspects of the Mineral Nutrition of Plants*, ed. Rorison I.H. Oxford: Blackwell Scientific.

BARBER D.A. & ROVIRA A.D. (1975) Rhizosphere microorganisms and the absorption of phosphate by plants. *Annual Report of the ARC Letcombe Laboratory, 1974*, pp. 27–28.

BARBER S.A. (1962) A diffusion and mass-flow concept of soil nutrient availability. *Soil Sci.* **93**, 39–49.

BARBER S.A. (1978) Growth and nutrient uptake of soyabean roots under field conditions. *Agron. J.* **70**, 457–461.

BARBER S.A. (1995) *Soil Nutrient Bioavailability*, 2nd Edition. New York: Wiley.

BARBER S.A. & CUSHMAN J.H. (1981) Nutrient uptake model for agronomic crops. In *Modelling Wastewater Renovation Land Treatment*, ed. Iskander I.K., pp. 382–409. New York: Wiley.

BARBER S.A., WALKER J.M. & VASEY E.H. (1962) Principles of ion movement through the soil to the plant root. In *Transactions of the International Society of Soil Science, Commissions IV and V*, pp. 121–124. International Conference, Soil Bureau, P.B. Lower Hutt, New Zealand, 1963.

BARBER S.A., WALKER J.M. & VASEY E.H. (1963) Mechanisms for the movement of plant nutrients from the soil and fertiliser to the plant root. *J. Agric. Food Chem.* **11**, 204–207.

BAREA J.M., AZCON R. & AZCON-AGUILAR C. (1992) Vesicular-arbuscular mycorrhizal fungi in nitrogen-fixing systems. *Methods Microbiol.* **24**, 391–416.

BARLEY K.P. (1970) The configuration of the root system in relation to nutrient uptake. *Adv. Agron.* **22**, 159–201.

BARLEY K.P. & GREACEN E.L. (1967) Mechanical resistance as a soil factor influencing the growth of roots and underground shoots. *Adv. Agron.* **19**, 1–43.

BARLEY K.P. & ROVIRA A.D. (1970) The influence of root hairs on the uptake of phosphate. *Commun. Soil Sci. Plant Anal.* **1**, 287–292.

BARNES A., GREENWOOD D.J. & CLEAVER T.J. (1976) A dynamic model for the effects of potassium and nitrogen fertilizers on the growth and nutrient uptake of crops. *J. Agric. Sci.* **86**, 225–244.

BARRACLOUGH D. (1976) The diffusion of macromolecules in the soil pore space. D.Phil. Thesis, Oxford.

BARRACLOUGH D. & NYE P.H. (1979) The effect of molecular size on diffusion characteristics in soil. *J. Soil Sci.* **30**, 29–42.

BARRACLOUGH P.B. (1984) The growth and activity of winter wheat roots in the field: root growth of high-yielding crops in relation to shoot growth. *J. Agric. Sci., Camb.* **103**, 419–422.

BARRACLOUGH P.B. (1986a) The growth and activity of winter wheat roots in the field: nutrient uptakes of high-yielding crops. *J. Agric. Sci., Camb.* **106**, 45–52.

BARRACLOUGH P.B. (1986b) The growth and activity of winter wheat roots in the field: nutrient inflows of high-yielding crops. *J. Agric. Sci., Camb.* **106**, 53–59.

BARRACLOUGH P.B. (1989a) Root growth and nutrient uptake by field crops under temperate conditions. *Asp. Appl. Biol.* **22**, 227–233.

BARRACLOUGH P.B. (1989b) Root growth, macro-nutrient dynamics and soil fertility requirements of a high-yielding winter oilseed rape crop. *Plant Soil* **119**, 59–70.

BARRACLOUGH P.B. & LEIGH R.A. (1984) The growth and activity of winter wheat roots in the field: the effects of sowing date and soil type on root growth of high yielding crops. *J. Agric. Sci., Camb.* **103**, 59–74.

BARRACLOUGH P.B. & TINKER P.B. (1981) The determination of ionic diffusion coefficients in field soils. I. Diffusion coefficients in sieved soils in relation to water content and bulk density. *J. Soil Sci.* **32**, 225–236.

BARRACLOUGH P.B. & WEIR A.H. (1988) Effects of a compacted subsoil layer on root and shoot growth, water use and nutrient uptake of winter wheat. *J. Agric. Sci., Camb.* **110**, 207–216.

BARRACLOUGH P.B., KUHLMANN H. & WEIR A.H. (1989) The effects of prolonged drought and nitrogen fertilizer on root and shoot growth and water uptake by winter wheat. *J. Agron. Crop Sci.* **163**, 352–360.

BARRACLOUGH P.B., WEIR A.H. & KUHLMANN H. (1991) Factors affecting the growth and distribution of winter wheat roots under U.K. field conditions. In *Plant Roots and their Environment,* eds. McMichael M. & Persson H., pp. 410–419. Amsterdam: Elsevier.

BARRETT-LEONARD E.G., DRACUP M. & GREENWAY H. (1993) Role of extra-cellular phosphatases in the phosphorus nutrition of clover. *J. Exp. Bot.* **44**, 1595–1600.

BARROW N.J. (1979) Description of desorption of phosphate from soil. *J. Soil Sci.* **30**, 259–270.

BARROW N.J. (1980) Differences among a wide-ranging collection of soils in the rate of reaction with phosphate. *Aust. J. Soil Res.* **3**, 423–483.

BARROW N.J. (1983) A mechanistic model for describing the sorption and desorption of phosphate by soil. *J. Soil Sci.* **34**, 733–750.

BARROW N.J. (1987) *Reactions with Variable-Charged Soils.* Developments in Soil and Plant Sciences, Vol. 31. Dordrecht: Martinus Nijhoff.

BARROW N.J. & SHAW T.C. (1975a) The slow reaction between soils and anions. 2: Effect of time and temperature on the decrease in phosphate concentration in the soil solution. *Soil Sci.* **119**, 311–320.

BARROW N.J. & SHAW T.C. (1975b) The slow reaction between soil and anions. 3: The effect of time and temperature on the decrease in the isotopically exchangeable phosphate. *Soil Sci.* **119**, 190–197.

BARROW N.J. & SHAW T.C. (1975c) The slow reaction between soil and anions. 4: Effect of time and temperature of contact between soil and molybdate on the uptake of molybdenum by plants and on the molybdate concentration in the soil solution. *Soil Sci.* **119**, 301–310.

BARROW N.J. & SHAW T.C. (1977) The slow reaction between soil and anions 6: effect of time and temperature of contact with fluoride. *Soil Sci.* **124**, 265–278.

BAR-TAL A., BAR-YOSEF B. & CHEN Y. (1991). Validation of a model of the transport of zinc to an artificial root. *J. Soil Sci.* **42**, 399–411.

BARTHOLOMEW W.V. & CLARK F.E. (1965) *Soil Nitrogen*, Agronomy Monograph 10. Madison, WI: American Society of Agronomy.

BARTHOLOMEW W.V. & KIRKHAM D. (1960) Mathematical descriptions and interpretation of culture-induced soil nitrogen changes. *Trans. 7th Int. Cong. Soil Sci.* **2**, 471–477.

BAR-YOSEF B., FISHMAN S. & TALPAZ H. (1980) A model of zinc movement to single roots in soils. *Soil Sci. Soc. Am. J.* **44**, 1272–1279.

BATES T.R. & LYNCH J.P. (1996) Stimulation of root hair elongation in *Arabidopsis thaliana* by low phosphorus availability. *Plant Cell Environ.* **19**, 529–538.

BAVER L.D., GARDNER W.H. & GARDNER W.R. (1972) *Soil Physics*. New York: Wiley & Sons.

BAYLIS G.T.S. (1970) Root hairs and phycomycetous mycorrhizas in phosphorus-deficient soil. *Plant Soil* **33**, 713–716.

BAZIN M.J., MARKHAM P., SCOTT E.M. & LYNCH J.M. (1990) Population dynamics and rhizosphere interactions. In *The Rhizosphere*, ed. Lynch J.M., pp. 99–128. Chichester: Wiley & Sons.

BECKETT P.H.T. (1964) Potassium–calcium exchange equilibria in soils: specific adsorption sites of potassium. *Soil Sci.* **97**, 376–383.

BECKETT P.H.T. (1969) Residual potassium and magnesium: a review. *Tech. Bull. No. 20*, pp. 183–196. London: Ministry of Agriculture, Fisheries & Food, H.M.S.O.

BECKETT P.H.T. & CRAIG J.B. (1964) The determination of potassium potentials. *Trans. 8th Int. Cong. Soil Sci.* **2**, 249–255.

BECKETT P.H.T. & NAFADY M.H.M. (1967a) Studies on soil potassium. VI. The effect of K fixation and release on the form of the K: (Ca + Mg) exchange isotherm. *J. Soil Sci.* **18**, 244–262.

BECKETT P.H.T. & NAFADY M.H.M. (1967b) Potassium–calcium exchange equilibria in soils: the location of non-specific (Gapon) and specific exchange sites. *J. Soil Sci.* **18**, 263–281.

BEERLING D.J., WOODWARD F.I., HEATH J. & MANSFIELD T.A. (1996) Drought–CO_2 interactions in trees: observations and mechanisms. *New Phytol.* **134**, 235–242.

BEGG C.M.B., KIRK G.J.D., MACKENZIE A.F. & NEUE H.-U. (1994) Root-induced iron oxidation and pH changes in the lowland rice rhizosphere. *New Phytol.* **128**, 469–477.

BEGON M., HARPER J.L. & TOWNSEND C.R. (1986) *Ecology*. Oxford: Blackwell.

BENGOUGH A.G. & MULLINS C.E. (1990) Mechanical impedance to root growth: a review of experimental techniques and root growth responses. *J. Soil Sci.* **41**, 341–358.

BENGOUGH A.G., MULLINS C.E. & WILSON G. (1997) Estimating soil frictional resistance to metal probes and its relevance to the penetration of soil by roots. *Eur. J. Soil Sci.* **48**, 603–612.

BERNTSON G.M. (1994) Modelling root architecture: are there tradeoffs between efficiency and potential of resource acquisition? *New Phytol.* **127**, 483–493.

BETHLENFALVAY G., REYES-SOLIS M.G., CAMEL S.B. & FERRERA-CERRATO L. (1991) Nutrient transfer between roots of soybean and maize plants connected by a common mycorrhizal mycelium. *Physiol. Plant* **82**, 423–432.

BHADORIA P.B.S., KASELOWSKY J., CLAASEN N. & JUNGK A. (1991) Impedance factor for chloride diffusion in soil as affected by bulk density and water content. *Z. Pflanzenernähr. Bodenk.* **154**, 69–72.

BHAT K.K.S. & NYE P.H. (1973) Diffusion of phosphate to plant roots in soil. I. Quantitative autoradiography of the depletion zone. *Plant Soil* **38**, 161–175.

BHAT K.K.S. & NYE P.H. (1974a) Diffusion of phosphate to plant roots in soil. II. Uptake along the roots at different times and the effect of different levels of phosphorus. *Plant Soil* **41**, 365–382.

BHAT K.K.S. & NYE P.H. (1974b) Diffusion of phosphate to plant roots in soil. III. Depletion around onion roots without root hairs. *Plant Soil* **41**, 383–394.

BHAT K.K.S., BALDWIN J.P. & NYE P.H. (1976) Diffusion of phosphate to plant roots in soil. IV. The concentration–distance profile in the rhizosphere of roots with root hairs in a low-P soil. *Plant Soil* **44**, 63–72.

BIELESKI R.L. & FERGUSON I.B. (1983) Physiology and metabolism of phosphate and its compounds. In *Inorganic Plant Nutrition*, eds. Lauchli A. & Bieleski R.L., Encyclopedia of Plant Physiology 15A, pp. 422–449. Berlin: Springer-Verlag.

BIGGAR J.W. & NIELSEN D.R. (1962) Miscible displacement. II. Behaviour of tracers. *Soil Sci. Soc. Am. Proc.* **26**, 125–128.

BIGGAR J.W. & NIELSEN D. R. (1976) Spatial variability of the leaching characteristics of a field soil. *Water Resour. Res.* **12**, 78–94.

BIRD R.B., STEWART W.E. & LIGHTFOOT E.N. (1960) *Transport Phenomena.* New York: Wiley.

BLACK C.A. (1968) *Soil–Plant Relationships.* London: Wiley.

BLACK R. & TINKER P.B. (1979) The development of endomycorrhizal root systems. II — Effects of agronomic factors and soil conditions on the development of vesicular-arbuscular mycorrhizal infection in barley and on the endophyte spore density. *New Phytol.* **83**, 401–413.

BLAND W.L. (1991) Root length density from core-break observations — sources of error. In *Plant Roots and their Environment*, eds. McMichael M. & Persson H., pp. 565–569. Amsterdam: Elsevier.

BLOOM A.J., CHAPIN F.S. & MOONEY H.A. (1985) Resource limitation in plants — an economic analogy. *Annu. Rev. Ecol. Syst.* **16**, 361–392.

BLOOMFIELD J., VOGT K.A. & WARGO P.M. (1991) Tree root turnover and senescence. In *Plant Roots — The Hidden Half*, eds. Waisel Y., Eshel A. & Kafkafi U., pp. 363–382. New York: Marcel Dekker.

BLUM A., JOHNSON J.W., RAMSEUR E.L. & TOLLNER E.W. (1991) The effect of a drying top soil and a possible non-hydraulic root signal on wheat growth and yield. *J. Exp. Bot.* **42**, 1225–1231.

BODMAN G.B. & COLMAN E.A. (1943) Moisture and energy conditions during downward entry of water into soils. *Soil Sci. Soc. Am. Proc.* **8**, 116–122.

BOEHM W. (1979) *Methods of Studying Root Systems.* New York: Springer-Verlag.

BOEUF-TREMBLAY V., PLANTUREUX S. & GUCKERT A. (1995) Influence of mechanical impedance on the root exudation of maize seedlings at two developmental stages. *Plant Soil* **172**, 279–287.

BOLE J.B. (1973) Influence of root hairs in supplying soil phosphorus to wheat. *Can. J. Soil Sci.* **53**, 169–175.

BOLT G.H. (1955) Ion adsorption by clays. *Soil Sci.* **79**, 267–276.

BOLT G.H. (1982) *Soil Chemistry B. Physico-Chemical Models*, ed. Bolt G.H. Amsterdam: Elsevier.

BOLT G.H. & GROENEVELT P.H. (1972) Coupling between transport processes in porous media. *Proceedings of the 2nd IAHR-ISSS Symposium on Transport Phenomena in Porous Media*, pp. 630–652. Guelph.

BOLTON J. (1971) Quantity–intensity relationships for labile sodium in field soils. *J. Soil Sci.* **22**, 417–429.

BOND W.J. & VERBURG K. (1997) Comparison of methods for predicting ternary exchange from binary isotherms. *Soil Sci. Soc. Am. J.* **61**, 444–454.

BOOKER F.L., BLUM U. & FISCUS E.L. (1992) Short-term effects of ferulic acid on ion uptake and water relations in cucumber seedlings. *J. Exp. Bot.* **43**, 649–655.

BOOT R.G.A. & MENSINK M. (1991) The influence of nitrogen availability on growth parameters of fast and slow growing perennial grasses. In *Plant Root Growth*, ed. Atkinson D., pp. 161–170. Oxford: Blackwell.

BOTTOMLEY P.A., ROGERS H.H. & FOSTER T.H. (1986) NMR imaging shows water distribution and transport of plant roots *in situ*. *Proc. Natl. Acad. Sci. USA* **83**, 87–89.

BOUGER N.L., GROVE T.S. & MALAJCZUK N. (1990) Growth and phosphorus acquisition of karri (*Eucalyptus diversicolor* F. Muell.) seedlings inoculated with ecto-mycorrhizal fungi in relation to phosphorus supply. *New Phytol.* **114**, 77–85.

BOULDIN D.R. (1961) Mathematical description of diffusion processes in the soil–plant system. *Soil Sci. Soc. Am. Proc.* **25**, 475–480.

BOULDIN D.R. (1989) A multiple ion uptake model. *J. Soil Sci.* **40**, 309–319.

BOULDIN D.R., MIYASAKA S.C. & GRUNES D.L. (1992) Cation accumulation by winter wheat forage. II Correlation with multi-ion model. *J. Plant Nutr.* **15**, 1081–1097.

BOUMAN B.A.M., VAN KEULEN H., VAN LAAR H.H. & RABBINGE R. (1996) The 'school of de Wit' crop growth simulation models: a pedigree and historical overview. *Agric. Syst.* **52**, 171–198.

BOWDEN J.W., POSNER A.M. & QUIRK J.P. (1977) Ionic adsorption on variable charge surfaces. Theoretical charge development and titration curves. *Aust. J. Soil Res.* **15**, 121–126.

BOWEN G.D. & ROVIRA A.D. (1961) The effects of microorganisms on plant growth. I. Development of roots and root hairs in sand and agar. *Plant Soil* **15**, 166–188.

BOWEN G.D. & ROVIRA A.D. (1991) The rhizosphere: the hidden half of the hidden half. In *Plant Roots — The Hidden Half*, eds. Waisel Y., Eshel A. & Kafkafi U., pp. 641–670. New York: Marcel Dekker.

BOWER C.A., GARDNER W.R. & GOERTZEN J.O. (1957) Dynamics of cation exchange in soil columns. *Soil Sci. Soc. Am. Proc.* **21**, 20–24.

BOWERS J.H. & PARKE J.L. (1993) Colonization of pea taproots by *Pseudomonas fluorescens*: effect of soil temperature and bacterial motility. *Soil Biol. Biochem.* **25**, 1693–1701.

BOX J.E. (1996) Modern methods for root investigation. In *Plant Roots — The Hidden Half*, eds. Waisel Y., Eshel A. & Kafkafi U., pp. 193–237. New York: Marcel Dekker.

BOX J.E. & HAMMOND L.C. (1990) *Rhizosphere Dynamics*, AAAS Selected Symposium 113. Washington, DC: American Association for the Advancement of Science.

BOYER J.S. (1985) Water transport. *Ann. Rev. Plant Physiol.* **36**, 473–516.

BRADLEY P.M. & MORRIS J.T. (1991) Relative importance of ion exclusion, secretion and accumulation in *Spartina alterniflora* Loisel. *J. Exp. Bot.* **42**, 1525–1532.

BRADY D.J., EDWARDS D.G., ASHER C.J. & BLAMEY F.P.C. (1993a) Calcium amelioration of Al toxicity effects on root hair development in soybean (*Glycine max* (L.) Merr). *New Phytol.* **123**, 531–538.

BRADY D.J., GREGORY P.J. & FILLERY I.R.P. (1993b) The contribution of different regions of the seminal roots of wheat to uptake of nitrate from soil. *Plant Soil* **155**, 155–158.

BRAGG P.L., RUBINO P., HENDERSON F.K.G., FIELDING W.J. & CANNELL R.Q. (1984) A comparison of the root and shoot growth of winter barley and winter wheat and the effects of an early application of chlormequat. *J. Agric. Sci., Camb.* **103**, 257–264.

BRAUM S.M. & HELMKE P.A. (1995) White lupin utilizes soil phosphorus that is unavailable to soybean. *Plant Soil* **176**, 95–100.

BRAY R.H. (1954) A nutrient mobility concept of soil–plant relationships. *Soil Sci.* **78**, 9–22.

BRESLER E. (1972) Control of soil salinity. In *Optimizing the Soil Physical Environment towards Greater Crop Yields*, ed. Hillel D., pp. 101–132. New York: Academic Press.

BRESLER E. (1973) Anion exclusion and coupling effects in non-steady transport through unsaturated soils. I. Theory. *Soil Sci. Soc. Am. Proc.* **37**, 663–669.

BRESLER E. & HANKS R.J. (1969) Numerical method for estimating simultaneous flow of water and salt in unsaturated soils. *Soil Sci. Soc. Am. Proc.* **33**, 827–840.

BREWSTER J.L. (1971) Some factors affecting the uptake of plant nutrients from the soil. D.Phil. Thesis, Oxford.

BREWSTER J.L. & TINKER P.B. (1970) Nutrient cation flow in soil around plant roots. *Soil Sci. Soc. Am. Proc.* **34**, 421–426.

BREWSTER J.L. & TINKER P.B. (1972) Nutrient flow rates into roots. *Soils Fert.* **35**, 355–359.

BREWSTER J.L., BHAT K.K.S. & NYE P.H. (1975a) The possibility of predicting solute uptake and plant growth response from independently measured soil and plant characteristics. II. The growth and uptake of onions in solutions of constant phosphate concentration. *Plant Soil* **42**, 171–195.

BREWSTER J.L., BHAT K.K.S. & NYE P.H. (1975b) The possibility of predicting solute uptake and plant growth response from independently measured soil and plant characteristics. III. The growth and uptake of onions in a soil fertilized to different initial levels of phosphate and a comparison of the results with model predictions. *Plant Soil* **42**, 197–226.

BREWSTER J.L., BHAT K.K.S. & NYE P.H. (1976) The possibility of predicting solute uptake and plant growth response from independently measured soil and plant characteristics. V. The growth and phosphorus uptake of rape in soil at a range of phosphorus concentrations and a comparison of results with the predictions of a simulation model. *Plant Soil* **44**, 295–328.

BRISKIN D.P. & HANSON J.B. (1992) How does the plant plasmalemma membrane H^+ATPase pump protons? *J. Exp. Bot.* **43**, 269–289.

BROECKER W.S. & OLSON E.A. (1960) Radiocarbon from nuclear tests II. *Science* **132**, 712–721.

BROUDER S.M. & CASSMAN K.G. (1994) Evaluation of a mechanistic model of potassium uptake by cotton in vermiculitic soil. *Soil Sci. Soc. Am. J.* **58**, 1174–1183.

BROUWER R. (1965) Water movement across the root. In *The State and Movement of Water in Living Organisms*, pp. 121–150, Society of Experimental Biology Symposium No. XIX.

BROWN D.A., FULTON B.E. & PHILLIPS R.E. (1964) Ion diffusion. I. A quick-freeze method for the measurement of ion diffusion in soil and clay systems. *Soil Sci. Soc. Am. Proc.* **28**, 628–631.

BROWN K.F. & BISCOE P.V. (1985) Fibrous root growth and water use of sugar beet. *J. Agric. Sci., Camb.* **105**, 679–691.

BROWN K.F., GREGORY P.J., COOPER P.J.M. & KEATINGE J.D.H. (1989) Root and shoot growth and water use of chickpea (*Cicer arietinum*) grown in dryland conditions; effects of sowing date and genotype. *J. Agric. Sci., Camb.* **113**, 41–49.

BROWN M.E. (1974) Seed and root bacterization. *Annu. Rev. Phytopath.* **35**, 443–451.

BROWN V.K. & GANGE A.C. (1990) Insect herbivory below ground. *Adv. Ecol. Res.* **20**, 1–59.

BROWN V.K. & GANGE A.C. (1991) Effects of root herbivory on vegetation dynamics. In *Plant Root Growth*, ed. Atkinson D., pp. 453–470. Oxford: Blackwell.

BRUGGENWERT M.G.M. & KAMPHORST A. (1979) Survey of experimental information on cation exchange systems. In *Soil Chemistry B. Physico-Chemical Models*, ed. Bolt G.H., pp. 141–203. Amsterdam: Elsevier.

BRUNDRETT M.C. (1991) Mycorrhizae in natural ecosystems. *Adv. Ecol. Res.* **21**, 171–315.

BRUSSEAU M.L. & RAO P.S.C. (1989) Sorption kinetics of organic chemicals: methods, models and mechanisms. In *Reactions and Movement of Organic Chemicals in Soils*, eds. Sawhney B.L. & Brown K., Spec. Pub. No. 22, pp. 281–299. Madison, WI: Soil Science Society of America.

BRYANT R.B. & ARNOLD R.W., eds. (1994) *Quantitative Modeling of Soil Forming Processes*, SSSA Spec. Pub. No. 39. Madison, WI: Soil Science Society of America.

BUCKINGHAM E. (1904) Contributions to our knowledge of the aeration of soils. *U.S. Bur. Soils Bull.* **25**.

BUCKLAND S.T., CAMPBELL C.D., MACKIE-DAWSON L.A., MORGAN G.W. & DUFF E.I. (1993) A method for counting roots observed in minirhizotrons and their theoretical conversion to root length density. *Plant Soil* **153**, 1–9.

BURD J.S. & MARTIN J.C. (1923) Water displacement of soils and soil solution. *J. Agric. Sci.* **13**, 265–295.

BURKERT B. & ROBSON A. (1994) Zn uptake in subterranean clover (*Trifolium subterraneum* L.) by the vesicular-arbuscular mycorrhizal fungi in a root-free sandy soil. *Soil Biol. Biochem.* **26**, 1117–1124.

BURNS G.R. & DEAN L.A. (1964) The movement of water and nitrate around bands of sodium nitrate in soils and glass beads. *Soil Sci. Soc. Am. Proc.* **28**, 470–474.

BURNS I.G. (1974) A model for predicting the redistribution of salts applied to fallow soils after excess rainfall or evaporation. *J. Soil Sci.* **25**, 165–178.

BURNS I.G. (1980) Influence of the spatial distribution of nitrate on the uptake of nitrate by plants: a review and a model for rooting depth. *J. Soil Sci.* **31**, 155–173.

BUSH D.R. (1993) Proton-coupled sugar and amino acid transporters in plants. *Ann. Rev. Plant Physiol. Plant Mol. Biol.* **44**, 513–542.

BUWALDA J.G., ROSS G.J.S., STRIBLEY D.P. & TINKER P.B. (1982) The development of endomycorrhizal root systems. III — The mathematical representation of the spread of vesicular-arbuscular mycorrhizal infection in root systems. *New Phytol.* **91**, 669–682.

BUWALDA J.G., STRIBLEY D.P. & TINKER P.B. (1983) Increased uptake of bromide and chloride by plants infected with vesicular-arbuscular mycorrhizas. *New Phytol.* **93**, 217–225.

BYRNE J.M. (1974) Root morphology. In *The Plant Root and its Environment*, ed. Carson E.W., pp. 3–27. Charlottesville: University of Virginia Press.

CAILLOUX M. (1972) Metabolism and absorption of water by root hairs. *Can. J. Bot.* **50**, 557–575.

CAKMAK I., HENGELER C. & MARSCHNER H. (1994) Partitioning of root and shoot dry matter and carbohydrate in bean plants suffering from phosphorus, potassium and magnesium deficiency. *J. Exp. Bot.* **45**, 1245–1250.

CALDWELL M.M. (1987) Competition between root systems in natural communities. In *Root Development and Function*, eds. Gregory P.J., Lake V.J. & Rose D.A., pp. 167–186. Cambridge: Cambridge University Press.

CALDWELL M.M., EISENSTAT D.M., RICHARDS J.H. & ALLEN M.F. (1985) Competition for phosphorus: differential uptake from dual-isotope-labelled soil interspaces between shrub and grass. *Science* **229**, 384–386.

CALDWELL M.M., RICHARDS J.H. & BEYSCHLAG W. (1991) Hydraulic lift: ecological implications of water efflux from roots. In *Plant Root Growth*, ed. Atkinson D., pp. 423–436. Oxford: Blackwell.

CALLAWAY R.M. (1990) Effects of soil water distribution on the lateral root development of three species of California oaks. *Am. J. Bot.* **77**, 1469–1475.

CALOIN M. & YU O. (1984) Analysis of the time course change in nitrogen content in *Dactylis glomerata* L. using a model of plant growth. *Ann. Bot.* **54**, 69–76.

CAMPBELL C.D., MACKIE-DAWSON L.A., REID E.J., PRATT S.M., DUFF E.I. & BUCKLAND S.T. (1994) Manual recording of minirhizotron data and its application to study the effect of herbicide and nitrogen fertilizer on tree and pasture root growth in a silvopastoral system. *Agroforest. Systems* **26**, 75–87.

CAMPBELL D.J., KINNIBURGH D.G. & BECKETT P.H.T. (1989) The solution chemistry of some Oxfordshire soils. Spatial and temporal variability. *J. Soil Sci.* **40**, 321–339.

CAMPBELL G.S. (1991) Simulation of water uptake by plant roots. In *Modelling Plant and Soil Systems*, eds. Hanks J. & Ritchie J.T., pp. 273–284. Agronomy Series 31. Madison, WI: American Society of Agronomy.

CAMPBELL R. & GREAVES M.P. (1990). Anatomy and community structure of the rhizosphere. In *The Rhizosphere*, ed. Lynch J.M., pp. 11–34. Chichester: Wiley & Sons.

CANNELL M.G.R. & DEWAR R.C. (1994) Carbon allocation in trees: a review of concepts for modelling. *Adv. Ecol. Res.* **29**, 59–104.

CANNY M.J. (1991) *Phloem Translocation*. Cambridge: Cambridge University Press.

CANNY M.J. & HUANG C.X. (1994) Rates of diffusion into roots of maize. *New Phytol.* **126**, 11–19.

CARADUS J.R. (1982) Genetic differences in the length of root hairs in white clover and their effect on phosphorus uptake. In *Proceedings of the 9th International Plant Nutrition Colloquium, Warwick, England*, ed. Scaife, A., pp. 84–88. Farnham Royal, Bucks: Commonwealth Agricultural Bureau.

CARRODUS B.B. (1967) Absorption of nitrogen by mycorrhizal roots of beech. II. Ammonia and nitrate as sources of nitrogen. *New Phytol.* **66**, 1–4.

CARSLAW M.S. & JAEGER J.C. (1959) *Conduction of Heat in Solids*, 2nd Edition. Oxford: Clarendon Press.

CASSEL D.K., NELSON D.R. & BIGGAR J.W. (1969) Soil water movement in response to imposed temperature gradients. *Soil Sci. Soc. Am. Proc.* **33**, 493–500.

CHAPIN F.S. (1980) The mineral nutrition of wild plants. *Ann. Rev. Ecol. Systemat.* **11**, 233–260.

CHAPIN F.S. (1988) Ecological aspects of plant mineral nutrition. *Advances in Plant Nutrition*, Vol. 3, eds. Tinker P.B. & Lauchli A., pp. 161–192. New York: Praeger.

CHAPMAN P.J., SHAND C.A., EDWARDS A.C. & SMITH S. (1997) Effect of storage and sieving on the phosphate composition of soil solution. *Soil Sci. Soc. Am. J.* **61**, 315–321.

CHAPMAN S. & COWLING T.G. (1951) *Mathematical Theory of Non-Uniform Gases*, 2nd Edition. Cambridge: Cambridge University Press.

CHARLTON W.A. (1996) Lateral root initiation. In *Plant Roots — The Hidden Half*, eds. Waisel Y., Eshel A. & Kafkafi U., pp. 149–174. New York: Marcel Dekker.

CHARNEY C.P., TORKINGTON R. & HOLL F.B. (1991) Ecological implications of specificity between plants and rhizosphere microorganisms. *Adv. Ecol. Res.* **21**, 121–169.

CHEN W. & WAGENET R.J. (1997) Description of atrazine transport in soil with heterogeneous non-equilibrium sorption. *Soil Sci. Soc. Am. J.* **61**, 360–371.

CHILDS E.C. & COLLIS-GEORGE N. (1950) The permeability of porous materials. *Proc. R. Soc. London, Ser.* A, **201**, 392–405.

CHIOU C.T. (1989) Theoretical considerations of the partition and uptake of nonionic organic compounds by soil organic matter. In *Reactions and Movement of Organic Chemicals in Soils*, eds. Sawhney B.L. & Brown K., Spec. Pub. No. 22, pp. 1–30. Madison, WI: Soil Science Society of America

CLAASSEN N. & BARBER S.A. (1974) A method for characterizing the relation between nutrient concentration and the flux into roots of intact plants. *Plant Physiol.* **54**, 564–568.

CLAASSEN N. & BARBER S.A. (1976) Simulation model for nutrient uptake from soil by a growing plant root system. *Agron. J.* **68**, 961–964.

CLAASSEN N. & BARBER S.A. (1977) Potassium influx characteristics of corn roots and interaction with N, P, Ca and Mg influx. *Agron. J.* **69**, 860–864.

CLAASSEN N., SYRINGK M. & JUNGK A. (1986) Verification of a mathematical model by simulating potassium uptake from soil. *Plant Soil* **95**, 209–220.

CLARHOLM M. (1981) Protozoan grazing of bacteria in soil – impact and importance. *Microb. Ecol.* **7**, 343–350.

CLARHOLM M. (1985) Possible roles for roots, bacteria, protozoa and fungi in supplying nitrogen to plants. In *Ecological Interactions in Soil*, eds. Fitter A.H., Atkinson D., Read D.J. & Usher M.B. Special Bulletin British Ecological Society No. 4, pp. 355–365.

CLARK L.J., WHALLEY W.R., DEXTER A.R., BARRACLOUGH P.B. & LEIGH R.A. (1996) Complete mechanical impedance increases the turgor of cells in the apex of pea roots. *Plant Cell Environ.* **19**, 1099–1102.

CLARK R.B. (1982) Plant genotype differences to uptake, translocation, accumulation and use of mineral elements. In *Genetic Specificity of Mineral Nutrition of Plants*, ed. Saric M., pp. 41–55. Belgrade: Serbian Academy of Sciences and Arts.

CLARK R.B. & DUNCAN P.R. (1991) Improvement of plant mineral nutrition through breeding. *Field Crops Res.* **27**, 219–240.

CLARK R.B., ROMHELD V. & MARSCHNER H. (1988) Iron uptake and phytosiderophore release by roots of sorghum genotypes. *J. Plant Nutr.* **11**, 663–676.

CLARKE A.L. & BARLEY K.P. (1968) The uptake of nitrogen from soils in relation to solute diffusion. *Aust. J. Soil Res.* **6**, 75–92.

CLARKSON D.T. (1985) Factors affecting mineral acquisition by plants. *Ann. Rev. Plant Physiol.* **36**, 77–115.

CLARKSON D.T. (1993) Roots and the delivery of solutes to the xylem. *Phil. Trans. R. Soc. London*, B **341**, 5–17.

CLARKSON D.T. (1996) Root structure and sites of ion uptake. In *Plant Roots — The Hidden Half*, eds. Waisel Y., Eshel A. & Kafkafi U., pp. 417–453. New York: Marcel Dekker.

CLARKSON D.T. & HANSON J.B. (1980) The mineral nutrition of higher plants. *Ann. Rev. Plant Physiol.* **31**, 239–298.

CLARKSON D.T. & HAWKESFORD M.J. (1993) Molecular biological approaches to plant nutrition. *Plant Soil* **156**, 21–31.

CLARKSON D.T. & LUTTGE U. (1991) Mineral nutrition: inducible and repressible nutrient transport systems. *Prog. Bot.* **52**, 61–83.

CLARKSON D.T., SANDERSON J. & RUSSELL R.S. (1968) Ion uptake and root age. *Nature* **220**, 805–806.

CLARKSON D.T., GRAHAM J. & SANDERSON J. (1974) Water uptake by the roots of marrow and barley plants. *Letcombe Laboratory Annual Report, 1973*, pp. 9–11.

CLARKSON D.T., SANDERSON J. & SCATTERGOOD C.B. (1978) Influence of phosphate stress on phosphate absorption and translocation by various parts of the root system of *Hordeum vulgare* L. (barley). *Planta* **139**, 47–53.

CLARKSON D.T., EARNSHAW M.J., WHITE P.J. & COOPER H.D. (1988) Temperature-dependent factors influencing nutrient uptake: an analysis of responses at different levels of organization. In *Plants and Temperature*, eds. Long S.P. & Woodward F.I., pp. 281–310. Cambridge: Company of Biologists.

CLAUSNITZER V. & HOPMANS J.W. (1994) Simultaneous modelling of transient three dimensional root growth and soil water flow. *Plant Soil* **164**, 299–314.

COCKCROFT B., BARLEY K.P. & GREACEN E.L. (1969) The penetration of clays by fine probes and root tips. *Aust. J. Soil Res.* **7**, 333–348.

COGLIATTI D.H. & CLARKSON D.T. (1983) Physiological changes in phosphate uptake by potato plants during development of and recovery from phosphorus deficiency. *Physiol. Plant.* **58**, 287–294.

COMERFORD N.B., PORTER P.S. & ESCAMILLA J.A. (1994a) Use of Theissen areas in models of nutrient uptake in forested ecosystems. *Soil Sci. Soc. Am. J.* **58**, 210–215.

COMERFORD N.B., SMETHURST P.J. & ESCAMILLA J.A. (1994b) Nutrient uptake by woody root systems. *N.Z. J. Forest. Sci.* **24**, 195–212.

COMINS H.M. & MCMURTRIE R.E. (1993) Long-term response of nutrient-limited forests to CO_2 environment: equilibrium behaviour of plant–soil models. *Ecol. Appl.* **4**, 666–681.

COOK A., MARRIOTT C.A., SEEL W. & MULLINS C.E. (1997) Does the uniform packing of sand in a cylinder provide a uniform penetration resistance? *Plant Soil* **190**, 279–287.

COOKE G.W. (1967) *The Control of Soil Fertility*, Chapter 15. London: Crosby Lockwood.

COOKE G.W. (1969) Plant nutrient cycles. *Proceedings of the VIIth Colloquium of the International Potash Institute*, pp. 75–95. Berne.

COOPER A.J. (1973) Root temperature and plant growth. Research review B No. 4, *Commonwealth Bureau of Horticulture and Plantation Crops.* Farnham Royal: Commonwealth Agricultural Bureau.

COOPER H.D. & CLARKSON D.T. (1989) Cycling of amino-nitrogen and other nutrients between shoots and roots in cereals — a possible mechanism integrating shoot and root in the regulation of nutrient uptake. *J. Exp. Bot.* **40**, 753–762.

COOPER K.M. & TINKER P.B. (1978) Translocation and transfer of nutrients in vesicular-arbuscular mycorrhizas. II — Uptake and translocation of phosphorus, zinc and sulphur. *New Phytol.* **81**, 43–52.

COOPER K.M. & TINKER P.B. (1981) Translocation and transfer of nutrients in vesicular-arbuscular mycorrhizas. IV — Effect of environmental variables on movement of phosphorus. *New Phytol.* **88**, 327–339.

CORMACK R.H.H., LEMOY P. & MACLACHLAN G.A. (1963) Calcium in the root hair wall. *J. Exp. Bot.* **14**, 311–315.

COTTENIE A. & KIEKENS L. (1972) Exchange of Zn, Mn, Cu and Fe in relation to saturation of the soil complex. *Proceedings of the 9th Colloquium of the International Potash Institute*, pp. 113–123. Berne.

COWAN I.R. (1965) Transport of water in the soil–plant atmosphere system. *J. Appl. Ecol.* **2**, 221–239.

COX G. & SANDERS F.E.T. (1974) Ultrastructure of the host–fungus interface in a vesicular-arbuscular mycorrhiza. *New Phytol.* **73**, 901–912.

COX G. & TINKER P.B. (1976) Translocation and transfer of nutrients in vesicular-arbuscular mycorrhizas. I — The arbuscule and phosphorus transfer; a quantitative ultrastructural study. *New Phytol.* **77**, 371–378.

COX G., SANDERS F.E., TINKER P.B. & WILD J.A. (1975) Ultrastructural evidence relating to host–endophyte transfer in vesicular–arbuscular mycorrhiza. In *Endomycorrhizas*, eds. Sanders F.E., Mosse B. & Tinker P.B., pp. 297–307. London: Academic Press.

CRAM W.J. (1973) Internal factors regulating nitrate and chloride influx in plant cells. *J. Exp. Bot.* **24**, 328–341.

CRANK J. (1975) *Mathematics of Diffusion*, 2nd Edition. Oxford: Clarendon Press.

CROWLEY D.E., ROMHELD V., MARSCHNER H. & SZANISZLO P.J. (1992) Root-microbial effects on plant iron uptake from siderophores and phytosiderophores. *Plant Soil* **142**, 1–7.

CRUZ R.T., JORDAN W.R. & DREW M.C. (1992) Structural changes and associated reduction of hydraulic conductance in roots of *Sorghum vulgare* L. following exposure to water deficits. *Plant Physiol.* **99**, 203–212.

CUMBUS I.P. & NYE P.H. (1982) Root zone temperature effects on growth and nitrate absorption in rape (*Brassica napus* cv. Emerald). *J. Exp. Bot.* **33**, 1138–1146.

CUNNINGHAM R.K. (1964) Cation–anion relationships in crop nutrition: III Relationships between the ratios of sum of cations: sum of the anions and nitrogen concentrations in several species. *J. Agric. Sci.* **63**, 109–111.

CUNNINGHAM R.K. & COOKE G.W. (1957) *Inorganic nitrogen in soils.* Report of the Rothamsted Experimental Station, 1956, pp. 53–54.

CURL E.A. & TRUELOVE B.T. (1986) *The Rhizosphere.* Berlin: Springer-Verlag.

CURRIE J.A. (1960) Gaseous diffusion in porous media. Pt. II. Dry granular materials. *Br. J. Appl. Phys.* **11**, 318–324.

CURRIE J.A. (1961) Gaseous diffusion in porous media. Pt. III. Wet granular materials. *Br. J. Appl. Phys.* **12**, 275–281.

CURRIE J.A. (1970) Movement of gases in soil respiration. In *Sorption and Transport Processes in Soils*, pp. 152–169, SCI Monograph No. 37. London: Society of Chemical Industry.

CURTISS C.F. & HIRSCHFELDER J. (1949) Transport properties of multicomponent gas mixtures. *J. Chem. Phys.* **17**, 550–555.

CUSHMAN J.H. (1982) Nutrient transport inside and outside the root rhizosphere: theory. *Soil Sci. Soc. Am. J.* **46**, 704–709.

CUSHMAN J.H. (1984) Numerical study of some age-dependent parameters in root nutrient uptake. *Plant Soil* **79**, 123–141.

CUTTER E.G. (1978) *Plant Anatomy: Experiment and Interpretations.* London: Edward Arnold.

CZAPEK F. (1896) Zur Lehre von den Wurzelausscheidungen. *Jahrb. wiss. Bot.* **29**, 321–390.

DALTON F.N., RAATS P.A.C. & GARDNER W.R. (1975) Simultaneous uptake of water and solutes by plant roots. *Agron. J.* **67**, 334–339.

DARRAH P.R. (1991) Models of the rhizosphere II. A quasi three-dimensional simulation of the microbial population dynamics around a growing root releasing soluble exudates. *Plant Soil* **138**, 147–158.

DARRAH P.R. (1993) The rhizosphere and plant nutrition; a quantitative approach. *Plant Soil* **156**, 1–20.

DARRAH P.R. (1996) Rhizodeposition under ambient and elevated CO_2 levels. *Plant Soil* **187**, 265–275.

DAUBENY C.G.B. (1846) On the distinction between the dormant and active ingredients of the soil. *J. R. Agric. Soc. Eng.* **7**, 237–245.

DAVIDSON R.L. (1969) Effect of root/leaf/temperature differentials on root/shoot ratios in a vegetative legume — a model. *Ann. Bot.* **33**, 561–566.

DAVIES C.W. (1962) *Ion Association*. London: Butterworth.

DAVIES M.S. (1991) Effects of toxic concentrations of metals on root growth and development. In *Plant Root and Development*, ed. Atkinson D., pp. 211–228. Oxford: Blackwell.

DAVIES W.J. & ZHENG J. (1991) Root signals and the regulation of growth and development of plants in a drying soil. *Ann. Rev. Plant Physiol. Plant Mol. Biol.* **42**, 55–76.

DAVIES W.J., TARDIEU F. & TREGO C. (1994) How do chemical signals work in plants that grow in drying soil? *Plant Physiol.* **104**, 309–314.

DEGENHARDT J., LARSON P.B., HOWELL S.H. & KOCHIAN L.V. (1998) Aluminium resistance in the *Arabidopsis* mutant *alr*-104 is caused by an aluminium-induced increase in rhizosphere pH. *Plant Physiol.* **117**, 19–27.

DE HAAN F.A.M., BOLT G.H. & PIETERS B.G.M (1965) Diffusion of ^{40}K into an illite during prolonged shaking. *Soil Sci. Soc. Am. Proc.* **29**, 528–530.

DEIST J. & TALIBUDEEN O. (1967) Ion exchange in soils from the ion pairs K–Ca, K–Rb, and K–Na. *J. Soil Sci.* **18**, 125–137.

DE JAGER A. (1982) Effects of localized supply of H_2PO_4, NO_3, SO_4, Ca and K on the production and distribution of dry matter in young maize plants. *Neth. J. Agric. Sci.* **30**, 193–203.

DE SAUSSURE TH. (1804) *Recherches Chimiques sur la Vegetation*, p. 327. Paris.

DEVAUX H. (1916) Action rapide des solutions salines sur les plantes vivantes: deplacement reversible d'une partie des substances basiques centenues dans la plante. *C. R. Acad. Sci., Paris* **162**, 561–563.

DEVIENNE F., MARY B. & LAMAZE T. (1994) Nitrate transport in intact wheat roots. II Long-term effects of nitrate concentration in the nutrient solution on NO_3^- unidirectional fluxes and distribution within the tissues. *J. Exp. Bot.* **45**, 677–684.

DEWAR, R.C. (1993) A root–shoot partitioning model based on carbon–nitrogen water interactions and Munch phloem flow. *Func. Ecol.* **7**, 356–368.

DE WEGER L.A., VAN DER VLUGT L., BAKKER A.P.A.H.M., SCHIPPERS B. & LUGTENBERG B. (1987) Role of flagella of the plant growth stimulating *Pseudomonas fluorescens* isolate WCS374 in the colonization of potato roots. In *Recognition of Microbe–Plant Symbiotic and Pathogenic Interactions*, ed. Lugtenberg B., NATO ASI Series, pp. 409–412. Berlin: Springer-Verlag.

DE WEGER L., DEKKERS L.C., VAN DERBIJL A.J. & LUGTENBERG B.J.J. (1994) Use of phosphate reporter genes to study phosphate limitation in the rhizosphere and in bulk soil. *Mol. Plant-Microbe Interact.* **7**, 32–38.

DE WILLIGEN P. (1991) Nitrogen turnover in the soil–crop system: a comparison of fourteen simulation models. *Fertil. Res.* **27**, 141–149.

DE WILLIGEN P. & VAN NOORDWIJK M. (1987) Uptake potential of non-regularly distributed roots. *J. Plant Nutr.* **10**, 1273–1280.

DE WILLIGEN P. & VAN NOORDWIJK M. (1989) Model calculations on the relative importance of internal longitudinal diffusion for aeration of roots of non-wetland plants. *Plant Soil* **113**, 111–119.

DE WILLIGEN P. & VAN NOORDWIJK M. (1994a) Diffusion and mass flow to a root with constant nutrient demand or behaving as a zero-sink. 1. Constant uptake. *Soil Sci.* **157**, 162–170.

DE WILLIGEN P. & VAN NOORDWIJK M. (1994b) Diffusion and mass flow to a root with constant nutrient demand or behaving as a zero sink. 2. Zero sink. *Soil Sci.* **157**, 171–175.

DE WILLIGEN P., HEINEN M. & VAN DEN BROEK B.J. (1995) Modelling water and nitrogen uptake of a potato crop growing on a ridge. In *Potato Ecology*, eds. Haverkort A.J. & MacKerron D.J.L., pp. 75–88. Dordrecht: Kluwer Academic Press.

DE WIT C.T. (1960) On competition. *Veersl. landbouwk. Onder.* **66**, 1–88.

DE WIT C.T. & VAN KEULEN H. (1972) *Simulation of Transport Processes in Soils.* Wageningen: Centre for Agricultural Publishing and Documentation.

DEXTER A.R. (1986a) Model experiments on the behaviour of roots at the interface between a tilled seed-bed and a compacted sub-soil. I Effects of seed-bed aggregate size and sub-soil strength on wheat roots. *Plant Soil* **95**, 123–133.

DEXTER A.R. (1986b) Model experiments on the behaviour of roots at the interface between a tilled seed-bed and a compacted sub-soil. II Entry of pea and wheat roots into sub-soil cracks. *Plant Soil* **95**, 135–147.

DEXTER A.R. (1986c) Model experiments on the behaviour of roots at the interface between a tilled seed-bed and a compacted sub-soil. III Entry of pea and wheat roots into cylindrical biopores. *Plant Soil* **95**, 149–161.

DIGGLE A.J. (1988) ROOTMAP — A model in three-dimensional coordinates of the growth and structure of fibrous root systems. *Plant Soil* **105**, 169–178.

DINKELAKER B., ROMHELD V. & MARSCHNER H. (1989) Citric acid excretion and precipitation of calcium citrate in the rhizosphere of white lupin (*Lupinus albus* L.). *Plant Cell Environ.* **12**, 285–292.

DITTMER H.J. (1940) A quantitative study of the subterranean members of soyabean. *Soil Cons.* **6**, 33–34.

DITTMER H.J. (1949) Root hair variation in plant species. *Am. J. Bot.* **36**, 152–155.

DIXON J.B. & WEED S.B., eds. (1989) *Minerals in Soil Environments,* 2nd Edition, Book Series No. 1. Madison, WI: Soil Science Society of America.

DONALD C.M. (1958) The interaction of competition for light and for nutrients. *Aust. J. Agric. Res.* **9**, 421–432.

DONALD C.M. (1963) Competition among crop and pasture plants. *Adv. Agron.* **15**, 1–118.

DORMAAR D.F. & SAUERBECK D.R. (1983) Seasonal effects of photoassimilated carbon-14 in the root system of blue grama and associated soil organic matter. *Soil Biol. Biochem.* **15**, 475–479.

DREW M.C. (1975) Comparison of the localised effects of phosphate, nitrate, ammonium and potassium on the growth of the seminal root system and shoot of barley. *New Phytol.* **75**, 479–490.

DREW M.C. (1988) Effects of flooding and oxygen deficiency on plant mineral nutrition. In *Advances in Plant Nutrition*, 3, eds. Tinker P.B. & Lauchli A., pp. 115–159. New York: Praeger.

DREW M.C. & GOSS M.J. (1973) Effect of soil physical factors on root growth. *Chem. Ind.* 679–684.

DREW M.C. & NYE P.H. (1969) The supply of nutrient ions by diffusion to plant roots in soil. II. The effect of root hairs on the uptake of potassium by roots of rye-grass (*Lolium multiflorum*). *Plant Soil* **31**, 407–424.

DREW M.C. & NYE P.H. (1970) The supply of nutrient ions by diffusion to plant roots in soil. III. Uptake of phosphate by roots of onion, leek and ryegrass. *Plant Soil* **33**, 545–563.

DREW M.C. & SAKER L.R. (1975a) Nutrient supply and the growth of the seminal root system in barley. II. Localized compensatory increases in lateral growth and rates of nitrate uptake. *J. Exp. Bot.* **26**, 79–90.

DREW M.C. & SAKER, L.R. (1975b) Further studies on the modification to root growth and ion uptake caused by localized enrichment of phosphate in the rooting zone of barley. *Annual Report of the ARC Letcombe Laboratory, 1974*, pp. 8–10.

DREW M.C. & SAKER L.R. (1984) Uptake and long-distance transport of phosphate, potassium and chloride in relation to internal ion concentrations in barley: evidence for non-allosteric regulation. *Planta* **160**, 500–507.

DREW M.C. & STOLZY L.H. (1996) Growth under oxygen stress. In *Plant Roots — The Hidden Half*, eds. Waisel Y., Eshel A. & Kafkafi U., pp. 397–414. New York: Marcel Dekker.

DREW M.C., SAKER L.R., BARBER S.A. & JENKINS W. (1984) Changes in the kinetics of phosphate and potassium absorption in nutrient-deficient barley roots measured by a nutrient-depletion technique. *Planta* **160**, 490–499.

DREW M.C., COBB B.G., JOHNSON J.R., ANDREWS D., MORGAN P.W., JORDAN W. & HE C.J. (1994) Metabolic acclimation of root tips to oxygen deficiency. *Ann. Bot.* **74**, 281–286.

DUDDRIDGE J.A., MALIBARI A. & READ D.J. (1980) Structure and function of mycorrhizal rhizomorphs with special reference to their role in water transport. *Nature* **287**, 834–836.

DUDDRIDGE J.A., FINLAY R.D., READ D.J. & SODERSTROM B. (1988) The structure and function of the vegetative mycelium of ectomycorrhizal plants III. Ultrastructural and autoradiographic analysis of inter-plant carbon distribution through intact mycelial systems. *New Phytol.* **108**, 183–188.

DULLIEN F.A.L. (1979) *Porous Media Fluid Transport and Pore Structure*. New York: Academic Press.

DUNHAM R.J. & NYE P.H. (1973) The influence of soil water content on the uptake of ions by roots. I. Soil water content gradients near a plane of onion roots. *J. Appl. Ecol.* **10**, 585–598.

DUNHAM R.J. & NYE P.H. (1974) The influence of soil water content on the uptake of ions by roots. II. Chloride uptake and concentration gradients in soil. *J. Appl. Ecol.* **11**, 581–596.

DUNHAM R.J. & NYE P.H. (1976) The influence of soil water content on the uptake of ions by roots. III. Phosphate, potassium, calcium and magnesium uptake and concentration gradients in soil. *J. Appl. Ecol.* **13**, 967–984.

DUNLOP J. & GARDENER S. (1993) Phosphate uptake, proton extrusion and membrane electropotentials of phosphorus-deficient *Trifolium repens*. *J. Exp. Bot.* **44**, 1801–1808.

DURRALL D.M., JONES M.D. & TINKER P.B. (1995) Carbon allocation in ectomycorrhizal willow. *New Phytol.* **128**, 109–114.

DYER B. (1894) On the analytical determination of probably available 'mineral' plant food in soil. *J. Chem. Soc.* **65**, 115–167.

DYER B. (1901) A chemical study of the phosphoric acid and potash contents of wheat soils of Broadbalk Field, Rothamsted. *Phil. Trans. R. Soc. B* **194**, 235–290.

EATON F.M., HARDING R.B. & GANGE T.J. (1960) Soil solution extractions at 1/10 bar moisture percentages. *Soil Sci.* **90**, 253–258.

EAVIS B.W., TAYLOR H.M. & HUCK M.G. (1971) Radicle elongation of pea seedlings as affected by oxygen concentration and gradients between shoot and root. *Agron. J.* **63**, 770–772.

ECKERSTEN H. (1994) Modelling daily growth and nitrogen turnover for a short-rotation forest over several years. *Forest Ecol. Manage.* **69**, 57–72.

EDWARDS C.A. (1974) Factors affecting the persistence of pesticides in the soil. *Chem. Ind.* **5**, 190–193.

EL BASSAM N., DAMBROTH M. & LOUGHMAN B.C., eds. (1990) *Genetic Aspects of Plant Mineral Nutrition.* Dordrecht: Kluwer Academic.

ELLIOTT G.C., LYNCH, J. & LAUCHLI, A. (1984) Influx and efflux of P in roots of intact maize plants. *Plant Physiol.* **76**, 336–341.

ENGELAAR W.M.H.G., VAN BRUGGEN M.H., VAN HOEK W.P.M., HERYSEN M.A.H. & BLOM C.W.P.M. (1993) Root porosities and radical oxygen losses of *Rumex* and *Plantago* species as influenced by soil pore diameter and soil aeration. *New Phytol.* **125**, 565–574.

ENNIK G.C. & BAAN HOFMAN T. (1983) Variation in the root mass of ryegrass types and its ecological consequences. *Neth. J. Agric. Sci.* **31**, 325–334.

EPSTEIN E. (1972) *Mineral Nutrition of Plants: Principles and Perspectives.* New York: Wiley.

ERICSSON T. (1995) Growth and shoot:root ratio of seedlings in relation to nutrient availability. *Plant Soil* **169**, 205–214.

ERIKSSON E. (1952) Cation exchange equilibria on clay minerals. *Soil Sci.* **74**, 103–113.

ESAU K. (1965) *Plant Anatomy*, 2nd Edition. New York: John Wiley & Sons.

ESCAMILLA J.A., COMERFORD N.B. & NEARY D.G. (1991a) Spatial pattern of slash pine roots and its effects on nutrient uptake. *Soil Sci. Soc. Am. J.* **55**, 1716–1722.

ESCAMILLA J.A., COMERFORD N.B. & NEARY D.G. (1991b). Soil core break method to estimate pine root distribution. *Soil Sci. Soc. Am. J.* **55**, 1722–1726.

ESHEL A. & WAISEL Y. (1996) Multiforms and multifunctions of various constituents of one root and system. In *Plant Roots — The Hidden Half*, eds. Waisel Y., Eshel A. & Kafkafi U., pp. 175–193. New York: Marcel Dekker.

EVANS D.E., BRIERS S.A. & WILLIAMS L.E. (1991) Active Ca transport by plant cell membranes. *J. Exp. Bot.* **42**, 285–303.

EVERARD J.D. & DREW M.C. (1987) Mechanism of inhibition of water movement in anaerobically treated roots of *Zea mays* L. *J. Exp. Bot.* **38**, 1154–1165.

EVERETT D.H. & WHITTON W.I. (1952) A general approach to hysteresis. I. *Trans. Faraday Soc.* **48**, 749–757.

FAHN A. (1982) *Plant Anatomy*, 4th Edition. Oxford: Pergamon.

FAIZ S.M.A. & WEATHERLEY P.E. (1982) Root contraction in transpiring plants. *New Phytol.* **92**, 333–343.

FARR E., VAIDYANATHAN L.V. & NYE P.H. (1969) Measurement of ionic concentration gradients in soil near roots. *Soil Sci.* **107**, 385–391.

FARR E., VAIDYANATHAN L.V. & NYE P.H. (1970) The measurement and mechanisms of ion diffusion in soils. V. Diffusion of hydrogen ion in soils. *J. Soil Sci.* **21**, 1–14.

FARRAR J.F. (1992) The whole plant: carbon partitioning during development. In *Carbon Partitioning within and between Organisms*, eds. Pollock, C.J., Farrar J.F. & Gordon A.J., pp. 163–179. Oxford: Bios Scientific.

FARRAR J.F. (1993) Sink strength: What is it and how are we to measure it? Summary. *Plant Cell Environ.* **16**, 1045–1046.

FARRAR J.F. (1996) Regulation of root weight ratio is mediated by sucrose: opinion. *Plant Soil* **185**, 13–19.

FARRELL D.A., GREACEN E.L. & GURR C.G. (1966) Vapour transfer in soil due to air turbulence. *Soil Sci.* **102**, 305–313.

FAXEN H. (1922) Der Widerstand gegen die Bewegung einen starren Kugel in einer zahen Flussigkeit die zwischen zwei parallellen ebenen Wanden eingeschlossen ist. *Ann. Physik*. **68**, 89–119.

FEDDES R.A., KOWALIK P.J. & ZARADNY H. (1978) *Simulation of Field Water Use and Crop Yield*. Wageningen: Centre for Agricultural Publishing and Documentation.

FIELD C. & MOONEY H.A. (1986) The photosynthesis–nitrogen relationship in wild plants. In *On the Economy of Plant Form and Function*, ed. Givnish T.J., pp. 25–55. Cambridge: Cambridge University Press.

FINLAY R.D. (1992) Uptake and translocation of nutrient by ectomycorrhizal fungal mycelia. In *Mycorrhizas in Ecosystems*, eds. Read D.J., Lewis D., Fitter A.H. & Alexander I.J., pp. 91–97. Wallingford: CAB International.

FINLAY R.D. & READ D.J. (1986) The structure and function of the vegetative mycelium of ectomycorrhizal plants. II The uptake and distribution of phosphorus by mycelial strands interconnecting host plants. *New Phytol.* **103**, 157–165.

FISCUS E.L. & MARKHART A.H. III (1979) Relationships between root system water transport properties and plant size in *Phaseolus*. *Plant Physiol.* **64**, 770–773.

FISHER, R.A. (1925) *Statistical Methods for Research Workers*. Edinburgh: Oliver & Boyd.

FITTER A.H. (1988) Water relations of red clover *Trifolium pratense* L. as affected by VA mycorrhizal infection and phosphorus supply before and during drought. *J. Exp. Bot.* **39**, 595–603.

FITTER A.H. (1991a) Costs and benefits of mycorrhizas: implications for functioning under natural conditions. *Experientia* **47**, 350–355.

FITTER A.H. (1991b) The ecological significance of root systems architecture; an economic approach. In *Plant Root Growth*, ed. Atkinson D., pp. 229–246. Oxford: Blackwell.

FITTER A.H. (1996) Characteristics and functions of root systems. In *Plant Roots — The Hidden Half*, eds. Waisel Y., Eshel A. & Kafkafi U., pp. 1–20. New York: Marcel Dekker.

FITTER A.H. & GARBAYE J. (1994) Interactions between mycorrhizal fungi and other soil organisms. *Plant Soil* **159**, 123–132.

FITTER A.H. & HAY R.K.M. (1981) *Environmental Physiology of Plants*. London: Academic Press.

FITTER A.H. & MERRYWEATHER J.W. (1992) Why are some plants more mycorrhizal than others? An ecological enquiry. In *Mycorrhizas in Ecosystems*, eds. Read D.J., Lewis D.H., Fitter A.H. & Alexander I.J., pp. 26–36. Wallingford: CAB International.

FITTER A.H. & STICKLAND T.R. (1991) Architectural analysis of plant root systems 2. Influence of nutrient supply on architecture in contrasting plant species. *New Phytol.* **118**, 383–388.

FITTER A.H. & STICKLAND T.R. (1992) Fractal characterization of root system architecture. *Func. Ecol.* **6**, 632–635.

FITTER A.H., STICKLAND T.R., HARVEY M.L. & WILSON G.W. (1991) Architectural analysis of plant root systems I. Architectural correlates of exploration efficiency. *New Phytol.* **118**, 375–382.

FITTER A.H., GRAVES J.D., WATKINS N.K., ROBINSON D. & SCRIMGEOUR C. (1998) Carbon transfer between plants and its control in networks of arbuscular mycorrhizas. *Func. Ecol.* **12**, 406–412.

FOGEL R. (1991) Root system demography and production in forest ecosystems. In *Plant Root Growth*, ed. Atkinson D., pp. 89–101. Oxford: Blackwell.

FOHSE D., CLAASSEN N. & JUNGK A. (1991) Phosphorus efficiency of plants. II Significance of root radius, root hairs and cation–anion balance for phosphorus influx in seven plant species. *Plant Soil* **132**, 261–272.

FORRESTER S.D. & GILES C.H. (1971) From manure heaps to monolayers: the earliest development of solute–solid adsorption studies. *Chem. Ind.* 1314–1321.

FOY C.D. (1992) Soil chemical factors limiting plant root growth. *Adv. Soil Sci.* **19**, 97–150.

FRANCIS R. & READ D.J. (1984) Direct transfer of carbon between plants connected by vesicular-arbuscular mycorrhizal mycelium. *Nature* **307**, 53–56.

FREY B. & SCHUEPP H. (1992a) Transfer of symbiotically fixed nitrogen from berseem (*Trifolium alexandreum* L.) to maize via vesicular-arbuscular mycorrhizal hyphae. *New Phytol.* **122**, 447–454.

FREY B. & SCHUEPP H. (1992b) Nitrogen translocation through a root-free soil mediated by VA fungal hyphae. In *Mycorrhizas in Ecosystems*, eds. Read D.J., Lewis D.H., Fitter A.H. & Alexander I.J., pp. 378–379. Wallingford: CAB International.

FRIED J.J. & UNGEMACH P.O. (1971) Determination in situ de coefficient de dispersion longitudinale d'un milieu poreaux natural. *C. R. Acad. Sci., Paris,* **172A**, 1327.

FRIEND A.D., STEVENS, A.K., KNOX, R.G. & CANNELL M.G.R. (1997) A process-based terrestrial biosphere model of ecosystem dynamics (Hybrid v. 3.0). *Ecol. Modelling* **95**, 249–287.

FRISSEL M.J. & POELSTRA P. (1967) Chromatographic transport through soils. II. Column experiments with Sr and Ca isotopes. *Plant Soil* **27**, 20–32.

FRISSEL M.J., POELSTRA P. & REINIGER P. (1970a) Chromatographic transport through soils. III. A simulation model for the evaluation of the apparent diffusion coefficient in undisturbed soils with tritiated water. *Plant Soil* **33**, 161–176.

FRISSEL M.J., POELSTRA P. & REINIGER P. (1970b) Sorption and transport in soils. In *Sorption and Transport Processes in Soils*, pp. 135–151, SCI Monograph No. 37. London: Society of Chemical Industry.

FRISSEL M.J. *et al.* (1973) Tracing soil moisture migration with ^{36}Cl, ^{60}Co and tritium. *Proc. Symp. on Isotopes and Radiation Techniques in Studies of Soil Physics, Irrigation and Drainage in Relation to Crop Production.* IAEA, Vienna, pp. 145–151.

FUSSEDER A. (1987) The longevity and activity of the primary root of maize. *Plant Soil* **101**, 257–265.

GAHOONIA T.S. & NIELSEN N.E. (1992) The effects of root induced pH changes on the depletion of inorganic and organic phosphates in the rhizosphere. *Plant Soil* **143**, 185–191.

GAHOONIA T.S., CLASSEN N. & JUNGK A. (1992) Mobilization of phosphate in different soils by ryegrass supplied with ammonium or nitrate. *Plant Soil* **140**, 241–249.

GANNON J.T., MINGELGRIN U., ALEXANDER M. & WAGENET R.J. (1991) Bacterial transport through homogeneous soil. *Soil Biol. Biochem.* **23**, 1155–1160.

GAPON E.N. (1933) Theory of exchange adsorption in soils. *J. Gen. Chim., Moscow,* **3**, 144–163.

GARDNER W.R. (1960) Dynamic aspects of water availability to plants. *Soil Sci.* **89**, 63–73.

GARDNER W.R. (1965) Movement of nitrogen in soil. In *Soil Nitrogen*, eds. Bartholomew W.V. & Clark F.E., *Agronomy*, Monograph 10, pp. 550–572. Madison, WI: America Society of Agronomy.

GARDNER W.R. (1991) Modelling water uptake by roots. *Irrigation Science* **12**, 109–114.

GARDNER W.K. & BOUNDY K.A. (1983) The acquisition of phosphorus by *Lupinus albus* L. IV. The effect of interplanting wheat and white lupin on the growth and mineral composition of the two species. *Plant Soil* **70**, 391–402.

GARDNER W.K. & PARBERY D.C. (1982) The acquisition of phosphorus by *Lupinus albus* L. II. The effect of varying phosphorus supply and soil type on some characteristics of the soil–root interface. *Plant Soil* **68**, 33–41

GARDNER W.K., PARBERY D.C. & BARBER D.A. (1982) The acquisition of phosphorus by *Lupinus albus* L. I. Some characteristics of the soil–root interface. *Plant Soil* **68**, 19–32.

GARDNER W.K., BARBER D.A. & PARBERY D.C. (1983a) The acquisition of phosphorus by *Lupinus albus* L. III. The probable mechanism by which phosphorus movement in the soil–root interface is enhanced. *Plant Soil* **70**, 107–124.

GARDNER W.K., BARBER D.A. & PARBERY D.C. (1983b) Non-infecting rhizosphere microorganisms and the mineral nutrition of temperate cereals. *J. Plant Nutr.* **6**, 185–189.

GARNIER E. (1991) Resource capture, biomass allocation, and growth in herbaceous plants. *Trends Ecol. Evol.* **6**, 126–131.

GARWOOD E.A. & SINCLAIR J. (1979) Use of water by six grass species. 2. Root distribution and use of soil water. *J. Agric. Sci., Camb.* **93**, 25–35.

GAZZERI G. (1823) *A Textbook of Manuring*, quoted by Orth, A. (1873) *Landwirtsch. Vers-Sta.* **16**, 56.

GEERING H.R. (1967) M.Sci. Thesis, Cornell University, quoted in Olsen S.R. & Kemper W.D. (1968) *Adv. Agron.* **20**, 91–149.

GEORGE E., HAUSSLER K., KOTHARI S.K., LI X.L. & MARSCHNER H. (1992) Contribution of mycorrhizal hyphae to nutrient and water uptake of plants. In *Mycorrhizas in Ecosystems*, eds. Read D.J., Lewis D.H., Fitter A.H. & Alexander I.J., pp. 42–47. Wallingford: CAB International.

GERSTL Z., YARON B. & NYE P.H. (1979) Diffusion of biodegradable pesticide I. In a biologically inactive soil. *Soil Sci. Soc. Am. J.* **43**, 839–842.

GERWITZ A. & PAGE E.R. (1973) Estimation of root distribution in soil, by labelling with ^{86}Rb and counting with commercially available equipment. *Lab. Pract.* **22**, 35–36.

GERWITZ A. & PAGE E.R. (1974) An empirical mathematical model to describe plant root systems. *J. Appl. Ecol.* **11**, 773–781.

GIANINAZZI-PEARSON V., SMITH S.E., GIANINAZZI S. & SMITH V.A. (1991). Enzymatic studies on the metabolism of vesicular-arbuscular mycorrhizas. V. Is H^+-ATPase a component of ATP hydrolysing enzyme activities in plant–fungus interface? *New Phytol.* **117**, 61–76.

GIANINAZZI-PEARSON V., DUMAS-GAUDOT E., GOLLOTTE A., TAHIRI-ALAOUI A. & GIANINAZZI S. (1996) Cellular and molecular defence-related root responses to invasions by arbuscular mycorrhizal fungi. *New Phytol.* **133**, 45–57

GIFFORD R.M. & EVANS L.T. (1981) Photosynthesis, carbon partitioning and yield. *Ann. Rev. Plant Physiol. Plant Mol. Biol.* **32**, 485–509.

GILDON A. & TINKER P.B. (1983a) Interactions of vesicular-arbuscular mycorrhizal infection and heavy metals in plants. I — The effects of heavy metals on the development of vesicular-arbuscular mycorrhizas. *New Phytol.* **95**, 247–261.

GILDON A. & TINKER P.B. (1983b) Interactions of vesicular–arbuscular mycorrhizal infection and heavy metals in plants. II — The effects of infection on uptake of copper. *New Phytol.* **95**, 263–268.

GILLER K.E. & DAY J.M. (1985) Nitrogen fixation in the rhizosphere: significance in natural agricultural systems. In *Biological Interactions in Soils*, ed. Fitter A.H. Oxford: Blackwell.

GILLER K.E. & WILSON K.J. (1991) *Nitrogen Fixation in Tropical Cropping Systems.* Wallingford: CAB International.

GILLESPIE A.R. & POPE P.E. (1990a) Rhizosphere acidification increases phosphorus recovery of black locust: I Induced acidification and soil response. *Soil Sci. Soc. Am. J.* **54**, 534–537.

GILLESPIE A.R. & POPE P.E. (1990b) Rhizosphere acidification increases phosphorus recovery of black locust: II Model predictions and measured recovery. *Soil Sci. Soc. Am. J.* **54**, 538–541.

GLANDORF C.M., PETERE L.G.L., VAN DER SLUIS I., BAKKER P.A.H.M. & SCHIPPERS B. (1993) Crop specificity of rhizosphere *Pseudomonads* and the involvement of root agglutinins. *Soil Biol. Biochem.* **25**, 981–989.

GLASS A.D.M. (1976) Regulation of potassium absorption in barley roots: an allosteric model. *Plant Physiol.* **58**, 33–37.

GLASS A.D.M. (1983) Regulation of ion transport. *Ann. Rev. Plant Physiol. Plant Mol. Biol.* **34**, 311–326.

GLASS A.D.M. & SIDDIQUI M.Y. (1984) The control of nutrient uptake rates in relation to the inorganic composition of plants. *Advances in Plant Nutrition, 1*, eds. Tinker P.B. & Lauchli A., pp. 103–148. New York: Praeger.

GLASS R.J., STEENHUIS T.S. & PARLANGE J.-Y. (1989) Mechanism for finger persistence in homogeneous unsaturated porous media; theory and verification. *Soil Sci.* **148**, 60–70.

GLASSTONE S., LAIDLER K.J. & EYRING H. (1941) *Theory of Rate Processes.* New York: McGraw-Hill.

GLUEKAUF E. (1955) Theory of chromatography. Pt. 10. Formulae for diffusion into spheres and their application to chromatography. *Trans. Faraday Soc.* **51**, 1540–1551.

GODWIN D.C. & ALLAN JONES C. (1991) Nitrogen dynamics in soil–plant systems. In *Modelling Plant and Soil Systems*, eds. Hanks J. & Ritchie J.T., Agronomy Series 31, Agronomy Monograph No. 11, pp. 287–321. Madison, WI: American Society of Agronomy.

GOLDBERG S. & SPOSITO G. (1985) On the mechanism of specific phosphate adsorption by hydroxylated mineral surfaces: a review. *Comm. Soil Sci. Plant Anal.* **16**, 801–821.

GORE A.J.P. & OLSON J.S. (1967) Preliminary models for accumulation of organic matter in an *Eriophorum/Calluna* ecosystem. *Aquilo, Ser. Bot. Soc. Amic. Nat. Oulensis.* **6**, 297–313.

GORING C.A.I. (1968) The size, shape and origin of lignin macromolecules. In *Solution Properties of Natural Polymers.* Chem. Soc. London. Spec. Pub. No. 27, 115–134.

GORING C.A.I & HAMAKER J.W., eds. (1972) *Organic Chemicals in the Soil Environment*, 2 Vols. New York: Marcel Dekker.

GORING R.L. & CHURCHILL S.W. (1961) Thermal conductivity of heterogeneous materials. *Chem. Eng. Prog.* **57**, 53–59.

GOSS K.V. (1993) Effect of temperature and relative humidity on the sorption of organic vapors on clay minerals. *Environ. Sci. Technol.* **27**, 2127–2132.

GOUDRIAAN J. (1996) Predicting crop yields under global change. In *Global Change and Terrestrial Ecosystems*, eds. Walker B. & Steffen W., pp. 260–274. Cambridge: Cambridge University Press.

GOUDRIAAN J. & VAN LAAR H.H. (1994) *Modelling Potential Crop Growth Processes.* Dordrecht: Kluwer Academic.

GRACE J.B. & TILMAN D. (1990) *Perspectives on Plant Competition.* New York: Academic Press.

GRAF H., REICHENBACK V. & RICH C.I. (1968) Preparation of dioctahedral vermiculite from muscovite and subsequent exchange properties. *Trans. 9th Int. Congr. Soil Sci. Adelaide* **1**, 709–719.

GRAHAM J.H. & SYVERTSEN J.P. (1989) Vesicular-arbuscular mycorrhizas increase chloride concentration in citrus seedlings. *New Phytol.* **113**, 29–36.

GRAHAM R.D. (1984) Breeding for nutritional characteristics in cereals. *Advances in Plant Nutrition*, 1, eds. Tinker P.B. & Lauchli A., pp. 57–102. New York: Praeger.

GRAHAM-BRYCE I.J. (1969) Diffusion of organo-phosphorous insecticides in soil. *J. Sci. Food Agric.* **20**, 489–492.

GRAHAM-BRYCE I.J. & BRIGGS G.G. (1970) Pollution of soils. *R. Inst. Chem.* **3**, 87–104.

GRAVES J.D., WATKINS N.K., FITTER A.H., ROBINSON D. & SCRIMGEOUR C. (1997) Intraspecific transfer of carbon between plants linked by a common mycorrhizal network. *Plant Soil* **192**, 153–159.

GREACEN E.L. & OH J.S. (1972) Physics of root growth. *Nature* (Lond.) **235**, 24–25.

GREACEN E.L., FARRELL D.A. & COCKCROFT B. (1968) Soil resistance to metal probes and plant roots. *Trans. 9th Int. Congr. Soil Sci. Adelaide* **1**, 769–778.

GREAVES M.P. & DARBYSHIRE J.F. (1972) The ultrastructure of the mucilaginous layer on plant roots. *Soil Biol. Biochem.* **4**, 443–449.

GREEN J.S. (1976) An investigation of the self diffusion of chloride ions within soil aggregates. Thesis, Chemistry, Pt. 2, Oxford.

GREENLAND D.J. (1965) Interaction between clays and organic compounds in soils. Pt. I. Mechanisms of interaction between clays and defined organic compounds. *Soils Fert.* **28**, 415–425.

GREENLAND D.J. (1970) Sorption of organic compounds by clays and soils. In *Sorption and Transport Processes in Soils*, pp. 79–88, SCI Monograph No. 37. London: Society of Chemical Industry.

GREENLAND D.J. (1971) Changes in the nitrogen status and physical conditions of soil under pastures, with special reference to the maintenance of the fertility of Australian soils used for growing wheat. *Soils Fert.* **34**, 237–251.

GREENLAND D.J. & NYE P.H. (1959) Increases in the carbon and nitrogen contents of tropical soils under natural fallows. *J. Soil Sci.* **9**, 284–299.

GREENWAY H. & MUNNS R. (1980) Mechanisms of tolerance in non-halophytes. *Ann. Rev. Plant Physiol.* **31**, 149–190.

GREENWOOD D.J. & DRAYCOTT A. (1988) Quantitative relationships for growth and N content of different vegetable crops grown with and without ample fertilizer N in the same soil. *Fert. Res.* **18**, 153–174.

GREENWOOD D.J. & DRAYCOTT A. (1989) Experimental validation of an N-response model for widely different crops. *Fert. Res.* **18**, 153–174.

GREENWOOD D.J. & KARPINETS T.V. (1997) Dynamic model for the effects of K-fertilizer on crop growth, K-uptake and soil-K in arable cropping.1. Description of the model. *Soil Use Manage.* **13**, 178–183.

GREENWOOD D.J. & STONE D.A. (1998) Prediction and measurement of the decline in critical-K, maximum-K and total cation plant concentrations during the growth of field vegetable crops. *Ann. Bot.* **82**, 871–881.

GREENWOOD D.J., GERWITZ A., STONE D.A. & BARNES A. (1982) Root development of vegetable crops. *Plant Soil* **68**, 75–96.

GREENWOOD D.J., NEETESON J.J. & DRAYCOTT A. (1986) Quantitative relationships for the dependence of growth rate of arable crops on their nitrogen content, dry weight and aerial environment. *Plant Soil* **91**, 281–301.

GREENWOOD D.L., LEMAIRE G., GOSSE G., CRUZ P., DRAYCOTT A. & NEETESON J.J. (1990) Decline in percentage N of C3 and C4 crops with increasing plant mass. *Ann. Bot.* **66**, 425–436.

GREENWOOD D.J., GASTAL F., LEMAIRE G., DRAYCOTT A., MILLARD P. & NEETESON J.J. (1991) Growth rate and % N of field grown crops: theory and experiment. *Ann. Bot.* **67**, 181–190.

GREENWOOD D.J., RAHN C., DRAYCOTT A., VAIDYANATHAN L.V. & PATERSON C. (1996) Modelling and measurement of the effects of fertilizer-N and crop residue incorporation on N-dynamics in vegetable cropping. *Soil Use Manage.* **12**, 13–24.

GREGORY P.J. (1988) Growth and functioning of plant roots. In *Russell's Soil Conditions and Plant Growth.* ed. Wild A., p. 152. Harlow, UK: Longman Scientific and Technical.

GREGORY P.J. (1994a) Resource capture by root networks. In *Resource Capture by Crops*, eds. Monteith J.L., Scott R.K. & Unsworth M.H., pp. 77–87. Nottingham: Nottingham University Press.

GREGORY P.J. (1994b) Root growth and activity. In *Physiology and Determination of Crop Yield*, pp. 65–91. Madison, WI: American Society of Agronomy.

GREGORY P.J. & ATWELL B.J. (1991) The fate of carbon in pulse-labelled crops of barley and wheat. *Plant Soil* **136**, 205–213.

GREGORY P.J. & HINSENGER P. (1999) New approaches to studying chemical and physical changes in the rhizosphere. *Plant Soil* (in press).

GREGORY P.J., MCGOWAN M., BISCOE P.V. & HUNTER B. (1978) Water relations of winter wheat. I Growth of the root system. *J. Agric. Sci., Camb.* **91**, 91–102.

GREGORY P.J., CRAWFORD D.V. & MCGOWAN M. (1979) Nutrient relations of winter wheat. 2. Movement of nutrients to the root and their uptake. *J. Agric. Sci., Camb.* **93**, 495–504.

GREGORY P.J., TENNANT D. & BELFORD R.K. (1992) Root and shoot growth and water and light use efficiency of barley and wheat crops grown on a shallow duplex soil in a Mediterranean-type environment. *Austr. J. Agric. Res.* **43**, 555–573.

GREGORY P.J., PALTA J. & BATTS G.R. (1996) Root systems and root–shoot mass ratio: carbon allocation under current conditions in crops. *Plant Soil* **187**, 221–228.

GRIGNON C. & SENTENAC H. (1991) pH and ionic conditions in the apoplast. *Ann. Rev. Plant Physiol. Plant Mol. Biol.* **42**, 103–128.

GRIME J.P., MACKAY J.M.L., HILLIER S.H. & READ D.J. (1987) Floristic diversity in a model system using experimental microcosms. *Nature* **328**, 420–422.

GRIME J.P., HODGSON J.G. & HUNT R. (1988) *Comparative Plant Ecology*. London: Unwin Hyman.

GRIMES D.W., MILLER R.J. & WILEY P.L. (1975) Cotton and corn development in two field soils of different strength characteristics. *Agron. J.* **67**, 519–523.

GRINSTED M.J., HEDLEY M.J., WHITE R.E. & NYE P.H. (1982) Plant induced changes in the rhizosphere of rape (*Brassica napus* var. *Emerald*) seedlings. I. pH change and the increase in P concentration in the soil solution. *New Phytol.* **91**, 19–29.

GUIDI G., POGGIO G. & PETRUZZELLI G. (1985) The porosity of soil aggregates from bulk soil and from soil adhering to roots. *Plant Soil* **87**, 311–314.

GUINEL F.C. & MCCULLY M.E. (1986) Some water-related physical properties of maize root-cap mucilage. *Plant Cell Environ.* **9**, 657–666.

GUNARY D. (1963) Behaviour of carrier free phosphorus-32 in natural soils in relation to the measurement of labile soil phosphorus. *J. Sci. Food Agric.* **14**, 319–324.

GUNARY D. (1970) A new adsorption isotherm for phosphate in soil. *J. Soil Sci.* **21**, 72–77.

GUR A. & SHULMAN Y. (1971) Influence of high root temperatures on the potassium nutrition and on certain organic constituents of apple-plants. In *Recent Advances in Plant Nutrition*. Vol. II, ed. Samish R.M., pp. 643–656. New York: Gordon & Breach.

GURR C.G., MARSHALL T.J. & HUTTON J.T. (1952) Movement of water in soil due to a temperature gradient. *Soil Sci.* **74**, 335–342.

HACKETT C. (1968) A study of the root system of barley. I. Effects of nutrition on two varieties. *New Phytol.* **67**, 287–299.

HACKETT C. & ROSE D.A. (1972a) A model of the extension and branching of a terminal root of barley, and its use in studying relations between root dimensions. I. The model. *Aust. J. Biol. Sci.* **25**, 669–679.

HACKETT C. & ROSE D.A. (1972b) A model of the extension and branching of a terminal root of barley, and its use in studying relations between root dimensions. II. Results and inferences from manipulation of the model. *Aust. J. Biol. Sci.* **25**, 681–690.

HAINSWORTH J.M. & AYLMORE, L.A.G. (1989) Non-uniform soil water extraction by plant roots. *Plant Soil* **113**, 121–124.

HALE M.G. & MOORE L.D. (1979) Factors affecting root exudation. II 1970–1978. *Adv. Agron.* **31**, 93–124.

HALLSWORTH E.G. & CRAWFORD D.V., eds. (1965) *Experimental Pedology*. London: Butterworth.

HAMAKER J.W. (1972) Diffusion and volatilization. In *Organic Chemicals in the Soil Environment*, eds. Goring C.A.I. & Hamaker J.W., pp. 341–397. New York: Marcel Dekker.

HAMAKER J.W. & THOMPSON J.M. (1972) Adsorption. In *Organic Chemicals in the Soil Environment*, Vol. 1, eds. Goring C.A.I. & Hamaker J.W., Chapter 2. New York: Marcel Dekker.

HAMAKER J.W., GORING C.A.I. & YOUNGSON C.R. (1966) Sorption and leaching of 4-amino-3,5,6-trichloropicolinic acid in soils. In *Organic Pesticides in the Environment*, pp. 23–37, Advances in Chemistry No. 60.

HAMBLIN A. (1985). The influence of soil structure on water movement, crop root growth and water uptake. *Adv. Agron.* **38**, 95–158.

HAMBLIN A.P. & HAMBLIN J. (1985) Root characteristics of some temperate legume species and varieties on deep free-draining entisols. *Austr. J. Agric. Res.* **36**, 63–72.

HAMEL C. & SMITH D.L. (1992) Mycorrhiza-mediated 15-N transfer from soybean to corn in field-grown intercrops: effect of component crop spatial relationships. *Soil Biol. Biochem.* **24**, 499–501.

HAMEL C., FYLES H. & SMITH D.L. (1990) Measurement of development of endomycorrhizal mycelium using 3 vital stains. *New Phytol.* **115**, 297–302.

HANCE R.J. (1970) Influence of sorption on the decomposition of pesticides. In *Sorption and Transport Processes in Soils*, pp. 92–104, SCI Monograph No. 37. London: Society of Chemical Industry.

HANKS J. & RITCHIE J.T. (1986) *Modelling Plant and Soil Systems*, Agronomy Series 31. Madison, WI: American Society of Agronomy.

HANNAPEL R.J., FULLER W.H. & FOX R.H. (1964a) Phosphorus movement in a calcareous soil. I. Predominance of organic forms of P in P movement. *Soil Sci.* **97**, 350–357.

HANNAPEL R.J., FULLER W.H. & FOX R.H. (1964b) Phosphorus movement in a calacareous soil. II. Soil microbial activity and organic P movement. *Soil Sci.* **97**, 421–427.

HANSEN S., JENSEN H.E., NIELSEN N.E. & SVENDSEN H. (1991) Simulation of nitrogen dynamics and biomass production in winter wheat using the Danish simulation model DAISY. *Fertil. Res.* **27**, 245–259.

HARDIE K. & LEYTON L. (1981) The influence of vesicular-arbuscular mycorrhiza on growth and water relations of red clover. 1. In phosphate deficient soil. *New Phytol.* **89**, 599–608.

HARLEY J.L (1969) *The Biology of Mycorrhiza.* London: Leonard Hill.

HARLEY J.L. & SMITH S.E. (1983) *Mycorrhizal Symbiosis.* London: Academic Press.

HARPER J.L., JONES M. & SACKVILLE-HAMILTON N.R. (1991) The evolution of roots and the problems of analysing their behaviour. In *Plant Root Growth*, ed. Atkinson D., pp. 3–22. Oxford: Blackwell.

HARRIS D., PACOVSKY R.S. & PAUL E.A. (1985) Carbon economy of soybean–*Rhizobium–Glomus* association. *New Phytol.* **101**, 427–440.

HARRISON A.F. & HELLIWELL D.R. (1979) A bioassay for comparing phosphorus availability in soils. *J. Appl. Ecol.* **16**, 497–505.

HARTLEY G.S. & GRAHAM-BRYCE I.J. (1980) *Physical Principles of Pesticide Behaviour*, Vol. 1. London: Academic Press.

HASELWANDTER K. (1995) Mycorrhizal fungi: siderophore production. *Crit. Rev. Biotechnol.* **15**, 287–291.

HASSETT J.J. & BANWART W.L. (1989) The sorption of nonpolar organics by soils and sediments. In *Reactions and Movement of Organic Chemicals in Soils*, eds. Sawney B.L. & Brown K., SSSA Spec. Pub. No. 22, pp. 31–44. Madison, WI: Soil Science Society of America.

HATTINGH M.J., GRAY L.E. & GERDEMANN J.W. (1973) Uptake and translocation of ^{32}P-labelled phosphate to onion roots by endomycorrhizal fungi. *Soil Sci.* **116**, 383–387.

HAUSSLING M., JORNS C.A., LEHMBECKER G., HECHT-BUCHHOLZ C.H. & MARSCHNER H. (1988) Ion and water uptake in relation to root development in Norway spruce. *J. Plant Physiol.* **133**, 486–491.

HAVERKORT A.J. & MACKERRON D.K.L. (1995) *Potato Ecology and Modelling of Crops under Conditions Limiting Growth.* Dordrecht: Kluwer Academic.

HAYNES R.J. (1980) Ion exchange properties of roots and ionic interactions within the root apoplasm: their role in ion accumulation by plants. *Bot. Rev.* **46**, 75–99.

HAYNES R.J. & GOH R.M. (1978) Ammonium and nitrate nutrition of plants. *Biol. Rev.* **53**, 465–510.

HAYWARD D.G. & TRAPNELL B.M.W. (1964) *Chemisorption*, 2nd Edition. London: Butterworth.

HEDLEY M.J., NYE P.H. & WHITE R.E. (1982a) Plant induced changes in the rhizosphere of rape (*Brassica napus* var. *Emerald*) seedlings. II. Origin of the pH change. *New Phytol.* **91**, 31–44.

HEDLEY M.J., WHITE R.E. & NYE P.H. (1982b) Plant induced changes in the rhizosphere of rape (*Brassica napus* var. *Emerald*) seedlings. III. Changes in L value, soil phosphate fractions and phosphatase activity. *New Phytol.* **91**, 45–56.

HEDLEY M.J., WHITE R.E. & NYE P.H. (1983) Plant-induced changes in the rhizosphere of rape (*Brassica napus* var. *Emerald*) seedlings. IV. The effect of rhizosphere phosphorus status on the pH, phosphatase activity and depletion of soil phosphorus fractions in the rhizosphere and on the cation exchange balance in the plants. *New Phytol.* **95**, 69–82.

HEISEY R.M. (1990) Allelopathic and herbicidal effects of extracts from tree of heaven (*Ailanthus altissima*). *Am. J. Bot.* **77**, 662–670.

HELAL H.M. & SAUERBECK D.R. (1991a) Short-term determination of the actual respiration rate of plant roots. In *Plant Roots and Their Environments*, eds. McMichael B.L. & Persson H., pp. 88–92. Amsterdam: Elsevier.

HELAL H.M. & SAUERBECK D. (1991b) Soil and root phosphatase activity and the utilization of inositol phosphates as dependent upon phosphorus supply. In *Plant Roots and Their Environment*, eds. McMichael B.L. & Persson H., pp. 93–97. Amsterdam: Elsevier.

HELFFERICH F. (1962) *Ion Exchange*. New York: McGraw-Hill.

HENDRICK R. & PREGITZER K.S. (1996) Temporal and depth-related patterns of fine root dynamics in northern hardwood forests. *J. Ecol.* **84**, 167–176.

HETRICK B.A.D., HARTNETT D.C., WILSON G.W.T. & GIBSON D.J. (1994) Effects of mycorrhizae, phosphorus availability and plant density on yield relationships among competing tallgrass prairie grasses. *Can. J. Bot.* **72**, 168–176.

HIGUCHI T., YODA K. & TENSHO K. (1984) Further evidence for gaseous CO_2 transport in relation to the uptake of CO_2 in rice plant. *Soil Sci. Plant Nutr.* **30**, 125–136.

HILL-COTTINGHAM D.G. & LLOYD JONES H.P. (1965) The behaviour of ion chelating agents with plants. *J. Exp. Bot.* **16**, 233–242.

HILLOCKS R.J. & WALKER J.M. (1997) *Soilborne Diseases of Tropical Crops*. Wallingford: CAB International.

HINGSTON F.J., ATKINSON R.J., POSNER A.M. & QUIRK J.P. (1968) Specific adsorption of anions on goethite. *Proc. 9th Int. Congr. Soil Sci. Adelaide* **1**, 669–678.

HINSINGER P. & GILKES R.J. (1996) Mobilization of phosphate from phosphate rock and aluminium-sorbed phosphate by the roots of ryegrass and clover as related to rhizosphere pH. *Eur. J. Soil Sci.* **47**, 533–544.

HIRREL M.C. & GERDEMANN J.W. (1979) Enhanced carbon transfer between onions infected with a vesicular-arbuscular mycorrhizal fungus. *New Phytol.* **83**, 731–738.

HIRTH J.R., MCKENZIE B.M. & TISDALL J.M. (1997) Do the roots of perennial ryegrass elongate to biopores filled with the casts of endogenic earthworms? *Soil Biol. Biochem.* **20**, 529–531.

HOAGLAND D.R. (1944) *Lectures on the Inorganic Nutrition of Plants*. Waltham, MA: Chronica Botanica.

HODGE A.M., PATERSON E., THORNTON B., MILLARD P. & KILLHAM K. (1997) Effects of photon flux density on carbon partitioning and rhizosphere carbon flow of *Lolium perenne*. *J. Exp. Bot.* **48**, 1797–1805.

HODGE A., STEWART J., ROBINSON B., GRIFFITHS D.S. & FITTER A.H. (1998) Root proliferation, soil fauna and plant nitrogen capture from nutrient-rich patches in soil. *New Phytol.* **139**, 479–494.

HODGSON J.F. (1963) Chemistry of the micronutrient elements in soils. *Adv. Agron.* **15**, 119–159.

HODGSON J.F. (1969) Contribution of metal–organic complexing agents to the transport of metals to roots. *Soil Sci. Soc. Am. Proc.* **33**, 68–75.

HOFER R.M. (1996) Root hairs. In *Plant Roots — The Hidden Half*, eds. Waisel Y., Eshel A. & Kafkafi U., pp. 111–126. New York: Marcel Dekker.

HOOKE J.E., BLACK K.E., PERRY R.L. & ATKINSON D. (1995) Arbuscular mycorrhizal fungi induced alteration to root longevity of poplar. *Plant Soil* **172**, 327–329.

HORNBY D. (1990) *Biological Control of Soil-Borne Plant Pathogens*. Wallingford: CAB International.

HORST W.J., WAGNER A. & MARSCHNER H. (1982) Mucilage protects root meristems from aluminium injury. *Zeits. Pflanzenphysiolog.* **105**, 435–444.

HOZORE E. & ALEXANDER M. (1991) Bacterial characteristics important to rhizosphere competence. *Soil Biol. Biochem.* **23**, 717–723.

HRI (1994) *WELL_N Nitrogen Advisory Model.* Wellesbourne: Horticultural Research International.

HSIAO T.C. (1973) Plant response to water stress. *Ann. Rev. Plant Physiol.* **24**, 519–570.

HSIAO T.C., ACEVEDO E., FERERES E. & HENDERSON D.W. (1976) Water stress, growth and osmotic adjustment. *Phil. Trans. R. Soc. London B* **273**, 479–500.

HUBEL F. & BECK E. (1995) *In situ* determination of the P relations around the primary root of maize with respect to inorganic and phytate P. *Plant Soil* **170**, 1–9.

HUFFMAN E.O. (1962) Reactions of phosphate in soils: recent research by TVA. *Fertil. Soc. Proc.* **71**, 1–48.

HUNT P.G. (1990) Microbial responses in the rhizosphere of agricultural plants. In *Rhizosphere Dynamics*, eds. Box J.E. & Hammond L.E. AAAS Symposium 113, pp. 116–143. Washington, DC: American Association for the Advancement of Science.

HUNT R. (1973) A method of estimating root efficiency. *J. Appl. Ecol.* **10**, 157–164.

HUTSON J.L. & WAGENET R.J. (1995) The application of chemical equilibrium in solute transport models. In *Chemical Equilibrium and Reaction Models*, eds. Loeppert R.H., Schwab A.P. & Goldberg S., pp. 97–112. Spec. Pub. No. 42. Madison, WI: Soil Science Society of America.

HUXLEY J. & ONG C. (1996) *Tree–Crop Interactions.* Wallingford: CAB International.

IBRIKCI H., COMERFORD N.B., HANLON E.A. & RECHCIGL J.E. (1994) Phosphorus uptake by bahiagrass from spodosols: modelling of uptake from different horizons. *Soil. Sci. Soc. Am. J.* **58**, 139–143.

IKEDA T.M., SASAKI R.D. & YASUNAGA T. (1984) Kinetic behaviour of alkali metal ion on zeolite 4A surface using stopped-flow method. *J. Colloid Interface Sci.* **97**, 278–283.

INGESTAD T. & AGREN G.I. (1988) Nutrient uptake and allocation at steady state nutrition. *Physiol. Plant.* **72**, 450–459.

INGESTAD T. & AGREN G.I. (1992) Theories and methods on plant nutrition and research. *Physiol. Plant.* **84**, 177–184.

INGRAM J. (1997) Food security in the face of global change: the GCTE Rice Network as a framework for international collaborative research. *J. Agric. Meteorol.* **52**, 759–768.

IPCC (1996) *Climate Change 1995: The Science of Climate Change.* Cambridge: Cambridge University Press.

ISCHTSCHERIKOW W. (1907) Die Gewinnung der Bodenlosung in unveranderten Zustande. *Russ. J. Exp. Agric.* **8**, 147–166.

ISHIDA T., CAMPBELL G.S. & CALISSENDORFF C. (1991) Improved heat balance method for determining sap flow rate. *Agric. Forest Meteorol.* **56**, 35–48.

ISRAEL D.W. & JACKSON W.A. (1978) The influence of nitrogen nutrition on ion uptake and translocation in leguminous plants. In *Mineral Nutrition of Legumes in Tropical and Subtropical Soils*, eds. Andrews C.S. & Kamprath E.J., pp. 113–129. Melbourne, Australia, 1972.

ITOH S. & BARBER S.A. (1983a) Phosphorus uptake by 6 plant species as related to root hairs. *Agron. J.* **75**, 457–461.

ITOH S. & BARBER S.A. (1983b) A numerical solution of whole plant nutrient uptake for soil root systems with root hairs. *Plant Soil* **70**, 403–413.

IWASAKI T. (1937) Some notes on sand filtration. *J. Am. Water Works Assoc.* **9**, 1591.

JACKSON M.B. (1991) Ethylene in root growth and development. In *The Plant Hormone Ethylene*, eds. Mattoo A.K. & Suttle J.C., pp. 159–181. Boca Raton, FL: CRC Press.

JACKSON, M.B. (1993) Are plant hormones involved in root to shoot communication? *Adv. Bot. Res.* **19**, 104–189.

JAEGER J.C. & CLARK M.A. (1942) A short table of I(01; x). *Proc. R. Soc. Edin.* **61A**, 229–230.

JAKOBSEN I. (1986) Phosphorus inflow into roots of mycorrhizal and non-mycorrhizal peas under field conditions. In *Physiological and Genetic Aspects of Mycorrhizae*, eds. Gianinazzi-Pearson V. & Gianinazzi S., pp. 317–322. Paris: INRA.

JAMIESON P.D., PORTER J.R., GOUDRIAAN J., RITCHIE J.T., VAN KEULEN H. & STOL W. (1998) A comparison of the models AFRCWHEAT2, CERES-Wheat, Sirius, SUCROS2, and SWHEAT with measurements from wheat grown under drought. *J. Field Crop Res.* **55**, 23–44.

JARVIS P.G. & MCNAUGHTON K.G. (1986) Stomatal control of transpiration: scaling up from leaf to region. *Adv. Ecol. Res.* **15**, 1–49.

JAYACHANDRAN K., SCHWAB A.P. & HETRICK B.A.D. (1992) Mineralization of organic phosphorus by vesicular-arbuscular mycorrhizal fungi. *Soil Biol. Biochem.* **24**, 897–903.

JENKINSON D.S. (1966) The turnover of organic matter in soil. In *The Use of Isotopes in Soil Organic Matter Studies*, FAO, pp. 187–197. Oxford: Pergamon.

JENKINSON D.S. (1990) The turnover of organic carbon and nitrogen in soils. *Phil. Trans. R. Soc. London B* **329**, 361–368.

JENKINSON D.S., HARKNESS D.D., VANCE E.D., ADAMS D.E. & HARRISON A.F. (1992) Calculating net primary production and annual input of organic matter to soil from the amount and radiocarbon content of soil organic matter. *Soil Biol. Biochem.* **24**, 295–308.

JENNY H. (1960) *Growth in Living Systems*, ed. Zarrow M.X., International Symposium on Growth. New York: Basic Books.

JENNY H. & GROSSENBACHER K. (1963) Root–soil boundary zones as seen in the electron microscope. *Soil Sci. Soc. Am. Proc.* **27**, 273–277.

JENNY H. & OVERSTREET R. (1939a) Cation interchange between plant roots and soil colloids. *Soil Sci.* **47**, 257–272.

JENNY H. & OVERSTREET R. (1939b) Surface migration of ions and contact exchange. *J. Phys. Chem.* **43**, 1185–1196.

JENSEN A. & JAKOBSEN I. (1980) The occurrence of vesicular-arbuscular mycorrhiza in barley and wheat grown in some Danish soils with different fertilizer treatments. *Plant Soil* **55**, 403–414.

JOHNSON I.R. & THORNLEY J.H.M. (1987) A model of root–shoot partitioning with optimal growth. *Ann. Bot.* **60**, 133–142.

JOHNSTON A.E. (1994). The Rothamsted Classical Experiments. In *Long-Term Experiments in Agricultural and Ecological Sciences*, eds. Leigh R.A. & Johnston A.E., pp. 9–38. Wallingford: CAB International.

JONES D.L. (1998) Organic acids in the rhizosphere: a critical review. *Plant Soil* **205**, 25–44.

JONES D.L. & DARRAH P.R. (1992) Resorption of organic compounds of roots by *Zea mays* L. and its consequences in the rhizosphere. I. Resorption of ^{14}C-labelled glucose, mannose and citric acid. *Plant Soil* **143**, 259–266.

JONES D.L. & DARRAH P.R. (1993) Re-absorption of organic compounds by roots of *Zea mays* L. and consequences in the rhizosphere. II. Experimental and model evidence for simultaneous exudation and re-absorption of soluble C compounds. *Plant Soil* **153**, 47–59.

JONES D.L. & DARRAH P.R. (1994) Amino-acid inflow at the soil–root interface of *Zea mays* L. and its implications in the rhizosphere. *Plant Soil* **163**, 1–12.

JONES D.L., DARRAH P.R. & KOCHIAN L.V. (1996a) Critical evaluation of organic acid mediated iron dissolution in the rhizosphere and its potential role in root iron uptake. *Plant Soil* **180**, 57–66.

JONES D.L., PRABOWO A.M. & KOCHIAN L.V. (1996b) Kinetics of malate transport and decomposition in acid soils and isolated bacterial populations: the effect of microorganisms on root exudation of malate under A1 stress. *Plant Soil* **182**, 239–247.

JONES H., TOMOS A.D., LEIGH R.A. & WYN JONES R.G. (1983) Water relation parameters of epidermal and cortical cells in the primary root of *Triticum aestivum* L. *Planta* **158**, 230–236.

JONES L.H.P. (1957) The relative content of manganese in plants. *Plant Soil* **8**, 328–336.

JONES L.H.P. & HANDRECK K.A. (1967) Silica in soil, plants and animals. *Adv. Agron.* **19**, 107–149.

JONES M.D., DURRALL D.M. & TINKER P.B. (1990) Phosphorus relationships and production of extramatrical hyphae by two types of willow ectomycorrhizas at different soil phosphorus levels. *New Phytol.* **115**, 259–267.

JONES M.D., DURRALL D.M. & TINKER P.B. (1991) Fluxes of carbon and phosphorus between symbionts in willow mycorrhizas and their changes with time. *New Phytol.* **119**, 99–106.

JONES M.D., DURRALL D.M. & TINKER P.B. (1998) A comparison of arbuscular and ectomycorrhizal *Eucalyptus coccifera*: growth response, phosphorus uptake efficiency and external hyphal production. *New Phytol.* **140**, 125–134.

JOPONY M. & YOUNG S.D. (1994) The solid–solution equilibria of lead and cadmium in polluted soils. *Eur. J. Soil Sci.* **45**, 59–70.

JUNGK A. (1996) Dynamics of nutrient movement at the soil–root interface. In *Plant Roots — The Hidden Half*, eds. Waisel Y., Eshel A. & Kafkafi U., pp. 529–556. New York: Marcel Dekker.

JUNGK A. & BARBER S.A. (1975) Plant age and the phosphorus uptake characteristics of trimmed and untrimmed corn root systems. *Plant Soil* **42**, 227–239.

JUNGK A., CLAASSEN N. & KUCHENBUCH R. (1982) Potassium depletion of the soil–root interface in relation to soil parameters and root properties. In *Proceedings of the 9th International Plant Nutrition Colloquium*, Warwick, England, ed. Scaife A., pp. 250–255. Farnham Royal, Bucks: Commonwealth Agricultural Bureau.

JUNGK A., ASCHER C.J., EDWARDS D.G. & MEYER D. (1990) Influence of phosphate status on phosphate uptake kinetics of maize (*Zea mays*) and soybean (*Glycine max*). *Plant Soil* **124**, 175–182.

JURY W.A. (1982) Simulation of solute leaching using a transfer function model. *Water Resour. Res.* **18**, 363–368.

JURY W.A. (1983) Chemical transport modeling: current approaches and unresolved problems. In *Chemical Mobility and Reactivity in Soil Systems*, SSSA Spec. Pub. No. 11, pp. 49–64. Madison, WI: Soil Science Society of America.

JUSTES E., MARY B., MEYNARD J.M., MACHET J.M. & THELIER-HUCHE L. (1994) Determination of a critical dilution curve for winter wheat crops. *Ann. Bot.* **74**, 397–407.

KAGE H. (1997) Is low rooting density of faba beans a cause of high residual nitrate content of soil at harvest? *Plant Soil* **190**, 47–60.

KAGE H. & EHLERS W. (1996) Does transport of water to roots limit water uptake of field crops? *Z. Pflanzenernahr. Bodenk.* **159**, 583–590.

KAPULNIK Y. (1996a) Nonsymbiotic nitrogen-fixing soil microorganisms. In *Plant Roots — The Hidden Half*, eds. Waisel Y., Eshel A. & Kafkafi U., pp. 757–768. New York: Marcel Dekker.

KAPULNIK Y. (1996b) Plant growth promotion by rhizosphere bacteria. In *Plant Roots — The Hidden Half*, eds. Waisel Y., Eshel A. & Kafkafi U., pp. 769–782. New York: Marcel Dekker.

KAUFMAN M.R. & LANDSBERG J.J., eds. (1991). Advancing towards closed models of forest ecosystems. IUFRO meeting. *Tree Physiol.* **9**, 1–324.

KAUFMANN D.D., STILL G.G., PAULSON G.D. & BANDAL S.K. (1976) Bound and conjugated pesticide residues. ACS Symposium Series 29. Washington DC: American Chemical Society.

KEITH H. (1998) Calibration of the ^{32}P bioassay for eucalyptus trees in the field. *Soil Biol. Biochem.* **30**, 651–660.

KEITH H., OADES J.M. & MARTIN J.K. (1986) Input of carbon to soil from wheat plants. *Soil Biol. Biochem.* **18**, 445–449.

KEMPER W., MAASLAND D.E.L. & PORTER L.K. (1964) Mobility of water adjacent to mineral surfaces. *Soil Sci. Soc. Am. Proc.* **28**, 164–167.

KEMPER W.D., OLSEN J. & HODGSON A. (1975) Fertilizer or salt leaching as affected by surface sloping and placement of fertilizer and irrigation water. *Soil Sci. Soc. Am. Proc.* **39**, 115–119.

KERSHAW K.A. (1985) *Quantitative and Dynamic Plant Ecology*, 3rd Edition. London: Arnold.

KETSCHMAR A., ed. (1992) Fourth International Symposium on Earth Worm Ecology. Special issue. *Soil Biol. Biochem.* **24**, 1193–1773.

KIERKEGAARD J.A., SO H.B. & TROEDSEN R.G. (1992) The effect of soil strength on the growth of pigeon pea radicles and seedlings. *Plant Soil* **140**, 65–74.

KINNIBURGH D.G. & MILES D.C. (1983) Extraction and chemical analysis of interstitial water from soils. *Environ. Sci. Technol.* **17**, 362–368.

KIRCHHOFF G. (1992) Measurement of root length and thickness using a handheld counterscanner. *Field Crops Res.* **29**, 79–88.

KIRK G.J.D & BAJITA J.B. (1995) Root-induced iron oxidation, pH changes and zinc solubilization in the rhizosphere of lowland rice. *New Phytol.* **131**, 129–137.

KIRK G.J.D. & NYE P.H. (1985) The dissolution and dispersion of dicalcium phosphate dihydrate in soil. I. A predictive model for a planar source. II. Experimental evaluation of the model. *J. Soil Sci.* **36**, 446–468.

KIRK G.J.D. & NYE P.H (1986) The dissolution and dispersion of dicalcium phosphate dihydrate in soil. III. A predictive model for regularly distributed particles. IV. Experimental evaluation of the model for particles. *J. Soil Sci.* **37**, 511–528.

KIRK G.J.D. & NYE P.H. (1991) A model of ammonia volatilization from applied urea. V. The effects of steady-state drainage and evaporation. VI. The effects of transient state water evaporation. *J. Soil Sci.* **42**, 103–126.

KIRK G.J.D. & NYE P.H. (1996) A simple model for predicting the rates of dissolution of sparingly soluble calcium phosphates in soil. I. The basic model. II Applications of the model. *J. Soil Sci.* **37**, 529–554.

KIRK G.J.D. & SALEQUE M.A. (1995) Solubilization of soil phosphate by lowland rice. Measurements of solubilization in a range of reduced soils and prediction of the resultant increase in uptake. *Eur. J. Soil Sci.* **46**, 247–256.

KIRSCH B.H. (1992) Solute movement in soil under conditions of evaporating water. D. Phil. Thesis, University of Oxford.

KLEPPER B. (1991) Root–shoot relationships. In *Plant Roots — The Hidden Half*, eds. Waisel Y., Eshel A. & Kafkafi U., pp. 265–286. New York: Marcel Dekker.

KLEPPER B. (1992) Development and growth of crop root systems. *Adv. Soil Sci.* **19**, 1–25.

KLEPPER B. & RICKMAN R.W. (1990) Modelling root growth and function. *Adv. Agron.* **44**, 113–131.

KLEPPER B., BELFORD R.K. & RICKMAN R.W. (1984) Root and shoot development in winter wheat. *Agron. J.* **76**, 117–122.

KLOEPPER J.W., LIFSCHITZ R. & ZABLOTOWICZ R.M. (1989) Free-living bacterial inocula for enhancing crop productivity. *Trends Biotechnol.* **7**, 39–44.

KLOEPPER J.W., RODRIGUEZ-KABANA R., MCINTAY J.A. & COLLINS J.D. (1991) Analysis of population and physiological characterization of microorganism in rhizospheres of plants with antagonistic properties to phytopathogenic nematodes. *Plant Soil* **136**, 95–102.

KNIGHT B.A.G. & TOMLINSON T.E. (1967) The interaction of paraquat (1:1 dimethyl 4:4 dipyridylium dichloride) with mineral soil. *J. Soil Sci.* **18**, 233–243.

KNOWLES C.J. & BUNCH A.W. (1986) Microbial cyanide metabolism. *Adv. Microbial Physiol.* **27**, 73–110.

KOCHIAN L.V. (1995) Cellular mechanisms of aluminium toxicity and resistance in plants. *Ann. Rev. Plant Physiol. Plant Mol. Biol.* **46**, 237–260.

KOIDE R.T. (1985) The effect of mycorrhizal infection and phosphorus status on sunflower hydraulic and stomata properties. *J. Exp. Bot.* **36**, 1087–1098.

KOIDE R.T. (1993) The physiology of the mycorrhizal plant. *Adv. Plant Pathol.* **9**, 33–50.

KOIDE R. & ELLIOTT G. (1989) Cost benefit and efficiency of the vesicular-arbuscular mycorrhizal symbiosis. *Func. Ecol.* **3**, 4–7.

KOIDE R.T. & LI M. (1990) On host regulation of the vesicular-arbuscular mycorrhizal symbiosis. *New Phytol.* **114**, 59–64.

KOOISTRA M.J., SCHOONDERBEEK D., BOONE F.R. & VEEN B.W. (1992) Root–soil contact of maize, as measured by a thin-section technique. II Effects of soil compaction. *Plant Soil* **139**, 119–129.

KORHONEN T.K., HAAHTELA K., ROMANTSCHUK M. & BAMFORD D. (1986) Role of fimbriae and pili in the attachment of *Klebsiella, Enterobacter* and *Pseudomonas* to plant surfaces. In *Recognition of Microbe Plant Symbiotic and Pathogenic Interactions*, ed. Lugtenberg B., NATO ASI Series, pp. 229–242. Berlin: Springer-Verlag.

KOTHARI S.K., MARSCHNER H. & GEORGE E. (1990) Effect of VA mycorrhiza and rhizosphere microorganisms on root and shoot morphology, growth and water relations of maize. *New Phytol.* **116**, 303–311.

KOVDA V.A., VAN DEN BURG C. & HAGAN R.M. (1973) *Irrigation, Drainage, and Salinity*. FAO/UNESCO. London: Hutchinson.

KOZLOWSKI, T.T., KRAMER T.T. & PALLARDY S.G. (1991) *The Physiological Ecology of Woody Plants*. San Diego: Academic Press.

KRAFFCZYK L., TROLLDENIER G. & BERINGER H. (1984) Soluble root exudates of maize: influence of potassium supply and rhizosphere microorganisms. *Soil Biol. Biochem.* **16**, 315–322.

KRAMER P.J. & BOYER J.S. (1995) *Water Relations of Plants and Soils*. San Diego: Academic Press.

KRAMER P.J. & COILE J.S. (1940) An estimation of the volume of water made available by root extension. *Plant Physiol.* **15**, 743–747.

KRAMER U., COTTER-HOWELLS J.D., CHARNOCK J.M., BAKER A.J. & SMITH A.J. (1996) Free histidine as a metal chelator in plants that accumulate nickel. *Nature* **379**, 635–638.

KRAUS M., FUSSEDER A. & BECK E. (1987a) In situ determination of the phosphate gradient around a root by autoradiography of frozen soil. *Plant Soil* **97**, 407–418.

KRAUS M., FUSSEDER A. & BECK E. (1987b) Development and replenishment of the P-depletion zone around the primary root of maize during the vegetation period. *Plant Soil* **101**, 247–255.

KROPFF M.J. (1993a) Mechanisms for competition for nitrogen. In *Modelling Crop–Weed Interactions*, eds. Kropff M.J. & Van Laar H.H., pp. 77–82. Wallingford: CAB International.

KROPFF M.J. (1993b) Mechanisms of competition for water. In *Modelling Crop–Weed Interactions*, eds. Kropff M.J. & van Laar H.H., pp. 63–76. Wallingford: CAB International.

KROPFF M.J. & VAN LAAR H.H., eds. (1993) *Modelling Crop–Weed Interactions*. Wallingford: CAB International.

KUCHENBUCH R. & JUNGK A. (1982) A method for determining ion concentration profiles at the root soil interface by thin slicing rhizospheric soil. *Plant Soil* **68**, 391–394.

KUCKE M., SCHMID H. & SPIESS A. (1995) A comparison of four methods for measuring roots of field crops in three contrasting soils. *Plant Soil* **172**, 63–71.

KUHLMANN H. & BARRACLOUGH P.B. (1987) Comparison between the seminal and the nodal root systems of winter wheat in their activity for N and K uptake. *Z. Pflanzenernahr. Bodenk.* **150**, 24–30.

KUTILEK M. & NIELSEN D.R. (1994) *Soil Hydrology*. Cremlingen-Destedt, Germany: Catena-Verlag.

KUTSCHERA L. (1960) *Wurzelatlas mitteleuropäischer Ackerunkrauter und Kulturpflanzen*. Frankfurt a.M.: DLG Verlag GMBH.

KUO S. & LOTSE E.G. (1974) Kinetics of phosphate adsorption and desorption by lake sediments. *Soil Sci. Soc. Am. Proc.* **38**, 50–54.

LAGERWERFF J.V. (1960) The contact-exchange theory amended. *Plant Soil* **13**, 253–264.

LAI T.M. & MORTLAND M.M. (1962) Self-diffusion of exchangeable cations in bentonite. *Clays and Clay Min.* **9**, 229–247.

LAI T.M. & MORTLAND M.M. (1968) Cationic diffusion in clay minerals. I. Homogeneous and heterogeneous systems. *Soil Sci. Soc. Am. Proc.* **32**, 56–61.

LAMBERS H. (1987) Growth, respiration, exudation and symbiotic associations. In *Root Development and Function*, eds. Gregory P.J., Lake J.V. and Rose D.A., pp. 125–145. Cambridge: Cambridge University Press.

LAMBERS H., ATKIN O.K. & SCHEURWATER I. (1996) Respiratory patterns in roots in relation to their functioning. In *Plant Roots — The Hidden Half*, eds. Waisel Y., Eshel A. & Kafkafi U., pp. 323–362. New York: Marcel Dekker.

LANG A.R.G. & GARDNER W.R. (1970) Limitation to water flux from soils to plants. *J. Agron.* **62**, 693–695.

LARSEN S. (1967) Soil phosphorus. *Adv. Agron.* **19**, 151–210.

LARSEN S., GUNARY D. & SUTTON C.D. (1965) The rate of immobilisation of applied phosphate in relation to soil properties. *J. Soil Sci.* **16**, 141–148.

LARSSON S. (1986) Drought resistance index. In *Research and Results in Plant Breeding*, ed. Olsson G., pp. 241–251. Sweden: Svalof AB. (Quoted in O'Toole & Bland, 1987.)

LAST P.J. & TINKER P.B. (1968) Nitrate nitrogen in leaves and petioles of sugar beet in relation to yield of sugar and juice purity. *J. Agric. Sci. Camb.* **71**, 383–392.

LAST F.T., DIGHTON J. & MASON P.A. (1987) Successions of sheathing mycorrhizal fungi. *Trends Ecol. Evol.* **2**, 157–161.

LAUDELOUT H., VAN BLADEL R., BOLT G.H. & PAGE A.L. (1968) Thermodynamics of heterovalent cation exchange reactions in a montmorillonite clay. *Trans. Faraday Soc.* **64**, 1477–1488.

LAUTER F.R., NINNEMAN O., BUCHER M., RIESMEIER J.W. & FROMMER W.B. (1996) Preferential expression of an ammonium transporter and of two putative nitrate transporters in root hairs of tomato. *Proc. Natl. Acad. Sci. USA* **93**, 8139–8144.

LAVY T.L. & BARBER S.A. (1964) Movement of molybdenum in the soil and its effect on availability to the plant. *Soil Sci. Soc. Am. Proc.* **28**, 93–97.

LAWLOR D.W. (1972) Growth and water use of *Lolium perenne*. I. Water transport. *J. Appl. Ecol.* **9**, 79–98.

LEE J.A. & STEWART G.R. (1978) Ecological aspects of nitrogen assimilation. *Adv. Bot. Res.* **6**, 1–43.

LEE R.B. & RATCLIFFE R.G. (1983) Phosphorus nutrition and the intracellular distribution of inorganic phosphate in pea root tips: a quantitative study using NMR. *J. Exp. Bot.* **34**, 1222–1244.

LEEDS-HARRISON P.B. (1995) The movement of water and solutes to surface and groundwaters. In *Pesticide Movement to Water*, eds. Walker A. and Allen R., Monograph No. 62, pp. 3–12. Farnham, Surrey: British Crop Protection Council.

LEGGEWIE G., WILLMIZER L. & RIESMAIER J. (1997) Two cDNAs from potato are able to complement a phosphate uptake-deficient yeast mutant: identification of phosphate transporters from higher plants. *Plant Cell* **9**, 381–392.

LEHR J.R., BROWN W.E. & BROWN E.H. (1959) Chemical behaviour of monocalcium phosphate monohydrate in soils. *Soil Sci. Soc. Am. Proc.* **23**, 3–7.

LEIGH R.A. & WYN JONES R.G. (1986) Cellular compartmentation in plant nutrition: the selective cytoplasm and the promiscuous vacuole. In *Advances in Plant Nutrition*, 2, eds. Tinker P.B. & Lauchli A., pp. 249–280. New York: Praeger.

LEISTRA M., SMELT J.H., VERLAAT J.G. & ZANDVOORT R. (1974) Measured and computed concentration patterns of propyzamide in field soils. *Weed Res.* **14**, 87–95.

LE TACON F., ALVAREZ I.F. *et al.* (1992) Variations in field response of forest trees to nursery ectomycorrhizal inoculation in Europe. In *Mycorrhizas in Ecosystems*, eds. Read D.J., Lewis D.H., Fitter A.H. & Alexander I.J., pp. 119–135. Wallingford: CAB International.

LETEY J., KEMPER W.D. & NOONAN L. (1969) The effect of osmotic pressure gradients on water movement in unsaturated soil. *Soil Sci. Soc. Am. Proc.* **33**, 15–18.

LEWIS D.G. & QUIRK J.P. (1967) Phosphate diffusion in soil and uptake by plants. III ^{31}P movement and uptake by plants as indicated by ^{32}P auto-radiography. *Plant Soil* **26**, 445–453.

LI X.-L. GEORGE E. & MARSCHNER H. (1991) Phosphorus depletion and pH decrease at the root–soil and the hyphae–soil interface of VA mycorrhizal white clover fertilized with ammonium. *New Phytol.* **119**, 397–404.

LINDER S. (1995) Foliar analysis for detecting and correcting nutrient imbalances in Norway spruce. *Ecol. Bull.* **44**, 178–190.

LINDER S. & AXELSSON B. (1982) Changes in carbon uptake and allocation patterns as a result of irrigation and fertilization in a young *Pinus sylvestris* stand. In *Carbon Uptake and Allocation in Subalpine Ecosystems as a Key to Management*, ed. Waring R.H., pp. 38–44. Corvallis, OR: Forest Research Laboratory, Oregon State University.

LINDSAY W.L. (1974) Role of chelation in micronutrient availability. In *The Plant Root and its Environment*, ed. Carson E.W., pp. 507–524. Charlottesville: University Press of Virginia.

LINDSAY W.L. & STEPHENSON H.F. (1959) Nature of the reactions of monocalcium phosphate monohydrate in soils. I. The solution that reacts with the soil. *Soil Sci. Soc. Am. Proc.* **23**, 12–17.

LITAV M. & HARPER J.L. (1967) A method of studying spatial relationships between the root systems of two neighbouring plants. *Plant Soil* **26**, 390–391.

LONERAGAN J.F. & ASHER C.J. (1967) Response of plants to phosphate concentration in solution culture: II. Rate of phosphate absorption and its relation to growth. *Soil Sci.* **103**, 311–318.

LÖSCH R. (1995) Plant water relations. *Prog. Bot.* **56**, 56–59.

LOW A.J. (1972) Some aspects of soil structure. *Chem. Ind.* 373–378.

LU S. & MILLER M.H. (1989) The role of VA mycorrhizas in the absorption of P and Zn by maize in field and growth chamber experiments. *Can. J. Soil Sci.* **14**, 931–939.

LUDLOW M.M. & MUCHOW R.C. (1990) A critical evaluation of the traits for improving crop yields in water-limited environments. *Adv. Agron.* **43**, 107–153.

LUNGLEY D.R. (1973) The growth of root systems — a numerical computer simulation model. *Plant Soil* **38**, 145–159.

LUTTGE G. (1983) Import and export of mineral nutrients in plant roots. In *Inorganic Plant Nutrition*, eds. Lauchli A. & Bieleski R.L., Encyclopedia of Plant Physiology 15A, pp. 179–211. Berlin: Springer-Verlag.

LUXMOORE R.J., KING A.J. & THARP M.L. (1991) Approaches to scaling up physiologically-based soil–plant models in space and time. In *Advancing towards Closed Models of Forest Ecosystems*, eds. Kaufman M.R. & Landsberg J.J., *Tree Physiol.* **9**, 281–292 (special issue).

LYNCH J.M., ed. (1990a) *The Rhizosphere*. Chichester: John Wiley & Sons.

LYNCH J.M. (1990b) Microbial metabolites. In *The Rhizosphere*, ed. Lynch J.M., pp. 177–206. Chichester: John Wiley & Sons.

LYNCH J. & NIELSEN K.L. (1996) Simulation of root system architecture. In *Plant Roots — The Hidden Half*, eds. Waisel Y., Eshel A. & Kafkafi U., pp. 247–257. New York: Marcel Dekker.

LYNCH, J. & WHIPPS, J.M. (1990) Substrate flow in the rhizosphere. *Plant Soil* **129**, 1–10.

MAATHUIS F.J.M. & SANDERS D. (1997) Regulation of K^+ absorption in plant root cells by external K^+: interplay of different plasma membrane K^+ transporters. *Ann. Exp. Bot.* **48**, 451–458.

MACKAY A.D. & BARBER S.A. (1987) Effect of cyclic wetting and drying of a soil on root hair growth of maize roots. *Plant Soil* **104**, 291–293.

MACKIE-DAWSON L.B. & ATKINSON D. (1991) Methodology for the study of roots in field experiments and the interpretation of results. In *Plant Root Growth*, ed. Atkinson D., pp. 25–48. Oxford: Blackwell.

MACKLON A.E.S. & SIM A. (1992) Modifying effects of a non-toxic level of aluminium on phosphate fluxes and compartmentation in root cortex cells of intact ryegrass seedlings. *J. Exp. Bot.* **43**, 1483–1490.

MAKELA A. (1986) Partitioning coefficients in plant models with turn-over. *Ann. Bot.* **57**, 291–297.

MAKELA A.A. & SIEVANEN R.P. (1987) Comparison of two root/shoot partitioning models with respect to substrate utilization and functional balance. *Ann. Bot.* **59**, 129–140.

MALAMY J.E. & BENFEY P.N. (1997) Down and out in *Arabidopsis*: the formation of lateral roots. *Trends Plant Sci.* **2**, 390–396.

MALCOLM R.L. & KENNEDY V.C. (1969) Rate of cation exchange on clay minerals as determined by specific-ion electrode techniques. *Soil Sci. Soc. Am. Proc.* **33**, 247–253.

MARRIOTT F.H.C. (1972) Buffons problem for non-random distribution. *Biometrics* **28**, 621–624.

MARSCHNER H. (1995) *Mineral Nutrition of Higher Plants*, 2nd Edition. London: Academic Press.

MARSCHNER H. & ROMHELD V. (1983) *In vivo* measurement of root-induced pH changes at the soil–root interface. Effect of plant species and nitrogen source. *Z. Pflanzenphysiol.* **111**, 241–251.

MARSCHNER H. & ROMHELD V. (1994) Strategies for plants for the acquisition of iron. *Plant Soil* **165**, 261–274.

MARTIN H. & LAUDELOUT H. (1963) Thermodynamique de l'échange des cations alcalins dans les argiles. *J. Chim. Phys.* **60**, 1086–1099.

MARTIN J.T. & JUNIPER B.E. (1970) *The Cuticles of Plants*. London: Edward Arnold.

MARTIN P., GLATZLE A., KOLB W., OMAY H. & SCHMIDT W. (1989) N_2-fixing bacteria in the rhizosphere: quantification and hormonal effects on root development. *Z. Pflanzenernahr. Bodenk.* **152**, 237–245.

MARX D.H., MAUL S.B. & CORDELL C.E. (1989) Application of specific ectomycorrhizal fungi in world forestry. In *Frontiers in Industrial Mycology*, pp. 78–98. American Mycological Society. New York: Chapman & Hall.

MARX D.H., RUEHLE J.L. & CORDELL C.E. (1991) Methods for studying nursery and field responses of trees to specific ectomycorrhiza. *Methods Microbiol.* **23**, 383–411.

MASSE J., TARDIEU F. & COLNENNE C. (1991) Rooting depth and spatial arrangement of roots in winter wheat. In *Plant Roots and their Environment*, eds. McMichael B.L. & Persson H., pp. 480–486. Amsterdam: Elsevier.

MATAR A.E., PAUL J.L. & JENNY H. (1967) Two phase experiments with plants growing in phosphate-treated soil. *Soil Sci. Soc. Am. Proc.* **31**, 235–237.

MAY E., LEWIS F.J., PEREIRA S., TAYLER S., SEAWARD M.R.D. & ALLSOP D. (1993) Microbial deterioration of building stone — a review. *Biodeterior. Abstr.* **7**, 109–123.

MCAULIFFE C.D., HALL N.S., DEAN L.A. & HENDRICKS S.B. (1947) Exchange reactions between phosphate and soils: hydroxylic surfaces of soil minerals. *Soil Sci. Soc. Am. Proc.* **12**, 119–123.

MCCOLL J.G. & COLE D.W. (1968) A mechanism of cation transport in forest soil. *North-West Sci.* **42**, 134–140.

MCCULLY M.E. (1987) Selected aspects of the structure and development of field-grown roots. In *Root Development and Function*, eds. Gregory P.J., Lake J.V. & Rose D.A., pp. 53–70. Cambridge: Cambridge University Press.

MCCULLY M.E. & CANNY M.J. (1989) Pathways and processes of water and nutrient movement in roots. In *Structural and Functional Aspects of Transport in Roots*, eds. Loughman B. C. *et al.*, pp. 3–14. Dordrecht: Kluwer Academic.

MCGRATH S.P. (1998) Phytoextraction for soil remediation. In *Plants that Hyperaccumulate Heavy Metals*, ed. Brooks R.R., pp. 261–287. Wallingford: CAB International.

MCINTYRE B.D., RIHA S.J. & ONG C.K. (1996) Light interception and evaporation in the hedgerow forestry systems. *Agric. Forest Meteorol.* **81**, 31–40.

MCLAREN A.D. (1970) Temporal and vectorial reactions of nitrogen in soil: a review. *Can J. Soil Sci.* **50**, 97–109.

MCMAHON M.A. & THOMAS G.W. (1974) Chloride and tritiated water flow in disturbed and undisturbed soil cores. *Proc. Soil Soc. Am.* **38**, 707–732.

MCMICHAEL B.L. & BURKE J.L. (1996) Temperature effects on root growth. In *Plant Roots–The Hidden Half*, eds. Waisel Y., Eshel A. & Kafkafi U., pp. 383–396. New York: Marcel Dekker.

MCMICHAEL B.L. & QUISENBERRY J.E. (1993) The impact of the soil environment on the growth of root systems. *Environ. Exp. Bot.* **33**, 53–61.

MCMURTRIE R.A., COMINS H.N., KIRSCHBAUM M.U.F. & WANG Y.-P. (1992) Modifying existing forest growth models to take account of effects of elevated CO_2. *Aust. J. Bot.* **40**, 657–677.

MEAD J.A. (1981) A comparison of the Langmuir, Freundlich and Temkin equations to describe phosphate adsorption properties of soils. *Aust. J. Soil Res.* **19**, 333–342.

MEHARG A.A. (1994) A critical review of labelling techniques used to quantify rhizosphere carbon flow. *Plant Soil* **166**, 55–62.

MEHARG A.A. & BLATT M.R. (1995) NO_3^- transport across the plasma membrane of *Arabidopsis thaliana* root hairs: kinetic control by pH and membrane voltage. *J. Membr. Biol.* **145**, 49–66.

MEHARG A.A. & KILLHAM K. (1995) Loss of exudates from the roots of perennial ryegrass inoculated with a range of microorganisms. *Plant Soil* **170**, 345–349.

MELHUISH F.W. & LANG A.R.G. (1968) Quantitative studies of roots in soil. I. Length and diameters of cotton roots in a clay loam soil by analysis of surface-ground blocks of resin-impregnated soil. *Soil Sci.* **106**, 16–22.

MELHUISH F.W. & LANG A.R.G. (1971) Quantitative studies of roots in soil. Analysis of non-random populations. *Soil Sci.* **112**, 161–166.

MENGE J.A. (1983) Utilization of vesicular-arbuscular mycorrhiza in agriculture. *Can. J. Bot.* **61**, 1015–1024.

MENGEL D.B. & BARBER S.A. (1974) Rate of nutrient uptake per unit of corn root under field conditions. *J. Agron.* **66**, 391–402.

MERCKX R., DEN HARTOG A. & VAN VEEN J.A. (1985) Turnover of root-derived material and related microbial biomass formation in soils of different texture. *Soil. Biol. Biochem.* **17**, 565–569.

MEREDITH R.E. & TOBIAS C.W. (1962) Conduction in heterogeneous systems. *Adv. Electro-chem. Eng.* **2**, 15–47.

MEYER W.S. & BARRS H.D. (1991) Roots in irrigated clay soils: measurement techniques and responses to root zone conditions. *Irrig. Sci.* **12**, 125–134.

MEYERHOFF G. & SCHULTZ G.V. (1952) Molecular weight determination of polymethacrylate esters by means of sedimentation in an ultracentrifuge and diffusion. *Makromol. Chem.* **7**, 294–319.

MICHEL B.E. (1971) Further comparisons between Carbowax 6000 and mannitol as suppressants of cucumber hypocotyl elongation. *Plant Physiol.* **48**, 513–516.

MILCHUNAS D.G., LEE C.A., LEUWENROTH W.K. & COFFIN D.P. (1992) A comparison of ^{14}C, ^{86}Rb and total excavation for determination of root distribution of individual plants. *Plant Soil* **144**, 125–132.

MILLER A.J. & SMITH S.J. (1996) Nitrate transport and compartmentation in cereal root cells. *J. Exp. Bot.* **47**, 843–854.

MILLER M.H. & MCGONIGLE T.P. (1992) Soil disturbance and the effectiveness of arbuscular mycorrhizas in an agricultural ecosystem. In *Mycorrhizas in Ecosystems*, eds. Read D.J., Lewis D.H., Fitter A.H. & Alexander I.J., pp. 156–163. Wallingford: CAB International.

MILLER M.H., MAMARIL C.P. & BLAIR G.J. (1970) Ammonium effects on phosphorus absorption through pH changes and phosphorus precipitation, at the soil–root interface. *J. Agron.* **62**, 524–527.

MILLINGTON R.J. (1959) Gas diffusion in porous media. *Science* **130**, 100–102.

MILLINGTON R.J. & QUIRK J.P. (1961) Permeability of porous solids. *Trans. Faraday Soc.* **57**, 1–8.

MILLINGTON R.J. & SHEARER R.C. (1971) Diffusion in aggregated porous media. *Soil Sci.* **111**, 372–378.

MINCHIN P.E.H., THORPE M.R. & FARRAR J.F. (eds.) (1994) Short-term control of root–shoot partitioning. *J. Exp. Bot.* **45**, 615–622.

MISRA R.K., ALSTON A.M. & DEXTER A.R. (1988) Role of root hairs in phosphorus depletion from a microstructured soil. *Plant Soil* **107**, 11–18.

MITCHELL A.R. (1969) *Computational Methods in Partial Differential Equations.* London: Wiley.

MITCHELL R.L. & RUSSELL W.J. (1971) Root development and rooting patterns of soya bean (*Glycine max* L. Merill) evaluated under field conditions. *J. Agron.* **63**, 313–316.

MOBBS D.C. & CANNELL M.G.R. (1995) Optimal tree-fallow rotation: some principles revealed by modelling. *Agroforest. Systems* **29**, 113–132.

MOGHIMI A., TATE M.E. & OADES J.M. (1978) Characterization of rhizosphere products, especially 2-keto-gluconic acid. *Soil Biol. Biochem.* **10**, 283–287.

MOHREN G.M.J., BARTELINK H.H. & JANSEN J.J. (eds.) (1994) Contrasts between bio-logically-based process models and management oriented growth and yield models. *Forest Ecol. Manage.* **69**, 1–350.

MOKADY R.S. & ZASLAVSKY D. (1967) Movement and fixation of phosphates applied to soils. *Soil Chemistry and Fertility, Meeting Comm. II & IV International Society of Soil Science* (Aberdeen), pp. 329–334.

MOORBY H. & NYE P.H. (1984) The effect of temperature variations over the root system on root extension and phosphate uptake by rape. *Plant Soil* **78**, 283–293.

MOORBY H., NYE P.H. & WHITE R.E. (1985) The influence of nitrate nutrition on the H ion efflux by young rape plants. *Plant Soil* **84**, 403–415.

MOREN J., KEREN R., BENJAMINI Y., BEN-HUR M. & SHAINBERG I. (1989) Water infiltration as affected by soil crust and moisture profile. *Soil Sci.* **148**, 53–59.

MORGAN J.M. (1984) Osmoregulation and water stress in higher plants. *Annu. Rev. Plant Physiol.* **35**, 299–319.

MOSS P. (1963) Some aspects of the cation status of soil moisture. I. The ratio law and moisture content. *Plant Soil* **18**, 99–113.

MOSS P. (1969) A comparison of potassium activity ratios derived from equilibration procedures and from measurements on displaced soil solution. *J. Soil Sci.* **20**, 297–306.

MOSSE B. (1973) Advances in the study of vesicular-arbuscular mycorrhiza. *Annu. Rev. Phytopathol.* **11**, 171–196.

MOSSE B., STRIBLEY D.P. & LE TACON F. (1981) Ecology of mycorrhizae and mycorrhizal fungi. *Adv. Microbial Ecol.* **2**, 137–210.

MOTT C.J. (1967) Cationic mobility in orientated bentonite. D.Phil. Thesis, Oxford University.

MOTT C.J. (1988) Surface chemistry of soil particles. In *Russell's Soil Conditions and Plant Growth*, ed. Wild A., pp. 239–281. Harlow, UK: Longman Scientific and Technical.

MOUNTNEY A.W. & WILLIAMS D.R. (1992) Computer simulation of metal ion humic and fulvic acid interactions. *J. Soil Sci.* **43**, 679–688.

MUGNIER J. & MOSSE B. (1987) Vesicular-arbuscular mycorrhizal infection in transformed root-inducing T-DNA roots grown axenically. *Phytopathology* **77**, 1045–1050.

MULJADI D., POSNER A.M. & QUIRK J.P. (1966) The mechanism of phosphate adsorption by kaolinite gibbsite and pseudo boehmite. I. The isotherms and the effect of pH on adsorption. *J. Soil Sci.* **17**, 212–247.

MULLINS C.E., MCKENZIE B. & TISDALL J.M. (1997) How roots penetrate soil: the importance of cell abscission and mucilage. *J. Exp. Bot.* **48**, 35 (abstract).

MULLINS G.L. & EDWARDS J.H. (1989) A comparison of two methods for measuring potassium influx kinetics by intact corn seedlings. *J. Plant Nutr.* **12**, 485–496.

MUNNS D.N. (1978) Soil acidity and nodulation. In *Mineral Nutrition* of *Legumes in Tropical and Sub-tropical Soils*, eds. Andrews C.S. & Kamprath E.J., pp. 243–263. Melbourne, Australia.

MUNNS R. & PASSIOURA J.B. (1984) Effect of prolonged exposure to NaCl on the osmotic pressure of leaf xylem sap from intact transpiring barley plants. *Aust. J. Plant Physiol.* **11**, 497–507.

NASSAR I.N. & HORTON R. (1992) Simultaneous transfer of heat, water and solutes in porous media. I Theoretical development. *Soil Sci. Soc. Am. J.* **56**, 1350–1356.

NASSAR I.N., HORTON R. & GLOBUS A.M. (1992) Simultaneous transfer of heat, water and solutes in porous media. II Experiment and analyses. *Soil Sci. Soc. Am. J.* **56**, 1357–1365.

NEARPASS D.C. (1965) Effects of soil acidity on the adsorption, penetration and persistence of simazine. *Weeds* **13**, 341–346.

NEILANDS J.B. (1984) Siderophores of bacteria and fungi. *Microbiol. Sci.* **1**, 9–14.

NEILANDS J.B. & LEONG S.A. (1986) Siderophores in relation to plant growth and disease. *Annu. Rev. Plant Physiol. Plant Mol. Biol.* **37**, 187–208.

NELSON L.M., YARON B. & NYE P.H. (1982) Biologically-induced hydrolysis of parathion in soil: kinetics and modelling. *Soil Biol. Biochem.* **14**, 223–237.

NEPSTED D.C., CARVALHO C.R., DAVIDSON E.A., JUPP P.H., LAFEBVRE P.A., NEGRERES G.H., DA SILVA ELSAN, STOW T.A., TRUMBORE S.E. & VICERA S. (1994) The role of deep roots in the hydrological and carbon cycles of the Amazon forests and pastures. *Nature* **372**, 666–669.

NEWMAN A.C.D. (1970) Discussion in: *Sorption and Transport Processes in Soils*, S.C.I. Monograph No. 37, p. 32. London: Society of Chemical Industry.

NEWMAN E.I. (1966) A method of estimating the total root length in a sample. *J. Appl. Ecol.* **3**, 139–145.

NEWMAN E.I. (1974) Root and soil–water relations. In *The Plant Root and its Environment*, ed. Carson E.W., pp. 363–440. Charlottesville, VA: University of Virginia Press.

NEWMAN E.I. (1976) Water movement through root systems. *Phil. Trans. R. Soc. London B* **273**, 463–478.

NEWMAN E.I. (1988) Mycorrhizal links between plants: their functioning and ecological significance. *Adv. Ecol. Res.* **18**, 243–270.

NEWMAN E.I. & EASON W.R. (1993) Rates of phosphorus transfer within and between ryegrass (*Lolium perenne*) plants. *Func. Ecol.* **7**, 242–248.

NEWMAN E.I. & WATSON A. (1977) Microbial abundance in the rhizosphere: a computer model. *Plant Soil* **48**, 17–56.

NEWMAN E.I., EASON W.R., EISENSTADT D.M. & RAMOS M.I.R.F. (1992) Interactions between plants: the role of mycorrhizas. *Mycorrhiza* **1**, 47–53.

NICHOLS B.E., OLIVEIRA L.A., GLASS A.D.M. & SIDDIQUI M.Y. (1993) The effects of aluminium on the influx of calcium, potassium, ammonium, nitrate and phosphate in an Al-sensitive cultivar of barley (*Hordeum vulgare* L.). *Plant Physiol.* **101**, 1263–1266.

NIELSEN D.R. & BIGGAR J.W. (1961) Miscible displacement. III. Theoretical considerations. *Soil Sci. Soc. Am. Proc.* **26**, 216–221.

NIELSEN D.R., VAN GENUCHTEN M.TH. & BIGGAR J.W. (1986) Water flow and solute transport processes in the unsaturated zones. *Water Resour. Res.* **22**, 895–1085.

NIELSEN K.L., LYNCH J.P., JABLOKOV A.G. & CURTIS P.S. (1994) Carbon cost of root systems: an architectural approach. *Plant Soil* **165**, 161–169.

NIELSEN N.E. (1972) A transport kinetic concept of ion uptake from soil by plants. I. A method for isolating soil solution from soils with or without plant cover. *Plant Soil* **36**, 505–520.

NIELSEN N.E. & BARBER S.A. (1978) Differences among genotypes of corn in the kinetics of phosphate uptake. *Agron. J.* **70**, 695-698.

NIJHUIS E.H., MAT M.J., ZOEGERS I.W., WAALWIJK C. & VAN VEEN J.A. (1993) Selection of bacteria suitable for introduction into the rhizosphere of grass. *Soil Biol. Biochem.* **25**, 885–895.

NIMAH M.N. & HANKS R. (1973a) Model for estimating soil–water, plant and atmospheric inter-relations. I. Description and sensitivity. *Soil Sci. Soc. Am. Proc.* **37**, 522–527.

NIMAH M.N. & HANKS R. (1973b) Model for estimating soil–water, plant and atmospheric inter-relations. II. Field test of model. *Soil Sci. Soc. Am. Proc.* **37**, 528–532.

NISSEN P. (1974) Uptake mechanism: organic and inorganic. *Ann. Rev. Plant Physiol.* **25**, 38–80.

NISSEN P. (1996) Uptake mechanisms. In *Plant Roots — The Hidden Half*, eds. Waisel Y., Eshel A. & Kafkafi U., pp. 511–528. New York: Marcel Dekker.

NOBEL P.S. (1989) Temperature, water availability and nutrient levels at various soil depths: consequences for shallow-rooted desert succulents, including nurse plant effects. *Am. J. Bot.* **76**, 1486–1492.

NOBEL P.S. (1991) *Physicochemical and Environmental Plant Physiology*. San Diego, CA: Academic Press.

NOBEL P.S. & CUI M. (1992) Hydraulic conductance of the soil, the root–soil air gap and the root: changes for desert succulents in drying soil. *J. Exp. Bot.* **43**, 319–326.

NYE P.H. (1966a) The measurement and mechanism of ion diffusion in soil. The relation between self-diffusion and bulk diffusion. *J. Soil Sci.* **17**, 16–23.

NYE P.H. (1966b) The effect of nutrient intensity and buffering power of a soil, and the absorbing power, size and root-hairs of a root, on nutrient absorption by diffusion. *Plant Soil* **25**, 81–105.

NYE P.H. (1966c) Changes in the concentration of nutrients in the soil near planar absorbing surfaces when simultaneous diffusion and mass flow occur. *Trans. Comm. II & IV. Int. Soc. Soil Sci.* (Aberdeen), 317–327.

NYE P.H. (1968a) The use of exchange isotherms to determine diffusion coefficients in soil. *9th Int. Cong. Soil Sci. Trans.* (Adelaide) **1**, 117–126.

NYE P.H. (1968b) Processes in the root environment. *J. Soil Sci.* **19**, 205–215.

NYE P.H. (1972) The measurement and mechanism of ion diffusion in soils. VIII. A theory for the propagation of changes of pH in soils. *J. Soil Sci.* **23**, 82–92.

NYE P.H. (1973) The relation between the radius of a root and its nutrient absorbing power. *J. Exp. Bot.* **24**, 783–786.

NYE P.H. (1979) Diffusion of ions and uncharged solutes in soils and soil clays. *Adv. Agron.* **31**, 225–272.

NYE P.H. (1981) Changes of pH across the rhizosphere induced by roots. *Plant Soil* **61**, 7–26.

NYE P.H. (1984) On estimating the uptake of nutrients solubilized near roots or other surfaces. *J. Soil Sci.* **35**, 439–446.

NYE P.H. (1986) Acid–base changes in the rhizosphere. In *Advances in Plant Nutrition, 2*, eds. Tinker P.B. & Lauchli A., pp. 129–154. New York: Praeger.

NYE P.H. (1992a) Towards the quantitative control of crop production and quality. I. The role of computer models in soil and plant research. *J. Plant Nutr.* **15**, 1131–1150.

NYE P.H. (1992b) Towards the quantitative control of crop production and quality. 2. The scientific basis for guiding fertilizer and management practice, particularly in poorer countries. *J. Plant Nutr.* **15**, 1151–1173.

NYE P.H. (1994) The effect of root shrinkage on soil water influx. *Phil. Trans. R. Soc. Lond.* **B345**, 395–402.

NYE P.H. & AMELOKO A. (1986) A comparison of measured and theoretical soil acidity diffusion coefficients over a wide range of pH. *J. Soil Sci.* **37**, 191–196.

NYE P.H. & AMELOKO A.Y. (1987) Predicting the rate of dissolution of lime in soil. *J. Soil Sci.* **38**, 641–650.

NYE P.H. & GREENLAND D.J. (1960) *The Soil under Shifting Cultivation.* Farnham, England: Commonwealth Agricultural Bureau.

NYE P.H. & MARRIOTT F.H.C. (1969) A theoretical study of the distribution of substances around roots resulting from simultaneous diffusion and mass flow. *Plant Soil* **30**, 459–472.

NYE P.H. & SPIERS J.A. (1964) Simultaneous diffusion and mass flow to plant roots. *8th Int. Congr. Soil Sci. (Bucharest)*, **11**, 535–542.

NYE P.H. & STAUNTON S. (1994) The self-diffusion of strongly adsorbed anions in soil: a two-path model to simulate restricted access to exchange sites. *Eur. J. Soil Sci.* **45**, 145–152.

NYE P.H. & TINKER P.B. (1969) The concept of a root demand coefficient *J. Appl. Ecol.* **6**, 293–300.

NYE P.H. & TINKER P.B. (1977) *Solute Movement in the Soil–Root System.* Oxford: Blackwell.

NYE P.H., CRAIG D., COLEMAN N.T. & RAGLAND J.L. (1961) Ion exchange equilibria involving aluminium. *Soil Sci. Soc. Am. Proc.* **25**, 14–17.

NYE P.H., BREWSTER J.L. & BHAT K.K.S. (1975a) The possibility of predicting solute uptake and plant growth response from independently measured soil and plant characteristics. I. The theoretical basis of the experiments. *Plant Soil* **42**, 161–170.

NYE P.H., BREWSTER J.L. & BHAT K.K.S. (1975b) The possibility of predicting solute uptake and plant growth response from independently measured soil and plant characteristics. III. The growth and uptake of onions in a soil fertilized to different initial levels of phosphate and a comparison of the results with model predictions. *Plant Soil* **42**, 197–226.

NYE P.H., GERSTL Z. & GALIN T. (1994a) Prediction of sorption by soils of volatile hydrocarbon mixtures. *J. Environ. Qual.* **23**, 1031–1037.

NYE P.H., YARON B., GALIN T. & GERSTL Z. (1994b) Volatilization of a multicomponent liquid through dry soils: testing a model. *Soil Sci. Soc. Am. J.* **58**, 269–277.

OADES J.M. (1978) Mucilages at the root surface. *J. Soil Sci.* **29**, 1–7.

OERTLI J.J. (1996) Transport of water in the rhizosphere and in roots. In *Plant Roots — The Hidden Half*, eds. Waisel Y., Eshel A. & Kafkafi U., pp. 607–633. New York: Marcel Dekker.

OFORI F. & STERN W.R. (1987) Cereal–legume intercropping systems. *Adv. Agron.* **41**, 41–90.

OLAFSDOTTER M., NAVAREZ D. & MOODY K. (1995) Allelopathic potential in rice germplasm. *Ann. Appl. Biol.* **127**, 543–560.

O'LEARY G.J. & CONNOR D.J. (1996a) A simulation model of the wheat crop in response to water and nitrogen supply: I. Model construction. *Agric. Syst.* **52**, 1–29.

O'LEARY G.J. & CONNOR D.J. (1996b) A simulation model of the wheat crop in response to water and nitrogen supply: II. Model validation. *Agric. Syst.* **52**, 31–55.

OLIVER S. & BARBER S.A. (1966) An evaluation of the mechanisms governing the supply of Ca, Mg, K and Na to soyabean roots. *Soil Sci. Soc. Am. Proc.* **30**, 82–86.

OLSEN R.A. & PEECH M. (1960) The significance of the suspension effect in the uptake of cations by plants from soil-water systems. *Soil Sci. Soc. Am. Proc.* **24**, 257–261.

OLSEN S.R. & KEMPER W.D. (1968) Movement of nutrients to plant roots. *Adv. Agron.* **20**, 91–151.

OLSEN S.R. & WATANABE F.S. (1957) A method to determine a phosphorus adsorption maximum of soils as measured by the Langmuir isotherm. *Soil Sci. Soc. Am. Proc.* **21**, 144–149.

OLSEN S.R., KEMPER W.D. & JACKSON R.D. (1962) Phosphate diffusion to plant roots. *Soil Sci. Soc. Am. Proc.* **26**, 222–227.

ONDERDONK J.J. & KETCHESON J.W. (1973) Effect of soil and temperature on direction of corn root growth. *Plant Soil* **39**, 177–186.

ONG C.K. (1995) The 'dark side' of intercropping: manipulation of soil resources. In *Ecophysiology of Tropical Intercropping*, INRA Science Update, pp. 45–65. Paris: INRA.

ONG C.K. & BLACK C.R. (1994) Complementarity of resource use in intercropping and agroforestry systems. In *Resource Capture by Crops*, eds. Monteith J.L., Scott R.K. & Unsworth M.H., pp. 255–278. Loughborough: Nottingham University Press.

ONG C.K., CORLETT J.E., SINGH R.P. & BLACK C.R. (1991) Above and below ground interactions in agroforestry. *Forest Ecol. Manage.* **45**, 45–47.

ONG C.K., BLACK C.R., MARSHALL F.M. & CORLETT J.E. (1996) Principles of resource capture and utilization of light and water. In *Tree–Crop Interactions*, eds. Ong C.K. & Huxley P., pp. 74–158. Wallingford: CAB International.

O'TOOLE J.C. & BLAND W.L. (1987) Genotypic variations in the root systems. *Adv. Agron.* **41**, 91–140.

OYANAGI A., TAKAHASHI H. & SUGE H. (1995) Interactions between hydrotropism and gravitropism in the primary seminal roots of *Triticum aestivum* L. *Ann. Bot.* **75**, 229–235.

PAGES L. & SERRA V. (1994) Growth and branching of the taproot of young oak trees — a dynamic study. *J. Exp. Bot.* **45**, 1327–1334.

PAGES L., JORDAN M.O. & PICARD D. (1989) A simulation model of the three-dimensional architecture of the maize root system. *Plant Soil* **119**, 147–154.

PAIRUNAN A.K., ROBSON A.D. & ABBOTT L.K. (1980) The effectiveness of vesicular-arbuscular mycorrhizas in increasing growth and phosphorus uptake of subterranean clover from phosphorus sources of different solubilities. *New Phytol.* **84**, 327–338.

PAPAVIZAS G.C. & DAVEY C.B. (1961) Extent and nature of the rhizosphere of *Lupinus*. *Plant Soil*, **14**, 215–236.

PARLANGE J.Y. (1973) Movement of salt and water in relatively dry soils. *Soil Sci.* **116**, 249–255.

PARR J.F., GARDNER W.R. & ELLIOTT L.F., eds. (1981) *Water Potential Relations in Soil Microbiology*. SSSA Spec. Pub. No. 9. Madison. WI: Soil Science Society of America.

PARTON W.J., SCHIMEL D.S., COLE C.V. & OJIMA D.S. (1987) Analysis of factors controlling soil organic matter levels in Great Plains grasslands. *Soil Sci. Soc. Am J.* **51**, 1173–1179.

PASSIOURA J.B. (1963) A mathematical model for the uptake of ions from the soil solution. *Plant Soil* **18**, 225–238.

PASSIOURA J.B. (1966) *The Amounts of Nutrients Contacted by Roots*. Technical Report Series No. 65, pp. 82–84. Vienna: International Atomic Energy Agency.

PASSIOURA J.B. (1971) Dispersion in aggregated media. I. Theory. *Soil Sci.* **111**, 339–344.

PASSIOURA J.B. (1988) Water transport in and to roots. *Ann. Rev. Plant Physiol. Plant Mol. Biol.* **39**, 245–265.

PASSIOURA J.B. & COWAN I.R. (1968) On solving the non-linear diffusion equation for the radial flow of water to roots. *Agric. Meteorol.* **5**, 129–134.

PASSIOURA J.B. & FRERE M.H. (1967) Numerical analysis of the convection and diffusion of solute to roots. *Aust. J. Soil Res.* **5**, 149–159.

PASSIOURA J.B. & ROSE D.A. (1971) Hydrodynamic dispersion in aggregated media: 2. Effects of velocity and aggregate size. *Soil Sci.* **111**, 345–351.

PATE J.S. (1994) The mycorrhizal association: just one of many nutrient acquiring specializations in natural ecosystems. *Plant Soil* **159**, 1–10.

PATRIQUIN D.G., DOBEREINER J. & JAIN D.K. (1983) Sites and processes of association between diazotrophs and grasses. *Can. J. Microbiol.* **29**, 900–915.

PAUL E.A. & CLARK F.E. (1996) *Soil Microbiology and Biochemistry*. London: Academic Press.

PAUL J.L. (1965) Influence of soil moisture on chloride uptake by wheat seedlings at low rates of transpiration. *Agrochimica* **9**, 368–379.

PAUL N.D. & AYRES P.G. (1990) Root form and function after foliar infection. *Br. Soc. Plant Growth Reg.* **21**, 351–352.

PEARSON C.J. & JACOBS B.C. (1985) Root distribution in space and time of *Trifolium subterraneum*. *Aust. J. Agric. Res.* **36**, 601–614.

PEARSON J.N. & JAKOBSEN I. (1993) The relative contribution of hyphae and roots to phosphorus uptake by arbuscular mycorrhizal plants, measured by dual labelling with ^{32}P and ^{33}P. *New Phytol.* **124**, 489–494.

PEARSON J.N., ABBOTT L.K. & JASPER D.A. (1993) Mediation of competition between two colonizing VA mycorrhizal fungi by the host plant. *New Phytol.* **123**, 93–98.

PEARSON V. & TINKER P.B. (1975) Measurements of fluxes in the external hyphae of endomycorrhizas. In *Endomycorrhizas*, eds. Sanders F.E., Mosse B. & Tinker P.B., pp. 277–288. London: Academic Press.

PEEL A.J. (1974) *Transport of Nutrients in Plants*. London: Butterworth.

PENNING DE VRIES F.W.T. & RABBINGE R. (1995) Models in research and education, planning and practice. In *Potato Ecology and Modelling of Crops under Conditions Limiting Growth*, eds. Haverkort A.J. & Mackerron D.K.L., pp. 1–18. Dordrecht: Kluwer.

PERSSON H.A. (1983) The distribution and productivity of fine roots in boreal forests. *Plant Soil* **71**, 87–101.

PERSSON H. (1990) Methods of studying root dynamics in relation to nutrient cycling. In *Nutrient Cycling in Terrestrial Ecosystems*, eds. Harrison A.F., Ineson P. & Heal O.W., pp. 198–217. London: Elsevier.

PETERSEN C.A. (1988) Exodermal Casparian bands: their significance for ion uptake by roots. *Physiol. Plant.* **72**, 204–208.

PETERSEN C.A., EMANUAL M.E. & HUMPHREYS G.B. (1981) Pathways of movement of apoplastic fluorescent dye tracers through the endodermis at the site of secondary root formation in corn (*Zea mays*) and broad bean (*Vicia faba*). *Can. J. Bot.* **59**, 618–625.

PETERSEN P.J. (1993) Metal pollutant tolerance. In *Plant Adaptation to Environmental Stress*, eds. Mansfield T.A., Fowden L. & Stoddard J., pp. 171–188. London: Chapman & Hall.

PETERSON R.L. & FARQUHAR M.L. (1996) Root hairs: specialised tubular cells extending root surfaces. *Bot. Rev.* **62**, 2–40.

PHILIP J.R. (1957) The physical principles of soil-water movement during the irrigation cycle. *3rd Congr. Inter. Comm. Irrig. Drain. Quest.* **8**, 125–154.

PHILIP J.R. (1969) *Theory of Flow and Transport Processes in Pores and Porous Media*, eds. Wolstenholme G.E.W. & Knight J., Ciba Foundation Symposium on Circulatory and Respiratory Mass Transport, pp. 24–44. London: Churchill.

PHILIP J.R. (1973) On solving the unsaturated flow equation. I The flux–concentration relation. *Soil Sci.* **116**, 328–335.

PHILIP J.R. & DE VRIES D.A. (1957) Moisture movement in porous materials under temperature gradients. *Trans. Am. Geophys. Union* **38**, 222–232.

PIELOU E.C. (1979) *Mathematical Ecology*, 2nd Edition. Chichester: Wiley.

PIGNATELLO J.J. (1989) Sorption dynamics of organic compounds in soils and sediments. Nonpolar organics by soils and sediments. In *Reactions and Movement of Organic Chemicals in Soils*, eds. Sawney B.L. & Brown K. SSA Spec. Pub. No. 22, pp. 45–80. Madison, WI: Soil Science Society of America.

PINNER A. & NYE P.H. (1982) A pulse method for studying effects of dead-end pores, slow equilibration and soil structure on diffusion of solutes in soil. *J. Soil Sci.* **33**, 25–35.

PINTER P.J., KIMBALL B.A., GARCIA R.L., WALL G.W., HUNSAKER D.J. & LAMORTE R.L. (1996) Free-air CO_2 enrichment: responses of cotton and wheat crops. In *Terrestrial Ecosystems Responses to Elevated Carbon Dioxide*, eds. Koch G.W. & Mooney H.A., pp. 215–249. New York: Academic Press.

PITMAN M.G. (1965) Ion exchange and diffusion in roots of *Hordeum vulgare*. *Aust. J. Biol. Sci.* **18**, 541–546.

PLENCHETTE C., FORTIN J.A. & FURLAN V. (1983) Growth responses of several plant species to mycorrhizae in a soil of low fertility. I Mycorrhizal dependency under field conditions. *Plant Soil* **70**, 199–209.

POELSTRA P., FRISSEL M.J., VAN DER KLUYT N. & TAO W. (1974) Behaviour of mercury compounds in soils: accumulation and evaporation. FAO/IAEA/WHO. *Symposium on Comparative Aspects of Food and Environmental Contamination*, pp. 281–292. Helsinki.

PONNAMPERUMA F.N. (1972) The chemistry of submerged soils. *Adv. Agron.* **24**, 29–96.

PORTER J.R. (1984) A model of canopy development in winter wheat. *J. Agric. Sci., Camb.* **102**, 383–392.

PORTER J.R. (1993) AFRCWHEAT 2: a model of the growth and development of wheat incorporating responses to water and nitrogen. *Eur. J. Agron.* **2**, 69–82.

PORTER J., KLEPPER B. & BELFORD R.K. (1986) A model (WHTROOT) which synchronizes growth and development with shoot development for winter wheat. *Plant Soil* **92**, 133–145.

PORTER L.K., KEMPER W.D., JACKSON R.D. & STEWART B.A. (1960) Chloride diffusion in soils as influenced by moisture content. *Soil Sci. Soc. Am. Proc.* **24**, 460–463.

PREGITZER K.S., HENDRICKS R.L. & FOGEL R. (1993) The demography of fine roots in response to patches of water and nitrogen. *New Phytol.* **125**, 575–580.

PRESTON G., HAUBOLD B. & RAINEY P.B. (1998) Bacterial genomics and adaptation to life on plants: implications for the evolution of pathogenicity and symbiosis. *Curr. Opinion Microbiol.* **1**, 589–597.

PROSSER J.I., KILLHAM K., GLOVER L.A. & RATTRAY E.A.S. (1996) Luminescence-based systems for detection of bacteria in the environment. *Crit. Rev. Biotechnol.* **16**, 157–183.

RABBINGE R., GOUDRIAAN R.J., VAN KEULEN H., PENNING DE VRIES F.W.T. & VAN LAAR H.H., eds. (1990) *Theoretical Production Ecology, Reflections and Prospects.* Wageningen: PUDOC.

RACHE K.D. & LICHTENSTEIN E.P. (1985) Effects of soil microorganisms on the release of bound ^{14}C residues from soils previously treated with [^{14}C] parathion. *J. Agric. Food Chem.* **33**, 938–943.

RACHPAL-SINGH & NYE P.H. (1986) A model of ammonia volatilization from applied urea. Pt. I. Development of the model. Pt. II. Experimental testing. Pt. III. Sensitivity analysis, mechanisms and applications. *J. Soil Sci.* **37**, 9–40.

RACHPAL-SINGH & NYE P.H. (1988) The volatilization of ammonia from surface supplied urea. IV. Effect of method of urea application. *J. Soil Sci.* **39**, 9–14.

RADIN J.W. & MATTHEWS M.A. (1989) Water transport properties of cells in the root cortex of N and P deficient cotton seedlings. *Plant Physiol.* **89**, 264–268.

RADOSEVICH S.R. & ROUSCH M.L. (1990) The role of competition in agriculture. In *Perspectives on Competition*, eds. Grace J.B. & Tilman D., pp. 341–366. New York: Academic Press.

RAINEY P. (1999) Adaptation of *Pseudomonas fluorescens* to the plant rhizosphere. *Environ. Microbiol.* **1**, 243–257.

RAJU P.S., CLARK R.B., ELLIS J.R., DUNCAN R.R. & MARANVILL E.J.W. (1990) Benefit and cost analysis of phosphorus efficiency of VA mycorrhizal fungi colonization with *Sorghum bicolor* genotypes grown at varied phosphorus levels. *Plant Soil* **124**, 199–204.

RATNAYAKE R.T., LEONARD R.T. & MENGE J.A. (1978) Root exudation in relation to supply of phosphorus and its possible relevance to mycorrhizal infection. *New Phytol.* **81**, 543–552.

RATTRAY E.A.S., PROSSER J.I., GLOVER L.A. & KILLHAM K. (1995) Characterization of rhizosphere colonization by luminescent *Enterobacter cloaca* at the population and single-cell level. *Appl. Environ. Microbiol.* **61**, 2950–2957.

RAVEN J.A. & SMITH F.A. (1976) Nitrogen assimilation and transport in vascular land plants in relation to intracellular pH regulation. *New Phytol.* **76**, 415–431.

RAVEN K.P. & HORSNER L.R. (1993) Phosphorus desorption quantity intensity re-lationships in soils. *Soil Sci. Soc. Am. J.* **57**, 1501–1508.

RAY T.B., CALLOW J.A. & KENNEDY J.F. (1988) Composition of root mucilage polysaccharides from *Lepidium sativum. J. Exp. Bot.* **39**, 1249–1261.

READ D.J., FRANCIS R. & FINLAY R.D. (1985) Mycorrhizal mycelia and nutrient cycling in plant communities. In *Ecological Interactions in Soil*, ed. Fitter A.H., pp. 193–217. Oxford: Blackwell.

REDDY M.S. & WILLEY R.W. (1981) Growth and resource use studies in an intercrop of pearl millet-groundnut. *Field Crops Res.* **4**, 13–24.

REID C.P.P. (1990) Mycorrhizas. In *The Rhizosphere*, ed. Lynch J.M., pp. 281–316. Chichester: John Wiley & Sons.

REITEMEIER R.F. & RICHARDS L.A. (1944) Reliability of pressure membrane method for extraction of soil solution. *Soil Sci.* **57**, 119–136.

RENGEL Z. (1993) Mechanistic simulation models of nutrient uptake: a review. *Plant Soil* **152**, 161–173.

RENGEL Z. & WHEAL M.S. (1997) Kinetic parameters of Zn uptake by wheat are affected by the herbicide chlorsulfuron. *J. Exp. Bot.* **48**, 935–941.

RENNIE P.J. (1955) The uptake of nutrients by mature forest growth. *Plant Soil* **7**, 249–95.

RENNIE R.J. & LARSON R.J. (1981) Dinitrogen fixation associated with disomic chromosome substitution lines of spring wheat in the phytotron and in the field. In *Associative N_2-Fixation*, Vol. I, eds. Vose P.B. & Ruschel A.P. Boca Raton, FL: CRC Press.

RENNIE R.J. & THOMAS J.B. (1987) ^{15}N-determined effect of inoculation with N-fixing bacteria on nitrogen assimilation in Western Canadian wheats. *Plant Soil* **100**, 213–223.

RIBBLE J.M. & DAVIS L.E. (1955) Ion exchange in soil columns. *Soil Sci.* **79**, 41–47.

RICE E.L. (1984) *Allelopathy*, 2nd Edition. New York: Academic Press.

RICHARDS L.A. (ed.) (1954) *Diagnosis of Saline and Alkaline Soil*. Riverside, CA: US Regional Salinity Laboratory.

RICHARDS R.A. & PASSIOURA J.B. (1989) A breeding programme to reduce the diameter of the major xylem vessels in the seminal roots of wheat and its effect on grain yield in rain-fed environments. *Aust. J. Agric. Res.* **40**, 943–950.

RILEY D. & BARBER S.A. (1969) Bicarbonate accumulation and pH changes at soyabean (*Glycine max* (L.) Merr.) root–soil interface. *Soil Sci. Soc. Am. Proc.* **33**, 905–908.

RILEY D. & BARBER S.A. (1970) Salt accumulation at the soyabean (*Glycine max* (L.) Merr.) root–soil interface. *Soil Sci. Soc. Am. Proc.* **34**, 154–155.

RILEY D. & BARBER S.A. (1971) Effect of ammonium and nitrate fertilization on phosphorus uptake as related to root-induced pH changes at the root–soil interface. *Soil Sci. Soc. Am. Proc.* **35**, 301–306.

RITCHIE J.T. & OTTER S. (1985) Description and performance of CERES-Wheat: a user-oriented wheat yield model. ARS Publication 38, pp. 159–175. USDA.

RITCHIE J.T., GODWIN D.C. & OTTER-NACKE S. (1988) *CERES Wheat. A Simulation Model of Wheat Growth and Development*. College Station, TX: Texas A&M University Press.

RIVIERE J. (1960) Etude de la rhizosphere du ble. *Annales Agronomiques* **11**, 397–440.

ROBARDS A.W. & LUCAS W.J. (1990) Plasmodesmata. *Ann. Rev. Plant Physiol. Plant Mol. Biol.* **41**, 369–419.

ROBERTSON M.J., FUKAI S., HAMER G.L. & LUDLOW M.H. (1993a) Water extraction by grain sorghum in a subhumid environment. I Analysis of water extraction patterns. *Field Crops Res.* **33**, 81–97.

ROBERTSON M.J., FUKAI S., HAMER G.L. & LUDLOW M.H. (1993b) Water extraction by grain sorghum in a subhumid environment. II Extraction in relation to root growth. *Field Crops Res.* **33**, 99–112.

ROBERTSON M.J., FUKAI S., HAMER G.L. & LUDLOW M.H. (1993c) Modelling root growth of sorghum using the CERES approach. *Field Crops Res.* **33**, 113–130.

ROBERTSON W.K., HAMMOND L.C., JOHNSON J.T. & BOOTE K.J. (1980) Effects of plant-water stress on root distribution of corn, soybean and peanuts in sandy soil. *Agron. J.* **72**, 548–550.

ROBINSON D. (1986) Limits to nutrient inflow rates in roots and root systems. *Physiol. Plant.* **68**, 551–559.

ROBINSON D. (1994) The responses of plants to non-uniform supplies of nutrients. *New Phytol.* **127**, 635–674.

ROBINSON D. & RORISON I.H. (1983) Relationships between root morphology and nitrogen availability in a recent theoretical model describing nitrogen uptake from soil. *Plant Cell Environ.* **6**, 641–647.

ROBINSON D., LINEHAN D.J. & CAUL S. (1991) What limits nitrate uptake from soil? *Plant Cell Environ.* **14**, 77–85.

ROBINSON R.A. & STOKES R.H. (1959) *Electrolyte Solutions*. London: Butterworth.

ROBSON A.D. & PITMAN M.G. (1983) Interactions between nutrients in higher plants. In *Inorganic Plant Nutrition*, eds. Lauchli A. & Bieleski R.L. Encyclopedia of Plant Physiology 15A, pp. 145–180. Berlin: Springer-Verlag.

RODIN L.E. & BAZILEVICH N.I. (1965) *Production and Mineral Cycling in Terrestrial Vegetation.* London: Oliver & Boyd.

ROGERS H.H., PETERSEN C.M., MCCRUMMEN J.N. & CURE J.D. (1992) Response of plant roots to elevated atmospheric carbon dioxide. *Plant Cell Environ.* **15**, 749–752.

ROGERS H.H., PRIOR S.A., RUNION G.B. & MITCHELL R.J. (1996) Root to shoot ratio of crops as influenced by CO_2. *Plant Soil* **187**, 229–248.

ROGERS W.S. (1939) Apple growth in relation to rootstock, soil, seasonal and climatic factors. *J. Pomol.* **17**, 99–130.

ROGERS W.S. & HEAD G.C. (1969) Factors affecting the distribution and growth of roots of perennial woody species. In *Root Growth*, ed. Whittington W.J., pp. 280–291. 15th Easter School, Nottingham. London: Butterworth.

ROMELL L.G. (1922) Luftvaxlingen i marken som okologisk faktor. *Medd Statens Skogsforsoks-anstalt* **19**, 125.

ROMHELD V. & MARSCHNER H. (1986) Mobilization of iron in the rhizosphere of different plant species. In *Advances in Plant Nutrition, 2*, eds. Tinker P.B. & Lauchli A., pp. 155–204. New York: Praeger.

ROSE D.A. (1963) Water movement in porous materials. 2. The separation of the components of water movement. *Br. J. Appl. Phys.* **14**, 491–496.

ROSE D.A. (1977) Hydrodynamic dispersion in porous materials. *Soil Sci.* **123**, 277–283.

ROSE D.A. (1983) The description of the growth of root systems. *Plant Soil* **75**, 405–415.

ROSENE H.F. (1943) Quantitative measurement of the velocity of water absorption in individual root hairs by a microtechnique. *Plant Physiol.* **18**, 588–607.

ROUATT J.W. & KATZNELSON H. (1961) A study of the bacteria on the root surface and in the rhizosphere soil of crop plants. *J. Appl. Bacteriol.* **24**, 164–171.

ROVIRA A.D., NEWMAN E.I., BOWEN H.J. & CAMPBELL R. (1974) Quantitative assessment of the rhizosphere microflora by direct microscopy. *Soil Biol. Biochem.* **6**, 211–216.

ROVIRA A.D., FOSTER R.C. & MARTIN J.K. (1979) Note on terminology: origin, nature and nomenclature of the organic materials in the rhizosphere. In *The Soil–Root Interface*, eds. Harley J.L. & Scott-Russell R., pp. 1–4. London: Academic Press.

ROWELL D.L. (1981) Oxidation and reduction. In *The Chemistry of Soil Processes*, eds. Greenland D.J. & Hayes M.H.B., pp. 401–461. Chichester: Wiley.

ROWELL D.L., MARTIN M.W. & NYE P.H. (1967) The measurement and mechanism of ion diffusion in soils. III. The effect of moisture content and soil solution concentration on the self-diffusion of ions in soil. *J. Soil Sci.* **18**, 204–222.

ROY T.C., CALLOW J.A. & KENNEDY J.F. (1988) Composition of root mucilage polysaccharides from *Lepideum sativum*. *J. Exp. Bot.* **39**, 1249–1261.

RUNDEL P.W. & NOBEL P.S. (1991) Structure and function in desert root systems. In *Plant Root Growth*, ed. Atkinson D., pp. 349–380. Oxford: Blackwell.

RUSSELL Sir E.J. (1937) *Soil Conditions and Plant Growth*, 7th Edition. London: Longman.

RUSSELL Sir E.J. (1966) *A History of Agricultural Science in Great Britain.* London: Allen & Unwin.

RUSSELL Sir E.J. & RUSSELL E.W. (1950) *Soil Conditions and Plant Growth*, 8th Edition. London: Longman.

RUSSELL E.W. (1973) *Soil Conditions and Plant Growth*, 10th Edition. London: Longman.

RYAN P.R., DELHAIZE E. & RANDALL P.J. (1995) Malate efflux from root apices and tolerance to aluminium are highly correlated in wheat. *Aust. J. Plant Physiol.* **22**, 531–536.

SAAB I.N., SHARP R.E., PRITCHARD J. & VOETBERG, G.S. (1990) Increased endogenous abscisic acid maintains primary root growth and inhibited shoot growth of maize seedlings at low water potential. *Plant Physiol.* **93**, 1329–1336.

SABRY S.R.S., SALED A.S., BATCHELOR C.A., JONES J., JOTHAM J., WEBSTER G., KOTHARI S.L., DAVEY M.R. & COCKING E.C. (1997) Endophytic establishment of *Azorhizobium caulinodans* in wheat. *Proc. R. Soc. Lond.* **B264**, 341–346.

SACHS J. (1875) *Textbook of Botany* (English edition). Oxford: Oxford University Press.

SACKVILLE-HAMILTON C.A.G. & CHERRETT J.M. (1991) The development of clover and ryegrass root systems in a pasture and their interaction with the soil fauna. In *Plant Root Growth*, ed. Atkinson D., pp. 291–300. Oxford: Blackwell.

SAFIR G.R., BOYER J.S. & GERDEMANN J.W. (1972) Nutrient status and mycorrhizal enhancement of water transport in soybean. *Plant Physiol.* **49**, 700–703.

SALEQUE M.A. & KIRK G.D.J. (1995) Root-induced solubilization of phosphate in the rhizosphere of lowland rice. *New Phytol.* **129**, 325–336.

SALISBURY F.B. (1993) Gravitropism: changing ideas. *Hort. Rev.* **15**, 233–278.

SALLAM A., JURY W.A. & LETEY J. (1984) Measurement of gas diffusion coefficient under relatively low air filled porosity. *Soil Sci. Soc. Am. J.* **48**, 3–6.

SALMON R.C. (1964) Cation exchange reactions. *J. Soil Sci.* **15**, 273–283.

SAMTSEVICH S.A. (1968) Gel-like excretions of plant roots and their influence upon soil and rhizosphere microflora. In *Methods of Productivity Studies in Root Systems and Rhizosphere Organisms*, eds. Ghilasov M.S., Kanda V.A., Noichkova L.N., Rodin L.E. & Shveshnikova V.M. Soviet National Commission I.B.P.

SANCHEZ P.B. (1995) Science in agroforestry. *Agroforest. Syst.* **30**, 5–55.

SANDERS D. (1990) Kinetic modelling of plant and fungal membrane transport systems. *Ann. Rev. Plant Physiol. Plant Mol. Biol.* **41**, 77–107.

SANDERS F.E.T. (1971) Effects of root and soil properties on the uptake of nutrients by competing roots. D. Phil. Thesis, Oxford University.

SANDERS F.E. (1993) Modelling plant growth responses to vesicular-arbuscular mycorrhizal infection. *Adv. Plant Pathol.* **9**, 135–166.

SANDERS F.E. & SHEIKH N.A. (1983) The development of vesicular-arbuscular mycorrhizal infection in plant root systems. *Plant Soil* **71**, 223–246.

SANDERS F.E.T. & TINKER P.B. (1971) Mechanism of absorption of phosphate from soil by *Endogone* mycorrhizas. *Nature* **233**, 278–279.

SANDERS F.E. & TINKER P.B. (1973) Phosphate flow into mycorrhizal roots. *Pestic. Sci.* **4**, 385–395.

SANDERS F.E., TINKER P.B. & NYE P.H. (1971) Uptake of solutes by multiple root systems from soil. I. An electrical analog of diffusion to root systems. *Plant Soil* **34**, 453–466.

SANDERS F.E., TINKER P.B., BLACK R.L.B. & PALMERLEY S.M. (1977) The development of endomycorrhizal root systems. I. Spread of infection and growth promoting effects with four species of vesicular-arbuscular endophyte. *New Phytol.* **78**, 257–268.

SANDERS F.E.T., BUWALDA J.G. & TINKER P.B. (1983) A note on modelling methods for studies of ectomycorrhizal systems. *Plant Soil* **71**, 507–512.

SANDERS J.L. & BROWN D.A (1977) Measurement of rooting patterns for determinate and indeterminate soybean genotypes with a fiber-optic scope. In *The Soil–Root Interface*, eds. Harley J.L. & Scott Russell R., pp. 369–380. London: Academic Press.

SANTANTONIO D. & GRACE J.C. (1987) Estimating fine root production and turnover from biomass and decomposition data: a compartment-flow model. *Can. J. Forest Res.* **17**, 900–908.

SANWAR M. & KREMER R.J. (1995) Enhanced suppression of plant growth through production of L-tryptophan-derived compounds by deleterious rhizobacteria. *Plant Soil* **172**, 261–269.

SARKAR A.N. & WYN JONES R.G. (1982) Influence of rhizosphere on the nutrient status of dwarf French beans. *Plant Soil* **64**, 369–380.

SCHACHTMANN D.P. & SCHROEDER J.I. (1994) Structure and transport mechanism of a high-affinity potassium uptake transporter from higher plants. *Nature* **370**, 655–658.

SCHANDER H. (1941) The displacement of optimum soil reaction during development of *Lupinus lutens*. *Bodenk. Pfl. Ernah.* **20**, 129–151.

SCHENK M.K. & BARBER S.A. (1979a) Root characteristics of corn genotypes as related to phosphorus uptake. *Agron. J.* **71**, 921–924.

SCHENK M.K. & BARBER S.A. (1979b) Phosphate uptake by corn as affected by soil characteristics and root morphology. *Soil Sci. Soc. Am. J.* **43**, 880–883.

SCHIPPERS B. & VAN WURDE J.W.L. (1978) Studies of microbial colonization of wheat roots and the manipulation of the rhizosphere microflora. In *Microbial Ecology*, eds. Loutit M.W. & Miles J.A.R., pp. 295–298. New York: Wiley.

SCHIPPERS B., BAKKAR A.W. & BAKKAR P.A.H.M. (1987) Interaction of deleterious and beneficial rhizosphere microorganisms and the effects of cropping practice. *Ann. Rev. Phytopathol.* **25**, 339–358.

SCHLIEFF U. (1980) Chloride content of onion roots and their adhering soil under irrigation with saline drainage water. *Z. Pflanzen. Boden.* **143**, 638–644.

SCHOBERT C. & KOMOR E. (1987) Amino acid uptake by *Ricinus communis* roots: characterization and physiological significance. *Plant Cell Environ.* **10**, 493–500.

SCHOFIELD R.K. (1955) Can a precise meaning be given to available soil phosphorus? *Soils Fert.* **18**, 373–375.

SCHOFIELD R.K. & GRAHAM-BRYCE I.J. (1960) Diffusion of ions in soils. *Nature* **188**, 1048–1049.

SCHOFIELD R.K. & TAYLOR A.W. (1955) The measurement of soil pH. *Soil Sci. Soc. Am. Proc.* **19**, 164–167.

SCHULZE, E.D. (1986) Carbon dioxide and water vapour exchange in response to drought in the atmosphere and in the soil. *Ann. Rev. Plant Physiol.* **37**, 247–274.

SCHULZE, E.D. (1993) Soil water deficits and atmospheric humidity as environmental signals. In *Water Deficits in Plants*, eds. Smith J.A.C. & Griffiths H., pp. 129–145. Oxford: Bios Scientific.

SCHWAB S.M., MENGE J.A. & TINKER P.B. (1991) Regulation of nutrient transfer between host and fungus in VA mycorrhizae. *New Phytol.* **117**, 387–398.

SCOTT RUSSELL R. (1977) *Plant Root Systems: Their Function and Interaction with the Soil.* Maidenhead, UK: McGraw-Hill Book Co.

SCOTTER D.R. & RAATS P.A.C. (1970) Movement of salt and water near crystalline salt in relatively dry soil. *Soil Sci.* **109**, 170–178.

SELKER J.S., STEENHUIS T.S. & PARLANGE J.-Y. (1992) Wetting front instability in homogeneous sandy soils under continuous infiltration. *Soil Sci. Soc. Am. J.* **56**, 1346–1350.

SERRANO R. & GAXIOLA R. (1994) Microbial models and salt stress tolerance in plants. *Crit. Rev. Plant Sci.* **13**, 121–138.

SEWARD P., BARRACLOUGH P.B. & GREGORY P.J. (1990) Modelling potassium uptake by wheat (*Triticum aestivum*) crops. *Plant Soil* **124**, 303–307.

SHACHAR-HILL Y., PFEFFER P.E., DOUDS D., OSMAN S.F., DONER L.W & RATCLIFFE R.G. (1995) Partitioning of intermediary carbon metabolism in VAM colonised leek. *Plant Physiol.* **108**, 7–15.

SHARPLEY A.N., MEISINGER J.J., POWER J.F. & SUAREZ D.L. (1992) Root extraction of nutrients associated with long-term soil management. *Adv. Soil Sci.* **15**, 151–217.

SHONE M.G.T. (1966) The initial uptake of ions by barley roots. II. Applications of measurements of adsorption of ions to elucidate the structure of free space. *J. Exp. Bot.* **17**, 89–95.

SHONE M.G.T. & WOOD A.V. (1972) Factors affecting absorption and translocation of simazine by barley. *J. Exp. Bot.* **23**, 141–151.

SIDDIQUI M.Y. & GLASS A.K.M. (1982) Simultaneous consideration of tissue and substrate potassium concentration in K^+ uptake kinetics; a model. *Plant Physiol.* **69**, 283–285.

SIEVERS A. & BRAUN M. (1996) The root cap: structure and function. In *Plant Roots — The Hidden Half*, eds. Waisel Y., Eshel A. & Kafkafi U., pp. 31–50. New York: Marcel Dekker.

SILBERBUSH M. (1996) Simulation of ion uptake from the soil. In *Plant Roots — The Hidden Half*, eds. Waisel Y., Eshel A. & Kafkafi U., pp. 643–658. New York: Marcel Dekker.

SILBERBUSH M. & BARBER S.A. (1983) Sensitivity analyses of parameters used in simulating potassium uptake with a mechanistic mathematical model. *Agron. J.* **75**, 851–854.

SILBERBUSH M. & BARBER S.A. (1984) Phosphorus and potassium uptake by field-grown soy bean cultivars predicted by a simulation model. *Soil Sci. Soc. Am. J.* **48**, 592–596.

SILBERBUSH M., SOREK S. & YAKIREVICH A. (1993) K^+ uptake by root systems grown in soil under salinity. I. A mathematical model. *Transport in Porous Media* **11**, 101–116.

SIMARD W., PERRY D.A., JONES M.D., MYROLD D.D., DURALL D.M. & MOLINA R. (1997) Net transfer of carbon between ectomycorrhizal tree species in the field. *Nature* **388**, 579–582.

SIMPSON R.J., LAMBERS H. & DALLING M.J. (1982) Translocation of nitrogen in a vegetative wheat plant (*Triticum aestivum*). *Physiol. Plant.* **56**, 11–17.

SINGH R.P., ONG C.K. & SAHARAN N. (1989) Above- and below-ground interactions in alley cropping in semi-arid India. *Agroforest. Syst.* **9**, 259–274.

SINHA B.K. & SINGH N.T. (1974) Effect of transpiration rates on salt accumulation around roots in saline soil. *Agron. J.* **66**, 557–560.

SINHA B.K. & SINGH N.T. (1976a) Salt distribution around roots of wheat under different transpiration rates. *Plant Soil* **44**, 141–147.

SINHA B.K. & SINGH N.T. (1976b) Root uptake coefficient for chloride ion in corn as affected by transpiration rates and solution concentration. *Plant Soil* **44**, 521–525.

SKENE K.R., KIERANS M., SPRENT J.I. & RAVEN J.A. (1996) Structural aspects of cluster root development and their possible significance for nutrient acquisition in *Grevillea robusta* (Protaceae). *Ann. Bot.* **77**, 443–451.

SLATYER R.O. (1967) *Plant-Water Relationships.* London: Academic Press.

SMART C.J., GARVIN D.F., PRINCE J.P., LUCAS W.J. & KOCHIAN L.V. (1996) The molecular basis of potassium nutrition in plants. *Plant Soil* **187**, 81–89.

SMETHURST P.J. & COMERFORD N.B. (1993a) Simulating nutrient uptake by single or competing and contrasting root systems. *Soil Sci. Soc. Am. J.* **57**, 1361–1367.

SMETHURST P.J. & COMERFORD N.B. (1993b) Potassium and phosphorus uptake by competing pine and grass: observations and model verification. *Soil Sci. Soc. Am. J.* **57**, 1602–1610.

SMETHURST P.J., COMERFORD N.B. & NEARY D.G. (1993) Predicting the effects of weeds on K and P uptake by young slash pine on a spodosol. *Forest Ecol. Manage.* **60**, 27–39.

SMILEY R.W. (1974) Rhizosphere pH as influenced by plant, soils and nitrogen fertilizer. *Soil Sci. Soc. Am. Proc.* **38**, 795–799.

SMITH F.A. & RAVEN J.A. (1979) Intracellular pH and its regulation *Ann. Rev. Plant Physiol.* **30**, 289–311.

SMITH F.A., SMITH S.E., ST. JOHN B.J. & NICHOLAS D.J.D. (1986) Inflows of N and P into roots of mycorrhizal and non-mycorrhizal onions. In *Physiological and Genetic Aspects of Mycorrhizae*, eds. Gianinazzi-Pearson V. & Gianinazzi S., pp 371–378. Paris: INRA.

SMITH F.W., EALING P.M., HAWKESFORD M.J. & CLARKSON D.T. (1995) Plant members of a family of sulphate transporters reveal functional subtypes. *Proc. Natl. Acad. Sci. USA* **92**, 9373–9377.

SMITH G.D. (1978) *Numerical Solution of Partial Differential Equations: Finite Difference Methods*, 2nd Edition. Oxford: Clarendon Press.

SMITH S.E. (1982) Inflow of phosphate into mycorrhizal and non-mycorrhizal plants of *Trifolium subterraneum* at different levels of soil phosphate. *New Phytol.* **90**, 293–303.

SMITH S.E. & GIANINAZZI-PEARSON V. (1988) Physiological interactions between symbionts in vesicular-arbuscular mycorrhizal plants. *Ann Rev. Plant Physiol. Plant Mol. Biol.* **39**, 221–244.

SMITH S.E. & READ D.J. (1997) *Mycorrhizal Symbiosis.* San Diego, CA: Academic Press.

SMITH S.E. & WALKER N.A. (1981) A quantitative study of mycorrhizal infection in *Trifolium*: separate determination of the rates of infection and of mycelial growth. *New Phytol.* **89**, 225–240.

SMITH S.E., DICKSON S., MORRIS N. & SMITH F.A. (1994a) Transfer of phosphate from fungus to plant in VA mycorrhizas: calculation of the area of symbiotic interface and of fluxes of P from two different fungi to *Allium porrum* L. *New Phytol.* **127**, 93–99.

SMITH S.E., GIANINAZZI-PEARSON V., KOIDE R. & CAIRNEY J.W.G. (1994b) Nutrient transport in mycorrhizas: structure, physiology and consequences for efficiency of the symbiosis. *Plant Soil* **159**, 103–113.

SMITH W.H. (1969) Release of organic materials from the roots of tree seedlings. *Forest Sci.* **15**, 138–143.

SMITH W.H. (1970) Root exudates of seedlings and mature sugar maple. *Phytopathology* **60**, 701–703.

SNELLGROVE R.C., SPLITTSTOESSER W.E., STRIBLEY D.P. & TINKER P.B. (1982) The distribution of carbon and the demand of the fungal symbiont in leek plants with vesicular-arbuscular mycorrhizas. *New Phytol.* **92**, 75–87.

SNELLGROVE R.C., STRIBLEY D.P., TINKER P.B. & LAWLOR D.W. (1986) The effect of vesicular-arbuscular mycorrhizal infection in photosynthesis and carbon distribution in leek plants. In *Mycorrhiza: Physiology and Genetics*, eds. Gianinazzi-Pearson V. & Gianinazzi S., pp. 421–424. Paris: INRA.

SO H.B. & NYE P.H. (1989) The effect of bulk density, water content and soil type on the diffusion of chloride in soil. *J. Soil Sci.* **40**, 743–749.

SOON Y.K. (1988) Nutrient uptake by barley roots under field conditions. *Plant Soil* **109**, 171–179.

SOON Y.K. & MILLER M.H. (1977) Changes in the rhizosphere due to NH_4^+ and NO_3^- fertilization and phosphorus uptake by corn seedlings (*Zea mays* L.). *Soil Sci. Soc. Am. J.* **41**, 77–80.

SPARLING G.P. (1976) Effects of arbuscular vesicular mycorrhizas on Pennine grassland vegetation. Ph.D. Thesis, University of Leeds.

SPARLING G.P. & TINKER P.B. (1978) Mycorrhizal infection in Pennine grassland. II Effects of mycorrhizal infection on some upland grasses on γ-irradiated soils. *J. Appl. Ecol.* **15**, 951–958.

SPEK L.Y. & VAN NOORDWIJK M. (1994) Proximal root diameter as predictor of total root size for fractal branching. II Numerical model. *Plant Soil* **164**, 119–127.

SPOLLEN W.G., SHARP R.E., SAABI N. & WU Y. (1993) Regulation of cell expansion in roots and shoots at low water potential. In *Water Deficits*, eds. Smith J.A.C. & Griffiths H., pp. 37–52. Oxford: Bios Scientific Publications.

SPONCHIADO B.N., WHITE J.W., CASTILLO J.A. & JONES P.G. (1989) Root growth of four common bean cultivars in relation to drought tolerance in environments with contrasting soil types. *Exp. Agric.* **25**, 249–257.

SPOSITO G. (1981) *The Thermodynamics of Soil Solutions.* Oxford: Clarendon Press.

SPOSITO G. (1982) On the use of the Langmuir equation in the interpretation of adsorption phenomena. *Soil Sci. Soc. Am. J.* **46**, 1147–1152.

SPOSITO G. (1984) *The Surface Chemistry of Soils.* New York: Oxford University Press.

SPOSITO G. & MATTIGOD S.V. (1979) A computer program for the calculation of chemical equilibria in soil solutions and other natural water systems. Kearney Foundation of Soil Science. Riverside, CA: University of California.

SQUIRE G.R., GREGORY P.J., MONTEITH J.L., RUSSELL M.B. & PRAIA SINGH (1984) Control of water use by pearl millet (*Pennisetum typhoides* S & H). *Exp. Agric.* **20**, 135–139.

STADELMANN E.J. (1977) Passive transport parameters of plant cell membranes. In *Regulation of Cell Membrane Activities in Plants*, eds. Marre E. & Cifferi O., pp. 3–18. Amsterdam: North-Holland.

STANFORD G. & PIERRE W.H. (1947) The relation of potassium fixation to ammonium fixation. *Soil Sci. Soc. Am. Proc.* **11**, 155–160.

STAPLE W.J. (1965) Moisture tension, diffusivity and conductivity of a loam soil during wetting and drying. *Can. J. Soil Sci.* **45**, 78–86.

STARR J.L., BROADBENT F.E. & NIELSEN D.R. (1974) Nitrogen transformations during continuous leaching. *Soil Sci. Soc. Am. Proc.* **38**, 282–289.

STAUNTON S. (1986) The self-diffusion of sodium in soil: factors affecting the surface mobility. *J. Soil Sci.* **37**, 373–377.

STAUNTON S. (1990) A comparison of the surface impedance factors of Ca, Na, Rb and Cs derived from their self-diffusion coefficients in various soils. *J. Soil Sci.* **41**, 643–653.

STAUNTON S. (1995) Diffusion processes. In *The Encyclopaedia of Soil Science*, Part 2, eds. Fairbridge R.W. & Finkl C.H., Stroudsburg, PA: Dowden Hutchinson & Ross.

STAUNTON S. & NYE P.H. (1989a) The effect of non-instantaneous exchange on the self-diffusion of phosphate in soil. *J. Soil Sci.* **40**, 751–760.

STAUNTON S. & NYE P.H. (1989b) Three approaches to the simulation of the self-diffusion and non-instantaneous isotopic exchange of phosphate in soil. *J. Soil Sci.* **40**, 761–771.

STAUNTON S., CLAY P.G. & REES L.V.C. (1990) Diffusion of neptunium(V) in clays. *Radiochim. Acta* **49**, 147–153.

STEEN E. (1991) Usefulness of the mesh bag method in quantitative root studies. In *Plant Root Growth.*, ed. Atkinson D., pp. 75–88. Oxford: Blackwell.

STEIN W.D. (1986) *Transport and Diffusion across Cell Membranes.* Orlando, FL: Academic Press.

STEUDLE E. (1993) Pressure probe techniques: basic principles and applications to studies of water and solute relations at cell, tissue and organ level. In *Water Deficits*, eds. Smith, J.A.C. and Griffiths, H.C., pp. 5–36. Oxford: Bios Press.

STEUDLE E. (1994) The regulation of plant water at the cell, tissue and organ level: role of active processes and compartmentation. In *Flux Control in Biological Systems*, ed. Schulze E.-D., pp. 237–299. San Diego, CA: Academic Press.

STEUDLE E. & JESCHKE W.D. (1983) Water transport in barley roots. *Planta* **158**, 237–248.

STEUDLE E., MURRMANN M. & PETERSEN C. (1993) Transport of water and solutes across maize roots modified by puncturing the endodermis. *Plant Physiol.* **103**, 335–349.

STEVENSON F.J., ed. (1982) *Nitrogen in Agricultural Soils.* Agronomy Monograph No. 22. Madison, WI: American Society of Agronomy.

STRAUSS R., BRUMMER G.W. & BARROW N.J. (1997) Effects of crystallinity of goethite: II Rate of sorption and desorption of phosphate. *Eur. J. Soil Sci.* **48**, 101–114.

STRIBLEY D.P., TINKER P.B. & RAYNER J.H. (1980a) Relation of internal phosphorus concentration and plant weight in plants infected by vesicular-arbuscular mycorrhizas. *New Phytol.* **86**, 261–266.

STRIBLEY D.P., TINKER P.B. & SNELLGROVE R.C. (1980b) Effect of vesicular-arbuscular mycorrhizal fungi on the relations of plant growth, internal phosphorus concentration and soil phosphate analyses. *J. Soil Sci.* **31**, 655–672.

STROM L., OLSSON T. & TYLER G. (1995) Differences between calcifuge and acidifuge plants in root exudates of low-molecular weight organic acids. *Plant Soil* **167**, 239–245.

SUSLOW T. (1982) Role of root-colonizing bacteria in plant growth. In *Phytopathogenic Prokaryotes, I*, eds. Mount M.S. and Lacey G.H., pp. 187–223. New York: Academic Press.

SUTCLIFFE J.F. (1962) *Mineral Salts Absorption in Plants.* Oxford: Pergamon.

SWINBURNE T.R. ed. (1986) *Iron, Siderophores and Plant Diseases.* New York: Plenum Press.

SWINNEN J., VAN VEEN J.A. & MERCKX R. (1994a) ^{14}C pulse-labelling of field-grown spring wheat: an evaluation of its use in rhizosphere carbon budget estimations. *Soil Biol. Biochem.* **26**, 161–170.

SWINNEN J., VAN VEEN J.A. & MERCKX R. (1994b) Rhizosphere carbon fluxes in field-grown spring wheat: model calculations based on ^{14}C partitioning after pulse-labelling. *Soil Biol. Biochem.* **26**, 171–182.

SYLVIA D.M. (1990) Distribution, structure and function of external hyphae of vesicular-arbuscular mycorrhizal fungi. In *Rhizosphere Dynamics*, eds. Box J.E. & Hammond L.C., Selected Symposium 113, pp. 144–167. American Association for the Advancement of Science.

TACKETT J.L. & PEARSON R.W. (1964) Oxygen requirements of cotton seedling roots for penetration of compacted soil cores. *Soil Sci. Soc. Am. Proc.* **28**, 200–205.

TAN Y., BUD W.J., ROVIRA A.D., BRISBANE P.G. & GRIFFIN D.M. (1991) Movement through soil of a biological control agent *Pseudomonas fluorescens. Soil Biol. Biochem.* **23**, 821–826.

TANFORD C. (1961) In *Physical Chemistry of Macromolecules*, Chapter 6. Transport processes. New York: Wiley.

TARAFDAR J.C. & CLAASSEN N. (1988) Organic phosphorus compounds as a phosphorus source for higher plants through the activity of phosphatases produced by plant roots and microorganisms. *Biol. Fert. Soils* **5**, 308–312.

TARAFDAR J.C. & JUNGK A. (1987) Phosphatase activity in the rhizosphere and its relation to the depletion of soil organic phosphorus. *Biol. Fert. Soils* **3**, 199–204.

TARDIEU F. (1993) Will progress in understanding soil–root relations and root signalling substantially alter water flux models? *Phil. Trans. R. Soc. London* **341**, 57–66.

TARDIEU F. & DAVIES W.J. (1993) Root–shoot communication and whole-plant regulation of water flux. In *Water Deficits*, eds. Smith J.A.C. & Griffiths H.C., pp. 147–162. Oxford: Bios Press.

TARDIEU F. & PELLERIN S. (1990) Trajectory of the nodal roots of maize in fields with low mechanical constraints. *Plant Soil* **124**, 39–45.

TAYLOR H.M. & BOEHM W. (1976) Use of acrylic plastic as rhizotron windows. *Agron. J.* **68**, 693–694.

TAYLOR S.A. & CARY J.W. (1964) Linear equations for the simultaneous flow of matter and energy in a continuous soil system. *Soil Sci. Soc. Am. Proc.* **28**, 167–172.

TEPE W. & LEIDENFROST E. (1958) Ein Vergleich zwischen pflanzenphysiologischen, kinetischen und statischen Bodenuntersuchung-werten. *I. Mitt. Landw. Forsch.* **11**, 217–229.

TESTER M. (1990) Plant ion channels: whole cell and single channel studies. *New Phytol.* **114**, 305–340.

TESTER M. (1997) Techniques for studying ion channels: an introduction. *J. Exp. Bot.* **48**, 353–359.

TEZERA TSEGAYE & MULLINS C.E. (1994) Effect of mechanical impedance on root growth and morphology of two varieties of pea. *New Phytol.* **126**, 707–713.

TEZERA TSEGAYE, MULLINS C.E. & DIGGLE A.J. (1995a) An experimental procedure for obtaining input parameters for the ROOTMAP simulation program for peas (*Pisum sativum* L.). *Plant Soil* **172**, 1–16.

TEZERA TSEGAYE, MULLINS C.E. & DIGGLE A.J. (1995b) Modelling pea (*Pisum sativum*) root growth in drying soil. A comparison between observations and model predictions. *New Phytol.* **131**, 179–189.

THENG B.K.G., GREENLAND D.J. & QUIRK J.P. (1967) Adsorption of alkylammonium cations by montmorillonite. *Clay Miner.* **7**, 1–17.

THOMPSON J.P. (1994) Inoculation with vesicular-arbuscular mycorrhizal fungi from cropped soil overcomes long-fallow disorder of linseed (*L. usitatissimum* L.) by improving P and Zn uptake. *Soil Biol. Biochem.* **26**, 1133–1143.

THOMPSON L., THOMAS C.D., RADLEY J.M.A., WILLIAMSON S. & LAWTON J.H. (1993) The effect of earthworms and snails in a simple plant community. *Oecologia* **95**, 171–178.

THORNLEY J.H.M. (1972) A balanced quantitative model for root/shoot ratios in vegetative plants. *Ann. Bot.* **36**, 431–441.

THORNLEY J.H.M. (1991) A transport-resistance model of forest growth and partitioning. *Ann. Bot.* **68**, 211–226.

THORNLEY J.H.M. (1995) Shoot:root allocation with respect to C, N and P: an investigation and comparison of resistance and teleonomic models. *Ann. Bot.* **75**, 391–405.

THORNLEY J.H.M. & CANNELL M.G.R. (1992) Nitrogen relations in a forest plantation — soil organic matter ecosystem model. *Ann. Bot.* **70**, 137–151.

THORNLEY J.H.M. & JOHNSON I.R. (1990) *Plant and Crop Modelling — A Mathematical Approach to Plant and Crop Physiology*. Oxford: Oxford University Press.

THORNLEY J.H.M., BERGELSON J. & PARSONS A.J. (1995) Complex dynamics in a carbon–nitrogen model of a grass–legume pasture. *Ann. Bot.* **75**, 79–94.

THORNTON B. & MILLARD P. (1996) Effects of severity of defoliation on root functioning in grasses. *J. Range Manag.* **49**, 443–447.

THORUP R.M. (1969) Root development and P uptake by tomato plants under controlled soil moisture conditions. *J. Agron.* **61**, 808–811.

TIKTAK A. & VAN GRINSVEN J.M. (1995) A review of sixteen forest soil–atmosphere models. *Ecol. Modelling* **83**, 17–53.

TILMAN D. (1987) On the measuring of competition and the mechanisms of competitive superiority. *Functional Ecol.* **1**, 304–315.

TINKER P.B. (1964a) Cation activity ratios in acid Nigerian soils. *J. Soil Sci.* **15**, 24–34.

TINKER P.B. (1964b) Equilibrium cation activity ratios and responses to potassium fertilizer of Nigerian oil palms. *J. Soil Sci.* **15**, 35–41.

TINKER P.B. (1968) Changes in soil sodium following its addition in fertilizers. *MAFF Tech. Bull.* **20**, 318–325.

TINKER P.B. (1970) Some problems in the diffusion of ions in soils. *Soc. Chem. Ind. Monogr.* **37**, 120–134.

TINKER P.B. (1975a) Effects of vesicular-arbuscular mycorrhizas on higher plants. *Symp. Soc. Exp. Biol.* **29**, 325–349.

TINKER P.B. (1975b) The soil chemistry of phosphorus and mycorrhizal effects on plant growth. In *Endomycorrhizas*, eds. Sanders F.E., Mosse B. & Tinker P.B., pp. 353–372. London: Academic Press.

TINKER P.B. (1976) Roots and water. Transport of water to plant roots in soil. *Phil. Trans. R. Soc. Lond.* **B273**, 445–461.

TINKER P.B. (1980) The role of rhizosphere microorganisms in the uptake of phosphorus in plants. In *The Role of Phosphorus in Agriculture*, eds. Kwasaneh F. & Sample E., pp. 617–654. Madison, WI: American Society of Agronomy.

TINKER P.B. (1985) Site-specific yield potentials in relation to fertilizer use. In *Nutrient Balances and Fertilizer Needs in Temperate Agriculture*, ed. von Peter A., pp. 193–208. 18th Coll. International Potash Institute, Bern, 1984.

TINKER P.B. (1986) Trace elements in arable agriculture. Hills Bequest Lecture of the Royal Agricultural Society of England. *J. Soil. Sci.* **37**, 585–601.

TINKER P.B. (1990) Nutrient uptake by plant roots in natural systems. In *Nutrient Cycling in Terrestrial Ecosystems*, eds. Harrison A.F., Ineson P. & Heal O.W., pp. 322–334. Elsevier: London.

TINKER P.B. & BOLTON J. (1966) Exchange equilibria of sodium on some British soils. *Nature* (Lond.) **212**, 548.

TINKER P.B. & GILDON A. (1983) Mycorrhizal fungi and ion uptake. In *Metals and Micronutrients: Uptake and Utilisation by Plants*, eds. Robb D. & Pierpoint W.S., pp. 21–32. London: Academic Press.

TINKER P.B. & INGRAM J.S.I. (1996) Agriculture, forestry and soils: the work of Focus 3. In *Global Change and Terrestrial Ecosystems*, eds. Walker B. & Steffen W., pp. 207–228. Woods Hole Conference 1994. Cambridge: Cambridge University Press.

TINKER P.B. & SANDERS F.T. (1975) Rhizosphere microorganisms and plant nutrition. *Soil Sci.* **119**, 363–368.

TINKER P.B., JONES M.D. & DURRALL D.M. (1991) Principles in use of radioisotopes in mycorrhizal studies. In *Methods in Microbiology, Vol. 23: Techniques for the Study of Mycorrhizae*, eds. Varma A.K., Norris J.R. & Read D.J., pp. 295–308. London: Academic Press.

TINKER P.B., JONES M.D. & DURRALL D.M. (1992) A functional comparison of ecto- and endo-mycorrhizas. In *Mycorrhizas in Ecosystems*, eds. Read D.J., Lewis D.H., Fitter A.H. & Alexander I.J., pp. 303–310. Wallingford: CAB International.

TINKER P.B., DURRALL D.M. & JONES M.D. (1994) Carbon use efficiency in mycorrhizas: theory and sample calculations. *New Phytol.* **128**, 115–122.

TINKER P.B., GREGORY P.J., INGRAM J.S.I. & CANADELL J., eds. (1996) Plant-soil carbon below ground: the effects of elevated CO_2. *Plant Soil* **187**, 107–401.

TOTH R. & MILLER R.M. (1984) Dynamics of arbuscule development and degeneration in a *Zea mays* mycorrhiza. *Am. J. Bot.* **71**, 449–460.

TOURAINE B., CLARKSON D.T. & MULLER B. (1994) Regulation of nitrate uptake at the whole plant level. In *A Whole Plant Perspective on Carbon–Nitrogen Interactions*, eds. Roy J. & Garnier E., pp. 11–30. The Hague: SPB Academic.

TROUGHTON A. (1957) The underground organs of herbage grasses. *Bull 44*. Farnham Royal: Imperial Bureau of Plant Genetics, Commonwealth Agricultural Bureau.

TRUEMAN L.J., RICHARDSON A.R. & FORDE B.G. (1996) Molecular cloning of higher plant homologues of the high-affinity nitrate transporters of *Chlamydomonas reinhardtii* and *Aspergillus nidulans*. *Gene* **175**, 223–231.

TULL J. (1731) *Horse Hoeing Husbandry*. London.

TURNBULL M.H., GOODALL R. & STEWART G.R. (1995) The impact of mycorrhizal colonization upon nitrogen source utilization, metabolism in seedlings of *Eucalyptus grandis* Hill ex Marden and *Eucalyptus maculata* Hook. *Plant Cell Environ.* **18**, 1386–1394.

TYREE M.T., PATINO S., BENNINK J. & ALEXANDER J. (1995) Dynamic measurements of root hydraulic conductance using a high-pressure flowmeter in the laboratory and field. *J. Exp. Bot.* **46**, 83–94.

UREN N.C. (1993) Mucilage secretion and its interaction with soil, and contact reduction. *Plant Soil* **156**, 79–82.

UREN N.C. & REISENAUER H.M. (1988) The role of root exudates in nutrient acquisition. *Advances in Plant Nutrition 3*, eds. Tinker P.B. & Lauchli A., pp. 79–114. New York: Praeger.

VAIDYANATHAN L.V., DREW M.C. & NYE P.H. (1968) The measurement and mechanism of ion diffusion in soils. IV. The concentration dependence of diffusion coefficients of potassium in soils at a range of moisture levels and a method for the estimation of the differential diffusion coefficient at any concentration. *J. Soil Sci.* **19**, 94–107.

VAN DE GEIJN S.S., VOS J., GROENWOLD L., GOUDRIAAN J. & LEFFELAAR P.A. (1994) The Wageningen Rhizolab — a facility to study soil–root–shoot–atmosphere interactions in crops. *Plant Soil* **161**, 275–287.

VAN GENUCHTEN (1981) Non-equilibrium transport parameters from miscible displacement experiments. Research Report 119. Riverside, CA: U.S. Salinity Laboratory.

VAN KEULEN H. & SELIGMAN N.G. (1987) *Simulation of Water Use, Nitrogen Nutrition and Growth of a Spring Wheat Crop*. Simulation Monographs. Wageningen: Pudoc.

VAN NOORDWIJK M. & BROUWER G. (1991) Review of quantitative root length data in agriculture. In *Plant Roots and their Environment*, eds. McMichael B.L. & Persson H., pp. 515–525. Amsterdam: Elsevier.

VAN NOORDWIJK M. & DE WILLIGEN P. (1979) Calculation of the root density required for growth in soils of different status. In *The Soil–Root Interface*, eds. Harley J.L. & Scott Russell R., pp. 381–390. London: Academic Press.

VAN NOORDWIJK M. & DE WILLIGEN P. (1987) Agricultural concepts of roots: from morphogenetic to functional equilibrium. *Neth. J. Agric. Sci.* **35**, 487–496.

VAN NOORDWIJK M. & DE WILLIGEN P. (1991) Root functions in agricultural systems. In *Plant Roots and their Environment*, eds. McMichael M. & Persson H., pp. 381–395. Amsterdam: Elsevier.

VAN NOORDWIJK M. & VAN DE GEIJN S.C. (1996) Root, shoot and soil parameters required for process-oriented models of crop growth limited by water and nutrients. *Plant Soil* **183**, 1–25.

VAN NOORDWIJK M., DE WILLIGEN P., EHLERT P.A.I. & CHARDEN W.J. (1990) A simple model of P uptake as a possible basis for P fertilizer recommendations. *Neth. J. Agric. Sci.* **38**, 317–332.

VAN NOORDWIJK M., SCHOONDERBEEK D. & KOOISTRA M.J. (1993) Root–soil contact of field-grown winter wheat. *Geoderma* **56**, 277–286.

VAN OLPHEN H. (1957) Surface conductance of various ion forms of bentonite in water as the electrical double layer. *J. Phys. Chem.* **61**, 1276–1280.

VAN RAIJ B. & VAN DIEST A. (1979) Utilization of phosphate from different sources by six different plant species. *Plant Soil* **51**, 577–589.

VAN REES K.C.J., COMERFORD N.D. & MCFEE W.W. (1990) Modelling potassium uptake by slash pine seedlings from low-potassium-supplying soils. *Soil Sci. Soc. Am. J.* **54**, 1413–1421.

VAN REES K.C.J., HOSKINS J.A. & HOSKINS W.D. (1994) Analysing root competition with Dirichlet tesselations. *Soil Sci. Soc. Am. J.* **58**, 423–432.

VARNEY G.T. & CANNY M.J. (1993) Rates of water uptake into the mature root system of maize plants. *New Phytol.* **123**, 775–786.

VEEN B.W., VAN NOORDWIJK M., DE. WILLIGEN P., BOONE F.R. & KOOISTRA M.J. (1992) Root–soil contact of maize, as measured by a thin-section technique. III. Effects on shoot growth, nitrate uptake and water uptake efficiency. *Plant Soil* **139**, 131–138.

VERMA D.S. (1998) Developing crops with resistance to salinity and water stress. In *Feeding a World Population of More than Eight Billion People: A Challenge to Science*, eds. Fowden L. & Riley R. Rank Prize Funds Meeting 1996, pp. 171–182. New York: Oxford University Press.

VINCENT C.D. & GREGORY P.J. (1989a) Effects of temperature on the development and growth of winter wheat roots. I. Controlled glasshouse studies of temperature, nitrogen and irradiance. *Plant Soil* **119**, 87–97.

VINCENT C.D. & GREGORY P.J. (1989b) Effect of temperature on the development and growth of winter wheat roots. II. Field studies of temperature, nitrogen and irradiance. *Plant Soil* **119**, 99–110.

VINTEN A.J.A. & NYE P.H. (1985) Transport and deposition of dilute colloidal suspensions in soils. *J. Soil Sci.* **36**, 531–541.

VINTEN A., YARON B. & NYE P.H. (1983) Vertical transport of pesticides into soil when adsorbed on suspended particles. *Agric. Food Chem.* **31**, 662–664.

VLAMIS J. (1953) Acid soil infertility as related to soil-solution and solid phase effects. *Soil Sci.* **75**, 383–394.

VOGT K.A. & PERSSON H. (1989) Measuring growth and development of roots. In *Techniques and Approaches in Forest Tree Ecophysiology*, eds. Lassoie J.P. & Hinckley T.M., pp. 478–501. Boca Raton, FL: CRC Press.

VOGT K.A., GRIER C.C. & VOGT D.J. (1986) Production, turnover and nutrient dynamics of above- and below-ground detritus of world forests. *Adv. Ecol. Res.* **15**, 303–377.

VOGT K.A., VOGT D.J. & BLOOMFIELD J. (1991) Inputs of organic matter to the soil by tree roots. In *Plant Roots and their Environment*, eds. McMichael M. & Persson H., pp. 171–190. Amsterdam: Elsevier.

VOISARD C., KEEL C., HAAS D. & DEFAGO G. (1989) Cyanide production by *Pseudomonas fluorescens* helps suppress black root rot of tobacco under gnotobiotic conditions. *EMBO J.* **8**, 351–358.

VOLKMAR K.M. (1993) A comparison of mini-rhizotron techniques for estimating root length density in soils of different bulk density. *Plant Soil* **157**, 239–245.

VON WIREN N., ROMHELD V., MOREL J.L., GUCHERT A. & MARSCHNER H. (1993) Influence of microorganisms on iron acquisition in maize. *Soil Biol. Biochem.* **25**, 373–376.

VOS J. (1995) Nitrogen and the growth of potato crops. In *Potato Ecology*, eds. Haverkort J.A. & MacKerron D.K.L., pp. 115–128. Dordrecht: Kluwer Academic.

VOS J. & GROENWOLD J. (1986) Root growth of potato crops on a marine clay soil. *Plant Soil* **94**, 17–33.

WADLEIGH C.H. (1968) *Wastes in Relation to Agriculture and Forestry*. Misc. Pub. No. 1065. Washington, DC: U.S. Department of Agriculture.

WAGENET R.J. (1990) Quantitative prediction of the leaching of organic and inorganic solutes in soil. *Phil. Trans. R. Soc. Lond.* **B329**, 321–330.

WALKER A. (1971) Effects of soil moisture content on availability of soil-applied herbicides to plants. *Pestic. Sci.* **2**, 56–59.

WALKER A. & CRAWFORD D.V. (1968) The role of organic matter in adsorption of the triazine herbicides by soil. In *Isotopes and Radiation in Soil Organic Matter Studies*, pp. 91–108. Vienna: International Atomic Energy Authority.

WALKER A. & FEATHERSTONE R.M. (1973) Absorption and translocation of atrazine and linuron by plants with implications concerning linuron selectivity. *J. Exp. Bot.* **24**, 450–458.

WALKER A., ALLEN R., BAILEY S.W., BLAIR C.D., GUNTHER P., LEAKE C.R. & NICHOLLS P.H., eds. (1995) *Pesticide Movement to Water*. Monograph Series No. 62. Farnham, Surrey: British Crop Protection Council.

WALKER B. & STEFFEN W., eds. (1996) *Global Change and Terrestrial Ecosystems*. Cambridge: Cambridge University Press.

WALKER J.M. & BARBER S.A. (1961) Ion uptake by living plant roots. *Soil Sci.* **133**, 881–882.

WALWORTH J.L.A. & SUMNER M.E. (1988) Foliar diagnosis: a review. *Advances in Plant Nutrition, 2*, eds. Tinker P.B. & Lauchli A., pp. 193–242. New York: Praeger.

WANG G.M., STRIBLEY D.P., TINKER P.B. & WALKER C. (1993) Effects of pH on arbuscular mycorrhiza. I — Field observations on the long-term liming experiments at Rothamsted and Woburn. *New Phytol.* **124**, 465–472.

WARNCKE D.D. & BARBER S.A. (1972) Diffusion of zinc in soil: I. The influence of soil moisture. *Soil Sci. Soc. Am. Proc.* **36**, 39–42.

WATKINS N.K., FITTER A.H., GRAVES J.D. & ROBINSON D. (1996) Carbon transfer between C3 and C4 plants linked by a common mycorrhizal network quantified using stable carbon isotopes. *Soil Biol. Biochem.* **28**, 471–477.

WATT D.A., AMAY A.M. & CRESSWELL C.F. (1992) Effect of N supply on the kinetics and regulation of nitrate assimilation in *Chlamydomonas reinhardtii* Dangeard. *J. Exp. Bot.* **43**, 605–616.

WAY J.T. (1850) On the power of soils to absorb manure. *J. R. Agric. Soc. Eng.* **11**, 313–379; *J. R. Agric. Soc. Eng.* (1852) **13**, 123–143.

WEATHERLEY P.J. (1982) Uptake and flow in roots. In Encyclopedia of Plant Physiology 12B, *Physiological Plant Ecology*, eds. Lange O.L., Osmond C.B. & Ziegler H., pp. 79–109. Heidelberg: Springer-Verlag.

WEAVER J.E. (1926) *Root Development of Field Crops*. New York: McGraw-Hill.

WEBER J.B. & MILLER C.T. (1989) Organic chemical movement over and through soil. In *Reactions and Movement of Organic Chemicals in Soils*, eds. Sawney B.L. & Brown K., pp. 31–44, SSSA Spec. Pub. No. 22. Madison, WI: Soil Science Society of America.

WEGNER L.H. & RASCHKE K. (1994) Ion channels in the xylem parenchyma of barley roots. *Plant Physiol.* **105**, 799–813.

WEIR A.H. & BARRACLOUGH P.B. (1986) The effect of drought on the root growth of winter wheat and on its water uptake from a deep loam. *Soil Use Manage.* **2**, 91–96.

WEIR A.H., BRAGG P.L., PORTER J.R. & RAYNER, J.H. (1984a) A winter wheat crop simulation model without water and nutrient limitations. *J. Agric. Sci., Camb.* **102**, 371–382.

WEIR A.H., RAYNER J.H., CATT J., SHIPLEY D.G. & HOLLIES J.D. (1984b) Soil factors affecting the yield of winter wheat: analyses of results from ICI surveys 1979–1980. *J. Agric. Sci., Camb.* **103**, 639–649.

WELBANK P.J., GIBB M.J., TAYLOR P.J. & WILLIAMS E.D. (1974) Root growth of cereal crops. *Annual Report of the Rothamsted Experimental Station, 1973*, pp. 26–66.

WETSELAAR R. (1962) The fate of nitrogenous fertilizer in a monsoonal climate. *Transactions of the International Soil Conference New Zealand*, pp. 588–595.

WHALLEY W.R. & DEXTER A.R. (1993) The maximum axial growth pressure of roots of spring and autumn cultivars of lupin. *Plant Soil* **157**, 313–318.

WHIPPS J.M. (1985) Effect of CO_2 concentration on growth, carbon distribution and loss of carbon from the roots of maize. *J. Exp. Bot.* **36**, 644–651.

WHIPPS J.M. (1987) Carbon loss from the roots of tomato and pea seedlings grown in soil. *Plant Soil* **103**, 95–100.

WHIPPS J.M. (1990) Carbon economy. In *The Rhizosphere*, ed. Lynch J.M., pp. 59–98. Chichester: John Wiley.

WHIPPS J.M. & LYNCH J.M. (1986) The influence of the rhizosphere on crop productivity. *Adv. Microbial Ecol.* **9**, 187–244.

WHISLER F.D., KLUTE A. & MILLINGTON R.J. (1970) Analysis of radial steady-state solution, and solute flow. *Soil Sci. Soc. Am. Proc.* **34**, 382–387.

WHITE P.J., BENFIELD J. & DIAZ M. (1992) Unidirectional Ca fluxes in roots of rye (*Secale cereale* L.). A comparison of excised roots with roots of an intact plant. *J. Exp. Bot.* **43**, 1061–1074.

WHITE R.E. (1979) *Introduction to the Principles and Practice of Soil Science*. Oxford: Blackwell Scientific.

WHITEHEAD A.G., DUNNING R.A. & COOKE D.A. (1971) Docking disorder and root ectoparasitic nematodes of sugar beet. *Annual Report of the Rothamsted Experimental Station 1970*, **2**, pp. 219–236.

WHITELEY G.M. & DEXTER A.R. (1984) Displacement of soil aggregates by elongating roots and emerging shoots of crop plants. *Plant Soil* **77**, 131–140.

WHITNEY M. & CAMERON F.K. (1903) *The Chemistry of Soil as Related to Crop Production*. Bull. 22. U.S. Department of Agriculture, Bureau of Soils.

WIERSUM L.K. (1957) The relationship of the size and the structural rigidity of pores to their penetration by roots. *Plant Soil* **9**, 75–85.

WIERSUM L.K. (1961) Utilization of soil by the plant root system. *Plant Soil* **15**, 189–192.

WILCOX H.E. (1996) Mycorrhizas. In *Plant Roots — The Hidden Half*, eds. Waisel Y., Eshel A. & Kafkafi U., pp. 689–722. New York: Marcel Dekker.

WILD A. (1964) Soluble phosphate in soil and uptake by plants. *Nature* **203**, 326–327.

WILD A., ed. (1988) *Russell's Soil Conditions and Plant Growth*, 11th edition. Harlow U.K.: Longman Scientific & Technical.

WILD A., ROWELL D.L. & OGUNFOWORA M.A. (1969) The activity ratio as a measure of the intensity factor in potassium supply to plants. *Soil Sci.* **108**, 432–439.

WILD A., SKARLOW V., CLEMENT C. & SNAYDON R.W. (1974) Comparison of potassium uptake data by four plant species grown in sand and flowing solution culture. *J. Appl. Ecol.* **11**, 801–812.

WILD A., JONES L.H.P. & MACDUFF J.H. (1987) Uptake of mineral nutrients and crop growth — the uses of flowing nutrient solution. *Adv. Agron.* **41**, 171–219.

WILLIAMS B.G., GREENLAND D.J. & QUIRK J.P. (1966) The adsorption of poly vinyl alcohol by natural soil aggregates. *Aust. J. Soil Res.* **4**, 131–143.

WILLIAMS B.G., GREENLAND D.J. & QUIRK J.P. (1967) The effect of poly vinyl alcohol on the nitrogen area and pore structure of soils. *Aust. J. Soil Res.* **5**, 77–83.

WILLIAMS R.F. (1946) The physiology of plant growth with special reference to the concept of net assimilation rate. *Ann. Bot. N.S.* **10**, 41–72.

WOLF O., MUNNS R., TONNET M.L. & JESCHKE W.D. (1991) The role of the stem in the partition of Na and K in salt-treated barley. *J. Exp. Bot.* **42**, 697–704.

WOOD J.T. & GREENWOOD D.J. (1971) Distribution of carbon dioxide and oxygen in the gas phase of aerobic soils. *J. Soil Sci.* **22**, 281–288.

WOODRUFF C.M. (1955) The energies of replacement of calcium by potassium in soils. *Soil Sci. Soc. Am. Proc.* **19**, 167–171.

WOODWARD J.M.D. (1699) Some thoughts and experiments concerning vegetation. *Phil. Trans. R. Soc.* **21**, 381–398.

WORRAL F., PARKER A., RAE J.E. & JOHNSON A.C. (1995) A study of suspended and colloidal matter in the leachate from lysimeters: implications for pollution and lysimeter studies. In *Pesticide Movement to Water*, eds. Walker A., Allen R., Bailey S.W., Blair C.D., Gunther P., Leake C.R. & Nicholls P.H., Monograph Series No. 62, pp. 129–136. Farnham, Surrey: British Crop Protection Council.

WRAY F.J. (1971) Changes in the ionic environment around plant roots. D.Phil. Thesis, Oxford.

WRAY F.J. & TINKER P.B. (1969) A scanning apparatus for detecting concentration gradients around single plant roots. In *Root Growth*, ed. Whittington W.J., pp. 418–422. 15th Easter School, Nottingham. London: Butterworth.

WULLSCHLAGER S.D., LYNCH J.P. & BERNTSEN G.M. (1994) Modelling the belowground response of plants and soil biota to edaphic and climatic change — what can we expect to gain? *Plant Soil* **165**, 149–160.

YADAV R., FLOWERS T.J. & YEO A.R. (1996) The involvement of the transpirational bypass flow in sodium uptake by high- and low-sodium-transporting lines of rice developed through intravarietal selection. *Plant Cell Environ.* **19**, 329–336.

YAN H.T., BEYROUTY C.A., NORMAN R.J. & GBUR E.E. (1995) Nutrient uptake relationships of root characteristics in rice. *Plant Soil* **171**, 297–302.

YANAI R.D. (1994) A steady-state model of nutrient uptake accounting for newly grown roots. *Soil Sci. Soc. Am. J.* **58**, 1562–1571.

YEATES G. & DARRAH P.R. (1991) Microbial changes in a model rhizosphere. *Soil Biol. Biochem.* **23**, 963–971.

YOST R.S. & FOX R.L. (1979) Contribution of mycorrhiza to P nutrition of crops growing on an oxisol. *Agron. J.* **71**, 903–908.

YOUNGS E.G. & GARDNER W.R. (1963) A problem of diffusion in the infinite hollow cylinder. *Soil Sci. Soc. Am. Proc.* **27**, 475–476.

ZHANG J.C. & BARBER S.A. (1992) Maize root distribution between P-fertilized and unfertilized soil. *Soil Sci. Soc. Am. J.* **56**, 819–822.

ZHEN R.G. & LEIGH R.A. (1989) Nitrate accumulation by wheat (*Triticum aestivum*) in relation to growth and tissue N concentration. In *Plant Nutrition: Physiology and Application*, ed. Van Beusichem, pp. 17–20. Dordrecht: Kluwer Academic.

ZHU J.K., HASEGAWA P.M. & BRESSAN R.A. (1997) Molecular aspects of osmotic stress in plants. *Crit. Rev. Plant Sci.* **16**, 253–277.

ZHU W. & EHRENFELD J.G. (1996) Effects of mycorrhizal roots on litter decomposition, soil biota and nutrients in a spodosolic soil. *Plant Soil* **179**, 109–118.

ZOBEL R.W. (1992) Soil environment constraints to root growth. *Adv. Soil Sci.* **19**, 27–52.

ZOBEL R.W. (1996) Genetic control of root systems. In *Plant Roots — The Hidden Half*, eds. Waisel Y., Eshel A. & Kafkafi U., pp. 21–30. New York: Marcel Dekker.

ZRIEK R.A. (1987) Concentration changes of sodium and chloride associated with roots in saline environments. D.Phil. Thesis. Bodlean Library, Oxford.

ZUBERER D.A. (1990) Soil and rhizosphere aspects of N-fixing plant–microbe associations. In *The Rhizosphere*, ed. Lynch J.M., pp. 317–354. Chichester: John Wiley.

Index